Neurotoxicology

TARGET ORGAN TOXICOLOGY SERIES

Series Editors

A. Wallace Hayes, John A. Thomas, and Donald E. Gardner

Neurotoxicology, Third Edition. *G. Jean Harry and Hugh A. Tilson, editors, 324 pp. 2010.*
Adrenal Toxicology. *Philip W. Harvey, David J. Everett, and Christopher J. Springall, editors, 336 pp., 2008*
Cardiovascular Toxicology, Fourth Edition. *Daniel Acosta, Jr., editor, 712 pp., 2008*
Toxicology of the Gastrointestinal Tract. *Shayne C. Gad, editor, 384 pp., 2007*
Immunotoxicology and Immunopharmacology, Third Edition. *Robert Luebke, Robert House, and Ian Kimber, editors, 676 pp., 2007*
Toxicology of the Lung, Fourth Edition. *Donald E. Gardner, editor, 696 pp., 2006*
Toxicology of the Pancreas. *Parviz M. Pour, editor, 720 pp., 2005*
Toxicology of the Kidney, Third Edition. *Joan B. Tarloff and Lawrence H. Lash, editors, 1200 pp., 2004*
Ovarian Toxicology. *Patricia B. Hoyer, editor, 248 pp., 2004*
Cardiovascular Toxicology, Third Edition. *Daniel Acosta, Jr., editor, 616 pp., 2001*
Nutritional Toxicology, Second Edition. *Frank N. Kotsonis and Maureen A. Mackey, editors, 480 pp., 2001*
Toxicology of Skin. *Howard I. Maibach, editor, 558 pp., 2000*
Neurotoxicology, Second Edition. *Hugh A. Tilson and G. Jean Harry, editors, 386 pp., 1999*
Toxicant–Receptor Interactions: Modulation of Signal Transductions and Gene Expression. *Michael S. Denison and William G. Helferich, editors, 256 pp., 1998*
Toxicology of the Liver, Second Edition. *Gabriel L. Plaa and William R. Hewitt, editors, 444 pp., 1997*
Free Radical Toxicology. *Kendall B. Wallace, editor, 454 pp., 1997*
Endocrine Toxicology, Second Edition. *Raphael J. Witorsch, editor, 336 pp., 1995*
Carcinogenesis. *Michael P. Waalkes and Jerrold M. Ward, editors, 496 pp., 1994*
Developmental Toxicology, Second Edition. *Carole A. Kimmel and Judy Buelke-Sam, editors, 496 pp., 1994*
Nutritional Toxicology. *Frank N. Kotsonis, Maureen A. Mackey, and Jerry J. Hjelle, editors, 336 pp., 1994*
Ophthalmic Toxicology. *George C. Y. Chiou, editor, 352 pp., 1992*
Toxicology of the Blood and Bone Marrow. *Richard D. Irons, editor, 192 pp., 1985*
Toxicology of the Eye, Ear, and Other Special Senses. *A. Wallace Hayes, editor, 264 pp., 1985*
Cutaneous Toxicity. *Victor A. Drill and Paul Lazar, editors, 288 pp., 1984*

Neurotoxicology

Third Edition

Edited by

G. Jean Harry
National Institute of Environmental Health Sciences
National Institutes of Health
Department of Health and Human Services
Research Triangle Park, North Carolina, U.S.A.

and

Hugh A. Tilson
Durham, North Carolina, U.S.A.

informa
healthcare

New York London

Informa Healthcare USA, Inc.
52 Vanderbilt Avenue
New York, NY 10017

10 9 8 7 6 5 4 3 2 1

International Standard Book Number-10: 1-4200-5487-2(Hardcover)
International Standard Book Number-13: 978-1-4200-5487-3 (Hardcover)

Library of Congress Cataloging-in-Publication Data

Neurotoxicology / edited by G. Jean Harry, Hugh A. Tilson. — 3rd ed.
p. ; cm. — (Target organ toxicology series ; 28)
Tilson's name appears first on the earlier ed.
Includes bibliographical references and index.
ISBN-13: 978-1-4200-5487-3 (hb : alk. paper)
ISBN-10: 1-4200-5487-2 (hb : alk. paper) 1. Neurotoxicology. I.
Harry, G.
Jean (Gaylia Jean) II. Tilson, Hugh A. III. Series: Target organ toxicology
series ; v. 28.
[DNLM: 1. Nervous System Diseases—chemically induced.
2. Nervous System—
drug effects. 3. Neurobiology. WL 140 N4943 2010]
RC347.5.N483 2010
616.8--dc22

2009042987

For Corporate Sales and Reprint Permissions call 212-520-2700 or write to: Sales Department, 52 Vanderbilt Avenue, 7th floor, New York, NY 10017.

Visit the Informa Web site at
www.informa.com

and the Informa Healthcare Web site at
www.informahealthcare.com

Preface

The first edition of this series on the nervous system as a target organ was published in 1992. This edition reviewed methodological approaches to detect and quantify the effects of chemicals at various levels of neurobiological organization (i.e., behavioral, neurophysiological, neurochemical, and neuropathological). The first edition also included chapters on newly emerging areas such as developmental neurotoxicology, neurotoxicology risk assessment, and methods to evaluate neurotoxicity in humans. The evolution of neurotoxicology as a discipline could be seen clearly in the second edition published in 1999. The second edition contained chapters on emerging in vitro approaches to study effects of chemicals on the nervous system. Other chapters emphasized potential sites of action such as neuron–glial interactions, ion channels, the immune system, the endocrine hormone system, and neurotrophic factors during development. Other chapters focused on putative modes of action such as metabolic influences, apoptosis, and the formation of reactive oxygen species. The second edition also included a chapter on the use of brain imaging techniques to evaluate neurotoxicity in vivo. There was also a summary of approaches to evaluate learning and memory as neurotoxic endpoints and methods to assess neurotoxicity in children. This volume finished with a chapter on emerging concepts for quantitative neurotoxicology risk assessment.

The current edition on the nervous system as a target organ reflects the continued growth and refinement of the field of neurotoxicology over a 10-year period. Included in the current edition is a chapter devoted to principles of behavioral phenotyping of neonatal and adult mice, an important topic for those interested in working with inbred mouse strains, outbred mouse stocks, or genetically modified mice in neurotoxicological research. Recent developments concerning design-based stereology and video densitometry for the assessment of neurotoxicological damage are also included. The current edition pays considerable attention to the potential of newly developed methods such as network analysis and emerging molecular approaches to identify potential neurotoxicants and their modes of action. The current edition also describes new advances in understanding the long-term impact of chemicals on the development of the thyroid hormone system and the potential relationship between disruption of the endocrine system and reproductive senescence. As in the case of the second edition, there are chapters that expand upon potential sites of action for neurotoxicity, including adult neurogenesis and the dopaminergic neurotransmitter system. The current edition also describes exciting new developments concerning the role of inflammation in the expression of neurotoxicity and the complex compensatory changes that follow damage to the central nervous system both in vivo and in slice cultures. Chapters on the role of neurotoxicants and the manifestation of obesity and the effects of neurotoxins on brain tissue and function revealed through in vivo magnetic resonance imaging and spectroscopy round out the third edition.

The evolution of neurotoxicology as a discipline is clearly documented in the three editions spanning a period of nearly 20 years. The first edition focused primarily on how to measure neurotoxic effects on the nervous system and identified emerging trends such as developmental neurotoxicity. The second edition reflected the development of new methodologies to identify sites and modes of action of neurotoxicity and described approaches to understand risk to humans of exposure to neurotoxic chemicals. The current edition builds on the two previous editions by exploring the potential use of molecular approaches to screen chemicals for potential neurotoxicity, developing approaches to understand the complex interactions of chemicals with biological systems, and providing the basis to study the relationship between neurotoxicant exposure and the emergence of diseases such as obesity and reproductive senescence. In spite of the tremendous advances that have made over the last 20 years, the discipline of neurotoxicology is still in its infancy. Hopefully, the information provided in this and the previous editions will stimulate current and future students and researchers to study how the nervous system functions and how it responds to chemicals with potentially toxic effects.

G. Jean Harry
Hugh A. Tilson

Contents

Contributors

Lucyna Antkiewicz-Michaluk Department of Neurochemistry, Institute of Pharmacology, Polish Academy of Sciences, Krakow, Poland

Rana Awada Laboratoire de Biochimie et Génétique Moléculaire, Groupe d'Etude sur l'Inflammation Chronique et l'Obésité (GEICO), Faculté des Sciences, Université de La Réunion, France

Susan H. Brunssen Division of Family Health, School of Nursing, and Neurodevelopmental Disorders Research Center, University of North Carolina, Chapel Hill, North Carolina, U.S.A.

Christian Lefebvre d'Hellencourt Laboratoire de Biochimie et Génétique Moléculaire, Groupe d'Etude sur l'Inflammation Chronique et l'Obésité (GEICO), Faculté des Sciences, Université de La Réunion, France

Karen Dean Department of Animal and Avian Sciences, University of Maryland, College Park, Maryland, U.S.A.

Katarzyna M. Dziegielewska Department of Pharmacology, University of Melbourne, Parkville, Victoria, Australia

Francisco Fornai Department of Human Morphology and Applied Biology, University of Pisa, Pisa, and Laboratory of Neurobiology of Movement Disorders IRCCS, INM Neuromed Pozzilli, Isernia, Italy

Willard M. Freeman Department of Pharmacology, Penn State College of Medicine, Hershey, Pennsylvania, U.S.A.

Mary E. Gilbert Neurotoxicology Division, U.S. Environmental Protection Agency, Research Triangle Park, and Department of Psychology, University of North Carolina, Chapel Hill, North Carolina, U.S.A.

Julia M. Gohlke Laboratory of Molecular Toxicology, National Institute of Environmental Health Sciences, Research Triangle Park, North Carolina, U.S.A.

Marcia N. Gordon Byrd Alzheimer Institute and Department of Molecular Pharmacology and Physiology, School of Basic Biomedical Sciences, University of South Florida College of Medicine, Tampa, Florida, U.S.A.

Mark D. Habgood Department of Pharmacology, University of Melbourne, Parkville, Victoria, Australia

G. Jean Harry Neurotoxicology Group, Laboratory of Molecular Toxicology, National Institute of Environmental Health Sciences, National Institutes of Health, Department of Health and Human Services, Research Triangle Park, North Carolina, U.S.A.

Grace E. Kissling National Institute of Environmental Health Sciences, National Institutes of Health, Research Triangle Park, North Carolina, U.S.A.

Richard M. Kostrzewa Department of Pharmacology, Quillen College of Medicine, East Tennessee State University, Johnson City, Tennessee, U.S.A.

Andrew D. Kraft Laboratory of Molecular Toxicology, National Institute of Environmental Health Sciences, National Institutes of Health, Research Triangle Park, North Carolina, U.S.A.

Emma T. Lavoie Department of Animal and Avian Sciences, University of Maryland, College Park, Maryland, U.S.A.

Peter R. Mouton Byrd Alzheimer Center and Research Institute, University of South Florida College of Medicine, Tampa, Florida, and Stereology Resource Centers, Chester, Maryland, U.S.A.

Sheryl S. Moy Neurodevelopmental Disorders Research Center and Department of Psychiatry, University of North Carolina, Chapel Hill, North Carolina, U.S.A.

William R. Mundy Systems Biology Branch, Integrated Systems Toxicology Division, National Health and Environmental Effects Research Laboratory, U.S. Environmental Protection Agency, Research Triangle Park, North Carolina, U.S.A.

Mary Ann Ottinger Department of Animal and Avian Sciences, University of Maryland, College Park, Maryland, U.S.A.

Adolf Pfefferbaum Department of Psychiatry and Behavioral Sciences, Stanford University School of Medicine, Stanford, and Neuroscience Program, SRI International, Menlo Park, California, U.S.A.

Michael J. Quinn, Jr. U.S. Army Center for Health Promotion and Preventive Medicine, Directorate of Toxicology, Health Effects Research Program, Aberdeen Proving Ground, Maryland, U.S.A.

Ram Ramabhadran Genetic and Cellular Toxicology Branch, Integrated Systems Toxicology Division, National Health and Environmental Effects Research Laboratory, U.S. Environmental Protection Agency, Research Triangle Park, North Carolina, U.S.A.

Palaniyandi Ravanan* Laboratoire de Biochimie et Génétique Moléculaire, Group d'Etude sur l'Inflammation Chronique et l'Obésité (GEICO), Faculté des Sciences, Université de La Réunion, France

Margaret J. Rosenbloom Department of Psychiatry and Behavioral Sciences, Stanford University School of Medicine, Stanford, and Neuroscience Program, SRI International, Menlo Park, California, U.S.A.

Norman R. Saunders Department of Pharmacology, University of Melbourne, Parkville, Victoria, Australia

Steven O. Simmons Genetic and Cellular Toxicology Branch, Integrated Systems Toxicology Division, National Health and Environmental Effects Research Laboratory, U.S. Environmental Protection Agency, Research Triangle Park, North Carolina, U.S.A.

Brandon D. Sitzmann Department of Animal and Avian Sciences, University of Maryland, College Park, Maryland, U.S.A.

Edith V. Sullivan Department of Psychiatry and Behavioral Sciences, Stanford University School of Medicine, Stanford, California, U.S.A.

Kent E. Vrana Department of Pharmacology, Penn State College of Medicine, Hershey, Pennsylvania, U.S.A.

R. T. Zoeller Department of Biology, University of Massachusetts, Amherst, Massachusetts, U.S.A.

*Present address: Burnham Institute for Medical Research, La Jolla, California, U.S.A.

1 Molecular Approaches to Identify Specific Modes of Action for Neurotoxicity In Vitro

Ram Ramabhadran and Steven O. Simmons

Genetic and Cellular Toxicology Branch, Integrated Systems Toxicology Division, National Health and Environmental Effects Research Laboratory, U.S. Environmental Protection Agency, Research Triangle Park, North Carolina, U.S.A.

William R. Mundy

Systems Biology Branch, Integrated Systems Toxicology Division, National Health and Environmental Effects Research Laboratory, U.S. Environmental Protection Agency, Research Triangle Park, North Carolina, U.S.A.

INTRODUCTION

Advances in molecular biology in the last two decades have produced a powerful set of tools for the analysis of structure, function, and internal dynamics of eukaryotic cells, both in vitro and in vivo. The ability to clone genes, measure their expression, and to assess their function by modulating the expression of the native gene or their mutant versions introduced into cells, are the central driving forces for this technological revolution. The confluence of these molecular techniques with the similar developments in cell biology, allowing the isolation and cultivation of cells in vitro in primary cultures, has greatly impacted our understanding of biology of differentiated cells from various mammalian organs including the brain. Despite the contributions of these complementary approaches to the study of basic cell biology and their successful applications to drug discovery, their use to address the questions of toxicology lags behind. A recent report released by the National Research Council of the National Academy of Sciences (U.S.A.) titled "Toxicity Testing in the 21st Century: A Vision and a Strategy" (1), places great emphasis on the use of in vitro assays, preferably using human cells in culture, in its vision of the future of toxicity testing as an alternative to the time- and resource-intensive methodologies currently used. This chapter highlights advances in molecular tools in the context of basic biology of neuronal cells, discusses their extant applications to neurotoxicology, and addresses their future applicability. While the application of the molecular tools in vitro forms the focus of this review, it should be recognized these same tools have been vital to studies at the whole organism level, notably through the creation of transgenic and knockout animal models (2–4), which will be discussed only as is relevant to the in vitro approaches or applications. Further, advances in recombinant DNA-based expression of proteins in heterologous organisms such as *Escherichia coli* and yeast have resulted in their use as antigens for the production of monoclonal antibodies highly specific to these proteins; these have been vital tools for immunohistochemical studies of cells in vitro as well as in vivo. However, this has been a well-established technology for well over a decade and will not be discussed in this review.

NEUROTOXIC MODES OF ACTION AT THE CELLULAR LEVEL

In vitro molecular approaches promise to have an increasing impact in neurotoxicology research as the focus moves away from apical tests of neurologic function in animals and towards an understanding of the sequence of key biological events that result in neurotoxicity (i.e., the *mode of action*). At the cellular level, mode of action may be defined as a series of key events starting with interaction of a chemical with a molecular target site and leading to perturbation of the normal physiological function of the cell. Mode of action data can be used to develop biologically-based dose response models and improve the ability to extrapolate data between species. The ability to manipulate and measure the interaction of genes and proteins at the cellular level will contribute to our understanding of how chemicals with known actions at neuronal targets (e.g., receptors, ion channels, cytoskeletal proteins) alter critical cellular pathways, ultimately resulting in adverse effects on the nervous system. It can also facilitate the neurobiological study of cellular networks and signaling pathways underlying normal nervous system function, and allow the identification of new modes of action for neurotoxicity.

Elucidating the mode of action of a neurotoxicant at the cellular level can be a daunting task considering that the nervous system consists of a heterogeneous population of cells exhibiting diverse phenotypes. This complexity, however, serves one major function: the transmission of information between and within cells. Thus, neurotoxicity at the cellular level can be considered in light of how chemicals interfere with the molecular mechanisms of neural signaling. To carry out this role, cells of the nervous system are specialized in terms of their morphology and biochemistry. Unique morphological features of neurons include relatively long axonal processes ending with pre-synaptic terminals and complex dendritic processes containing post-synaptic densities, both of which support the pre- and post-synaptic machinery for neurotransmission. This morphology is supported by neuron-specific proteins that make up the cytoskeleton and transport system required to move cellular components to and from these relatively distant sites (5). Biochemically, neurons are highly enriched in the enzymes necessary for neurotransmitter synthesis, release, uptake, and degradation. Neurotransmitters act at cell surface receptors to activate ion channels that are responsible for fast synaptic transmission between cells. In addition, neurotransmitter- and trophic factor-activation of cell surface receptors engage second messengers and intracellular signaling pathways that regulate normal physiological processes including neurotransmitter release, neuronal survival and growth, and synaptic plasticity (6).

The morphological and biochemical specialization of neurons that support the transmission of information provides a multitude of molecular target sites for chemicals (7). Ultimately, neurotoxicity results from the interaction of a chemical with one or more of these molecular targets. This interaction can be due to the recognition of the toxicant's chemical structure at some physiological binding site or to the reactive nature of the chemical structure itself. An example of the first case is the action of domoic acid as an agonist at the kainate and alpha-amino-3-hydroxy-5-methyl-4-isoxazole propionic acid subclasses of ionotropic glutamate receptors to result in excitotoxicity (8). In the second case, a chemical can interact with cellular macromolecules (protein, lipids, DNA) to result in catastrophic damage [e.g., adduct formation between 2,5-hexanedione and cytoskeletal proteins to leading to axonal degeneration (9)], or alter the molecular structure in such as way as to affect normal neuronal function [e.g., interaction of the pyrethroid insecticides with the α subunit of the voltage sensitive calcium channel resulting in prolonged depolarization (10)].

Neurotoxicants are structurally diverse, ranging from simple ions (10) or solvents (11) to complex naturally occurring substances (e.g., brevetoxins) (12). In light of the structural diversity of toxicants and the multitude of potential cellular target sites relating to neural signaling, it would appear to be an impossible task to identify the specific biochemical interactions of every chemical that leads to neurotoxicity. In contrast, an understanding of the key components of pathways that regulate normal neuronal functions will help to identify those that, when sufficiently perturbed by chemical exposure, produce neuronal damage or dysfunction. At one end of the spectrum are pathways that respond to levels of toxicant exposure that result in lethal cell injury. This would include the trophic factor pathways that promote cell survival (13), stress response pathways (14,15), and apoptotic pathways (16,17). At the other end of the spectrum are pathways that are more specific to the regulation of neural signaling. Examples of such pathways include those that regulate neuronal differentiation and growth (18–20), pathways that regulate the development and maintenance of axons and dendrites (5,21,22), pathways that regulate neurotransmitter release (23,24), and pathways involved in synaptogenesis and synaptic plasticity (25,26). Using an approach that focuses on normal regulatory pathways, it may be possible to identify general principles of dysregulation that are common to a diverse array of chemical structures. In other words, there are likely common modes of action at the molecular level that result in an adverse effect on cellular structure or function. For example, the neurotoxicant Pb^{2+} can substitute for other polyvalent cations (principally Ca^{2+} and Zn^{2+}) at multiple molecular sites in neurons such as ion channels and the signal transduction enzymes protein kinase C and calmodulin (27). These multiple sites of action, however, can converge on cellular pathways regulating axonal and dendritic growth and synaptic plasticity (28,29). Similarly, recent work has demonstrated that structurally diverse toxicants which cause subtle alterations in redox potential converge on a Fyn kinase pathway and alter neural progenitor cell division (30). These examples highlight an important aspect of mode of action studies at the cellular level: Toxicant-induced changes in critical biochemical pathways were associated with structural or functional effects in the cell that can be considered adverse. As discussed below, in vitro neural preparations provide models for both the molecular and genetic manipulation of regulatory pathways and cell-based assessment of neurotoxicity.

UTILITY OF IN VITRO MODELS

Faced with the immense complexity and cellular heterogeneity of the nervous system, in vitro model systems have been an indispensable tool for elucidating mechanisms governing both the normal development and function of the brain and those underlying disease states. The reasons for this include the ability to ask questions of a specific cell type in isolation and the ability to control the environmental conditions and variables of interest. Of course, the power of in vitro models is also the weakness since the nervous system does not consist of cells in isolation, and its very purpose is to communicate with other cells in a coordinated fashion. It is clear, however, that the responses of neural tissue in isolation can often predict that observed in vivo. Even with the constraints imposed by growth conditions in vitro (e.g., physical constraints of the culture vessel, artificial substrate and growth medium, lack of afferent and efferent connections), at the cellular level, neurons and glia in culture are remarkably similar to those in vivo. From differentiation and development of polarity to the organization of the cytoskeleton, the ion channels and receptors expressed, and the characteristics of synaptic communication, neural cells in culture resemble their counterparts in situ in both structure and function. Thus, neuronal cells in vitro provide a model system that is ideal for asking questions about those molecular pathways that regulate normal cell function and respond to perturbation by chemical exposure. The ability to finely control culture environment also facilitates the ability to manipulate these molecular pathways using genetic approaches.

A variety of in vitro preparations have been successfully developed and employed to address specific questions of cell biology and nervous system function (31–33). The use of in vitro techniques for the assessment of neurotoxicity has been the subject of recent reviews (34,35). Model systems of differing complexity provide the opportunity to examine cellular and molecular interactions at a number of levels. Developing brain tissue from different regions can be cut into slices that will survive in culture with preserved spatial, structural, and synaptic organization. These are termed organotypic slice cultures and can be maintained for several weeks. Organotypic slice cultures from hippocampus, cortex, and cerebellum have been used to study mechanisms for brain development, neurodegenerative disorders, and chemical neurotoxicity (36–38). In many cases organotypic cultures provide direct access to neurons for single-cell analysis, a feature that makes this model attractive for electrophysiological recording. Electrophysiological recording of long-term potentiation using hippocampal slice cultures have been used to understand chemical-induced effects on the cellular and molecular mechanisms of information storage (39). Primary neural cell cultures consist of dispersed cells that have been dissociated from nervous system tissue. When maintained under the appropriate culture conditions primary cells will acquire the properties of mature cells and attain a phenotype that is similar to that of cells from the brain region from which they were derived (40). Thus, neural cells with properties useful for assessing a particular molecular target or cell function can be obtained by choosing the appropriate source of cells. A comparison of properties exhibited by different neuronal culture systems can be found in (41). Although preparations of primary neuronal cultures can contain a predominant neuronal cell type (e.g., sympathetic neurons from the superior cervical ganglia, granule cell cultures from post-natal cerebellum) many preparations are a mixture of different neuronal populations. Pure neuronal cultures can be obtained using defined media which prohibit the growth of glial cells or by adding an anti-mitotic agent to the medium during the first days in culture. A mixed glial/neuronal co-culture will result if these steps are not employed. Primary cultures have a limited lifespan (typically days to weeks), and it is not possible to expand neuronal cell populations in culture. Thus, new cultures must be prepared from nervous system tissue on a regular basis. A continuous source of neural cells can be obtained by using cell lines, a term applied to a defined population of cells that can be maintained in culture for an extended period of time. Cell lines are usually clonal, meaning that the entire population originated from a single common ancestor cell. A number of neuronal cell lines are available (42) and many have been used as in vitro models in neurobiology (43,44). Neuronal cell lines have been derived from tumors, including pheochromocytomas (45) and neuroblastomas (46,47). The PC-12 cell line of rat origin and SH-SY5Y cell line of human origin are salient examples of such cell lines that have been routinely used in in vitro studies to elucidate basic neurobiologic principles and mechanism of chemical action (48–50). More recently, cell lines have been generated using oncogene-containing retroviruses (51). The introduction of oncogenes from the *myc* family into primary neural cells can result in neural cell lines that are immortalized while retaining many of the characteristics of the original cell population (43). Cell lines have a number of advantages as in vitro models. They provide a homogenous population of cells, are relatively easy to grow using standard tissue culture plastic and media, and can be induced to differentiate into a non-dividing cell with many of the characteristics of a neuron (52).

An emerging in vitro model with great potential is the stem cell, derived from rodents or preferably from human sources (53,54). Based on their functional properties of self renewal and multipotency (i.e., the ability to generate both neurons and glia), human neural stem cells are being widely considered as a potential therapy for neurodegenerative diseases (55–58). However, there is also great potential for their use in drug discovery and in toxicity screening assays (59,60), particularly those addressing developmental neurotoxicity. The use of human neuronal cells provides the opportunity to extend observations from rodent neurons to human neurobiology and thereby can be helpful in extrapolation of toxicity to humans. Stem cells are broadly divided into two types: embryonic, derived from developing human embryos, or somatic/adult, derived from adult tissues including the brain (54). There is, however, a great ethical debate regarding the use of human embryonic stem cells in research and therapy (61–63). The situation is relatively simpler with stem cells derived from adult tissues and these are available from commercial vendors. Although stem cells represent an ideal system for studying neurotoxicity in vitro, the realization of their full potential in this and other applications is hampered by experimental difficulties that remain to be resolved. For instance, despite their pluripotency or multipotency, the ability to derive homogenous population of a differentiated cell type remains difficult and involved, and the degree of desired differentiation is low and can vary between experiments (64–68). Maintenance of stem cells requires expensive substrates and complex media with growth factor supplementation (65,66,69). These requirements add to the cost and variability inherent to the cell system itself, making the use of stem cells in routine in vitro research somewhat daunting, especially for screening assays.

A caveat in the use of established cell lines is to be aware that the same cell line maintained under the same generic name in different laboratories often behaves very differently from each other due to differences in both the number of passages and the methodological differences between laboratories (70). While there is little that can be done to harmonize the numerous cell lines that are used under the same name in laboratories all around the world, it is prudent to minimally establish experimentally that the cells maintain the phenotypic features that led to their selection as the model for the phenomenon under study. A second problem is that, often slow growing cell lines may also be contaminated with other more robust cell lines grown alongside in the laboratory (70). Thus, where possible, cell lines should be obtained from established cell banks such as the American Type Culture Collection (ATCC: http://www.atcc.org/) or similar organizations dedicated to the archiving and distribution of cell lines.

The ability to genetically manipulate cells, be it to create screening assays or to define the mode of action of toxicants, provide many advantages that go far beyond the use of cells to biochemically measure molecular changes resulting from toxicant exposures. Genetically modified cells provide unique opportunities to address questions that would be difficult or impossible to answer through the traditional biochemical approach. Thus, the ability to introduce genes into cells in vitro has been the ramrod of many of the studies in cell biology for the last decade or more (71,72). However, from a neuroscience perspective, despite the availability of primary neuronal cells and established cell lines which exhibit neuronal phenotypes, application of many of the molecular tools to neurobiology was limited by the difficulty in genetically manipulating these cells. Several methods of gene delivery to cells in culture have been developed for transient and stable expression of exogenous genes (discussed further in the next section). These methods fall into two classes, namely, physio-chemical methods and viral vector methods (71,72). Both methods enable gene delivery to cells but at varying efficiencies depending on the method employed and the cell line of interest. These methods are collectively known as "transfection" generally for physicochemical methods, and as "transduction" or "infection" when using viral vectors. The exogenous gene construct and its components are referred to as "transgenes."

Cells of neuronal origin, primary or established, have proven to be some of the most recalcitrant to transfection and this has resulted in a lag in the advancement of the study of cellular neuroscience, which is in contrast to cells derived from most other organ systems where transfection techniques were relatively more effective (73). Many of these difficulties have now been overcome through recent advances in physicochemical and viral delivery methods and these are discussed below. Complementarily, the newer techniques such as transcriptomics and proteomics that do not require genetic manipulation of cells have been applied to neuronal systems on par with other organ systems. The power of genomic studies in vitro can be enhanced by the use of appropriate controls using genetically manipulated cells, and further genetic manipulation of cells can be vital to validation of genomics/proteomics-based hypotheses. Together, these techniques have contributed to spectacular advances in cellular neuroscience and these are being translated to in vitro neurotoxicology.

GENETIC MANIPULATION OF GENE EXPRESSION

Developments in molecular biology have provided a set of tools to alter cellular gene expression in targeted fashion by the expression of exogenous protein coding genes in cultured cells including neurons (73,74). Introduction of endogenously expressed genes under the transcriptional control of constitutive or inducible promoters, releases them from the natural regulators of their expression in the cell, permitting studies of protein function. In this context, elucidation of function is usually enhanced by the expression of mutant versions of the endogenous protein in a parallel experiment, which can also serve as a control in some cases (if the mutant version does not have dominant negative properties as described in a later section). The mutant versions of the protein may differ by single amino acid changes that alter some or all of a protein's activity, or by deletion of amino acids comprising one or more domains relevant to the function under study. Alternately, expression of genes foreign to the mammalian cell such as reporters (e.g., coral fluorescent proteins, firefly luciferase) (75) under the control of constitutive, inducible, or tissue and developmental stage-specific promoters (76) allows the development of screening assays as well as aids in mode of action studies. The end goal of the manipulations and the specifics of the genetic constructs used differ depending on the nature of the experiment. Promoter elements that direct the expression of the exogenously added gene are critical components of experimental design and the choice or promoter for a specific set of experiments is dictated by the question that is being addressed.

Gene Delivery into Neurons and Glia In Vitro

The techniques for gene delivery, known as transfection, fall into two basic categories: (*i*) physico-chemical methods, and (*ii*) viral vectors as shown in Table 1. Chemical methods involve the delivery of naked DNA or DNA complexed with agents such as charged polymers or cationic lipids to cells. Electroporation, microinjection, and biolistic (gene gun) delivery are the primary physical methods (71,77). Electroporation involves subjecting cells to an intense electric pulse leading to the creation of temporary "pores" in the cell membrane permitting the entry of DNA molecules from the surrounding medium into the intracellular space. Electroporation works efficiently only with single cell suspensions, and is a method of choice for transfection of mouse embryonic stem cells for creation of transgenic mice (4). Microinjection, as the name suggests, involves the injection of DNA into individual cells using a fine needle and a micromanipulator set up and is performed under a microscope. This technique is laborious and hence not practical for use in cell culture experiments where large numbers of cells need to be modified, but is used extensively in the injection of DNA into embryos to create transgenic animals (2,3). Biolistic gene delivery relies on the ability of accelerated metal particles to enter cells and deliver the DNA which has been pre-complexed onto the fine particles derived from inert metals, mainly gold. The biolistic approach can be used to deliver genes to cells in tissue explants, although effective gene delivery is limited to superficial layers of the explanted tissue slice (71,78,79).

The chemical-based methods originated with the discovery over two decades ago that complexation of DNA with cationic polymers such as diethylaminoethyl (DEAE) dextran increased its uptake by mammalian cells. Cationic compounds increase DNA uptake by masking the negatively-charged sugar-phosphate backbone of the DNA molecule. It was also shown that DNA incorporated into calcium phosphate precipitates could efficiently enter certain mammalian cell types. However, the toxicity of the DEAE dextran approach and the intra-laboratory and intra-experimental variability of the calcium phosphate technique prompted the search for more efficient materials to deliver DNA with limited toxicity (71). At this time, several commercially available reagents of the cationic lipid genre (e.g., Lipofectamine™, Fugene™) provide an easy-to-use means of introducing nucleic acids (both DNA and RNA) into mammalian

Table 1 Methods for Introducing DNA into Neuronal and Glial Cells

Class	Method	Advantages	Disadvantages	Comments
Chemical	Calcium phosphate	Rapid, inexpensive No DNA size limit	Transient expression Low, cell-type dependent efficiency Inter-experiment variability Cytotoxicity	Older method, less used
	Lipofection	Relatively expensive Simple, rapid No DNA size limit Less cytotoxic	Transient expression Low, cell-type dependent efficiency	Standardized commercial preparations available
Physical	Biolistic	Ideal for slice cultures No DNA size limit	Need for special equipment Transient expression Cytotoxic Restricted to superficial layers	Useful in plant cell transformation
	Microinjection	Directed to specific cells No DNA size limit	Laborious, Transient expression Needle induced cell death Need for special equipment Low throughput	Used in transgenic animal production for injecting embryos
	Electroporation	Works for many cell types No DNA size limit	Cells need to be in suspension Need for special equipment High level of cell death Optimization to cell type required	Used extensively in transducing mouse embryonic stem cells for knockout model production
Viral	Several (Table 2)	Access wide range of cell types Efficient delivery (~100%) Persistence of expression[a]	Laborious to produce Can be cytotoxic[a] Safety concerns[a]	Extremely useful for transducing primary neurons

[a]Depends on the specific viral vector.
Source: Compiled from Ref. 71, 73, 77.

Table 2 Comparison of Viral Vectors for Neuronal Transduction

Property	Retrovirus	Lentivirus	Adenovirus	Adeno-associated virus	Herpes
Neuronal transdution	–	++	++	++	+++
Ease of vector production	+++	+++	+	++	+
Gene payload (kb)	8	8	8[a]	4	30[b]
Persistence of expression	Long/short[c]	Very long	Transient	Long	Transient
Integration into genome	+	+	–	+/–	–
Expression of viral proteins	–	–	+	–	+
Cytotoxicity[d]	None	None	Cytotoxic	None	Cytotoxic
Insertional mutagenesis	+	+	–	+	–

[a]30 kb for "gutless" version which are considerably more difficult to produce.
[b]Amplicon version payload 15 kb, see Ref. 73 for details.
[c]Expression prone to silencing, hence can be shorter.
[d]At reasonable multiplicities of infection (vector particles/cell).

cells in culture with relatively low toxicity. These reagents permit transfection efficiencies upwards of 80% in many established and primary cell lines; however, the efficiency of transfection varies significantly among cell types, especially among primary cell cultures. The optimal ratio of DNA to the lipid reagent has to be empirically determined for each cell type line (71). One of the chief advantages of the physico-chemical methods of transfection is that there are no restrictions on the size of the DNA fragment, notably artificial chromosomes from yeast and bacteria of 100 to 1000 kb, that can be inserted (80). The limitations on the size of transgene termed "payload" is one of the restrictive features of viral vectors discussed below and is limited to about 50 kb at best or lower (72). The ability to analyze large fragments of DNA is critical for the delineation of regulatory regions of gene promoters that usually span many kilobases. Definition of the minimum required elements and the resultant shortening of the promoter permit their more efficent use in routine studies using physico-chemical or viral delivery methods.

Typically, although a higher percentage of cells transfected using physico-chemical methods take up DNA and express the transfected gene, this expression is transient for a period of 48 to 72 hours after which cells lose the transfected DNA. Analysis of gene expression during this time window is referred to as "transient transfection" experiment. In a low percentage of the cells however, (usually <1%) the exogenous DNA integrates into the chromosome (genome) and will remain stably integrated (71). Cells with integrated copies of the transgene are clonally selected using selection markers that are incorporated into the DNA construct. These markers, usually transcribed from a different promoter than the gene of interest, confer resistance to antibiotics such as neomycin, puromycin, or hygromycin, which kill the cells that do not harbor the transfected DNA. Isolated single clones or a pool of such clones that continue to express the transgene are referred to as "stably transfected" (71). While transient expression studies are rapid, establishment of stable clones is time consuming requiring weeks to months, but provide the opportunity to perform long term experiments without the loss of expression of the transfected gene and reduced experiment-to-experiment variability when compared with transiently transfected cells.

The cell type-dependent variability of efficiency physico-chemical methods of transfection makes them very difficult to use universally, especially with primary cell cultures where the method often causes toxicity. In contrast, viral vectors provide an efficient method for introducing genes into mammalian cells, particularly neurons (72,73). Viral vector systems capitalize on the facility of viruses to efficiently introduce and maintain their genomes in mammalian cells. The cytotoxic and disease-causing genes in the viral genome are deleted and replaced with regulatory and coding sequences needed for the expression of a foreign protein resulting in a "viral vector" that transfers the transgene to the transduced/infected cells, usually at high efficiencies (in many cases close to 100%). Depending on the viral vector used, the expression can be short-lived or long-lived. A number of DNA and RNA viruses have been modified along these lines to create viral vectors: however, from a neuronal cell perspective, vectors derived from adenovirus, adeno-associated virus (AAV), retrovirus (oncoretroviruses and lentiviruses), and herpes virus have the broadest utility (72,73,81–85). Each vector system has its advantages and limitations, chief among which are the size of the transgene or "payload" that can be encapsidated, its persistence in the cell and the cytotoxic perturbations caused by the viral-component-related gene expression. The properties of the various viral vectors are listed in Table 2. It is not possible to review all these systems in detail here and the references cited above should be consulted for details. The review by Craig (73) provides an extensive discussion of the adenoviral, AAV, and herpes viral vectors that have been used in many neuronal cell studies. Lentiviral vectors are a newer class of vectors that were in early stages of development at the time the review by Craig appeared. The lentiviral vector system is a welcome addition to the vector arsenal as it has properties overall well-suited for use in differentiated cells such as neurons. The use of lentiviral vectors in neurobiological studies, both in vitro and in vivo, has been on the rise significantly over the last five years. Established lentiviral vector system components can be obtained commercially for vector construction in the laboratory and several companies provide lentiviral production services.

Lentiviruses are members of the retroviridae family along with their cousins the oncoretroviruses such as Moloney leukemia virus which have been used extensively in human gene therapy trials over more than a decade (83,84). Retroviral particles contain a diploid RNA genome that upon entry into the host cell is converted to DNA by enzymes co-packaged into the viral capsid with the RNA genome. The resultant DNA is integrated into the host chromosome. Unfortunately, the oncoretroviral vectors have been of little use to studies in neuronal cells because of their inability to deliver genes to differentiated/non-dividing cells. In contrast, lentiviral vectors can transduce genes into differentiated/non-dividing cells with very high efficiency resulting in stably altered cells without the expression of any extraneous viral proteins that could perturb cellular functions, as commonly observed with both adenoviral and herpes viral vectors.

Lentiviruses are causative agents of diseases in many animal species, notably HIV-AIDS in humans. Despite this ominous property, the power of this class of viruses has been harnessed for gene delivery by deleting all of the disease-causing genes of the parent virus. Thus, while vectors have been constructed with equine and feline lentiviruses, the versions derived from HIV have been used most widely (84,86). Although safety concerns slowed the acceptance of this vector system, the issues with safety for laboratory in vitro applications (as opposed to use in human gene therapy or in vivo animal experiments) have significantly abated due to development of safer vector designs (87) and the use of the vectors in the laboratory for well over a decade in the absence of adverse events. Most in vitro work with lentiviral vectors can be carried out under NIH Biosafety level 2 (88), but caution should be exercised when working with vectors carrying toxic or oncogenic genes. Lentiviral vectors have been in use for in vitro and animal studies for nearly a decade and have contributed significantly to the advancement of neuroscience by permitting the genetic manipulation of neuronal cell lines and primary neurons.

Lentiviral genomes are around 9 kb in length, and all but ~2 kb of this can be replaced by foreign gene sequences which would include a promoter and the gene/cDNA of interest. The minimal viral sequences required for the vector to function are the two long terminal repeats (LTRs) of the virus and a short stretch of DNA, known as ψ (psi) sequence, required for encapsidation of the RNA transcribed from the vector into vector particles during production of the vector. Vector production involves the co-introduction of the foreign-gene-delivering vector (gene transfer vector) with the LTRs into transformed human kidney cells (293T) along with DNA constructs that direct the synthesis of the core and envelope proteins required for packaging the viral vector particles (84,86). The messenger RNAs (mRNAs) transcribed from the helper constructs are not packaged into particles because they lack the "ψ" sequence. A notable feature of the lentiviral vector particle is the replacement of the traditional envelope protein of HIV (or other lentiviruses) with the G protein of the vesicular stomatitis virus (VSV G protein), which enables entry through binding to phosphatidyl serine, a component of all cell membranes. In contrast to the Gp120 envelope protein of HIV that restricts the entry of HIV to a few cell types, VSV G protein, when present on the envelope of the vector particles, can permit their entry into (almost) any vertebrate cell, thereby conferring upon lentiviral vectors the ability to deliver DNA to any cell in any organ (89). Further details of lentiviral biology and the nuances of vector design and production can be found in the following references (84,86).

Lentiviral vectors offer significant advantages over traditional transfection methods for establishing stable cells lines. They are relatively easy to produce and concentrate and yield high titers with high infectivity. Lentiviral vectors transduce almost any cell type (including primary neurons) with 100% efficiency and stably integrate the transgene into the cellular genome. Long-term, stable expression of the transgene is obtained over extended periods in established cell lines of neuronal origin. Maintenance of expression is limited in the case of primary cultures only by the longevity of the culture itself. Transferred transgenes are integrated as whole cassettes, whereas with transfection/selection methods using physicochemical methods, random linearization of the circular plasmid carrying the transgene can result in permuted insertions, some of which are non functional because of breakage of the circular plasmid in a region crucial for gene expression. Furthermore, because of the inactivity of the lentiviral LTR promoter in the absence of the viral tat protein (90), transcription of the transgene occurs nearly exclusively form the internal promoter chosen to drive the gene. Lentiviral vectors permit the establishment of stable cells in two to three days, as opposed to several weeks with the traditional transfection and drug-selection approach used for the establishment of stable cell clones by physico-chemical transfection. Expression of the transgene is maintained over long periods without attenuation or silencing commonly seen with oncoretroviral vectors. Integration of lentiviral vectors is pseudorandom with under 1000 or so preferred sites in the human genome, with a bias towards the protein coding regions of transcriptionally active genes (91,92). Due the extreme stability of the integration in host chromatin, no drug selection is required to maintain transgenes delivered with lentiviral vectors. Lentiviral vectors offer the ability to modulate the average gene copy numbers of the transgene in transduced cells by varying amounts of input vector during the establishment of stable cell lines. Multiple vectors carrying different genes can be introduced simultaneously or sequentially to generate cell lines or primary cultures simultaneously expressing combinations of genes to study gene-gene interactions (93). Lastly, lentiviral vectors are practically nontoxic to all cell types in vitro including primary neurons [at the small number of vector copies/cell (5–20) used in most experiments] because they do not express any of the viral proteins and are practically immobile once they integrate into the genome. There are many variations in the design of lentiviral vectors used by many labs and available from commercial vendors.

Because of the ability of lentiviral vectors to transfer genes to differentiated cells, several early studies with lentiviral vectors were performed with primary neurons encompassing both in vitro and in vivo situations as well as in brain slices. Neurons of central nervous system and peripheral nervous system origin are efficiently transduced by lentiviral vectors at relatively low multiplicities of infection (moi: number of vector particles/cell) (93–98). They are also efficient in the transduction of human neuronal progenitor cells (99). In addition to primary neurons, several difficult-to-transfect cell lines such as the PC12 that is extensively used in neurotoxicological research are also transduced efficiently (98) as shown in Figure 1.

Promoter Elements: Determinants of Transgene Expression

In all the studies with engineered cells, in addition to the genes/proteins being expressed, the choice of the promoter used to direct their expression is a critical element that requires context-specific considerations (76,100,101). Each promoter is characterized by several factors such as transcriptional efficiency (strength), tissue specificity (limiting their use to specific cell types derived from the tissue), developmental stage specificity, and inducibility during normal biological processes as well as upon toxicant exposure. Some examples of promoters used

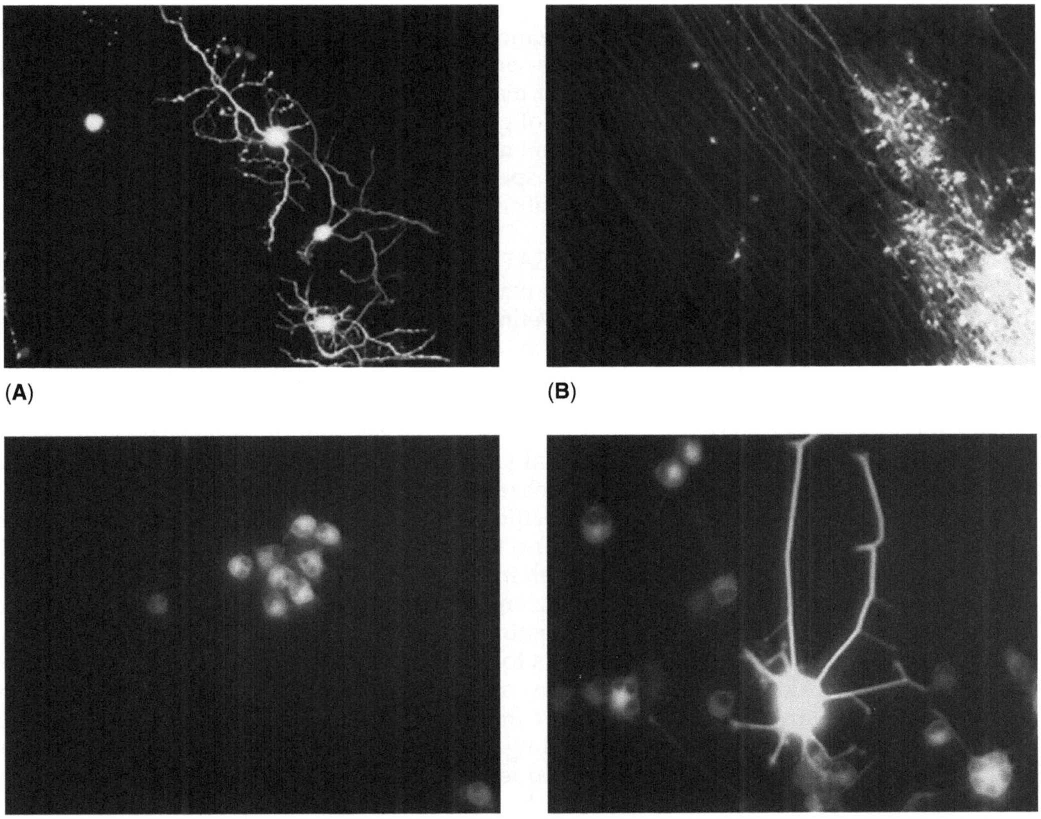

(A) (B) (C) (D)

Figure 1 (*See color insert*) Lentiviral transduction of neuronal cells. Lentiviral vectors expressing green fluorescent protein (GFP) under the constitutive cytomegalovirus or elongation factor alpha promoters were used to transduce cells and explants. Panel (**A**) isolated rat retinal ganglion cells in culture, three days after transduction; (**B**) porcine retinal ganglion cells in explant one week after transduction; (**C**) PC12 cells stably transduced and expressing GFP over several passages; (**D**) neuronal differentiation of cells in panel C, exposed to nerve growth factor. *Source*: Panel B reproduced from Ref. 93.

routinely and widely for constitutive expression in in vitro studies are the immediate early promoter derived from cytomegalovirus (CMV), the early promoter of SV40 virus, and cellular promoters of genes such as the elongation factor alpha (EF1-α), actin, or phosphoglycero kinase (PGK). Among these the CMV, SV40, and EF1-α drive high level expression of transgenes in most cell types and all of these elements are short, and therefore easy to engineer into DNA constructs or vectors with minimal impact on payload.

Tissue-specific or cell type specific promoters, which are highly transcribed in a particular tissue or a cell type within a tissue, are useful tools to restrict the expression of the transfected gene to specific organ or a cell type within the organ (100). In the study of neurotoxicology in vitro, these promoters are useful for determining the authenticity of the neuronal cell type being studied. In tissue explants and complex tissue culture systems, use of tissue specific promoters or their combinations enable focusing of the experiments specifically to neuronal/glial cell type(s) of interest. While the identification of tissue specific promoters is often easy (e.g., albumin promoter in the liver, neurofilament promoter in the brain, immunoglobulin promoter in B cells), the isolation of the minimum stretch of DNA required to maintain fidelity of expression when transferred as a part of the gene construct is difficult, and involves directed experiments as well as trial and error, notably by creating transgenic animals (2,3). Tissue-specific promoters are especially useful for in vitro studies of stem cell differentiation where their expression or lack thereof can enable the monitoring of the differentiation process (102). Over the past several years many neuron-specific and neuronal subclass-specific promoters have been identified and used in research (103,104). Further, a major effort by the Allen Brain Atlas Project (http://www.brainatlas.org) is focused on mapping the expression of known mRNAs across the brain by in situ hybridization using mRNA probes (105). This information is useful for the identification of cell type- or subregion-specific promoters for use in the study of specific cell types in vitro. Another major parallel effort known as GENSAT directed at the specificity and timing of expression of genes in the brain, has produced thousands of transgenic mouse lines expressing fluorescent marker protein under the control of various promoter fragments, thus defining the tissue specific promoter DNA fragments in vivo (105). These resources should aid the further identification of neuron-specific promoters for studies of neuronal cells and hence neurotoxicology.

The studies such as the ones in GENSAT (105) define large segments of DNA that contain the promoter elements, some of which may be too large to be useful in routine transfection experiments in tissue culture cells. Definition of the minimal promoter elements that maintain tissue- or cell type specificity is an arduous task (106). Normally, promoter fragments are of various lengths (usually hundreds of bases to kilobases) are isolated and characterized in vitro by laborious deletion analysis to define the minimum length of the DNA that retains the properties described for the promoter when resident in its normal chromosomal locus. However, these properties may not always translate to every cell type and/or condition beyond the cell system used for their characterization and it therefore it is advisable to verify their proper functioning in the specific cell system being used experimentally. Also analysis of families of promoters responding to the same or similar stimuli have led to the identification of short consensus sequence elements (10–30 bases) which when linked to other basic, minimal promoter elements are sufficient to qualitatively transfer the properties of the larger fragments (107). For instance, the hemoxygenase-1 (HO-1) promoter is contained in a complex 10 to 15 kb fragment and responds to various environmental stimuli including oxidative stress, but a 4 kb fragment appears to contain most of the sequences required for regulation by multiple stressors (108). A major component of the response to oxidative stress by the HO-1 promoter is directed by the binding of a transcription factor known as Nrf2 (108) that binds to a short, 10 base-pair DNA sequence element within the larger promoter fragment. A synthetic promoter comprising of the basic elements common to all eukaryotic promoters [i.e., TATA box and CAAT box (101)] and appropriately positioned Nrf2 binding consensus sequence results in a promoter that is responsive to oxidative stress. Multimerization of the transcription-factor binding consensus sequence in the synthetic promoter results usually in promoters that produce an enhanced response to the specific stimulus (107,109). However, their design involves trial and error for the optimum spacing between the elements and their placement relative to the basal promoter elements. Several such artificial promoter constructs that produce specific responses have been constructed in various laboratories and are also available from commercial vendors.

A novel and an extremely useful version of the synthetic promoters is the widely used tetracycline-regulated inducible promoter system that has provided unprecedented utility to the study of biology in vitro and notably in transgenic animals (110). Assembled using minimal binding elements of a bacterial promoter whose transcription in bacteria is induced by the antibiotic tetracycline, this promoter system works exquisitely in mammalian cells including neurons and neuronal cell types and its transcription is highly induced by addition or withdrawal of tetracycline (two designs respectively known as tet^{on} and tet^{off}). A detailed description of this promoter system is beyond the scope of this chapter and the reader should consult a number of comprehensive reviews in the scientific literature (110) or technical specification sheets provided by commercial vendors. However, for the tet system to function in mammalian cells, co-expression of the tet repressor protein (fused to a transcriptional activator) is obligatory. Other similar inducible promoter-related DNA elements from yeast such as the Gal 4 system as well as from mammalian genes such steroid receptor are useful for inducible expression of proteins and are useful to the study of neuronal cells (103,111). Major advantages of the use of inducible promoter systems are the ability to turn on the expression of the transgene at will and the ability to use a parallel culture of uninduced cells as a control.

Caveats in the Use of Genetically Modified Cells

There are several caveats that need to be considered in designing and interpreting the results of experiments with genetically modified cells. First, the expression/over-expression or untimely expression of an endogenous or foreign protein may compromise the health of the cell, although not overtly, or alter cells in some way not directly measured in the experiment. While appropriate controls using an "empty vector" that lacks the gene of interest or mutant forms of the protein under study are often helpful, they are by no means fool-proof and hence it is often difficult to engineer perfect controls. Use of inducible systems, as mentioned above, provides the best control for the effects of an exogenous protein on a cell, but with the assumption that the process of engineering the cell with the regulatory proteins required for induction itself does not alter the cell's properties in the un-induced state, particularity if the promoter being regulated has any extent of leakiness in the "off" state. Furthermore, as discussed later, the controls are dependent on the specific method used to introduce the genes into cells and the longevity of the engineered state (transient vs. stable expression). In stably transfected cell lines where the foreign gene and the promoter are integrated in the chromosome, the expression properties of the promoter can vary with the site of integration. Because integration of foreign DNA into mammalian chromosome is a random event, the use of a stable pool of cells resulting from independent transfection events, in contrast to the use of isolated single clones, will alleviate extreme deviations associated with a single integration site of the transgene construct. While well-characterized clonal lines are useful in situations like protein overproduction or screening assays, their use in mechanistic studies could be complicated by the chromosomal integration site of the transgene specific to the particular clone. The deviations could arise from the alterations in the properties of the promoter by the integration site or by the interruption and disruption of a key cellular DNA locus by the integration event. Although these caveats may appear so serious as to compromise studies using genetically modified cells, their increasing use in neuroscience and neurotoxicology as discussed below in detail, underscores that the power of genetically manipulated cells in providing insights far outweigh these problems.

EXPERIMENTAL APPROACHES TO GENETIC MODIFICATION OF CELLS

The approaches to genetic modification of cells, based on the desired endpoint, fall broadly into three categories: (*i*) exogenous gene expression/over-expression, (*ii*) inhibition of endogenous gene function using dominant negatives and small interfering RNAs (siRNAs), and (*iii*) reporter-based techniques for real time studies of transcriptional timing and protein localization. The details of the three approaches are discussed below.

Gene Expression/Over-Expression

Protein expression and over-production by genetic engineering of cells is one the earliest and medically beneficial application of genetic manipulation of cells ranging from bacteria to mammalian cells. These techniques form the basis for the production of biological therapeutics ranging from growth factors to antibodies which were then applied with necessary modifications to cell biology and pharmacology in vitro. While the focus of this review is on the applications of these methods to cells in vitro, it should be noted that availability of proteins of high purity created by genetic engineering have permitted the development of in vitro biochemical assays, such as those based on enzyme activity, for characterization of toxicants. For instance, in vitro assays using recombinant-expressed liver enzymes, such as members of the Cyp family, have been vital to the study of toxicant metabolism (112). Similarly, and more pertinent to neurotoxicology, rat, mouse, and human acetyl cholinesterases produced by recombinant methods have been used to characterize organophosphate pesticides, determine the relative species-specific sensitivities of the enzymes facilitating extrapolation of results obtained in rodents to human exposures. These recombinant enzymes are also being developed as antidotes for organophosphate poisoning (113–118).

In the context of cell-based in vitro studies, over-expression of native proteins (such as those normally expressed endogenously in neurons) or their mutated versions, often use promoter elements of higher strength or specificity. This approach results in the elevated production of the protein in the cell, thereby permitting the study of the role of the protein in normal cellular metabolism, in disease states (93), and/or in response to environmental insults (see next paragraph). In this context, it is also possible to express the constitutively active or non-activatable versions (such as by mutation of a post-translational modification site, e.g., phosphorylation) of a protein, permitting the study of its role in cell biology. This strategy is typically applied to proteins that are activated or silenced via phosphorylation where the replacement of neutral serine or threonine (which are normally phosphorylated by a kinase and acquire a negative charge) with a charged aspartic acid residue. Conversely, the serine or threonine in a protein that are substrates for phosphorylation can be replaced with alanine, a moiety that cannot serve as a phosphoacceptor, thereby rendering the mutant protein not activatable by phophorylation. There a large number of studies in literature analyzing kinase cascades using these approaches, however the following examples should be sufficient to illustrate the approach (103,119,120).

The current scientific literature abounds with examples of protein expression studies spanning a decade or more, but the following prototypic studies are cited as instances pertinent to neurological disease and neurotoxicology. Over-expression of glutathione peroxidase in cultured neurons has been shown to render the neurons relatively resistant to the toxic effects of the Alzheimer's disease associated A-beta peptides (121). Over-expression of the unphosphorylated form of the protein MARCKS (myristoylated alanine-rich C kinase substrate, which is normally phosphorylated) in the neuronal cell line SH-SY5Y cells has been shown to influence neurite initiation induced by insulin-like growth factor-1 (122). Heat shock protein over-expression has been shown to protect primary neuronal cells from apoptosis induced by nerve growth factor (NGF) withdrawal or by exposure to retinoic acid (123,124). Transfer of human Cu/Zn superoxide dismutase (SOD) protects dopaminergic neurons against damage by 6-hydroxydopamine (125). Lastly, pertinent to neurotoxicology, over-expression of the Parkinson's disease-associated protein, alpha synuclein, and its mutant forms has been shown to be toxic to neurons in culture and has been shown to increase the sensitivity of these neurons to apoptosis induced by the pesticide dieldrin (126). These and other such examples of relevance to neurotoxicology discussed in a later section, show that gene over-expression studies offer novel avenues to address the toxic mechanisms of compounds and their molecular targets in a way that was not possible without the use of genetic modification. Prior to the advent of genetically engineered cells, such studies depended on the availability of naturally isolated cell variants from human and animal sources that were obtained perchance by serendipity or isolated from diseased animals or humans (such as rat PC12 cells or human SH-SY5Y, discussed earlier under "Utility of in vitro models" on p. 3) and hence their applications are of limited in scope. Genetic manipulation of cells through the techniques described above affords the controlled and graded alteration of expression level of the protein under study that permits the design of more meaningful experiments with appropriate parallel controls.

Interference with Gene Expression or Function

Complementary to over-expression studies, inhibition of specific gene function in cultured neurons or neuronal cell lines offers a powerful tool for the study of gene function and the roles specific genes play in the cell's response to toxicants. Gene function can be inhibited by two broad mechanisms that produce a reduction of activity of the targeted endogenous gene. While achieving the same end, ablation of gene function, these methods differ greatly in the in their mechanics. The earlier of the two, termed "dominant negative expression," was developed based on the observation that the function of endogenous proteins can sometimes be disrupted by over-expression of mutant forms of the protein, resulting in a mutant cell phenotype. A significant number of such dominant negative proteins with specific applications to neuroscience have been described (103). The mechanism of action of dominant negative mutant proteins results from their binding to key targets of their normal counterparts resulting in little or no functionality of the complex. Similarly, the dominant negative protein may incorporate into multimeric complexes usually formed by their normal counterpart, rendering the multi-subunit complexes, such as ion channels, inoperative. Dominant negative version of proteins can be created through a variety of manipulations such as truncation, elimination of post translational medication sites (e.g., phosphorylation sites) either by deletion or point mutations and fusion of the protein to other proteins (103,127–130).

While dominant negative protein expression provided a novel strategy to study gene function, it is rather laborious from the perspective of designing and characterizing a dominant negative protein mutant. Additionally, it is not at all clear that dominant negative can be engineered for all proteins. In view of these limitations with the dominant negative approach and the recent emergence of RNA interference technology for inhibition of gene function, its use has decreased significantly. However, the dominant negative effect should be considered as a possible phenomenon when interpreting the results of experiments where mutant versions of protein are used as controls for their native counterpart.

About a decade ago, in a set of ground breaking experiments, it was observed that introduction of double stranded RNA representing the sequence of genes into a whole range of species from plants to worms to flies, resulted in the specific suppression of the expression of the homologous gene, through the phenomenon dubbed as RNA interference or RNAi (131,132). Frustratingly, this phenomenon could be demonstrated to work in most cells from most species other than vertebrates. This is because vertebrates mounted an interferon response when exposed to double stranded RNA, leading to cytotoxic consequences. It was soon discovered that this phenomenon of RNA interference worked in vertebrate cells, notably in mammalian cells, if the size of the double-stranded RNA was restricted to 18 to 21 nucleotides. This shortening permits RNA interference without eliciting an interferon response (133). Since this breakthrough discovery, a new field of research based on short interfering RNAs (siRNAs) has emerged with a growing number of research publications using the technique and a host of commercial vendors supplying siRNA reagents against practically every gene in the genome of wide range of species. siRNAs (short interfering RNAs) are believed to inhibit the expression of their homologous genes by either causing the degradation of the cognate mRNA or by inhibiting the translation of the mRNA into protein or both.

While space does not permit a detailed description of the siRNA technology, current scientific literature is replete with reviews that expound on the benefits and the caveats of this technology (133–135). Briefly, siRNAs can be delivered to a cell in culture by two methods. In the first method, chemically synthesized short double-stranded RNA molecules 18 to 21 nucleotides in length can be added to the cell culture to obtain specific inhibition of the corresponding genes. A fraction of the RNA from the extracellular space is taken up by the cell to cause specific inhibition of expression of the message cognate to the siRNA. Nucleic acids, owing to their charged nature, do not permeate cells easily. This necessitates the addition of high concentrations (high micromolar) of the siRNA, often with other facilitating chemical agents. The high concentration used often leads to off-target effects and other toxicities. Further, the effects of exogenously added siRNA are often transient prohibiting long-term analysis of gene suppression. Nonetheless, this approach has great value because of its ease and rapidity and has been used extensively to validate hypotheses.

An alternative to the use of exogenous RNA is the introduction of DNA constructs encoding the siRNA into cells. Stable integration of the siRNA construct into the genome of the cells produces long-lasting inhibition of the function of the cognate gene. Typically, the short RNAs are expressed as hairpins (short hairpin RNA:shRNA) from the transfected DNA that, upon intracellular expression, are processed by the cell's biochemical machinery to form double stranded siRNA similar to the moieties internalized upon external addition of synthetic double-stranded siRNA. In this case, however, the expression from the transfected DNA can be longer-lasting and permits the study of long-term consequences of inhibition of specific genes (134,136). Of note is the fact that lentiviral vectors described in the previous sections is a powerful delivery vector for shRNA application and affords the regulation of the expression of the shRNA through inducible promoters (89,137). Within the last five or so years, the siRNA technology has burgeoned into an industry with several vendors supplying siRNAs to practically every protein coding gene in the human and mouse genomes. Algorithms for selection of the target site within the target mRNA and modifications to the siRNA to improve their efficacy have been developed. High throughput methods for functional genomics based on loss of function and phenotypic suppression by siRNA have been developed (134,136). In addition to their use in research, several biotechnology companies are engaged in the development of siRNA as drugs for specific diseases. However, the difficulties with the systemic delivery of the siRNA drug through the biological barriers at the organism level and cellular level remains one of the major roadblocks to the fruition of siRNA therapeutics (138).

Both approaches of delivering siRNA to cells are currently used primarily based on the duration of inhibition required in experimentation. As the RNAi-based gene knockdown has become a routine component of in vitro neurobiological research, key examples of siRNA approaches to neurons are too numerous to review here, but a few salient examples are as follows: Krichevsky and Kosik (135) demonstrated that RNAi could be applied to primary neuronal cultures and demonstrated the knockdown of MAP2 expression. Calabrese and Halpain (139) showed that knockdown of endogenous MARCKS using RNAi reduced spine density and size in hippocampal neuronal cultures. Willard et al. (140) showed that siRNA-mediated knockdown of RGS12 expression also inhibits NGF-induced axonal growth in dissociated cultures of primary dorsal root ganglia neurons.

Exogenous siRNA methods have been used to deduce or confirm proposed regulatory mechanisms and for the validation of putative targets of inhibition by chemicals, notably drugs (141). This approach is based on the premise that sequela of inhibition of a protein target by a drug or toxin would resemble the consequences of inhibition of the production of that protein using siRNA. While this approach has not been applied significantly to toxicological studies, the siRNA approach represents a powerful tool for mechanistic studies in in vitro toxicology where the hypothesized mechanism of action of a toxicant through a target protein can be simulated using siRNA against the gene transcript encoding the target protein.

As with all technologies, there are caveats to be borne in mind with siRNA studies (142,143). It is well established that the inhibitory activity of siRNAs are variable ranging from no effect to >90% inhibition of the target; it is rare that 100% inhibition of the target is ever achieved as would be the case with a gene knockout (hence "knockdown"). siRNAs targeting different regions of the same target mRNA show differing efficiencies (144). Accordingly, it is necessary to test several siRNAs to identify an acceptable moiety. While mismatched (generally sequence scrambled) siRNA at the same concentration as the experimental version can be used to control for mass effects, it is always possible that the control RNA might affect another mRNA target in the cell. Chief among the difficulties in the application of siRNA are the off-target effects of the siRNAs directed against a specific target mRNA as well as global effects on the cell, primarily interferon response. These effects have been observed in experiments where the siRNAs are applied exogenously to the cells as well as when they are expressed endogenously in the cell using expression vectors. While the algorithms for the design of siRNAs are designed to minimize off target effects, it is critical to experimentally rule out off-target effects by use of two to three siRNAs directed to different sites within the same mRNA (144). Although the interferon response characteristic of vertebrate cells is minimized by the use of short siRNAs of 20 to 23 base-pair length, several studies have reported the activation of interferon response even with the small RNAs. Recent studies show that induction of the interferon response varies with both the length of the RNA as well as the cell type being treated (145). Thus, in experiments with siRNA, measurement of interferon response induction should be performed (143,145). Several commercial kits are available to measure interferon response using polymerase chain reaction (PCR) based assays.

While the siRNA methodology serves a powerful tool to analyze gene function in neurons, it would be inappropriate to omit the discussion of the centrality RNA interference phenomenon to cell biology because of its emergence as a newest frontier in biology. Although siRNA was discovered accidentally and developed as a research tool, it has become obvious that nature upstaged man and used the RNA interference as a regulatory phenomenon through much of the evolutionarily history of eukaryotic organisms. Since the detailed study of genomes of eukaryotes began, it has been assumed that much of the genome, except the DNA regions encoding proteins and regions regulating their production, was "junk." However, it was known that much of this "junk DNA" is transcribed to produce RNA (146). At least a fraction of this RNA is comprised of what is now recognized as microRNA (miRNA) that is now believed to be a vital to regulation in the eukaryotic organism through mechanisms analogous to that of siRNAs (131). Several miRNAs have been implicated in the development and maintenance of the nervous system (147,148). This is an unfolding area of biology whose full impact is yet to be felt. The possibility that a mammalian cell contains large classes of regulatory RNA is not inconceivable. Thus, it is possible that toxicants could elicit their adverse effects by interfering with miRNA production and metabolism.

Expression of Fluorescent Reported Proteins

The discovery of fluorescent reporter proteins from marine organisms, first green fluorescent protein (GFP) and subsequently GFP variants covering the entire visible spectrum, have advanced the study of cell biology, including that of neuronal cells into a new dimension (75,149–151). Ectopic expression of these proteins is achieved through the transfer of DNA encoding them. Cells expressing the fluorescent proteins (FPs) can be visualized for periods extending days under a microscope without the need to irreversibly fix the cells and hence permit the real-time monitoring of cellular events (149). The applications of the fluorescent proteins fall into two different categories: (*i*) as makers of cell type specificity and event timers and (*ii*) organelle marking and protein-protein interactions. While the focus here is on the use of FPs as reporters, it should be recognized that prior to the advent of FPs, a number of other reporters such as firefly luciferase, *E. coli* beta galactosidase and others with enzymatic activities were used very successfully as reporters in in vitro studies, primarily to monitor alterations in the activities of promoters of interest. In contrast to FPs that produce stoichiometric readouts (proportional to the amount of fluorescent protein) and hence are less sensitive, the enzymatic reporters are catalytic and hence provide a greater sensitivity and dynamic range to measurements (152).

Fluorescent protein expression driven by either a cell-type-specific or developmental stage-specific promoters respectively permits the identification of cell types or their stage in development. FPs driven by promoters encoding synaptic proteins such as synapsin (68,153) or post synaptic density protein 95 (psd95) (154) for instance, are expressed exclusively in neuronal cells, while FPs linked to the glial fibrillary acidic protein (GFAP) promoter are restricted to glial cells (155). Neuronal cell types could be further distinguished and studied by this approach, if promoters specific to those cell types could be identified and engineered. For instance, there are several promoters described that drive genes involved in neurotransmitter production and these would be specific for that particular class of neurons (103). A GFP protein inserted into the chromosomal tau gene coding sequences resulting in the expression of a GFP-tau fusion protein has been used to follow the differentiation of stem cells into neurons (156). When expressed in neuronal cells, FPs fill the entirety of the cytoplasmic and nucleoplasmic space within the neuron including long axonal processes and therefore provide a convenient and vivid method to assess changes in cell morphology caused by toxicants and other agents. Such experiments can be performed using dispersed neuronal cells or in explants such as brain slices using gene delivery approaches described in the previous sections (78,79).

The timing of appearance of FP expression can also be monitored as a real-time marker of the differentiation process as exemplified in their application in timing the differentiation of stem cells. Differentiation of pluripotent stem cells into cells of various lineages is accompanied by expression of lineage-specific genes. Expression of fluorescent proteins under the control of lineage-specific gene promoters "reports" the specific stage in the differentiation process (68,156). Multiplexing fluorescent proteins with distinct excitation/emission spectra permit the expression of different colored proteins under differentiation-stage-specific or lineage-specific promoters, permitting the real-time visualization of unfolding of developmental events (157). This is a task easily achieved using lentiviral vectors, which can deliver multiple genes sequentially or simultaneously (99). This approach is of particular significance to neurotoxicology, particularly to developmental neurotoxicology, as differentiation of pluripotent human stem cells into neuronal lineages has been proposed as system for screening developmental neurotoxicants (59,158). It is germane to note here that transgenic mouse strains expressing FPs in specific subsets of neurons have been established permitting the study of neuronal cells in dispersed cultures derived from animals treated with drugs or toxicants in vivo (159,160). More recently, transgenic mice with different populations of neurons labeled with different FPs have been generated (161) and the neuronal cells obtained from these mice will likely aid the parallel real-time studies of mixed neuronal populations in vitro.

While the FPs in their free form are useful in themselves, their ability to maintain their fluorescent properties when fused to other proteins has provided powerful avenues to study protein localization and protein-protein interactions in cells including neurons (162,163). With the judicious choice of the fusion partner, orientation (N- vs. C-terminus) and point of fusion between the two proteins, it is possible to maintain the functionality of the protein fused to the FP. While optimization of these factors must be performed empirically, once constructed, such fusion proteins are powerful tools for the real-time study of protein dynamics and localization within cells (163). A large number of studies using neuronal proteins such as receptors (164), synaptic proteins (154,165), and structural proteins (166) fused to GFP have enabled real time analysis of neuronal physiology in vitro, providing information on localization, physiological dynamics, and trafficking. However, it is always advisable to confirm these finding by other techniques such as immunolocalization to ensure that the fusion protein behaves like its native counterpart. The successful application of the fusion protein technology in the last several years shows that these fusion proteins and the cell lines expressing them can be very valuable tools in the study of mechanism of action of toxicants. We have engineered PC12 cells (167) to express a tau-GFP fusion for use in real-time assays of neurite outgrowth to measure effects of putative developmental neurotoxicants on this process . Fusion of GFP with a fragment of the tau proteins (166) enhances the localization of tau-GFP fusion to the neurites enabling their accurate visualization and quantification.

The availability of multiple FPs overlapping the excitation and emission spectra have enabled the study of protein co-localization and protein-protein interactions in real-time (162,168,169). Functional fusion proteins derived using two distinct FPs permit the detection of their proximity in cells through fluorescent resonance energy transfer (FRET). When the fusion proteins are separated spatially, they emit at their characteristic emission frequencies when excited at their respective excitation frequencies. However, when the fusion proteins are in close proximity, in the order of 5 nm (168), the emission of the first FPs (in response to its excitation wavelength) acts as the excitation for the second FP, resulting in non-radiative energy transfer and emission at the wavelength characteristic of the second, recipient FP. This technique has been used to design a number of real-time in vitro assays (170,171) and has found application in neurotoxicity studies such as that of the potent neurotoxin botulinum toxin (172).

GENOMICS, PROTEOMICS, METABOLOMICS

The last decade has seen the emergence of a group of technologies all with the suffix "-omics" that aim at parallel/global analysis of multiple biochemical processes in cells that are very relevant to the field of toxicology (173). These include but are not limited to: transcriptomics, the comparison of mRNA levels (often used interchangeably with genomics); proteomics, the comparison of protein levels; and metabolomics, the comparison of metabolites present in cells. The -omics technologies, particularly transcriptomics and proteomics, are well suited to in vitro studies as opposed to in vivo studies because of the degree of control the experimenter can exert over the cellular environment. Furthermore, unlike in vivo situations where tissues represent a mixture of cell types showing varying response to treatments, enriched populations used in in vitro studies provide data relevant to the particular cell type. We should note however, that although laborious, there are techniques such as laser capture microdissection that permit the isolation of specific cell populations from tissues generated in in vivo studies (174). One of the major benefits of the -omics techniques is the ability to compare the global patterns of changes produced by similar or related treatments. The two major objectives of the -omics approaches are: (*i*) identification of correlative biomarkers/signatures of disease, exposure, effect, or genetic susceptibility and (*ii*) hypothesis generation of causation for further investigation of the origins of diseases or mode of action of toxicants.

Genomics/Transcriptomics

Transcriptomics is the earliest of the -omics technologies that has now been largely reduced to highly automated "chip" based systems for parallel comparisons of the relative levels of thousands of mRNAs in cells subjected to different treatments, such as exposure to growth factors or toxicants (175–177). Cells respond to their environment by altering protein expression primarily through changes in expression of mRNAs. mRNAs are isolated and used to generate fluorescent dye-linked complimentary DNAs (cDNAs), which are then quantified by the degree of hybridization to small single-stranded DNA probes arrayed on solid-matrix chips (microarrays) using imprinting techniques such as photolithography, wherein each probe is specific for a particular gene. Due to the large number of data points generated using this technique, the procedures are highly automated at the wet lab and in silico levels and the evaluation of microarray data is laborious and complicated. Several competing procedures for data analysis have been developed and each differs with the particular algorithms used. Nonetheless, microarray data is useful for generating lists of genes whose mRNAs are up- or down-regulated in response to changes in cellular environment. Examination of the gene lists for clustering of the altered genes by function or pathways allows researchers to postulate hypotheses regarding the mode- or mechanism-of-action of the environmental stimulus (178).

Transcriptomics is a relatively mature -omics technology that has been used extensively in the last five or so years and several in-depth reviews discussing the technology itself and the methods of analysis have appeared. Originally started in individual laboratories, the technology has become an industry with several platforms competing for primacy (179). To address platform-dependent and operator-dependent variability which confounded the reproducibility of microarrays (180), a global consortium MicroArray Quality Control (MAQC) was formed (181). Minimum Information About a Microarray Experiment (MIAME) standards that should enable the harmonization of data obtained under different conditions have been proposed (182). Applications, extensively to cancer but also significantly to other area of biology in vitro and in vivo, have been described. An exemplary and successful application of microarrays from an in vitro research perspective is the analysis of the genomic profile of a panel of 60 cancer cell lines (NCI60) of different pathological origin (183) and their response to a library of thousands of compounds (184,185). This effort showed that expression patterns could be correlated with the histological origin of the cell line, and could also be used to identify pathways predominantly active in a tumor type. The chemical screening also provided information on the tumor type specific action of compounds, suggesting cancer-specific therapeutic approaches. More recently in 2007, FDA has approved the microarray-based cancer diagnostic test, MammaPrint, for determining the metastatic potential of breast cancers (186). Relevant to neuroscience, microarray analysis has examined the transcriptional changes associated with the differentiation of stem cells into neurons, providing stage-specific landmarks (187). Since trasnscriptomic analysis is a relatively mature field, several articles reporting its application to neurotoxicology have appeared in the past five years; these are described in a later section below.

Despite the power of the technology there are several issues that need to be considered when employing microarray analysis. The cost of the array chips and analysis technology renders microarray assays impractical for large experiments covering multiple doses and time points. Thus, microarray analysis is restricted to obtaining a "snap shot" at a limited set of time points or doses of a dynamic process of mRNA metabolism (188) that can yield an entirely different picture if the dose and time parameters are altered. Typically, before the generation of hypotheses, the mRNA changes observed in the microarray experiment are confirmed using complementary and more focused techniques such as quantitative PCR (qPCR) that may be relatively more reliable and cost-effective (189). Several qPCR-based commercial kits have also recently appeared permitting the study or smaller, defined gene sets related to an organ or a process or a pathway at higher reliability and lower noise and cost compared to the gene chips that interrogate thousands of genes. It should also be noted that depending on the nature of the question addressed, the analysis of microarray data can be very mathematically involved (190,191). Therefore microarray experiments should be performed with questions clearly defined and the analysis procedures chosen carefully.

One of the advantages of the in vitro genomics studies is the ease with which hypothesis generation as well as validation can be conducted relative to similar studies performed in vivo. The homogeneity of the cell samples and the ability to control the culture conditions and doses result in reproducible data which can permit efficient generation of mode of action (MOA) hypotheses. Validation of the hypotheses, leading to MOA information can be performed using genetic manipulations such as the gene over-expression and gene-silencing techniques described in detail in the earlier sections. Thus in vitro systems provide ideal substrates for genomic analysis to generate MOA information that can then be compared with in vivo genomics data to facilitate risk characterization.

Proteomics

While changes in the transcriptome of a cell generally result in changes in protein levels of the altered transcripts, the correlation between mRNA levels and the level of the proteins translated from them are not equivalent. Levels of a protein in a cell are influenced by factors such the level of its mRNA, mRNA stability, mRNA sequestration, translational efficiency, and the degradation rate of the protein (188). Additionally, a single pre mRNA can encode multiple protein isoforms through alternative mRNA splicing (192) which can be extreme for some genes (193) and also through the use of internal start codons (194). Furthermore, proteins undergo extensive posttranslational modifications such as proteolysis, glycosylation, phosphorylation, and others that control and/or profoundly alter their biological activities. Some modifications such as phosphorylation and acetylation are reversible and hence dynamic, strongly affecting the physiology of the cell (195,196). This diversification of function at the protein level is not reflected at the mRNA level. Thus, it is believed that direct comparison of protein isoforms and their levels might provide more relevant information regarding the state of a cell than the measurement of their mRNA levels. Therefore proteomics forms a powerful technology for the study of biology (197) and toxicology (198).

As proteins are composed of twenty amino acids, proteins exhibit a greater structural diversity than RNA, which is composed of only four bases. Thus the separation and identification of the protein complement of a cell poses a greater technical challenge than measuring mRNA levels. However, in the last few years there have been spectacular developments in proteomic methodologies such as two dimensional electrophoretic techniques, protein-labeling techniques, and mass spectrometric techniques to make proteomics studies reproducible and meaningful (196). A major advance in the reproducibility of two-dimensional electrophoresis through the development of two-dimensional differential gel electrophoresis (2D-DIGE) using a multi-dye system has led to rapid advances in the field (199,200). Differentially expressed proteins detected by DIGE are sequenced using mass spectrometric techniques to identify the protein (201,202). A technique that is particularly suited to proteomics of in vitro systems the differential labeling of cellular proteins with a heavy stable isotope followed by analysis by mass spectrometry, thus allowing the parallel comparison of the two proteomes (202). Because of the large numbers of posttranslational modifications in proteins and the special approaches need to study each of these, the field of proteomics is now splitting into subspeciality areas such as phosphoproteomics and glycoproteomics. The application of proteomics to the study of biology is only beginning to emerge and the application of this technique to the study of neurotoxicology is in its infancy, but has great potential (203).

Metabolomics

The rationale of metabolomics is that the action of thousands of genes and their protein products (30,000–100,000) are reflected at a cellular level in a much smaller number (a few thousands) of cellular metabolites. Therefore, changes in the levels of these metabolites represent the sum total of the actions of genes/proteins. Metabolomics analysis is currently focused primarily on in vivo studies, where body fluids such as serum, urine, or saliva, etc., are profiled for metabolites using mass spectrometric and nuclear magnetic resonance techniques (204). Such analysis is aimed at matching metabolite profiles to disease, drug treatment or environmental exposure (205). While powerful at a whole organism level, metabolomics is less-suited to in vitro studies because of the small molecule-rich media used in cell cultures and the difficulty in isolating the intracellular metabolites of a cell without causing significant alterations due to technical manipulations (206). However, developments are occurring in this field for successful application of metabolomics to microbial and mammalian cell bioengineering (206). Despite the demonstrated advances in the application of this technology at the organism level, the utility of metabolomics in vitro neurotoxicology are only beginning to emerge (207).

APPLICATIONS IN NEUROTOXICOLOGY

Recent advances in the molecular techniques discussed in the previous sections are already propelling research to identify modes of action in neurotoxicology. This section will review recent literature highlighting the utility of the previously described molecular techniques in uncovering toxicant modes of action.

Little has contributed as much to the recent understanding of toxicity mechanisms as functional gene analysis through such techniques as gene over-expression and targeted gene perturbation by siRNA knockdown, dominant negatives, and pharmacological agents. Used collectively, these techniques have allowed investigators to study the roles of specific genes and their gene products in neurotoxicity. Generally, studies have focused on identifying protective gene products by demonstrating how their over-expression can reduce the effects of neurotoxicants. Conversely, perturbation of these protective genes renders cells more sensitive to toxicant exposure. A recent study by Lee et al. (208) showed that over-expression of HO-1 increased reactive oxygen species (ROS) formation and cell death in mouse neuronal MN9D cells exposed to the commercial polychlorinated biphenyl (PCB) mixture, Aroclor 1254; additionally, HO-1 inhibition by RNAi knockdown blocked Aroclor 1254-induced ROS production and cell death. These results indicate that HO-1, an antioxidant protein that usually confers protection against electrophilic toxicants, can in some instances actually facilitate ROS-mediated toxicity. Overexpression of the yeast mitochondrial complex I subunit nicotinamide adenine dinucleotide dehydrogenase (NDI1) protected human epithelioma SK-N-MC cells from the toxic effects of rotenone, a powerful complex I inhibitor that produces a Parkinsonian syndrome in rats (209). In this case, the protective effect of NDI1 overexpression supports the mitochondria-specific mechanism that has been postulated for rotenone. Stable overexpression of free radical scavenging enzyme SOD-1 in murine N9 microglia decreased superoxide and nitric oxide production after lipopolysaccharide treatment; the altered ROS production reduced the ability of the N9 microglia to induce toxicity with co-cultured neurons (210) and these results could be reversed by inhibition of SOD-1 with disulfiram. Ectopic expression of neuroglobin, a vertebrate globin that protects against neuronal hypoxia and cerebral ischemia, rendered murine cortical neurons resistant to the toxic effects of the excitatory amino acid N-methyl-D-aspartate as well the Alzheimer's disease-related peptide Aβ (211).

As discussed earlier, over-expression experiments have caveats that often make interpretation using over-expression data alone problematic. The primary concern with over-expression analysis is that high levels of ectopic expression of a specific gene product will alter normal cellular function, creating a model that bears little resemblance to the actual biology in question. One of the more impactful gene analysis methods that has helped alleviate this concern with over-expression analysis is target gene knockdown

Dominant negative mutants, as discussed previously, provide a useful tool to interfere with normal gene function. The utility of dominant negative mutants stems from the partial functionality retained by the mutant. Newhouse et al. (212) used constitutively active MKK3, MMK4, and c-Jun dominant negative mutants to delineate which mitogen-activated protein kinase (MAPK) signaling cascades were involved with rotenone-induced apoptosis in SH-SY5Y cells. c-Jun and c-Fos dominant negatives were used in another study to examine the role of jun N-terminal kinase (JNK) signaling in thimerosal-mediated apoptosis in SK-N-SH cells (213). Expression of a dominant negative mixed lineage kinase 3 (MLK3) mutant attenuated the cytotoxicity of MPP+ in SH-SY5Y cells, revealing a protective role for MLK3/JNK pathways (214).

As discussed earlier, creation of dominant negative reagents is an arduous and often chancy task, and hence this approach is being supplanted by knockdown techniques using siRNA. Gene knockdown techniques, especially those using RNA interference that reduces the cell's capacity to express a particular gene product, allow researchers to systemically target components of pathways thought to be critical in the mechanism of action. The targeted knockdown of HO-1 using RNAi in the report by Lee et al. (208) has already been mentioned. In another study, siRNAs were used to inhibit expression of specific Bcl-2 family members to discern the role each member contributed to paraquat-induced cell death using human SK-N-SH cells (215). It was revealed that siRNA-mediated knockdown of BNip3, Noxa, and Bak protected cells from the effects of paraquat treatment, suggesting these Bcl-2 family members played a role in paraquat-mediated apoptosis. Antisense inhibition of the p53-activated gene PAG608 abated toxicity in methamphetamine-treated PC12 cells, suggesting a role for the p53 pathway in methamphetamine-induced neurotoxicity (216). Novitskya at al. (217) recently showed that targeted knockdown of endogenous prion protein (PrPc) abolished the toxic effect of recombinant mammalian prion protein (rPrP) in human SH-SY5Y cells, indicating a role for endogenous PrPc in prion-mediated toxicity.

Microarray technology provides researchers with a tool to quantify the simultaneous expression of thousands of genes. This has helped advance neurotoxicology in at least three important ways. First, microarray technology has increased our understanding of specific gene expression patterns associated with neurological diseases. Identification of key genes associated with neurological diseases provides insights into the mode of action for toxicants that elicit similar adverse health effects. A classic example of the overlap between neurological disease and neurotoxicity is that of the heroin analog 1-methyl-4-phenyl-1,2,5,6-tetrahydropyridine (MPTP) and Parkinson's disease. MPTP targets the dopaminergic neuron of the substantia nigra; these same neurons are degenerated in Parkinson's disease. Secondly, microarrays help illustrate how gene expression patterns change after exposure to toxicants, an application referred to as toxicogenomics. Altered gene expression patterns often highlight cellular pathways affected by a particular toxicant. Affected pathway components may be the targets of toxicity, such as cellular metabolism pathway components, or that are activated in response to toxicant exposure, such as heat shock response. Toxicogenomic data can thus be used to identify toxicant modes of action as the understanding of the pathways linked to regulating cellular activities increases. Lastly, gene expression changes identified using microarray experiments can be useful as biomarkers of exposure (218).

Toxicology as a field has not fully exploited microarrays due to a variety of factors such as cost, lack of familiarity with the required techniques, and difficulties with data analysis. However, there have been several studies using microarrays to analyze toxicity-induced changes in gene expression using cells derived from the nervous system. Gene expression profiling using primary human astrocytes highlighted that manganese chloride, long known to induce neurotoxicity, stimulated the interferon-γ inflammatory pathway, induced oxidative stress responses, and stymied cell cycle progression (219). Using microarray data from immortalized rat astrocytes, Bouton et al. identified several genes activated in response lead exposure, namely stress-responsive genes such as GFAP and heat shock protein 70 (220). A study using the MPTP-treated PC12 cells revealed several pathways likely to be involved in MPTP toxicity including oxidative stress, DNA and protein damage, cell cycle arrest, and apoptosis (221). A similar study using mouse dopaminergic neuronal MN9D cells also showed a role for oxidative stress, cell cycle arrest, and apoptosis in MPTP-induced toxicity but also suggested that cellular metabolism and iron homeostasis may also be disrupted by MPTP (222).

Proteomics is a relatively new field compared to genomics but recent work has highlighted the promise of proteomics in neurotoxicological research. Zhou et al. (223) used quantitative proteomics techniques to identify proteins involved in microglial activation following MPP+ exposure in rat primary neuron-microglia mixed cultures. Another study probed for proteins targeted by methylmercury exposure in mice cerebellar granule cells (224). The use of proteomics in neurotoxicology research will surely increase as the power of the applications develops. Mapping post-translational modifications, which have impact on protein stability, structure, and function are of keen interest to researchers interested in toxicant mechanisms of action. Protein expression profiling provides information about changes in quantity and pattern of protein expression resulting from toxicant exposure. Finally, a proteomics technique that isolates protein complexes for component identification known as protein network mapping would allow toxicologists to study how toxicants affect protein-protein interactions leading to cellular dysfunction (198).

Metabolomics emerged later than proteomics and while best suited to in vivo studies, applications to in vitro neurotoxicology studies are emerging. For instance van Vliet et al. (207) examined the alterations in the metabolomic profile of primary rat neuronal cells induced by the neurotoxicant, methyl mercury, raising the possibility that gamma-aminobutyric acid, choline, glutamine, creatine, and spermine could serve as metabolite biomarkers for methylmercury action. It is not unrealistic to anticipate that other studies in this vein would be forthcoming and will help expand the application of metabolomics to in vitro neurotoxicology with the simultaneous measurement in appropriate cell types of alterations in neurotransmitters, neuropeptides, and other metabolic markers of neurotoxicity.

ACKNOWLEDGMENTS

This chapter was reviewed by the National Health and Environmental Effects Research Laboratory, USEPA, and approved for publication. Approval neither signifies that the contents reflect the views of the Agency, nor does mention of trade names or commercial products constitute endorsement or recommendation for use.

REFERENCES

1. National Research Council. Toxicity Testing in the 21st Century: A Vision and a Strategy. Washington, DC: National Academies Press, 2007.
2. Pinkert CA. Introduction to transgenic animal technology. In: Pinkert CA, ed. Transgenic Animal Technology: A Laboratory Handbook, 2nd edn. New York: Academic Press, 2002: 3–12.
3. Houdebine LM. Transgenic animals. In: Encyclopedia of Life Sciences. Chichester: John Wiley and Sons, 2005: DOI: 10.1038/npg-els.0003839.
4. Mansouri A. Mouse knockouts: modifying genome by using embryonic stem cells. In: Encyclopedia of Life Sciences. Chichester: John Wiley and Sons, 2007: DOI: 10.1038/npg.els.0003840.
5. Setou M, Hayasaka T, Yao I. Axonal transport versus dendritic transport. J Neurobiol 2004; 58: 201–6.
6. Purves D, Augustine GJ, Fitzpatrick D, et al., eds. Neuroscience, 4th edn. Sunderland, MA: Sinauer Associates, Inc, 2007.
7. Lester DS, Slikker W, Lazarovici P, eds. Site-Selective Neurotoxicity. New York: Taylor and Francis, 2002.
8. Hampson DR, Manalo JL. The activation of glutamate receptors by kainic acid and domoic acid. Nat Toxins 1998; 6: 153–8.
9. DeCaprio AP. n-Hexane neurotoxicity: a mechanism involving pyrrole adduct formation in axonal cytoskeletal protein. Neurotoxicology 1987; 8: 199–210.
10. Narahashi T. Neuronal ion channels as the target sites of insecticides. Pharmacol Toxicol 1996; 79: 1–14.
11. Shafer TJ, Bushnell PJ, Benignus VA, Woodward JJ. Perturbation of voltage-sensitive Ca2+ channel function by volatile organic solvents. J Pharmacol Exp Ther 2005; 315: 1109–18.
12. Baden DG. Brevetoxins: unique polyether dinoflagellate toxins. FASEB J 1989; 3: 1807–17.
13. Reichardt LF. Neurotrophin-regulated signaling pathways. Philos Trans R Soc Lond B Biol Sci 2006; 361: 1545–64.
14. Mates JM, Segura JA, Alonso FJ, Marquez J. Pathways from glutamine to apoptosis. Front Biosci 2006; 11: 3164–80.
15. Morrison N, Wood AJ, Hancock D, et al. Annotation of environmental OMICS data: application to the transcriptomics domain. OMICS 2006; 10: 172–8.
16. Culmsee C, Landshamer S. Molecular insights into mechanisms of the cell death program: role in the progression of neurodegenerative disorders. Curr Alzheimer Res 2006; 3: 269–83.
17. Hara MR, Snyder SH. Cell signaling and neuronal death. Annu Rev Pharmacol Toxicol 2007; 47: 117–41.
18. Arevalo JC, Chao MV. Axonal growth: where neurotrophins meet Wnts. Curr Opin Cell Biol 2005; 17: 112–15.
19. Whitford KL, Dijkhuizen P, Polleux F, Ghosh A. Molecular control of cortical dendrite development. Annu Rev Neurosci 2002; 25: 127–49.
20. Yoshimura T, Arimura N, Kaibuchi K. Signaling networks in neuronal polarization. J Neurosci 2006; 26: 10626–30.
21. Keith CH, Wilson MT. Factors controlling axonal and dendritic arbors. Int Rev Cytol 2001; 205: 77–147.
22. Mattson MP. Establishment and plasticity of neuronal polarity. J Neurosci Res 1999; 57: 577–89.
23. Ghijsen WE, Leenders AG. Differential signaling in presynaptic neurotransmitter release. Cell Mol Life Sci 2005; 62: 937–54.
24. Rusakov DA. Ca2+-dependent mechanisms of presynaptic control at central synapses. Neuroscientist 2006; 12: 317–26.
25. Mayford M. Protein kinase signaling in synaptic plasticity and memory. Curr Opin Neurobiol 2007; 17: 313–17.
26. Waites CL, Craig AM, Garner CC. Mechanisms of vertebrate synaptogenesis. Annu Rev Neurosci 2005; 28: 251–74.
27. Garza A, Vega R, Soto E. Cellular mechanisms of lead neurotoxicity. Med Sci Monit 2006; 12: RA57–65.
28. Marchetti C. Molecular targets of lead in brain neurotoxicity. Neurotox Res 2003; 5: 221–36.

29. White LD, Cory-Slechta DA, Gilbert ME, et al. New and evolving concepts in the neurotoxicology of lead. Toxicol Appl Pharmacol 2007; 225: 1–27.
30. Li Z, Dong T, Proschel C, Noble M. Chemically diverse toxicants converge on Fyn and c-Cbl to disrupt precursor cell function. PLoS Biol 2007; 5: e35.
31. Boulton AA, Baker GB, Bateson AN, eds. Cell Neurobiology Techniques. Totowa, NJ: Humana Press, 1999.
32. Cestelli A, Savettieri G, Salemi G, Di Liegro I. Neuronal cell cultures: a tool for investigations in developmental neurobiology. Neurochem Res 1992; 17: 1163–80.
33. Federoff S, Richardson A, eds. Protocols for Neural Cell Culture, 2nd edn. Totowa, NJ: Humana Press, 1997.
34. Harry GJ, Billingsley M, Bruinink A, et al. *In vitro* techniques for the assessment of neurotoxicity. Environ Health Perspect 1998; 106(Suppl 1): 131–58.
35. Tiffany-Castiglioni E, ed. *In Vitro* Neurotoxicology. Totowa, NJ: Humana Press, 2004.
36. Lonchamp E, Dupont JL, Beekenkamp H, et al. The mouse cerebellar cortex in organotypic slice cultures: an *in vitro* model to analyze the consequences of mutations and pathologies on neuronal survival, development, and function. Crit Rev Neurobiol 2006; 18: 179–86.
37. Noraberg J, Poulsen FR, Blaabjerg M, et al. Organotypic hippocampal slice cultures for studies of brain damage, neuroprotection and neurorepair. Curr Drug Targets CNS Neurol Disord 2005; 4: 435–52.
38. Toyama K, Komatsu Y, Yamamoto N, et al. *In vitro* approach to visual cortical development and plasticity. Neurosci Res 1991; 12: 57–71.
39. Gilbert ME. *In vitro* systems as simulations of *in vivo* conditions: the study of cognition and synaptic plasticity in neurotoxicology. Ann N Y Acad Sci 2000; 919: 119–32.
40. Banker G, Goslin K, eds. Culturing of Nerve Cells, 2nd edn. Cambridge, MA: MIT Press, 1998.
41. Hollenbeck PJ, Bamburg JR. Comparing the properties of neuronal culture systems: a shopping guide for the cell biologist. Methods Cell Biol 2003; 71: 1–16.
42. James IF, Wood JN. In: James IF, Wood JN, eds. Neuronal Cell Lines: A Practical Approach. Oxford: IRL Press, 1992: 249–59.
43. Lendahl U, McKay RD. The use of cell lines in neurobiology. Trends Neurosci 1990; 13: 132–7.
44. Shastry P, Basu A, Rajadhyaksha MS. Neuroblastoma cell lines—a versatile *in vitro* model in neurobiology. Int J Neurosci 2001; 108: 109–26.
45. Greene LA, Tischler AS. Establishment of a noradrenergic clonal line of rat adrenal pheochromocytoma cells which respond to nerve growth factor. Proc Natl Acad Sci U S A 1976; 73: 2424–8.
46. Augusti-Tocco G, Sato G. Establishment of functional clonal lines of neurons from mouse neuroblastoma. Proc Natl Acad Sci U S A 1969; 64: 311–15.
47. Schubert D, Humphreys S, Baroni C, et al. I*n vitro* differentiation of a mouse neuroblastoma. Proc Natl Acad Sci U S A 1969; 64: 316–23.
48. Ehrich M. Using neuroblastoma cell lines to address differential specificity to organophosphates. Clin Exp Pharmacol Physiol 1995; 22: 291–2.
49. Fujita K, Lazarovici P, Guroff G. Regulation of the differentiation of PC12 pheochromocytoma cells. Environ Health Perspect 1989; 80: 127–42.
50. Shafer TJ, Atchison WD. Transmitter, ion channel and receptor properties of pheochromocytoma (PC12) cells: a model for neurotoxicological studies. Neurotoxicology 1991; 12: 473–92.
51. Bartlett PF, Reid HH, Bailey KA, et al. Immortalization of mouse neural precursor cells by the c-myc oncogene. Proc Natl Acad Sci U S A 1988; 85: 3255–9.
52. Banker GA, Goslin K, eds. Culturing Nerve Cells. Cambridge: MIT Press, 1998.
53. Behrstock S, Svendsen C. Neural stem cells. In: Encyclopedia of Life Sciences. Chichester: John Wiley and Sons, 2005: DOI: 10.1038/npg.els.0005765.
54. Terskikh AV, Bryant PJ, Schwartz PH. Mammalian stem cells. Pediatr Res 2006; 59(4 Pt 2): 13R–20R.
55. Allegrucci C, Young LE. Differences between human embryonic stem cell lines. Hum Reprod Update 2007; 13: 103–20.
56. Elkabetz Y, Panagiotakos G, Al Shamy G, et al. Human ES cell-derived neural rosettes reveal a functionally distinct early neural stem cell stage. Genes Dev 2008; 22: 152–65.
57. Lindvall O, Kokaia Z. Stem cells for the treatment of neurological disorders. Nature 2006; 441: 1094–6.
58. Nat R, Hovatta O. *In vitro* neural differentiation of human embryonic stem cells. J Cell Mol Med 2004; 8: 570–1.
59. Davila JC, Cezar GG, Thiede M, Strom S, Miki T, Trosko J. Use and application of stem cells in toxicology. Toxicol Sci 2004; 79: 214–23.

60. Rolletschek A, Blyszczuk P, Wobus AM. Embryonic stem cell-derived cardiac, neuronal and pancreatic cells as model systems to study toxicological effects. Toxicol Lett 2004; 149: 361–9.
61. Dalton R. California prepares to roll out stem-cell funding. Nature 2005; 437: 800–1.
62. Holden C. Biomedical research. States, foundations lead the way after Bush vetoes stem cell bill. Science 2006; 313: 420–1.
63. Wadman M. Stem-cell issue moves up the US agenda. Nature 2007; 446: 842.
64. Johe KK, Hazel TG, Muller T, et al. Single factors direct the differentiation of stem cells from the fetal and adult central nervous system. Genes Dev 1996; 10: 3129–40.
65. Lee G, Kim H, Elkabetz Y, et al. Isolation and directed differentiation of neural crest stem cells derived from human embryonic stem cells. Nat Biotechnol 2007; 25: 1468–75.
66. Lee H, Shamy GA, Elkabetz Y, et al. Directed differentiation and transplantation of human embryonic stem cell-derived motoneurons. Stem Cells 2007; 25: 1931–9.
67. Mayer-Proschel M, Liu Y, Xue H, et al. Human neural precursor cells—an *in vitro* characterization. Clin Neurosci Res 2002; 2: 58–69.
68. Pruszak J, Sonntag KC, Aung MH, et al. Markers and methods for cell sorting of human embryonic stem cell-derived neural cell populations. Stem Cells 2007; 25: 2257–68.
69. Flanagan LA, Rebaza LM, Derzic S, et al. Regulation of human neural precursor cells by laminin and integrins. J Neurosci Res 2006; 83: 845–56.
70. Hughes P, Marshall D, Reid Y, et al. The costs of using unauthenticated, over-passaged cell lines: how much more data do we need? Biotechniques 2007; 43: 575–86.
71. Chalberg TW, Phillips JE, Calos MP. Transfection of DNA into mammalian cells in culture. In: Encyclopedia of Life Sciences. Chichester: John Wiley and Sons, 2005; 10.1038/npg.els.0003928.
72. Ginsberg GS, Thyagarajan B, Phillips JE, et al. Gene delivery by viruses. In: Encyclopedia of Life Sciences. Chichester: John Wiley and Sons, 2005: DOI: 10.1038/npg.els.0003927.
73. Craig AM. Transfecting cultured neurons. In: Banker G, Goslin K, eds. Culturing Nerve Cells, 2nd edn. Cambridge, MA: MIT Press, 1998: 79–112.
74. Ghosh C, Song W, Lahiri DK. Efficient DNA transfection in neuronal and astrocytic cell lines. Mol Biol Rep 2000; 27: 113–21.
75. Hebshi LD, Angres BM, Li XL, et al Green fluorescent protein (GFP). In: Encyclopedia of Life Sciences. Chichester: John Wiley and Sons, 2007: DOI:10.1002/9780470015902.a0002663.pub2.
76. Ma J. Promoter fusions to study gene expression. In: Encyclopedia of Life Sciences. Chichester: John Wiley and Sons, 2007: DOI: 10.1002/97804700115902.a0000976.pub2.
77. Gromov PS, Cellis JE. Gene transfer and expression in tissue culture cells. In: Encylcopedia of Life Sciences. Chichester: John Wiley and Sons, 2001: DOI: 10.1038/npg.els.0002662.
78. O'Brien JA, Lummis SC. Biolistic transfection of neuronal cultures using a hand-held gene gun. Nat Protoc 2006; 1: 977–81.
79. Lo DC, McAllister AK, Katz LC. Neuronal transfection in brain slices using particle-mediated gene transfer. Neuron 1994; 13: 1263–8.
80. Roosen G, Velten FW, Vos J-MH. Artificial chromosomes. In: Encyclopedia of Life Sciences. Chichester: John Wiley and Sons, 2002: [Available from: http://www.mrw.interscience.wiley.com/emrw/9780470015902/els/article/a0001159/current/pdf].
81. Connelly S. Adenoviral vectors. In: Meager A, ed. Gene Therapy Technologies, Applications and Regulations. New York: John Wiley and Sons, 1999: 87–108.
82. Flotte TR, Carter BJ. Adeno-associated viral vectors. In: Meager A, ed. Gene Therapy Technologies, Applications and Regulations. New York: John Wiley and Sons, 1999: 109–26.
83. Gunzburg WH, Salmons B. Retroviral vectors. In: Meager A, ed. Gene Therapy Technologies, Applications and Regulations. New York: John Wiley and Sons, 1999: 35–60.
84. Lever AML. Lentiviral vectors. In: Meager A, ed. Gene Therapy Technologies, Applications and Regulations. New York: John Wiley and Sons, 1999: 61–86.
85. Soares MK, Goins WF, Glorioso JC, et al. Advances in engineering HSV vectors for gene transfer in the central nervous system. In: Meager A, ed. Gene Therapy Technologies, Applications and Regulations. New York: John Wiley and Sons, 1999: 127–64.
86. Cockrell AS, Kafri T. Gene delivery by lentivirus vectors. Mol Biotechnol 2007; 36: 184–204.
87. Wu X, Wakefield JK, Liu H, et al. Development of a novel trans-lentiviral vector that affords predictable safety. Mol Ther 2000; 2: 47–55.
88. Debyser Z. Biosafety of lentiviral vectors. Curr Gene Ther 2003; 3: 517–25.
89. Wiznerowicz M, Trono D. Harnessing HIV for therapy, basic research and biotechnology. Trends Biotechnol 2005; 23: 42–7.
90. Brigati C, Giacca M, Noonan DM, et al. Tat, its TARgets and the control of viral gene expression. FEMS Microbiol Lett 2003; 220: 57–65.

91. Mitchell RS, Beitzel BF, Schroder AR, et al. Retroviral DNA integration: ASLV, HIV, and MLV show distinct target site preferences. PLoS Biol 2004; 2: e234.
92. Schroder AR, Shinn P, Chen H, et al. HIV-1 integration in the human genome favors active genes and local hotspots. Cell 2002; 110: 521–9.
93. Shaughnessy L, Thomas MB, Wakefield J, et al. Lentiviral vector-based models of amyloid pathology: from cells to animals. Curr Alzheimer Res 2005; 2: 239–47.
94. Blomer U, Naldini L, Kafri T, et al. Highly efficient and sustained gene transfer in adult neurons with a lentivirus vector. J Virol 1997; 71: 6641–9.
95. Fleming J, Ginn SL, Weinberger RP, et al. Adeno-associated virus and lentivirus vectors mediate efficient and sustained transduction of cultured mouse and human dorsal root ganglia sensory neurons. Hum Gene Ther 2001; 12: 77–86.
96. Naldini L, Blomer U, Gage FH, et al. Efficient transfer, integration, and sustained long-term expression of the transgene in adult rat brains injected with a lentiviral vector. Proc Natl Acad Sci U S A 1996; 93: 11382–8.
97. Naldini L, Blomer U, Gallay P, et al. *In vivo* gene delivery and stable transduction of nondividing cells by a lentiviral vector. Science 1996; 272: 263–7.
98. Shaughnessy L, Chamblin B, McMahon L, et al. Novel approaches to models of Alzheimer's disease pathology for drug screening and development. J Mol Neurosci 2004; 24: 23–32.
99. Capowski EE, Schneider BL, Ebert AD, et al. Lentiviral vector-mediated genetic modification of human neural progenitor cells for ex vivo gene therapy. J Neurosci Methods 2007; 163: 338–49.
100. Bulger M, Groudine M. Enhancers. In: Encyclopedia of Life Sciences. Chichester: John Wiley and Sons, 2005: DOI: 10.1038/npg.els.0005034.
101. Smale ST. Promoters. In: Encyclopedia of Life Sciences. Chichester: John Wiley and Sons, 2005: DOI: 10.1038/npg.els.0005025.
102. Copray S, Balasubramaniyan V, Levenga J, et al. Olig2 overexpression induces the *in vitro* differentiation of neural stem cells into mature oligodendrocytes. Stem Cells 2006; 24: 1001–10.
103. Wells T, Carter DA. Genetic engineering of neural function in transgenic rodents: towards a comprehensive strategy? J Neurosci Methods 2001; 108: 111–30.
104. Luan H, White BH. Combinatorial methods for refined neuronal gene targeting. Curr Opin Neurobiol 2007; 17: 572–80.
105. Kuehn BM. Genomics proving a powerful tool for mapping the landscape of the brain. JAMA 2006; 295: 145–8.
106. Heintzman ND, Ren B. The gateway to transcription: identifying, characterizing and understanding promoters in the eukaryotic genome. Cell Mol Life Sci 2007; 64: 386–400.
107. Morella KK, Lai CF, Kumaki S, et al. The action of interleukin-2 receptor subunits defines a new type of signaling mechanism for hematopoietin receptors in hepatic cells and fibroblasts. J Biol Chem 1995; 270: 8298–310.
108. Ryter SW, Alam J, Choi AM. Heme oxygenase-1/carbon monoxide: from basic science to therapeutic applications. Physiol Rev 2006; 86: 583–650.
109. Nettelbeck DM, Curiel DT, Muller R. Promoter optimization and artificial promoters for transcriptional targeting in gene therapy. In: Curiel DT, Douglas JT, Valacarcel J, eds. Vector Targeting for Therapeutic Gene Delivery. Wilmington, DE: Wiley-Liss Inc, 2002: 481–503.
110. Rossi FM, Blau HM. Recent advances in inducible gene expression systems. Curr Opin Biotechnol 1998; 9: 451–6.
111. Burcin MM, O'Malley BW, Tsai SY. A regulatory system for target gene expression. Front Biosci 1998; 3: c1–7.
112. Wolf CR. The Gerhard Zbinden memorial lecture: application of biochemical and genetic approaches to understanding pathways of chemical toxicity. Toxicol Lett 2002; 127: 3–17.
113. Cohen O, Kronman C, Raveh L, et al. Comparison of polyethylene glycol-conjugated recombinant human acetylcholinesterase and serum human butyrylcholinesterase as bioscavengers of organophosphate compounds. Mol Pharmacol 2006; 70: 1121–31.
114. Kovarik Z, Radic Z, Berman HA, et al. Mutation of acetylcholinesterase to enhance oxime-assisted catalytic turnover of methylphosphonates. Toxicology 2007; 233: 79–84.
115. Kronman C, Cohen O, Raveh L, et al. Polyethylene-glycol conjugated recombinant human acetylcholinesterase serves as an efficacious bioscavenger against soman intoxication. Toxicology 2007; 233: 40–6.
116. Ma X, Tan J, Wei D, et al. High-level secretion and purification of recombinant acetylcholinesterase from human cerebral tissue in P. pastoris and identification by chromogenic reaction. Appl Microbiol Biotechnol 2006; 72: 316–22.
117. Sultatos LG. Concentration-dependent binding of chlorpyrifos oxon to acetylcholinesterase. Toxicol Sci 2007; 100: 128–35.

118. Taylor P, Reiner E, Kovarik Z, et al. Application of recombinant DNA methods for production of cholinesterases as organophosphate antidotes and detectors. Arh Hig Rada Toksikol 2007; 58: 339–45.
119. Rammes G, Steckler T, Kresse A, et al. Synaptic plasticity in the basolateral amygdala in transgenic mice expressing dominant-negative cAMP response element-binding protein (CREB) in forebrain. Eur J Neurosci 2000; 12: 2534–46.
120. Xu A, Wang Y, Xu LY, et al. Protein kinase C alpha-mediated negative feedback regulation is responsible for the termination of insulin-like growth factor I-induced activation of nuclear phospholipase C beta1 in Swiss 3T3 cells. J Biol Chem 2001; 276: 14980–6.
121. Barkats M, Millecamps S, Abrioux P, et al. Overexpression of glutathione peroxidase increases the resistance of neuronal cells to Abeta-mediated neurotoxicity. J Neurochem 2000; 75: 1438–46.
122. Shiraishi M, Tanabe A, Saito N, Sasaki Y. Unphosphorylated MARCKS is involved in neurite initiation induced by insulin-like growth factor-I in SH-SY5Y cells. J Cell Physiol 2006; 209: 1029–38.
123. Amin V, Cumming DV, Latchman DS. Over-expression of heat shock protein 70 protects neuronal cells against both thermal and ischaemic stress but with different efficiencies. Neurosci Lett 1996; 206: 45–8.
124. Kalwy SA, Akbar MT, Coffin RS, et al. Heat shock protein 27 delivered via a herpes simplex virus vector can protect neurons of the hippocampus against kainic-acid-induced cell loss. Brain Res Mol Brain Res 2003; 111: 91–103.
125. Barkats M, Millecamps S, Bilang-Bleuel A, et al. Neuronal transfer of the human Cu/Zn superoxide dismutase gene increases the resistance of dopaminergic neurons to 6-hydroxydopamine. J Neurochem 2002; 82: 101–9.
126. Sun F, Anantharam V, Latchoumycandane C, et al. Dieldrin induces ubiquitin-proteasome dysfunction in alpha-synuclein overexpressing dopaminergic neuronal cells and enhances susceptibility to apoptotic cell death. J Pharmacol Exp Ther 2005; 315: 69–79.
127. Ham J, Babij C, Whitfield J, et al. A c-Jun dominant negative mutant protects sympathetic neurons against programmed cell death. Neuron 1995; 14: 927–39.
128. Jia Y, Zhou J, Tai Y, Wang Y. TRPC channels promote cerebellar granule neuron survival. Nat Neurosci 2007; 10: 559–67.
129. Page KM, Heblich F, Davies A, et al. Dominant-negative calcium channel suppression by truncated constructs involves a kinase implicated in the unfolded protein response. J Neurosci 2004; 24: 5400–9.
130. Kolkova K, Novitskaya V, Pedersen N, et al. Neural cell adhesion molecule-stimulated neurite outgrowth depends on activation of protein kinase C and the Ras-mitogen-activated protein kinase pathway. J Neurosci 2000; 20: 2238–46.
131. Hannon GJ. RNA interference (RNAi) and microRNAs. In: Encyclopedia of Life Sciences. Chichester: John Wiley and Sons, 2006: DOI: 10.1002/97804700115902.a0006256.
132. Meister G, Tuschl T. Mechanisms of gene silencing by double-stranded RNA. Nature 2004; 431: 343–9.
133. Hannon GJ, Rossi JJ. Unlocking the potential of the human genome with RNA interference. Nature 2004; 431: 371–8.
134. Schepers U. RNAi in mammals. In: Schepers U, ed. RNA Interference in Practice. Wilmington, DE: Wiley-VCH, 2006: 101–88.
135. Krichevsky AM, Kosik KS. RNAi functions in cultured mammalian neurons. Proc Natl Acad Sci U S A 2002; 99: 11926–9.
136. Echeverri CJ, Perrimon N. High-throughput RNAi screening in cultured cells: a user's guide. Nat Rev Genet 2006; 7: 373–84.
137. Wiznerowicz M, Szulc J, Trono D. Tuning silence: conditional systems for RNA interference. Nat Methods 2006; 3: 682–8.
138. Grimm D, Kay MA. Therapeutic application of RNAi: is mRNA targeting finally ready for prime time? J Clin Invest 2007; 117: 3633–41.
139. Calabrese B, Halpain S. Essential role for the PKC target MARCKS in maintaining dendritic spine morphology. Neuron 2005; 48: 77–90.
140. Willard MD, Willard FS, Li X, et al. Selective role for RGS12 as a Ras/Raf/MEK scaffold in nerve growth factor-mediated differentiation. Embo J 2007; 26: 2029–40.
141. Xin H, Bernal A, Amato FA, et al. High-throughput siRNA-based functional target validation. J Biomol Screen 2004; 9: 286–93.
142. Bridge AJ, Pebernard S, Ducraux A, et al. Induction of an interferon response by RNAi vectors in mammalian cells. Nat Genet 2003; 34: 263–4.
143. Cullen BR. Enhancing and confirming the specificity of RNAi experiments. Nat Methods 2006; 3: 677–81.

144. Nair AR, Schliekelman M, Thomas MB, et al. Inhibition of p53 by lentiviral mediated shRNA abrogates G1 arrest and apoptosis in retinal pigmented epithelial cell line. Cell Cycle 2005; 4: 697–703.
145. Reynolds A, Anderson EM, Vermeulen A, et al. Induction of the interferon response by siRNA is cell type- and duplex length-dependent. RNA 2006; 12: 988–93.
146. Kapranov P, Willingham AT, Gingeras TR. Genome-wide transcription and the implications for genomic organization. Nat Rev Genet 2007; 8: 413–23.
147. Ashraf SI, Kunes S. A trace of silence: memory and microRNA at the synapse. Curr Opin Neurobiol 2006; 16: 535–9.
148. Kosik KS. The neuronal microRNA system. Nat Rev Neurosci 2006; 7: 911–20.
149. Giepmans BN, Adams SR, Ellisman MH, et al. The fluorescent toolbox for assessing protein location and function. Science 2006; 312: 217–24.
150. Shaner NC, Steinbach PA, Tsien RY. A guide to choosing fluorescent proteins. Nat Methods 2005; 2: 905–9.
151. Shaner NC, Patterson GH, Davidson MW. Advances in fluorescent protein technology. J Cell Sci 2007; 120(Pt 24): 4247–60.
152. Schenborn E, Groskreutz D. Reporter gene vectors and assays. Mol Biotechnol 1999; 13: 29–44.
153. Hioki H, Kameda H, Nakamura H, et al. Efficient gene transduction of neurons by lentivirus with enhanced neuron-specific promoters. Gene Ther 2007; 14: 872–82.
154. Zha XM, Green SH, Dailey ME. Regulation of hippocampal synapse remodeling by epileptiform activity. Mol Cell Neurosci 2005; 29: 494–506.
155. Jakobsson J, Ericson C, Jansson M, et al. Targeted transgene expression in rat brain using lentiviral vectors. J Neurosci Res 2003; 73: 876–85.
156. Wernig M, Tucker KL, Gornik V, et al. Tau EGFP embryonic stem cells: an efficient tool for neuronal lineage selection and transplantation. J Neurosci Res 2002; 69: 918–24.
157. Bilic-Curcic I, Kronenberg M, Jiang X, et al. Visualizing levels of osteoblast differentiation by a two-color promoter-GFP strategy: Type I collagen-GFPcyan and osteocalcin-GFPtpz. Genesis 2005; 43: 87–98.
158. Tamm C, Duckworth J, Hermanson O, et al. High susceptibility of neural stem cells to methylmercury toxicity: effects on cell survival and neuronal differentiation. J Neurochem 2006; 97: 69–78.
159. Feng G, Mellor RH, Bernstein M, et al. Imaging neuronal subsets in transgenic mice expressing multiple spectral variants of GFP. Neuron 2000; 28: 41–51.
160. Hechler D, Nitsch R, Hendrix S. Green-fluorescent-protein-expressing mice as models for the study of axonal growth and regeneration *in vitro*. Brain Res Rev 2006; 52: 160–9.
161. Livet J, Weissman TA, Kang H, et al. Transgenic strategies for combinatorial expression of fluorescent proteins in the nervous system. Nature 2007; 450: 56–62.
162. van Roessel P, Brand AH. Imaging into the future: visualizing gene expression and protein interactions with fluorescent proteins. Nat Cell Biol 2002; 4: E15–20.
163. White J, Stelzer E. Photobleaching GFP reveals protein dynamics inside live cells. Trends Cell Biol 1999; 9: 61–5.
164. Krapivinsky G, Krapivinsky L, Manasian Y, et al. The NMDA receptor is coupled to the ERK pathway by a direct interaction between NR2B and RasGRF1. Neuron 2003; 40: 775–84.
165. Chi P, Greengard P, Ryan TA. Synapsin dispersion and reclustering during synaptic activity. Nat Neurosci 2001; 4: 1187–93.
166. Yu JZ, Rasenick MM. Tau associates with actin in differentiating PC12 cells. FASEB J 2006; 20: 1452–61.
167. Greene LA, Farinelli SE, Cunningham ME, Park DS. Cultures and experimental use of the PC12 rat pheochromocytoma cell line. In: Banker G, Goslin K, eds. Culturing Nerve Cells, 2nd edn. Cambridge, MA: MIT Press, 1998: 161–88.
168. Piston DW, Kremers GJ. Fluorescent protein FRET: the good, the bad and the ugly. Trends Biochem Sci 2007; 32: 407–14.
169. Pollok BA, Heim R. Using GFP in FRET-based applications. Trends Cell Biol 1999; 9: 57–60.
170. Okamoto K, Hayashi Y. Visualization of F-actin and G-actin equilibrium using fluorescence resonance energy transfer (FRET) in cultured cells and neurons in slices. Nat Protoc 2006; 1: 911–19.
171. Okamoto K, Nagai T, Miyawaki A, et al. Rapid and persistent modulation of actin dynamics regulates postsynaptic reorganization underlying bidirectional plasticity. Nat Neurosci 2004; 7: 1104–12.
172. Dong M, Tepp WH, Johnson EA, et al. Using fluorescent sensors to detect botulinum neurotoxin activity *in vitro* and in living cells. Proc Natl Acad Sci U S A 2004; 101: 14701–6.

173. Heijne WH, Kienhuis AS, van Ommen B, et al. Systems toxicology: applications of toxicogenomics, transcriptomics, proteomics and metabolomics in toxicology. Expert Rev Proteomics 2005; 2: 767–80.
174. Craven RA, Banks RE. Laser capture microdissection and proteomics: possibilities and limitation. Proteomics 2001; 1: 1200–4.
175. Balbus JM. Ushering in the new toxicology: toxicogenomics and the public interest. Environ Health Perspect 2005; 113: 818–22.
176. Jaluria P, Konstantopoulos K, Betenbaugh M, et al. A perspective on microarrays: current applications, pitfalls, and potential uses. Microb Cell Fact 2007; 6: 4.
177. Pennie WD. Microarrays in toxicological research. In: Encyclopedia of Life Sciences. Chichester: John Wiley and Sons, 2005: DOI: 10.1038/npg.els.0005950.
178. Blalock EM, Chen KC, Stromberg AJ, et al. Harnessing the power of gene microarrays for the study of brain aging and Alzheimer's disease: statistical reliability and functional correlation. Ageing Res Rev 2005; 4: 481–512.
179. Woo Y, Affourtit J, Daigle S, et al. A comparison of cDNA, oligonucleotide, and Affymetrix GeneChip gene expression microarray platforms. J Biomol Tech 2004; 5: 276–84.
180. Eisenstein M. Microarrays: quality control. Nature 2006; 442: 1067–70.
181. Shi L, Reid LH, Jones WD, et al. The MicroArray Quality Control (MAQC) project shows inter- and intraplatform reproducibility of gene expression measurements. Nat Biotechnol 2006; 24: 1151–61.
182. Brazma A, Hingamp P, Quackenbush J, et al. Minimum information about a microarray experiment (MIAME)-toward standards for microarray data. Nat Genet 2001; 29: 365–71.
183. Scherf U, Ross DT, Waltham M, et al. A gene expression database for the molecular pharmacology of cancer. Nat Genet 2000; 24: 236–44.
184. Staunton JE, Slonim DK, Coller HA, et al. Chemosensitivity prediction by transcriptional profiling. Proc Natl Acad Sci U S A 2001; 98: 10787–92.
185. Lee JK, Havaleshko DM, Cho H, et al. A strategy for predicting the chemosensitivity of human cancers and its application to drug discovery. Proc Natl Acad Sci U S A 2007; 104: 13086–91.
186. Glas AM, Floore A, Delahaye LJ, et al. Converting a breast cancer microarray signature into a high-throughput diagnostic test. BMC Genomics 2006; 7: 278.
187. Chang HY, Thomson JA, Chen X. Microarray analysis of stem cells and differentiation. Methods Enzymol 2006; 420: 225–54.
188. Ji X, Liebhaber AA. mRNA stability and control of gene expression. In: Encyclopedia of Life Sciences. Chichester: John Wiley and Sons, 2007: [Available from: http://www.mrw.interscience.wiley.com/emrw/9780470015902/els/article/a0005972/current/pdf].
189. Provenzano M, Mocellin S. Complementary techniques: validation of gene expression data by quantitative real time PCR. Adv Exp Med Biol 2007; 593: 66–73.
190. Cobb K. Microarrays: search for meaning in a vast sea of data. Biomedical Computational Review 2006; 2: 16–23.
191. Clarke R, Ressom HW, Wang A, et al. The properties of high-dimensional data spaces: implications for exploring gene and protein expression data. Nat Rev Cancer 2008; 8: 37–49.
192. Valcarcel J, Smith CWJ. Alternative splicing: cell-type-specific and developmental control. In: Encyclopedia of Life Sciences. Chichester: John Wiley and Sons, 2005: DOI: 10.1038/npg.els.0005033.
193. Olson S, Blanchette M, Park J, et al. A regulator of Dscam mutually exclusive splicing fidelity. Nat Struct Mol Biol 2007; 14: 1134–40.
194. Sonnenberg N. Protein synthesis in eukaryotes: IRES-mediated internal initiation. Encyclopedia of Life Sciences. Chichester: John Wiley and Sons, 2005: [Available from: http://www.mrw.interscience.wiley.com/emrw/9780470015902/els/article/a0000544/current/pdf].
195. Wittmann-Liebold B, Choli-Papadopoulou T. Proteins: postsynthetic modification-function and physical analysis. In: Encyclopedia of Life Sciences. Chichester: John Wiley and Sons, 2007: [Available from: http://www.mrw.interscience.wiley.com/emrw/9780470015902/els/article/a0000551/current/pdf].
196. Meri S, Baumann M. Protein characterization in proteomics. In: Encyclopedia of Life Sciences. Chichester: John Wiley and Sons, 2006: [Available from: http://www.mrw.interscience.wiley.com/emrw/9780470015902/els/article/a0006217/current/pdf].
197. Zhu H, Bilgin M, Snyder M. Proteomics. Annu Rev Biochem 2003; 72: 783–812.
198. LoPachin RM, Jones RC, Patterson TA, et al. Application of proteomics to the study of molecular mechanisms in neurotoxicology. Neurotoxicology 2003; 4: 761–75.
199. Tannu NS, Hemby SE. Two-dimensional fluorescence difference gel electrophoresis for comparative proteomics profiling. Nat Protoc 2006; 1: 1732–42.

200. Carrette O, Burkhard PR, Sanchez JC, et al. DF. State-of-the-art two-dimensional gel electrophoresis: a key tool of proteomics research. Nat Protoc 2006; 1: 812–23.
201. Matt P, Fu Z, Fu Q, Van Eyk JE. Biomarker discovery: proteome fractionation and separation in biological samples. Physiol Genomics 2007; 33: 12–17.
202. Mirza SP, Olivier M. Methods and approaches for the comprehensive characterization and quantification of cellular proteomes using mass spectrometry. Physiol Genomics 2007; 33: 3–11.
203. Sheehan D. The potential of proteomics for providing new insights into environmental impacts on human health. Rev Environ Health 2007; 22: 175–94.
204. Fernie AR, Trethewey RN, Krotzky AJ, et al. Metabolite profiling: from diagnostics to systems biology. Nat Rev Mol Cell Biol 2004; 5: 763–9.
205. Kaddurah-Daouk R, Kristal BS, Weinshilboum RM. Metabolomics: a global biochemical approach to drug response and disease. Annu Rev Pharmacol Toxicol 2008; 48: 653–83.
206. Khoo SH, Al-Rubeai M. Metabolomics as a complementary tool in cell culture. Biotechnol Appl Biochem 2007; 47(Pt 2): 71–84.
207. van Vliet E, Morath S, Eskes C, et al. A novel *in vitro* metabolomics approach for neurotoxicity testing, proof of principle for methyl mercury chloride and caffeine. Neurotoxicology 2008; 29: 1–12.
208. Lee DW, Gelein RM, Opanashuk LA. Heme-oxygenase-1 promotes polychlorinated biphenyl mixture aroclor 1254-induced oxidative stress and dopaminergic cell injury. Toxicol Sci 2006; 90: 159–67.
209. Richardson JR, Caudle WM, Guillot TS, et al. Obligatory role for complex I inhibition in the dopaminergic neurotoxicity of 1-methyl-4-phenyl-1,2,3,6-tetrahydropyridine (MPTP). Toxicol Sci 2007; 95: 196–204.
210. Dimayuga FO, Wang C, Clark JM, et al. SOD1 overexpression alters ROS production and reduces neurotoxic inflammatory signaling in microglial cells. J Neuroimmunol 2007; 182: 89–99.
211. Khan AA, Mao XO, Banwait S, et al. Neuroglobin attenuates beta-amyloid neurotoxicity *in vitro* and transgenic Alzheimer phenotype *in vivo*. Proc Natl Acad Sci U S A 2007; 104: 19114–19.
212. Newhouse K, Hsuan SL, Chang SH, et al. Rotenone-induced apoptosis is mediated by p38 and JNK MAP kinases in human dopaminergic SH-SY5Y cells. Toxicol Sci 2004; 79: 137–46.
213. Herdman ML, Marcelo A, Huang Y, et al. Thimerosal induces apoptosis in a neuroblastoma model via the cJun N-terminal kinase pathway. Toxicol Sci 2006; 92: 246–53.
214. Mathiasen JR, McKenna BA, Saporito MS, et al. Inhibition of mixed lineage kinase 3 attenuates MPP+-induced neurotoxicity in SH-SY5Y cells. Brain Res 2004; 1003: 86–97.
215. Fei Q, McCormack AL, Di Monte DA, et al. Paraquat neurotoxicity is mediated by a Bak-dependent mechanism. J Biol Chem 2008; 283: 3357–64.
216. Asanuma M, Miyazaki I, Higashi Y, et al. Suppression of p53-activated gene, PAG608, attenuates methamphetamine-induced neurotoxicity. Neurosci Lett 2007; 414: 263–7.
217. Novitskaya V, Bocharova OV, Bronstein I, et al. Amyloid fibrils of mammalian prion protein are highly toxic to cultured cells and primary neurons. J Biol Chem 2006; 281: 13828–36.
218. Slikker W Jr, Bowyer JF. Biomarkers of adult and developmental neurotoxicity. Toxicol Appl Pharmacol 2005; 206: 255–60.
219. Sengupta A, Mense SM, Lan C, et al. Gene expression profiling of human primary astrocytes exposed to manganese chloride indicates selective effects on several functions of the cells. Neurotoxicology 2007; 28: 478–89.
220. Bouton CM, Hossain MA, Frelin LP, Laterra J, Pevsner J. Microarray analysis of differential gene expression in lead-exposed astrocytes. Toxicol Appl Pharmacol 2001; 176: 34–53.
221. Xu Z, Patterson TA, Wren JD, et al. A microarray study of MPP+-treated PC12 Cells: Mechanisms of toxicity (MOT) analysis using bioinformatics tools. BMC Bioinformatics 2005; 6(Suppl 2): S8.
222. Wang J, Duhart HM, Xu Z, et al. Comparison of the time courses of selective gene expression and dopaminergic depletion induced by MPP(+) in MN9D cells. Neurochem Int 2007; 52: 1037–43.
223. Zhou Y, Wang Y, Kovacs M, et al. Microglial activation induced by neurodegeneration: a proteomic analysis. Mol Cell Proteomics 2005; 4: 1471–9.
224. Vendrell I, Carrascal M, Vilaro MT, et al. Cell viability and proteomic analysis in cultured neurons exposed to methylmercury. Human Exp Toxiol 2007; 26: 263–72.

2 Bioinformatic and Computational Approaches for Analysis of Genomics Data in Neurotoxicology Research

Julia M. Gohlke

Laboratory of Molecular Toxicology, National Institute of Environmental Health Sciences, Research Triangle Park, North Carolina, U.S.A.

Willard M. Freeman and Kent E. Vrana

Department of Pharmacology, Penn State College of Medicine, Hershey, Pennsylvania, U.S.A.

INTRODUCTION

With the advent of microarray technology, the generation of a million data points from a single experiment is now commonplace. However, this explosion of data generation has not automatically equated to novel biological insights into neurotoxicology. Significant advancements since the initial introduction of microarray technology have come at the data generation stage through large-scale collaborations between industry, government, and academia. Significant advances at the data analysis stage have been made as well, in the form of novel computational and bioinformatic methods; yet, a consensus on the most appropriate method(s) for a given experimental design has not been reached. This chapter will provide a brief overview of microarray technology, its current applications in neurotoxicology research, common problems encountered with microarray data analysis, and novel methods developed for assessing the biological relevance of results from microarrays.

MICROARRAY TECHNOLOGIES IN NEUROTOXICOLOGY RESEARCH

Microarray technology evolved from the Southern blot, which detects the presence of specific a DNA sequence by attaching a fragmented mixture of DNA molecules to a solid substrate and then probing, through hybridization, with a known gene or fragment (1). The microarray reversed the process by immobilizing known sequences and then probing with labeled unknown nucleic acids. This technique was expanded such that hundreds of known DNA fragments could be interrogated in a single experiment. The first example was reported in 1987, when an array of complementary DNAs (cDNAs) was spotted onto filter paper to detect genes expression modulated by interferon (2).

Today, microarray technology has several applications. Microarrays can be used for DNA [e.g., in comparative genomic hybridization or single nucleotide polymorphism (SNP) analysis] or RNA [either messenger RNAs (mRNAs) or microRNAs (miRNAs)] analyses. In 1997, the first whole genome, that of *Saccharomyces cerevisiae*, was captured on a microarray (3). Currently, microarray-based, genome-wide scans of gene expression and/or SNPs are possible for several species commonly used in neurotoxicology research, such as mouse, rat, monkey, and human.

Specific microarray applications examining genomic DNA include studies of copy number variation through comparative genome hybridization, genotyping of millions of SNPs, and analysis of transcription factor binding to specific DNA sequences isolated via chromatin immunoprecipitation (ChIP-chip). The genetic basis of a wide variety of neuropsychiatric disorders has now been tested using genome-wide screening techniques that interrogate large patient cohorts for quantitative trait loci (4–7).

Detection of mRNA expression via microarray hybridization is often referred to as global gene expression analysis (whole genome analysis) or gene expression profiling, and is currently the most widely used application of microarray technology. Since its inception, the popularity of the gene expression profiling approach has grown enormously. In 2007, there were over 200,000 datasets deposited in the two major gene expression microarray data warehouses GEO and ArrayExpress (8,9). Expression profiling has now advanced to include transcriptional analyses where splicing patterns can be investigated using splice junction and exon tiling microarrays (10). In addition, microarrays are now being developed to also include examination of miRNAs (11). This chapter will focus on expression profiling applications of microarray technology where neurotoxicological research and advancements in data analysis methods are widespread.

Brief Overview of Gene Expression Microarray Technology

The appeal of the expression profiling array is simultaneous measurement of mRNA expression for tens of thousands of transcripts from a single biological sample. Indeed, the motivation is to use arrays to conduct unbiased assessments of the gene expression changes. At the inception of the technology, arrays were constructed by robotic spotting of cDNAs onto treated glass slides. This technology has generally been replaced by microarrays consisting of oligonucleotides. These oligonucleotides are either synthesized in situ on a slide, synthesized, and then spotted onto a slide, or synthesized on a bead and then deposited into microwells on a slide. No matter the type of microarray, databases of well characterized genes (e.g., RefSeq) and expressed sequence tags are used to select sequences complimentary to short segments of a gene for inclusion on the array. Oligonucleotide arrays have become the predominant technology due to the higher density and thus more genes and different probes for different regions of a gene can be included on an array.

For mRNA expression analysis, the experimental steps broadly involve: (*i*) isolating the mRNA; (*ii*) converting it to a labeled copy (through a combination of reverse transcription and amplification); (*iii*) hybridization to an array of known gene sequences in an addressable array; and (*iv*) signal collection, reduction, and analysis. In brief, mRNA is isolated from control and experimental samples, reverse transcribed to cDNAs, and then labeled with fluorescent dye(s), and hybridized to the microarray. In a two color design, both the control and experimental sample is hybridized to the same array; whereas, in a one color design, the control and experimental samples are hybridized to separate arrays. Due to the advances in technology which produce less costly, more consistent microarrays, the one-color experimental design has become the most common approach. After washing, the spot-bound fluorescent dyes are excited by lasers of appropriate wavelengths to generate images. Images are then analyzed to quantify the signal within each spot. Background hybridization intensity is removed (via quantification of mismatch probe binding, or measurement of signal intensity at "empty" addresses or negative control sequences). Therefore, the principal measurement is the signal minus the estimated background of each probe. Finally, a dataset of signal intensity measurements is generated usually expressed in a matrix whose columns correspond to samples and rows to genes.

Initially, microarrays were manufactured in-house using cloned polymerase chain reaction (PCR) products and led to data inconsistencies across users (12). Kothapalli et al. (2002) examined microarray data from two different systems (13). They reported inconsistencies in sequence fidelity for the spotted microarrays, variability of differential expression, low specificity of cDNA probes, discrepancy in fold-change calculations, and lack of probe specificity for different isoforms of a gene. In addition, Tan et al. (2003) examined gene expression measurements generated from identical RNA preparations that were obtained using three commercially available microarray platforms from Affymetrix, Amersham, and Agilent (14). Correlations in gene expression levels and comparisons for significant gene expression changes in this subset showed considerable divergence across the different platforms. These initial inconsistencies have largely been addressed through a variety of efforts to improve the reliability of microarrays. For example, introducing standards for microarray experiments and for data analysis (Minimum Information About a Microarray Experiment; MIAME) (15) establishing standardized RNA controls (16), and large scale collaborations between government, academia, and industry to assure quality control [MicroArray Quality Control (MAQC) Project] (17) have all led to dramatically increased reliability and reproducibility of microarray experiments. Recent studies comparing platforms across laboratories have concluded there is not a significant advantage of any one commercial design, but that researchers should pick the platform based on their individual needs. Although not covered in detail in this chapter, we refer readers interested in comparisons across platforms to several important reviews on this topic (18,19).

Applications in Neurotoxicology

The two broad applications of microarray technology in neurotoxicology research are the discovery of patterns of expression that can: (*i*) classify toxicity phenotypes and predict disease or (*ii*) identify novel molecular targets of a toxicant. The first use hinges on the hypothesis that mRNA expression can serve as an important classification system for molecular pathology. Indeed, microarray data can serve as an important subclinical biomarker of pathology. Research in this area includes reports of patterns of genetic expression that predict new classifications of central nervous system embryonal tumors and gliomas (20,21). Indeed, one of the first bioinformatics approaches to analyzing neurotoxicology microarray data [1-methyl-4-phenylpyridinium ion (MPP+) toxicity in PC12 cells] was published by Slikker and colleagues in 2005 (22). These studies provided validity for the approach as the analysis identified induction of oxidative stress, DNA/protein damage, cell cycle arrest, and apoptosis pathways as important sequelae to MPP+-induced neurotoxicity. A follow-up report extended these findings in MN9D dopaminergic cells (23). Additional neurotoxicology microarray studies include analysis of aryl hydrocarbon receptor-induction of gene expression changes (24) and Aroclor 1254-mediated developmental neurotoxicology (25).

The second use, and the focus of this chapter, addresses the hypothesis that neuropathological phenotypes are created by networks of molecular interactions. Therefore, analysis of global transcriptional responses can help to build hypotheses of the molecular system, including transcriptional regulation, signaling pathways, protein-protein, and protein–nucleic acid interactions, that are responsible for a particular phenotype. Most examples of microarray applications in systems biology are in non-mammalian systems such as regulation of the transcriptional responses when yeast cells encounter nutrients or the yeast galactose-utilization pathway (26). The translation of these methods to the mammalian nervous system is progressing. For example, Yonan et al. (2003) developed hypotheses of gene networks mediating autism (27) and Tropea et al. (2006) examined activity-dependent plasticity in the visual cortex (28) using bioinformatic evaluation of microarray results.

Primary Data Analysis Concerns

While the emphasis of this chapter is on the data analysis portion of microarray research, it is important to note the quality of the starting material will determine the success of any gene expression profiling experiment. Indeed, experimental design and sample collection and processing will frequently mean the difference between a successful experiment and a collection of uninterpretable data.

Some issues are of particular note. As we are focused, in the present context, on neurotoxicology, it is important to collect samples at a time point prior to frank cell death. Choose a condition at which altered gene expression is contributing to the pathology (as opposed to being the result of lost cells). With the widespread adoption of microarray technologies and the concomitant reduction in costs, it is no longer common practice to perform analyses on "pooled" samples. In fact, such practices deprive the investigator of potentially important insights into inter-individual variations. However, this does require that the experimental subjects or samples be processed in a consistent manner. Before hybridization, the quality of the RNA should always be verified either by gel electrophoresis or commercial products such as the Agilent RNA 6000 Nano or Pico kits for the 2100 Bioanalyzer. For most institutional (and commercial) core facilities, this is a requisite first step in determining whether to proceed with an array analysis on a given sample.

Developing methods that allow integration of previous research is an important area of growth. As mentioned above, repositories of microarray datasets are growing. Combined analysis of datasets as well as determining the biological context of results requires development of novel computational models to identify mechanisms of biological processes by integrating diverse sets of experimental data. Growth in this area of research has been initiated particularly in organisms where high throughput perturbation analyses are possible, including *S. cerevisiae*, *Caenorhabditis elegans*, and *Strongylocentrotus purpuratus* (29–33). However, finding and quantifying networks from datasets in mammalian species, particularly in the heterogeneous tissues of the brain, has been challenging. In fact, one of the most critical issues is the data requirements of current network based methods. The following sections will provide an overview of methods currently used at each step of microarray data analysis, starting with image analysis, normalization, and determination of differential expression. Next, methods for linking results with previous research will be discussed. Finally, approaches for developing novel hypotheses from microarray datasets are discussed.

GENERATING A LIST OF DIFFERENTIALLY EXPRESSED GENES

So far, we have reviewed mechanisms by which large-scale analysis of RNA can be conducted. These studies produce large databases of quantitative signals (generally of a fluorescent nature). Following this data generation step, a robust initial data analysis of the signals within and between chips is of primary concern. There are two areas of particular importance in the initial analysis of these large datasets: normalization and identification of significant differential expression. Both of these are complex and involved issues, therefore only a high-level overview will be provided here.

Initial Data Analysis and Normalization

With all microarray analyses of a functional genomic experiment, data normalization is required. This need derives from the fact that we are conducting complex analyses on independent platforms with several involved technical steps—each step of which provides opportunity for variability. Direct comparisons between microarrays are often not possible due to differences arising from technical variability in labeling efficiency, hybridization, or microarray fabrication.

The first step is background subtraction. Background subtraction corrects for non-specific background noise and permits comparison of specific signals. For illustration, if the signal intensities for the control and experimental spots are 200 and 300, respectively, it would appear that the experimental signal is 50% higher. However, if a background of 100 is subtracted from both signal intensities, the experimental value is actually 100% higher than control (100 vs. 200). Background is often taken from the blank areas on the array. A complication to background subtraction is that differences in background across the array can affect some spots more than others. An alternative is to use either a local background for the area around each spot or designate spots with the lowest signal intensities for background determination. The latter may be a more accurate determination of non-specific background because it represents the non-specific binding of targets to probe. Background intensities from blank areas (no nucleic acids) do not contain probe, and therefore are arguably a different form of background. Alternatively, nonsense probes (probes which should not be complimentary to any gene) can be used as a determinant of background signal intensity. In this regard, one innovative aspect of the Affymetrix platform is the use of the mismatch oligonucleotide. In this case, each hybridization oligonucleotide is accompanied by an oligonucleotide of the same sequence with a mismatch in the middle of the sequence. With oligonucleotides of such short sequence (25 residues), even such a single mismatch will disrupt specific hybridization signals.

The second step in initial data analysis is signal normalization. To overcome the variability in the labeling and hybridization steps, researchers frequently turn to a "sum" approach for normalization. This strategy is based on the precept that the total amount of labeled target should be the same in all samples. That is, even though individual genes will have selected increases and decreases, on balance, the total hybridization signal should be constant. Therefore, equilibrating the sum of the intensities for all control and experimental spots can be used to normalize arrays. In a similar vein, the median value of signal intensities can be used. This value is less susceptible to distortions caused by outlying signals. More complex methods of normalization have been developed such as those that apply Lowess intensity dependent normalization, which is even more impervious to skewing by outliers (34). Currently, there is no single standard method and generally the normalization strategy chosen is dependent on the type and degree of technical variance specific to an individual microarray technology. The ultimate point, however, is to convert all of the signal intensities from the various arrays to a "common scale" so as to permit statistical analyses with a reasonable level of confidence.

Significance Testing Methods

Perhaps the most obvious bioinformatics challenge is determining when a gene expression difference exists between samples. As with normalization, there is no single, commonly accepted method. In early microarray experiments, one method for "calling" a gene as differentially expressed between samples was based on the magnitude of change. Generally, a two-fold (i.e., 100% increase or 50% decrease) change was commonly accepted as a difference. This method has fallen from favor because it lacks statistical rigor and eliminates the possibility of detecting smaller magnitude changes (a common occurrence when examining brain tissue, in particular). As use of microarray technology has evolved, standard parametric tests (t-test and ANOVA) were applied to data analysis. The major limitation of this approach is that there are hundreds to tens of thousands of dependent measures in a microarray experiment. Standard methods for multiple testing correction (e.g., Bonferonni correction), however, are often too conservative. This led to further development of tools such as z-score analysis (35), statistical analysis of microarrays (SAM) (36), and false-discovery rates (37) to give just a few examples. These tools offer a better approach to determining significant changes because they are designed for the specific requirements of examining thousands of dependent measures. As with all statistical analyses, the appropriate balance between Type I and Type II error will depend on the goal of the experiment.

When examining thousands of mRNAs or proteins, it is almost a certainty that there will be false-positives (i.e., changes will be observed by microarray screens that in fact do not exist) as well as false-negatives. One of the best methods for dealing with this problem is to confirm changes by another method. Whenever possible, array results are used to identify targets for further validation using orthogonal technical approaches. This might involve quantitative reverse transcriptase-PCR (QRT-PCR) or western blot analysis of selected targets highlighted by the initial microarray screen. For microarray results, the most common method for confirmation is to use QRT-PCR. The current preferred method is to use real-time techniques. A range of techniques exists including Sybr green (38), fluorescent primers (LUX, light on extension) (39), and TaqMan fluorescent probes (40). Real time methods take advantage of the kinetics of the PCR reaction to accurately quantify transcript levels. For the specifics on quantitation using PCR there are several good resources (41,42). With microarrays now capable of examining the whole genome, there is a need for high-throughput confirmation by quantitative PCR. New instruments and protocols are making possible the PCR confirmation of a large number of genes (43). Other traditional methods such as in situ hybridization can be used for microarray confirmation (44) with the additional benefit being that the transcript is anatomically localized. However, this method is more time-consuming and requires more sample than quantitative PCR. Finally, if the functional endpoint of the mRNA in question is protein, then the gold standard for confirmation is the western blot or other protein quantitation method. Not only is this approach well suited to quantification, it also establishes that observed mRNA changes are in fact translated into changes in protein. The problem with this technology, however, is that it is not well-suited to high-throughput applications.

MOVING FROM DIFFERENTIALLY EXPRESSED GENE LISTS TO PATHWAYS

Initially, establishing a list of differentially expressed genes from a microarrray experiment was the endpoint. Now it is considered only the starting point of a research project as determining the biological significance of that list is a complex issue. As mentioned previously, these lists can be used to establish hypotheses and direct experimental research on the function of a single or couple of genes with a previously unknown role. In recent years, more sophisticated algorithms have been developed to analyze the implications of the complete set of differentially expressed genes from microarray results. Development of these algorithms has been an integral part of the resurgence of the field of systems biology, where the focus is the interactions between parts of a biological system, and not the individual parts themselves.

In the field of systems biology, there has been considerable discussion of "top-down" versus "bottom-up" strategies for delineating critical interactions in a system (45,46). In particular, a traditional top-down approach focuses on determining the functions of one or a few genes through incorporation of known information at several levels of organization, but may miss important novel discoveries gleaned from the integration of broad gene or protein expression scans. In contrast, several new computational methods have been recently developed that take a bottom-up approach, by constructing networks de novo through application of statistical or graph theory analyses to global gene or protein expression experiments (47,48). However, these bottom-up approaches have been criticized for utilizing data generated solely by novel high-throughput techniques at the gene, protein, or metabolite level, while excluding data generated using more traditional approaches at various levels of organization. Alleviation of the limitations at both ends is of paramount importance for the continued growth of systems biology approaches (49). Methods that combine top-down and bottom-up methodologies for elucidation of gene regulatory networks (GRNs) may provide the most useful information for experimentalists who want to balance previous research and novel insights from microarray experiments. An example of this methodology using the developing telencephalon as a model system will be discussed in greater detail in the next section.

In regards to top-down approaches, there are numerous ways in which microarray datasets can be analyzed within the context of previous research. Pathway enrichment algorithms are popular as they provide an efficient method for determining the most likely metabolic, signal transduction, and disease pathways affected by the experimental condition, thus summarizing results based on what is already known about the functions of gene products. The purpose of these algorithms is to determine how previously studied higher order processes (e.g., the interactions between gene products) contribute to the observed gene expression patterns.

In contrast, bottom-up strategies generate hypotheses about connectivity between gene products based on microarray results. These methods employ a wide variety of statistical and mathematical methods to determine connections between gene products that best explain the variation in expression results across samples and across genes. The reader is referred to several comprehensive reviews on these methods for further details (50–53). Because these methods generally require very large sample sizes to get reliable results, other methods have been developed to incorporate *a priori* information, thereby reducing the search space and thus reducing the sample size requirement (54). These methods utilize literature based information or sequence based information to require, prohibit, or weigh particular connections between a given set of genes. These methods are examined in more detail in the next section.

Pathway Enrichment Algorithms

Pathway or functional enrichment algorithms are the most widely used method to develop hypotheses of the global processes or pathways affected by the experimental condition in microarray analyses. These methods group biologically-relevant sets of genes/proteins based on known roles in biochemical, signaling, metabolic, or disease pathways. Some researchers have also extended the use of the methodology by including groupings of genes, or gene sets, based on chromosomal position and previous microarray experiments. The most important advantage of this methodology when compared to the conventional univariate statistical analyses is the ability to detect consistent, yet subtle changes in expression of a group of genes with related functions. In addition, these methods allow easy integration of genomic, metabolomic, and proteomic datasets, as known inter-relationships are annotated in many of the pathway and functional grouping databases. This section first describes sources of pathway information, then gives a brief overview of the statistical methods implemented in the most commonly used pathway enrichment algorithms, and ends with a few recent examples in neurotoxicological research. The reader is referred to several important reviews for a more detailed overview of these approaches (55–57).

Sources of Pathway or Functional Information

There are several sources of pathway or functional grouping information. Probably the most well known is the Gene Ontology (GO) annotation database, which was introduced in 1999 (58) and has been vastly expanded to incorporate several important sources of data as of 2008 (59). This database organizes genes into functional categories using a hierarchical system. Therefore, genes belong to very broad categories such as metabolism, but are also categorized further into more specific classes, such as pentose phosphate metabolism. The Kyoto Encyclopedia of Genes and Genomes (KEGG), Biocarta, and GenMAPP are other popular pathway annotation databases. Compared to the hierarchical system of GO, the advantage of KEGG, Biocarta, and GenMAPP is that gene products are organized into networks of biochemical processes such that connectivity between individual components of a pathway represents a known direct interaction. KEGG focuses on inter-relationships between classical metabolism, signal transduction, and disease based pathways (60). Biocarta develops fully annotated pathways from previous publications (www.biocarta.com/pathfiles/), whereas GenMAPP includes easy to use tools for creating user defined pathways and visualizing gene expression on the pathways (61). PANTHER (Protein ANalysis THrough Evolutionary Relationships) is another pathway resource that builds relationships through a protein functional classification system, either through published scientific experimental evidence or evolutionary relationships (62). In addition, chromosomal location datasets, as well as sets of genes found to be differentially expressed in previous experiments can serve as functional groupings or gene sets (63,64). Finally, a comprehensive guide to sources of pathway information has been developed (65).

Statistical Methods

The statistical methods vary widely across pathway enrichment algorithms. Sometimes referred to as overrepresentation analysis (ORA), the simplest form of pathway enrichment is to use the chi-squared test or the Fisher exact test to calculate how the observed number of matches (number of genes that are differentially expressed and are on a specific pathway) deviates from what is expected by random chance. Popular online software tools that fall into this category include EASE, GOMiner, and FatiGO (66–70). Other methods use the hypergeometric distribution or the binomial approximation to the hypergeometric distribution to determine the probability of there being a specific number of genes from one pathway or class within the list of differentially expressed genes from a microarray experiment. MAPPFinder and Onto-Tools employ this method and are widely used in microarray research (71,72). FatiGO and Onto-Tools have recently added additional functional groupings based on promoter region and SNP analyses (73,74).

It is important to note that many of the algorithms mentioned above require one to predetermine significance of differential expression for each gene in the experiment. Therefore, these methods still have the limitation of potentially missing a set of genes within a pathway whose gene expression levels change consistently across the pathway, but are not considered significant because the magnitude of change for individual genes within the pathway is not great enough to meet the pre-determined fold-change or p-value cutoff. In addition, these methods assume gene products act independently of each other as permutations are done across the genes to calculate the P-value for a given pathway.

To alleviate these limitations, other methods, sometimes referred to as functional class scoring (FCS) analysis, have been developed that do not require a fold change or p-value cutoff to determine differential gene expression (75). For example, gene set enrichment analysis, or GSEA, starts with the total set of genes in the microarray ranked by fold change or signal-to-noise ratio (63). The enrichment score for a given pathway is determined by the distribution of the positions of pathway genes within the ranked list. A weighted Kolmogorov-Smirnov test is then used to test whether the distribution differs from uniform. In addition, because the assumption of independence of expression across genes is likely not met (e.g., transcription of several genes may be regulated coordinately by the same set of transcription factors), the individual experiment can be used as the sampling unit for permutation to maintain the correlations between genes. Some popular online or downloadable tools that incorporate these methods include ErmineJ and SAM-GS (76,77). Finally, as with determining differential expression, the multiple testing problem needs to be addressed as the more pathways analyzed, the greater the chance of observing a false-positive result. As in determining significance of differential expression for individual genes, Bonferroni correction and false discovery rate calculations are popular methods to address this issue in pathway analysis as well (56).

Examples in Current Neurotoxicology Research

Although several applications of pathway analysis in neurotoxicology research exist, two studies are highlighted here to provide concrete examples for the neurotoxicologist. Wang et al. (2008) used pathway analysis techniques to identify strain and region specific responses to chronic nicotine treatment (78). Behavioral and pharmacological studies have shown that C3H/HeJ and C57BL/6 mice respond differently to nicotine treatment. Microarray analyses using tissue from amygdala, hippocampus, nucleus accumbens, prefrontal cortex, and ventral tegmental areas produced divergent gene expression patterns according to brain region and strain. Thirty key genes for each of five components were identified using principal components analysis to determine genes that explained the majority of strain specific variability. Using this set of key genes, ORA of GO categories using the EASE software was performed. Results suggested that nicotine perturbed cell cycle, organogenesis, and transmission of nerve impulse in both strains across all brain regions, but ubiquitin-dependent protein catabolism was specific to the

C57BL/6J strain whereas cell-surface receptor-linked signal transduction was specifically overrepresented in the datasets from the CeH/HeJ strain. Further validation of the predicted pathway changes using RT-PCR in the prefrontal region supported the hypothesis that the strain specific increased locomotor activity observed in the C3H/HeJ strain may be due to increased activation of the mitogen-activated protein kinase signaling cascade.

Looking at age-related changes in gene expression in the human prefrontal cortex, Pavlidis et al. (2004) compared ORA with FCS pathway analysis methods (79). They found that ORA methods requiring a cutoff to determine differential expression do not perform as well as FCS methods that use p-values, fold change, or an ANOVA result as a continuous variable across the whole gene expression dataset. For example, they found greater consistency in results across datasets from different brain regions when using a FCS approach. In addition, ORA methods were found to be less consistent, as results varied depending on the specific cutoff set for differential expression.

Many more examples exist in the neuroscience and neurotoxicology literature than can be covered in this chapter. In fact, a review produced in 2005 found over 502 original publications of microarray datasets in neural or glial tissue, 103 of which involved administration of a nonendogenous substance (80). This number has grown considerably since then (over 400 neuroscience related datasets deposited in GEO and ArrayExpress), and numerous efforts have been made to develop methods for re-analysis of these datasets to provide context for future studies (81). For example, GSEA provides easy to use tools to compare current microarray datasets with gene sets from previous microarray analyses to determine if similar patterns exist. In addition, novel analysis tools that use previous literature based hypotheses are described in further detail below.

Literature Based Search Algorithms

Although current pathway databases provide annotation of gene products that are part of the well known canonical pathways, several novel pathways are not included. For example, only 4000 of the 24,000 National Center for Biotechnology Information (NCBI) recognized human genes are represented in at least one KEGG pathway (60). One method to alleviate this limitation is to develop pathways based on mining of the literature. Two popular proprietary software programs that use expert knowledge and text mining algorithms to establish connections between gene products include Ariadne's Pathway Studio and Ingenuity's Pathway Analysis software packages (82,83), although neither emphasize analysis of neurosciences research. Other freeware text mining algorithms have been developed for retrieval of information specifically from the neurosciences literature, such as Textpresso for NeuroSciences (84) and NeuroExtract (85). The Comparative Toxicogenomics Database provides an easily searchable database of connections between genes, chemicals, and diseases based on text mining and expert annotation of the literature (86). In addition, a pathway can be built by traditional literature research and then tested against a microarray dataset. An example of this methodology applied to the regulatory cascade describing neurogenesis in the developing telencephalon is described below.

Testing Microarray Data Against Literature Based Networks: A Case Study in the Developing Telencephalon

The embryonic telencephalon gives rise to specialized structures such as the basal ganglia, the neocortex, and the hippocampus. The neocortex, mainly derived from the dorsolateral telencephalon, is the dominant structure of the adult mammalian brain and functions as the primary region for higher order processes such as language, decision making, and complex social behaviors (87). The embryonic ventral telencephalon also contributes to the neocortex, as it is the site of origin for a unique subset of neocortical GABAergic interneurons (88,89). The basal ganglia, also derived from the ventral telencephalon, is involved in behavioral control such as response inhibition, attention span, and overall executive functioning (90). The differential patterning of these regions begins early during the neurogenesis period, and perturbations of this process are linked to various neurological disorders often associated with developmental neurotoxicity including attention-deficit hyperactivity disorder, epilepsy, and schizophrenia. In addition, behavioral abnormalities associated with in utero exposures to substances such as ethanol, cocaine, and polychlorinated biphenyls (PCB) are thought to arise from perturbation of neurogenesis, creating an imbalance between inhibitory versus excitatory neurons (91–95).

During neurogenesis, which spans the second trimester in humans (approximately E11–E17 in the mouse), neural stem cells begin the differentiation process to become specialized, region-specific neuronal populations. This period is critical for correct telencephalon patterning in the adult, as perturbations during this period can result in massive shifts in positional identity (96,97). In general, neural stem cells in the dorsal telencephalon differentiate into stimulatory glutamatergic pyramidal neurons, whereas inhibitory GABAergic interneurons are derived from neural stem cells found in the ventral telencephalon (98). Furthermore, the regulation of cell fate choice is closely linked to the regulation of the cell cycle, thereby determining the final number of neurons in each region (99,100).

The proneural basic-helix-loop-helix (bHLH) transcription factors *Ngn1*, *Ngn2*, and *Mash1*, are particularly important for the correct differentiation of neuronal phenotypes in the developing telencephalic structures (101,102). Mice with null mutations of *Mash1* show severe defects in neurogenesis in the ventral telencephalon, whereas single and double mutants of *Ngn1* and *Ngn2* lack specific neuronal populations in the dorsal telencephalon (103,104). The generation of replacement mutations in mice in which the coding sequences of *Mash1* and *Ngn2* are swapped suggests *Mash1* is an instructive determinant of GABAergic neuronal differentiation, whereas *Ngn2* is a permissive factor that may act in combination with other factors to specify the dorsal glutamatergic neuronal phenotype (105).

The GRN shown in Figure 1 describes the current understanding of neuronal differentiation initiated by expression of proneural bHLH transcription factors in the developing telencephalon (106) and is based on extensive experimental research summarized in several important reviews (97,101,107).

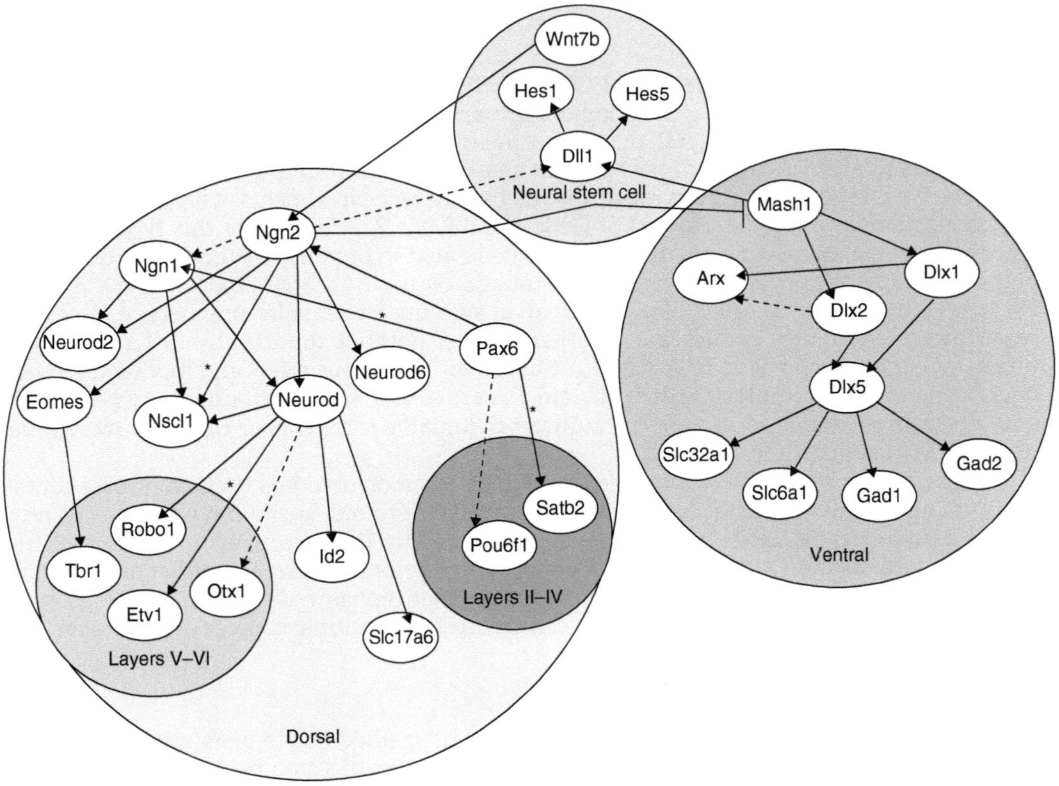

Figure 1 (*See color insert*) Literature-based gene regulatory network describing proneural bHLH regulation of telencephalon neurogenesis. Activations are identified with an arrow and repressions are identified with a barred line. Those connections that are non-significant based on the current microarray dataset are represented as dashed lines. Significant relationships were determined through analysis of the distribution of the strength of linkage parameter (β) after 500,000 MCMC simulations. If >95% of the simulations have values above zero they are considered significant. *Connections which were significant, but as inhibition. *Abbreviations*: bHLH, basic-helix-loop-helix; MCMC, Markov Chain Monte Carlo. *Source*: From Ref. 106 (originally published by Biomed Central).

A Bayesian based method was applied to global gene expression patterns in *Ngn2, Mash1*, and *Ngn1* gain of function (GOF) and loss of function (LOF) mice to test these literature based connections (106). This method removes the need to rely on fold change cutoffs by examining the strength of the predicted relationships based on concurrently evaluating variability in gene expression patterns across control and perturbation experiments. Significant connections based on this analysis [the 5th percentile of the posterior density for β is greater than zero (108)] are highlighted in Figure 1. This analysis predicted 86% (31/36) of the connections as significant based on the microarray dataset. Specificity of this method was estimated via permutation of gene labels, resulting in an estimated false positive rate of 8.2%. Therefore, the random chance of seeing 31 significant connections out of 36 is very low ($p < 0.001$), suggesting this method can provide an efficient way to test microarray datasets against literature-based network hypotheses.

One limitation of this approach is the assumption of an acyclic network (e.g., no feedback loops) and the assumption that gene expression relationships are linear. For example, notice that four of the significant interactions are considered inhibitory interactions instead of activations, which suggests possible feedback loops between *Pax6* and *Ngn1* and *Satb2, Ngn2*, and *Nscl*, as well as between *NeuroD* and *Etv1*. In addition, two of the connections originating from *Ngn2* (to *Ngn1* and *Dll1*) are not considered significant based on this analysis. As the algorithm relies on linear relationships between genes, its inability to detect these two connections may have resulted from non-linear variability between *Ngn2* and its targets created by the large increase in *Ngn2* transcript levels after the GOF electroporation. This highlights the importance of considering experimental design constraints in evaluating connectivity between gene products. Application of robust time-series datasets will allow for determination of acyclic networks and non-linear relationships (109).

GENERATING NOVEL PATHWAY HYPOTHESES FROM MICROARRAY DATA

As introduced above, another application of microarray datasets is generation of novel network or pathway hypotheses. In general, this type of analysis uses the variance structure across many experiments to estimate connections between gene products. In simple terms, if Gene 1 consistently increases when Gene 2 decreases across many different experiments, it would suggest there is a connection between the product of Gene 1 and Gene 2. Methods for this type of analysis were first developed using very large datasets generated in organisms where high throughput perturbation analyses have been developed, including *S. cerevisiae, C. elegans*, and *S. purpuratus* (29–33). Because of the higher complexity of mammalian systems and smaller numbers of microarray experiments available, application to other species without modification of the methods produced high false positive rates (48,110). Since then, numerous novel and modified methods have been developed to alleviate this problem. In particular, it has been found that incorporating datasets from several independent data sources to limit the search space results in more robust predictions of connectivity (54,111).

This section starts with a short description of common methods used to predict connectivity using microarray data. Next, databases available to aid in the development of novel networks in neurotoxicology research are explored. Finally, this section ends with a case study using a Bayesian based algorithm with an informative prior structure to predict novel connections in the GRN governing neurogenesis during telencephalon development. This type of algorithm is particularly useful because it can incorporate multiple sources of prior data.

Generating Networks from Microarray Data

Often referred to as "reverse engineering," numerous methods have been utilized to predict connections between gene products based on global gene expression patterns. These methods are detailed in several comprehensive reviews (50–53). Briefly, clustering is by far the simplest method employed to group genes by expression patterns. Different similarity metrics including Euclidean distance, correlation coefficient, ranked correlation coefficient, and mutual information based measure have been used to quantify the similarity in gene expression patterns. Multivariate unsupervised techniques including hierarchical clustering, k-means, and self-organizing maps have been used to predict functional groups of genes from microarray datasets. Supervised techniques such as support vector machine or decision tree analysis have been employed to classify genes of unknown function based on similarity in expression patterns of genes whose function is known. Neural network approaches have also been employed to predict connectivity between genes. Discreet, Boolean networks, where a gene is said to either be "on" or "off" based on a predetermined cutoff have been used to develop network hypotheses, whereas other methods use continuous functions to describe relationships between genes.

When compared to other methods, Bayesian approaches have performed particularly well for identifying connectivity networks based on global gene or protein expression datasets (112). Bayesian-based network analyses was first introduced in applications of large datasets from in vitro and invertebrate models (48,113); however, some attempts using datasets derived from mammalian species have been made (108). Through this research, it has been found that the sample size limitation can be eased through the use of an informative prior structure to incorporate data from other sources, such as from sequence-based transcription factor binding site (TFBS) information (54). In fact a wide variety of data can be used as prior knowledge to build a probability distribution for particular connections in the Bayesian framework, making it particularly amenable for prior knowledge data synthesis. Besides literature based search algorithms or traditional literature searches as described in the above section, there are several important sources of prior information. Some examples are explored below.

Prior Information Databases

A diverse set of information can be useful as prior knowledge for building a GRN. Some tools that are particularly suitable for neuroscience research are briefly discussed below. The reader is also referred to a recent review of informatics tools in neuroscience (114) and to a Neuroscience Database Gateway (http://ndg.sfn.org/) developed and managed by the Society for Neuroscience.

The National Institute of Mental Health Human Brain Project focuses on funding neuroinformatics research. One result is the SenseLab Project, which includes six linked databases bringing together neuroscience research in pharmacology, ion channels, membrane properties, olfactory pathways, and neuronal models (115). Functional magnetic resonance imaging datasets are deposited in the Brede database (116). This database includes software that allows searches and cross-referencing to genes, diseases, and receptors via SenseLab. In addition, compilation of electrophysiological datasets into databases are in the beginning stages (117), which when complete, will allow integration of these signal transmission datasets into other static datasets such as gene expression.

Cell to cell connections are an important part of neurophysiology and can help to interpret microarray datasets. Brain Architecture Management System (BAMS) (118,119) focuses on connectivity in the rat brain, whereas CoCoMac (120) compiles connectivity data in the Macaque monkey. Currently, *C. elegans* is the only species for which the complete wiring diagram of the nervous system is available (www.wormatlas.org).

GeneNetwork (http://www.genenetwork.org) provides easy to use query tools for gene expression datasets from several brain regions from well-studied recombinant inbred mouse strains (121). This allows researchers to treat expression levels as quantitative traits that can be related to morphological or behavioral traits (122). For example, this database has been used to show the link between genetic variation, gene expression variation, and adult neurogenesis in the hippocampus (123).

The highly heterogeneous nature of nervous system tissue contributes to the difficulty in interpretation of results from microarray analyses. Therefore, atlases of spatially restricted gene expression are ideal databases for supplementing microarray data to obtain morphological context. The Allen Brain Atlas contains high-resolution colorimetric in situ RNA hybridization data in the adult mouse brain for 21,000 genes (124) and has been used in regulatory network inference (125). For developmental expression patterns the Brain Gene Expression Map (BGEM), GenePaint, and Gene Expression Nervous System Atlas (GENSAT) provide searchable databases of radioactive in situ hybridization and fluorescent protein reporters (126–128).

Novel tools for integration of results from multiple microarray datasets are being developed. Gemma is one such project that focuses on the functional characterization of gene products via meta-analysis of variation across microarray experiments (129). Therefore, those genes that have correlated expression patterns are inferred to have a closer relationship than those with uncorrelated expression patterns. For example, by performing an analysis of 60 human data sets, Lee et al. (2004) was able to show more robust functional predictions were possible when results from several microarray experiments were combined (129).

Protein-protein interaction (PPI) databases can also be used to supplement microarray datasets in network generation based on the assumption that interactions between gene products at the protein level are predictive of relationships at the mRNA level. Although this is a simplistic assumption, it can be helpful to determine if the coordinated gene expression seen in a set of microarray experiments can be related to modules of protein interactions. Several PPI databases exist, some of which focus on particular species, yeast two hybrid screen results, or literature searches. Two well known freely accessible databases include database of interacting proteins (DIP) (130), which has no species restrictions and Human Protein Reference Database (HPRD) (131), which focuses on human protein interactions. Another popular database as well as a list of PPI resources can be found at the Mammalian Protein-Protein Interaction Database (MIPS) (132). Mathivanan et al. (2006) compares several PPI databases and evaluates their applicability based on the particular needs of a researcher (133).

The most popular method for supplementing gene expression data in network analysis is TFBS analyses. For example, if the binding site of a particular transcription factor is found in the upstream region of a gene, then this information can be used to increase confidence of a direct connection if correlation in gene expression exists between the transcription factor and the gene. Several important databases are useful in sequence searches for TFBS. TRANSFAC developed by Biobase provides the most comprehensive database of TFBS position weight matrices (PWM) (134). A given sequence can also be searched within TRANSFAC, although an initial sequence alignment and identification of evolutionary conserved regions is recommended. Because TFBS are very short sequences (usually between 6–15 bp in length) and degenerate in nature it is important to perform searches only within evolutionarily conserved regions to limit false positives. Several sequence alignment algorithms exist and are covered in detail elsewhere (135). A web-based platform developed at the Lawrence Livermore National Laboratory provides an integrated set of tools for analysis of TFBS in evolutionarily conserved sequence (136,137). MatInspector is part of a proprietary software package developed by Genomatix that includes up-to-date information on PWM and easy to use gene expression analysis tools (138). An example of how this type of information can be integrated with microarray datasets to generate hypotheses of GRNs is described in the following section.

A Case Study in the Developing Telencephalon

This example builds from the case study described in the previous section on testing of microarray datasets against a literature-based network. Briefly, microarray datasets were generated from dorsal and ventral telencephalon tissue derived from *Mash1*, *Ngn1*, and *Ngn2* GOF and LOF mice (106). To predict novel co-factors and co-regulators important for telencephalon specification, comparative genomics bioinformatics approaches were used to identify conserved TFBS surrounding the putative Ngn2 and Mash1 target genes identified from these global gene expression datasets (106). Subsequently, this sequence based information was incorporated into an informative prior structure for the TAO-Gen algorithm (139), which is a Bayesian based tool for finding the best network given a gene or protein expression dataset. Therefore, the resultant network incorporates prior information with microarray expression results to develop consensus predictions of connectivity between the predicted targets, co-factors, and co-regulators.

Sequence Analysis to Identify Potential Co-factors for Ngn2 and Mash1

It has been shown that developmentally expressed genes are often regulated via ultra conserved long-range enhancers (140–142). Therefore, comparative genomics and TFBS searches can be utilized to identify putative long-range enhancers. For example, in this study a minimum of 500 kb of sequence surrounding several predicted Ngn2 and Mash1 target genes was analyzed. In general, approximately 300 kb in front of and approximately 200 kb behind including UTRs and introns of the gene of interest was considered. A four step bioinformatic process was then implemented: (*i*) The ECR browser (www.ecr.dcode.org) was used to align human sequence with *Mus musculus*, *Gallus gallus*, *Xenopus tropicalis*, *Fugu rubripes*, and *Danio rerio* (137). (*ii*) Mulan (www.mulan.dcode.org) was used to perform a higher quality local multi-sequence alignment based on Blastz (137). (*iii*) Multitf (www.multitf.dcode.org) was applied to search for putative Mash1 (GCAGSTGK or CAGSTG) and Ngn2 (CANWTG) binding sites within the aligned conserved sequences. (4) All other TRANSFAC annotated TFBS within 30 bp upstream and downstream of the putative Ngn2 or Mash1 binding sites were then identified. In total, 114 conserved putative Ngn2 and Mash1 sites were identified. Finally, to identify the most likely co-factors for Ngn2 and Mash1, K-means clustering was carried out to predict those TFBS that most often co-occurred with Ngn2 or Mash1 binding sites within the Ngn2 versus Mash1 putative target genes, respectively.

This bioinformatics approach suggested that only the ubiquitously expressed E-proteins, Tcf12 and Tcfe2a, are specifically associated with conserved Mash1 binding sites in the Mash1 and common target genes, consistent with an instructive role for Mash1 in ventral cell fate determination (105). Alternatively, the analysis predicted several co-factors for Ngn2, again consistent with previous findings suggesting a permissive role for Ngn2 in dorsal fate specification (105). In addition, this analysis predicted CREB and Crebbp are the most likely candidates for a dorsally expressed Ngn2 co-factor, which is supported by similar interactions identified in other developing tissues (143–146). In addition, this analysis predicted transcriptional regulators downstream of Wnt signaling (Tcf4/Lef1) may bind to regulatory modules that also bind Ngn2. This result is consistent with research implicating the importance of Wnt signaling in the specification of the dorsal forebrain (147–149), and offers a hypothesis in which coordinated Wnt activation and Ngn2 expression act in concert to transcriptionally activate genes important for dorsal neuronal specification.

Identification of Putative Co-regulators of Ngn2 and Mash1 Targets

A second comparative genomics analysis was designed to identify important regulatory modules that bind transcription factors other than Ngn2 and Mash1. For this analysis, promoter region sequence (10,000 bp upstream of TSS) from mouse and human orthologs of Ngn2 predicted targets and Mash1 and Mash1/Ngn2 predicted common targets was automatically uploaded from the UCSC database via the CONFAC website (89). CONFAC then identified conserved TFBS from the TRANSFAC database version 7.0 in the human and mouse sequence alignments (150). As part of the CONFAC software, the Mann-Whitney statistical test was then applied to test for enrichment of TFBS in the given gene lists. Each list was compared with a list of 250 randomly picked genes available from the CONFAC website, as well as comparing the Ngn2 target list with the Mash1/common targets list and vice versa. Several transcription factors were identified that may act as co-regulators, including Hes1, Egr1, Nfy, Mef2a, Tef, and Sox9. Each of these transcription factors have previously been implicated in developmental regulation in neural and non-neural tissues (151–156) and based on the microarray analyses, showed gene expression consistent with a potential role in the developing forebrain.

Utilization of In Situ Hybridization and Protein-Protein Interaction Databases

Through interrogation of online databases of in situ hybridization in serial brain sections across development (126,127), the spatially distinct expression patterns of several predicted co-factors/co-regulators in the developing dorsal and ventral telencephalon was confirmed (106). In fact, this analysis was able to show that several of the predicted co-factors/co-regulators, including Sox9, Crebbp, Creb1, Tcf4, Lef1, Pou6f1, Pou2f1, Pou3f1, Tef, Hes1, and E2f1, have appreciable expression in the ventricular zone of dorsal and/or ventral telencephalon, where proneural bHLH proteins are specifically expressed. In addition, the HPRD (157) was interrogated to find if direct PPIs between proneural bHLH proteins and the predicted co-factors have been previously identified using a yeast two hybrid screen. Experimental evidence found in this database supported direct PPIs between proneural proteins and Crebbp, Tcf4, and Mef2a.

Bayesian Network Analysis with an Informative Prior Structure

Using Bayesian network analysis with an informative prior structure, the knowledge gained from the above experimental and computational approaches was synthesized to predict connectivity between the novel target genes, co-factors, and co-regulators (Fig. 2). An informative prior structure considered all significant literature-based connections as required (diagrammed using thicker lines) and used the TFBS information from the comparative genomics analyses to weight connections in which TFBS information was found. For example, a conserved *Sox9*

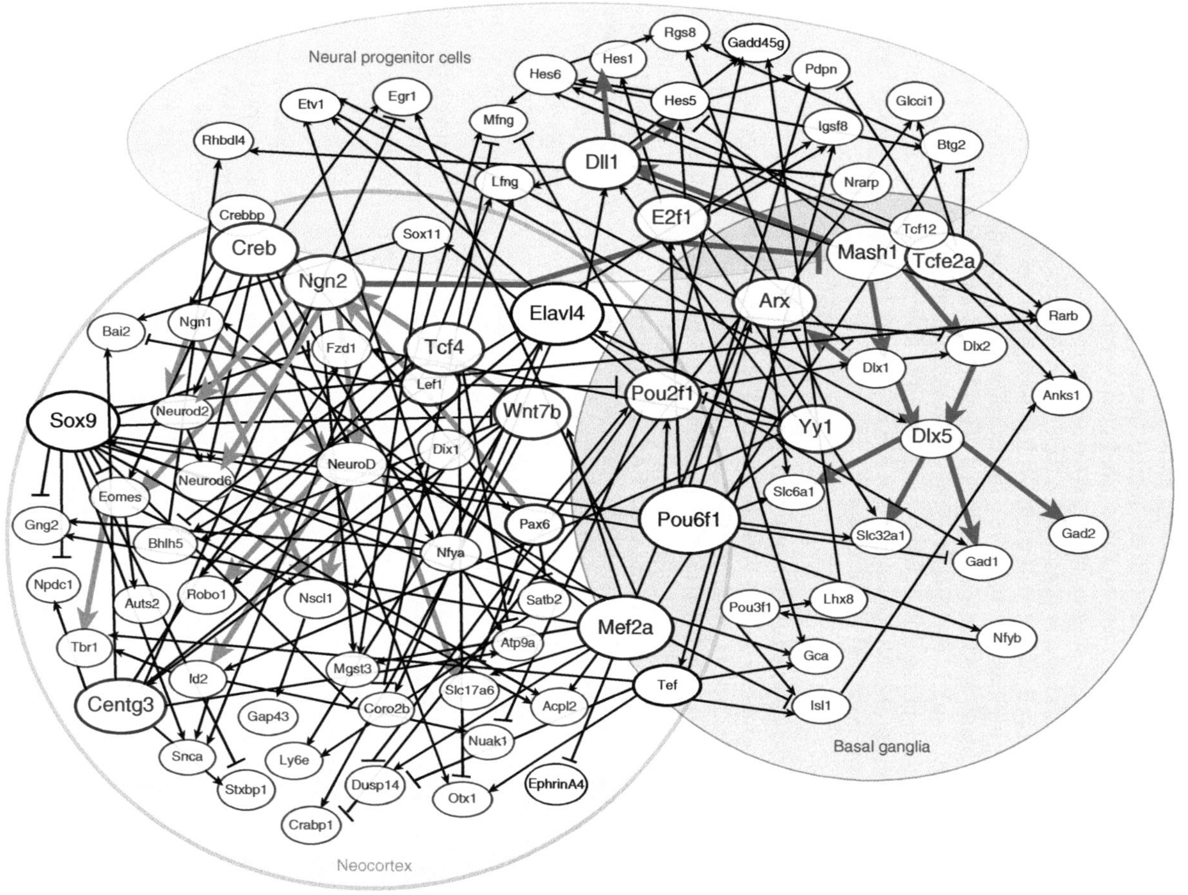

Figure 2 (*See color insert*) Algorithm-based gene regulatory network structure for dorsal and ventral telencephalon development. The Bayesian-based TAO-Gen algorithm (139) was implemented with an informative prior structure to predict the optimal network structure based on the LOF and GOF microarray datasets, evolutionarily conserved transcription factor binding site data, prior literature-based knowledge, and spatial and time specific expression patterns. To highlight the key regulators, the nodes representing genes predicted to be the parent of at least nine other genes are largest in size (Sox9, Mef2a, Elavl4, and Pou6f1), whereas those that are predicted to regulate at least five other genes are medium in size (Ngn2, Centg3, Tef, Tcf4, Wnt7b, Pou2f1, Yy1, Dll1, E2f1, Tcfe2a, Arx, and Creb). Arrows indicate activation and barred lines indicate inhibition. *Abbreviations*: GOF, gain of function; LOF, loss of function. *Source*: From Ref. 106 (originally published by Biomed Central).

TFBS was identified in the sequence surrounding *Lfng*, therefore a preference is given to *Sox9* being a parent to *Lfng*, which results in a slightly higher probability that a linear relationship between these two genes will be significant in the posterior distribution. The reliability of the TAO-Gen algorithm as well as the utility of the informative prior structure was tested through direct comparisons with results obtained from another Bayesian network algorithm (112,158), as well as with results obtained using TAO-Gen without an informative prior.

The final network structure predicts 174 linkages between the set of 82 genes, adding a considerable amount of knowledge to the previous literature based network. For example, Sox9, Mef2a, Elavl4, and Pou6f1 are predicted as prolific co-regulators of the target genes, with 14, 12, 9, and 12 children, respectively. Furthermore, our analysis predicts Creb1, Crebbp, and Yy1 as the most likely candidates for dorsally expressed Ngn2 co-factors, and supports a synergistic interaction between Pou-domain containing transcription factors and bHLH proneural proteins in the regulation of common target genes, which is consistent with previous studies (159). Several transcription factors were identified as candidate co-regulators of Ngn2 target genes, including Hes1, Egr1, Nfy, Mef2a, Tef, and Sox9, whereas Pou6f1 is a predicted co-regulator of Mash1 target genes.

Lessons Learned

One important application of a network analysis is the prioritization of the most useful experimental perturbation or ChIP experiment for resolving the overall network structure (160). This is particularly relevant to studies in mammalian species, in which perturbation analyses are much more time and resource intensive, although the development of RNAi strategies and ChIP-on-chip technologies in mammalian systems offers promising prospects for increasing the efficiency of network inference (161). For example, the network analysis presented here would suggest *Centg3* and *Elavl4* are important candidates for perturbation analysis and subsequent global gene expression analysis for further network resolution, as they are both highly connected nodes and little is known about their function in the developing telencephalon. Interestingly, variants in *Centg3* have been shown to modulate protection against neurodegeneration in Polyglutamine diseases (162). *Elavl4* regulates through mRNA stabilization (163), highlighting the importance of moving beyond cis regulatory binding and mRNA expression analysis to elucidate network relationships.

SUMMARY

With the maturation and widespread use of microarray technology, there is a critical need to understand how gene expression patterns can be utilized to understand gene regulation, the functional outcomes of sets of co-regulated genes, and ultimately phenotypic outcome. This chapter is intended to demonstrate the utility of bioinformatics approaches for elucidation of biological insights from omics technologies. Although this chapter focused on global gene expression analysis as it is the most developed omics field, the methods introduced here are relevant to all omics techniques. The analytical approaches are not restricted solely to genomic data, in fact the lessons learned from microarray studies over the past 15 years can aid in the more rapid development of proteomics, metabolomics, and other systems biology level analyses. This chapter covered initial data analysis to find differentially expressed genes, the placement of gene expression datasets into the context of known pathways, and ended with methods for developing novel network hypotheses by combining microarray datasets with prior knowledge. There is no doubt that as the technology for rapid experimental data generation continues to progress, the utilization of computational and bioinformatic tools will become an integral part of neurotoxicology research.

REFERENCES

1. Southern EM. Detection of specific sequences among DNA fragments separated by gel-electrophoresis. J Mol Biol 1975; 98: 503–17.
2. Kulesh DA, Clive DR, Zarlenga DS, Greene JJ. Identification of interferon-modulated proliferation-related cDNA sequences. Proc Natl Acad Sci U S A 1987; 84: 8453–7.
3. Lussier M, Sdicu AM, Winnett E, et al. Completion of the Saccharomyces cerevisiae genome sequence allows identification of KTR5, KTR6 and KTR7 and definition of the nine-membered KRE2/MNT1 mannosyltransferase gene family in this organism. Yeast 1997; 13: 267–74.

4. Craddock N, O'Donovan MC, Owen MJ. Genome-wide association studies in psychiatry: lessons from early studies of non-psychiatric and psychiatric phenotypes. Mol Psychiatry 2008; 13: 649–53.
5. Goldman D. Whole-genome association for complex psychiatric disorders. Neuropsychopharmacology 2008; 33: 201.
6. Almasy L, Gur RC, Haack K, et al. A genome screen for quantitative trait loci influencing schizophrenia and neurocognitive phenotypes. Am J Psychiatry 2008; 165: 1185–92.
7. Lewis CM, Ng MYM, Levinson DF, et al. Extended meta-analysis of genome-wide linkage studies in schizophrenia. Ann Hum Genet 2008; 72: 692–3.
8. Parkinson H, Kapushesky M, Shojatalab M, et al. ArrayExpress—a public database of microarray experiments and gene expression profiles. Nucleic Acids Res 2007; 35: D747–50.
9. Barrett T, Troup DB, Wilhite SE, et al. NCBI GEO: mining tens of millions of expression profiles—database and tools update. Nucleic Acids Res 2007; 35: D760–5.
10. Clark TA, Schweitzer AC, Chen TX, et al. Discovery of tissue-specific exons using comprehensive human exon microarrays. Genome Biol 2007; 8: R64.
11. Yin JQ, Zhao RC, Morris, KV. Profiling microRNA expression with microarrays. Trends Biotech 2008; 26: 70–6.
12. Knight J. When the chips are down. Nature 2001; 410: 860–1.
13. Kothapalli R, Yoder SJ, Mane S, Loughran TP. Microarray results: how accurate are they? BMC Bioinformatics 2002; 3: 22.
14. Tan PK, Downey TJ, Spitznagel EL, et al. Evaluation of gene expression measurements from commercial microarray platforms. Nucleic Acids Res 2003; 31: 5676–84.
15. Ball CA, Sherlock G, Parkinson H, et al. Standards for microarray data. Science 2002; 298: 539.
16. Baker SC, Bauer SR, Beyer RP. The external RNA controls consortium: a progress report. Nat Methods 2005; 2: 731–4.
17. Shi LM, Reid LH, Jones WD, et al. The MicroArray Quality Control (MAQC) project shows inter- and intraplatform reproducibility of gene expression measurements. Nat Biotechnol 2006; 24: 1151–61.
18. Shi L. MAQC: The microarray quality control project. Molecular and Cellular Toxicology 2007; 3: 32.
19. Fathallah-Shaykh HM. Microarrays—Applications and pitfalls. Arch Neurol 2005; 62: 1669–72.
20. Liang Y, Diehn M, Watson N. Gene expression profiling reveals molecularly and clinically distinct subtypes of glioblastoma multiforme. Proc Natl Acad Sci U S A 2005; 102: 5814–19.
21. Huang CJ, Liao WC. Application of probabilistic neural networks to the class prediction of leukemia and embryonal tumor of central nervous system. Neural Processing Letters 2004; 19: 211–26.
22. Xu Z, Patterson TA, Wren JD. A microarray study of MPP+-treated PC12 Cells: mechanisms of toxicity (MOT) analysis using bioinformatics tools. BMC Bioinformatics 2005; 6(Suppl 2): S8.
23. Wang J, Xu Z, Fang H, et al. Gene expression profiling of MPP+-treated MN9D cells: a mechanism of toxicity study. Neurotoxicology; 2007; 28: 979–87.
24. Maier MS, Legare ME, Hanneman WH. The aryl hydrocarbon receptor agonist 3,3′,4,4′,5-pentachlorobiphenyl induces distinct patterns of gene expression between hepatoma and glioma cells: chromatin remodeling as a mechanism for selective effects. Neurotoxicology 2007; 28: 594–612.
25. Royland JE, Kodavanti PR. Gene expression profiles following exposure to a developmental neurotoxicant, Aroclor 1254: pathway analysis for possible mode(s) of action. Toxicol Appl Pharmacol 2008; 231: 179–96.
26. Hohmann S. The yeast systems biology network: mating communities. Curr Opin Biotechnol 2005; 16: 356–60.
27. Yonan AL, Palmer AA, Smith KC, et al. Bioinformatic analysis of autism positional candidate genes using biological databases and computational gene network prediction. Genes Brain Behav 2003; 2: 303–20.
28. Tropea D, Kreiman G, Lyckman A, et al. Gene expression changes and molecular pathways mediating activity-dependent plasticity in visual cortex. Nat Neurosci 2006; 9: 660–8.
29. Vidal M. Interactome modeling. FEBS Lett 2005; 579: 1834–8.
30. Dearden P, Akam M. Segmentation in silico. Nature 2000; 406: 131–2.
31. Csete ME, Doyle JC. Reverse engineering of biological complexity. Science 2002; 295: 1664–9.
32. Kim JK, Gabel HW, Kamath RS. Functional genomic analysis of RNA interference in C-elegans. Science 2005; 308: 1164–7.
33. Davidson EH, Rast JP, Oliveri P, et al. A genomic regulatory network for development. Science 2002; 295: 1669–78.
34. Yang YH, Dudoit S, Luu P. Normalization for cDNA microarray data: a robust composite method addressing single and multiple slide systematic variation. Nucleic Acids Res 2002; 30: e15.

35. Cheadle C, Vawter MP, Freed WJ, Becker KG. Analysis of microarray data using Z score transformation. J Mol Diagn 2003; 5: 73–81.
36. Tusher VG, Tibshirani R, Chu G. Significance analysis of microarrays applied to the ionizing radiation response. Proc Natl Acad Sci U S A 2001; 98: 5116–21.
37. Shedden K, Chen W, Kuick R, et al. Comparison of seven methods for producing Affymetrix expression scores based on False Discovery Rates in disease profiling data. BMC Bioinformatics 2005; 6: 26.
38. Wittwer CT, Herrmann MG, Moss AA, Rasmussen RP. Continuous fluorescence monitoring of rapid cycle DNA amplification. Biotechniques 1997; 22: 130–8.
39. Nazarenko I, Lowe B, Darfler M, et al. Multiplex quantitative PCR using self-quenched primers labeled with a single fluorophore. Nucleic Acids Res 2002; 30: e37.
40. Heid CA, Stevens J, Livak KJ, Williams PM. Real time quantitative PCR. Genome Res 1996; 6: 986–94.
41. VanGuilder HD, Vrana KE, Freeman WM. Twenty-five years of quantitative PCR for gene expression analysis. Biotechniques 2008; 44: 619–26.
42. Bustin SA, ed. A-Z of Quantitative PCR. California: International University Line, 2004.
43. Pinhasov A, Mei J, Amaratunga D, et al. Gene expression analysis for high throughput screening applications. Comb Chem High Throughput Screen 2004; 7: 133–40.
44. Mirnics K, Middleton FA, Marquez A, et al. Molecular characterization of schizophrenia viewed by microarray analysis of gene expression in prefrontal cortex. Neuron 2000; 28: 53–67.
45. Bruggeman FJ, Westerhoff HV. The nature of systems biology. Trends Microbiol 2007; 15: 45–50.
46. Wilson I. Top-down versus bottom-up—rediscovering physiology via systems biology? Mol Syst Biol 2007; 3: 113.
47. Rhodes DR, Tomlins SA, Varambally S, et al. Probabilistic model of the human protein-protein interaction network. Nat Biotech 2005; 23: 951–9.
48. Basso K, Margolin AA, Stolovitzky G, et al. Reverse engineering of regulatory networks in human B cells. Nat Genet 2005; 37: 382–90.
49. Lee PH, Lee D. Modularized learning of genetic interaction networks from biological annotations and mRNA expression data. Bioinformatics 2005; 21: 2739–47.
50. Styczynski MP, Stephanopoulos G. Overview of computational methods for the inference of gene regulatory networks. Computers & Chemical Engineering 2005; 29: 519–34.
51. Ma'ayan A, Blitzer RD, Iyengar R. Toward predictive models of mammalian cells. Annu Rev Biophys Biomol Struct 2005; 34: 319–49.
52. Papin JA, Hunter T, Palsson BO, Subramaniam S. Reconstruction of cellular signalling networks and analysis of their properties. Nat Rev Mol Cell Biol 2005; 6: 99–111.
53. Barabasi AL, Oltvai ZN. Network biology: understanding the cell's functional organization. Nat Rev Genet 2004; 5: 101–15.
54. Bernard A, Hartemink AJ. Informative structure priors: joint learning of dynamic regulatory networks from multiple types of data. Pac Symp Biocomput 2005; (PSB05): 459–70.
55. Cary MP, Bader GD, Sander C. Pathway information for systems biology. FEBS Lett 2005; 579: 1815–20.
56. Curtis RK, Oresic M, Vidal-Puig A. Pathways to the analysis of microarray data. Trends Biotechnol 2005; 23: 429–35.
57. Werner T. Bioinformatics applications for pathway analysis of microarray data. Curr Opin Biotechnol 2008; 19: 50–4.
58. Shaw DR, Ashbumer M, Blake JA, et al. Gene Ontology: a controlled vocabulary to describe the function, biological process and cellular location of gene products in genome databases. Am J Hum Genet 1999; 65: A419.
59. Harris MA, Deegan JI, Lomax J, et al. The Gene Ontology project in 2008. Nucleic Acids Res 2008; 36: D440–4.
60. Kanehisa M, Goto S, Hattori M, et al. From genomics to chemical genomics: new developments in KEGG. Nucleic Acids Res 2006; 34: D354–7.
61. Salomonis N, Hanspers K, Zambon AC, et al. GenMAPP 2: new features and resources for pathway analysis. BMC Bioinformatics 2007; 8: 217.
62. Mi HY, Guo N, Kejariwal A, Thomas PD. PANTHER version 6: protein sequence and function evolution data with expanded representation of biological pathways. Nucleic Acids Res 2007; 35: D247–52.
63. Subramanian A, Tamayo P, Mootha VK, et al. Gene set enrichment analysis: a knowledge-based approach for interpreting genome-wide expression profiles. Proc Natl Acad Sci U S A 2005; 102: 15545–50.
64. Shi J, Walker MG. Gene set enrichment analysis (GSEA) for interpreting gene expression profiles. Curr Bioinform 2007; 2: 133–7.

65. Bader GD, Cary MP, Sander C. Pathguide: a pathway resource list. Nucleic Acids Res 2006; 34: D504–6.
66. Hosack DA, Dennis G, Sherman BT, et al. Identifying biological themes within lists of genes with EASE. Genome Biol 2003; 4: R70.
67. Dennis G, Sherman, BT, Hosack DA, et al. DAVID: Database for annotation, visualization, and integrated discovery. Genome Biol 2003; 4: P3.
68. Zeeberg BR, Qin HY, Narasimhan S, et al. High-throughput GoMiner, an "industrial-strength" integrative gene ontology tool for interpretation of multiple-microarray experiments, with application to studies of Common Variable Immune Deficiency (CVID). BMC Bioinformatics 2005; 6: 168.
69. Zeeberg BR, Feng WM, Wang G, et al. GoMiner: a resource for biological interpretation of genomic and proteomic data. Genome Biol 2003; 4: R28.
70. Al-Shahrour F, Diaz-Uriarte R, Dopazo J. FatiGO: a web tool for finding significant associations of Gene Ontology terms with groups of genes. Bioinformatics 2004; 20: 578–80.
71. Doniger SW, Salomonis N, Dahlquist KD, et al. MAPPFinder: using gene ontology and GenMAPP to create a global gene-expression profile from microarray data. Genome Biol 2003; 4: R7.
72. Khatri P, Bhavsar P, Bawa G, Draghici S. Onto-tools: an ensemble of web-accessible, ontology-based tools for the functional design and interpretation of high-throughput gene expression experiments. Nucleic Acids Res 2004; 32: W449–56.
73. Khatri P, Voichita C, Kattan K, et al. Onto-Tools: new additions and improvements in 2006. Nucleic Acids Res 2007; 35: W206–11.
74. Al-Shahrour F, Minguez P, Tarraga J. FatiGO+: a functional profiling tool for genomic data. Integration of functional annotation, regulatory motifs and interaction data with microarray experiments. Nucleic Acids Res 2007; 35: W91–6.
75. Goeman JJ, Buhlmann P. Analyzing gene expression data in terms of gene sets: methodological issues. Bioinformatics 2007; 23: 980–7.
76. Dinu I, Potter JD, Mueller T, et al. Improving gene set analysis of microarray data by SAM-GS. BMC Bioinformatics 2007; 8: 242.
77. Lee HK, Braynen W, Keshav K, et al. Tool for functional analysis of gene expression data sets. BMC Bioinformatics 2005; 6: 269.
78. Wang J, Gutala R, Hwang YY, et al. Strain- and region-specific gene expression profiles in mouse brain in response to chronic nicotine treatment. Genes Brain Behav 2008; 7: 78–87.
79. Pavlidis P, Qin J, Arango V, et al. Using the gene ontology for microarray data mining: a comparison of methods and application to age effects in human prefrontal cortex. Neurochem Res 2004; 29: 1213–22.
80. Aarnio V, Paananen J, Wong G. Analysis of microarray studies performed in the neurosciences. J Mol Neurosci 2005; 27: 261–8.
81. Wan X, Pavlidis P. Sharing and reusing gene expression profiling data in neuroscience. Neuroinformatics 2007; 5: 161–75.
82. Yuryev A, Mulyukov Z, Kotelnikova E, et al. Automatic pathway building in biological association networks. BMC Bioinformatics 2006; 7: 171.
83. Rajagopalan D, Agarwal P. Inferring pathways from gene lists using a literature-derived network of biological relationships. Bioinformatics 2005; 21: 788–93.
84. Muller HM, Kenny EE, Sternberg PW. Textpresso: an ontology-based information retrieval and extraction system for biological literature. PLoS Biol 2004; 2: 1984–98.
85. Crasto CJ, Masiar P, Miller PL. NeuroExtract: facilitating neuroscience-oriented retrieval from broadly-focused bioscience databases using text-based query mediation. J Am Med Inform Assoc 2007; 14: 355–60.
86. Mattingly CJ, Rosenstein MC, Davis AP, et al. The comparative toxicogenomics database: a cross-species resource for building chemical-gene interaction networks. Toxicol Sci 2006; 92: 587–95.
87. Mesulam M. From sensation to cognition. Brain 1998; 121: 1013–52.
88. Anderson SA, Eisenstat DD, Shi L, et al. Interneuron migration from basal forebrain to neocortex: dependence on Dlx genes. Science 1997; 278: 474–6.
89. Kriegstein AR, Noctor SC. Patterns of neuronal migration in the embryonic cortex. Trends Neurosci 2004; 27: 392–9.
90. Haber SN. The primate basal ganglia: parallel and integrative networks. J Chem Neuroanat 2003; 26: 317–30.
91. Slaght SJ, Paz T, Mahon S, et al. Functional organization of the circuits connecting the cerebral cortex and the basal ganglia: implications for the role of the basal ganglia in epilepsy. Epileptic Disord 2002; 4: S9–21.
92. LaMantia AS. Forebrain induction, retinoic acid, and vulnerability to schizophrenia: insights from molecular and generic analysis in developing mice. Biol Psychiatry 1999; 46: 19–30.

93. Voeller KKS. Attention-deflicit hyperactivity disorder (ADHD). J Child Neurol 2004; 19: 798–814.
94. Riley EP, McGee CL, Sowell ER. Teratogenic effects of alcohol: a decade of brain imaging. Am J Med Genet C Semin Med Genet 2004; 127C: 35–41.
95. Puglisi-Allegra S, Cabib S, Pascucci T, et al. Dramatic brain aminergic deficit in a genetic mouse model of phenylketonuria. Neuroreport 2000; 11: 1361–4.
96. Hirabayashi Y, Gotoh Y. Stage-dependent fate determination of neural precursor cells in mouse forebrain. Neurosci Res 2005; 51: 331–6.
97. Ross SE, Greenberg ME, Stiles CD. Basic helix-loop-helix factors in cortical development. Neuron 2003; 39: 13–25.
98. Letinic K, Zoncu R, Rakic P. Origin of GABAergic neurons in the human neocortex. Nature 2002; 417: 645–9.
99. Rakic P. A small step for the cell, a giant leap for mankind: a hypothesis of neocortical expansion during evolution. Trends Neurosci 1995; 18: 383–8.
100. Rubenstein JLR, Rakic P. Genetic control of cortical development. Cereb Cortex 1999; 9: 521–3.
101. Bertrand N, Castro DS, Guillemot F. Proneural genes and the specification of neural cell types. Nat Rev Neurosci 2002; 3: 517–30.
102. Schuurmans C, Guillemot F. Molecular mechanisms underlying cell fate specification in the developing telencephalon. Curr Opin Neurobiol 2002; 12: 26–34.
103. Casarosa S, Fode C, Guillemot F. Mash1 regulates neurogenesis in the ventral telencephalon. Development 1999; 126: 525–34.
104. Fode C, Ma QF, Casarosa S, et al. A role for neural determination genes in specifying the dorsoventral identity of telencephalic neurons. Genes Dev 2000; 14: 67–80.
105. Parras CM, Schuurmans C, Scardigli R, et al. Divergent functions of the proneural genes Mash1 and Ngn2 in the specification of neuronal subtype identity. Genes Dev 2002; 16: 324–38.
106. Gohlke JM, Armant O, Parham FM, et al. Characterization of the proneural gene regulatory network during mouse telencephalon development. BMC Biol 2008; 6: 15.
107. Grove EA, Fukuchi-Shimogori T. Generating the cerebral cortical area map. Annu Rev Neurosci 2003; 26: 355–80.
108. Toyoshiba H, Yamanaka T, Sone H, et al. Gene interaction network suggests dioxin induces a significant linkage between aryl hydrocarbon receptor and retinoic acid receptor beta. Environ Health Perspect 2004; 112: 1217–24.
109. Kim S, Imoto S, Miyano S. Dynamic Bayesian network and nonparametric regression for nonlinear modeling of gene networks from time series gene expression data. Biosystems 2004; 75: 57–65.
110. Hartemink AJ. Reverse engineering gene regulatory networks. Nat Biotechnol 2005; 23: 554–5.
111. Missal K, Cross MA, Drasdo D. Gene network inference from incomplete expression data: transcriptional control of hematopoietic commitment. Bioinformatics 2006; 22: 731–8.
112. Werhli AV, Grzegorczyk M, Husmeier D. Comparative evaluation of reverse engineering gene regulatory networks with relevance networks, graphical gaussian models and bayesian networks. Bioinformatics 2006; 22: 2523–31.
113. Segal E, Shapira M, Regev A, et al. Module networks: identifying regulatory modules and their condition-specific regulators from gene expression data. Nat Genet 2003; 34: 166–76.
114. French L, Pavlidis P. Informatics in neuroscience. Brief Bioinform 2007; 8: 446–56.
115. Crosto CJ, Marenco LN, Liu N, et al. SenseLab: new developments in disseminating neuroscience information. Brief Bioinform 2007; 8: 150–62.
116. Nielsen FA, Hansen LK, Balslev D. Mining for associations between text and brain activation in a functional neuroimaging database. Neuroinformatics 2004; 2: 369–79.
117. Gardner D, Erde SM. Database schemas for electrophysiological datasets may complement genomic and protein sequences. Biophys J 1997; 72, A118.
118. Bota M, Dong HW, Swanson LW. Brain architecture management system. Neuroinformatics 2005; 3: 15–47.
119. Bota M, Swanson LW. Online workbenches for neural network connections. J Comp Neurol 2007; 500: 807–14.
120. Kotter R. Online retrieval, processing, and visualization of primate connectivity data from the CoCoMac database. Neuroinformatics 2004; 2: 127–44.
121. Wu CC, Huang HC, Juan HF, et al. GeneNetwork: an interactive tool for reconstruction of genetic networks using microarray data. Bioinformatics 2004; 20: 3691–3.
122. Chesler EJ, Lu L, Shou SM, et al. Complex trait analysis of gene expression uncovers polygenic and pleiotropic networks that modulate nervous system function. Nat Genet 2005; 37: 233–42.
123. Kempermann G, Chesler EJ, Lu L, et al. Natural variation and genetic covariance in adult hippocampal neurogenesis. Proc Natl Acad Sci U S A 2006; 103: 780–5.

124. McCarthy M. Allen Brain Atlas maps 21000 genes of the mouse brain. Lancet Neurol 2006; 5: 907–8.
125. Olszewski PK, Cedernaes J, Olsson F, et al. Analysis of the network of feeding neuroregulators using the Allen Brain Atlas. Neurosci Biobehav Rev 2008; 32: 945–56.
126. Magdaleno S, Jensen P, Brumwell CL, et al. BGEM: An in situ hybridization database of gene expression in the embryonic and adult mouse nervous system. PLoS Biol 2006; 4: 497–500.
127. Visel A, Thaller C, Eichele G. GenePaint.org: an atlas of gene expression patterns in the mouse embryo. Nucleic Acids Res 2004; 32: D552–6.
128. Heintz N. Gene expression nervous system atlas (GENSAT). Nat Neurosci 2004; 7: 483.
129. Lee HK, Hsu AK, Sajdak J. Coexpression analysis of human genes across many microarray data sets. Genome Res 2004; 14: 1085–94.
130. Salwinski L, Miller CS, Smith AJ. The database of interacting proteins: 2004 update. Nucleic Acids Res 2004; 32: D449–51.
131. Gandhi TKB, Zhong J, Mathivanan S, et al. Analysis of the human protein interactome and comparison with yeast, worm and fly interaction datasets. Nat Genet 2006; 38: 285–93.
132. Pagel P, Kovac S, Oesterheld M, et al. The MIPS mammalian protein-protein interaction database. Bioinformatics 2005; 21: 832–4.
133. Mathivanan S, Periaswamy B, Gandhi TKB, et al. An evaluation of human protein-protein interaction data in the public domain. BMC Bioinformatics 2006; 7(Suppl 5): S19.
134. Wingender E, Chen X, Fricke E, et al. The TRANSFAC system on gene expression regulation. Nucleic Acids Res 2001; 29: 281–3.
135. Wei LP, Liu YY, Dubchak I. Comparative genomics approaches to study organism similarities and differences. J Biomed Inform 2002; 35: 142–50.
136. Ovcharenko I, Loots GG, Giardine BM, et al. Mulan: multiple-sequence local alignment and visualization for studying function and evolution. Genome Res 2005; 15: 184–94.
137. Loots GG, Ovcharenko I. Dcode.org anthology of comparative genomic tools. Nucleic Acids Res 2005; 33: W56–64.
138. Cartharius K, Frech K, Grote K. MatInspector and beyond: promoter analysis based on transcription factor binding sites. Bioinformatics 2005; 21: 2933–42.
139. Yamanaka T, Toyoshiba H, Sone H, et al. The TAO-Gen algorithm for identifying gene interaction networks with application to SOS repair in E-coli. Environ Health Perspect 2004; 112: 1614–21.
140. Bemstein BE, Mikkelsen TS, Xie XH, et al. A bivalent chromatin structure marks key developmental genes in embryonic stem cells. Cell 2006; 125: 315–26.
141. Hallikas O, Palin K, Sinjushina N, et al. Genome-wide prediction of mammalian enhancers based on analysis of transcription-factor binding affinity. Cell 2006; 124: 47–59.
142. Woolfe A, Goodson M, Goode DK, et al. Highly conserved non-coding sequences are associated with vertebrate development. PLoS Biol 2005; 3: 116–30.
143. Benjanirut C, Paris M, Wang WH. The cAMP pathway in combination with BMP2 regulates Phox2a transcription via cAMP response element binding sites. J Biol Chem 2006; 281: 2969–81.
144. Saxlund MA, Sadler-Riggleman I, Skinner MK. Role of basic helix-loop-helix (bHLH) and CREB transcription factors in the regulation of sertoli cell androgen-binding protein expression. Mol Reprod Dev 2004; 68: 269–78.
145. Mandolesi G, Gargano S, Pennuto M. NGF-dependent and tissue-specific transcription of vgf is regulated by a CREB-p300 and bHLH factor interaction. FEBS Lett 2002; 510: 50–6.
146. Kurisaki K, Kurisaki A, Valcourt U, et al. Nuclear factor YY1 inhibits transforming growth factor beta- and bone morphogenetic protein-induced cell differentiation. Mol Cell Biol 2003; 23: 4494–510.
147. Gunhaga L, Marklund M, Sjodal M, et al. Specification of dorsal telencephalic character by sequential Wnt and FGF signaling. Nat Neurosci 2003; 6: 701–7.
148. Backman M, Machon O, Mygland L, et al. Effects of canonical Wnt signaling on dorso-ventral specification of the mouse telencephalon. Dev Biol 2005; 279: 155–68.
149. Abu-Khalil A, Fu L, Grove EA, et al. Wnt genes define distinct boundaries in the developing human brain: implications for human forebrain patterning. J Comp Neurol 2004; 474: 276–88.
150. Karanam S, Moreno CS. CONFAC: automated application of comparative genomic promoter analysis to DNA microarray datasets. Nucleic Acids Res 2004; 32: W475–84.
151. Ohtsuka T, Ishibashi M, Gradwohl G. Hes1 and Hes5 as Notch effectors in mammalian neuronal differentiation. EMBO J 1999; 18: 2196–207.
152. James AB, Conway AM, Morris BJ. Regulation of the neuronal proteasome by Zif268 (Egr1). J Neurosci 2006; 26: 1624–34.
153. Gachon F, Fonjallaz P, Damiola F, et al. The loss of circadian PAR bZip transcription factors results in epilepsy. Genes Dev 2004; 18: 1397–412.

154. Zhu J, Zhang Y, Joe GJ, et al. NF-Ya activates multiple hematopoietic stem cell (HSC) regulatory genes and promotes HSC self-renewal. Proc Natl Acad Sci U S A 2005; 102: 11728–33.
155. Akiyama H, Kim JE, Nakashima K, et al. Osteo-chondroprogenitor cells are derived from Sox9 expressing precursors. Proc Natl Acad Sci U S A 2005; 102: 14665–70.
156. Shalizi A, Gaudilliere B, Yuan ZQ, et al. A calcium-regulated MEF2 surnoylation switch controls postsynaptic differentiation. Science 2006; 311: 1012–17.
157. Mishra GR, Suresh M, Kumaran K, et al. Human protein reference database—2006 update. Nucleic Acids Res 2006; 34: D411–14.
158. Geiger D, Heckerman D. Learning Gaussian networks. In: Proceedings of the Tenth Conference on Uncertainty in Artifical Intelligence. San Francisco, CA: Morgan Kaufmann, 1994: 235–43.
159. Castro DS, Skowronska-Krawczyk D, Armant O, et al. Proneural bHLH and Brn proteins coregulate a neurogenic program through cooperative binding to a conserved DNA motif. Dev Cell 2006; 11: 831–44.
160. Yeang CH, Mak HC, McCuine S, et al. Validation and refinement of gene-regulatory pathways on a network of physical interactions. Genome Biol 2005; 6: R62.
161. Gunsalus KC, Piano F. RNAi as a tool to study cell biology: building genome-phenome bridge. Curr Opin Cell Biol 2005; 17: 3–8.
162. Qin QY, Inatome R, Hotta A, et al. A novel GTPase, CRAG, mediates promyelocytic leukemia protein—associated nuclear body formation and degradation of expanded polyglutamine protein. J Cell Biol 2006; 172: 497–504.
163. Bolognani F, Tanner DC, Merhege M, et al. In vivo post-transcriptional regulation of GAP-43 mRNA by overexpression of the RNA-binding protein HuD. J Neurochem 2006; 96: 790–801.

3 Neurotoxicology of Barriers in the Developing Brain

Norman R. Saunders, Mark D. Habgood, and Katarzyna M. Dziegielewska
Department of Pharmacology, University of Melbourne, Parkville, Victoria, Australia

INTRODUCTION

This review will discuss the evidence explaining which blood-brain barrier mechanisms in the immature brain appear to be functional in relation to what is known about effects of neurotoxic agents in the fetus and newborn. The influence of neurotoxic agents on specific aspects of brain development have been reviewed in detail elsewhere (1–3) and will only be dealt with in outline here. There is a widespread belief amongst toxicologists (1,4–13) and neurobiologists (14–18) that "the" blood-brain barrier in the fetus and newborn is immature or even absent (3). This appears to be partly due to a teleological view that the developing brain would not "need" a specialized local environment for its development or if it does then this would be supplied by protective functions of the placenta (19). A more recent but similarly teleological view is that "the rapid growth of the cerebral cortex perhaps necessitates a 'leaky' structure of the blood-brain barrier to accommodate the high demand of blood-borne nutrients for brain growth" (20). The belief in the immaturity of the blood-brain barrier is often stated without citing evidence, or supported by uncritical acceptance and interpretation of results of three types of experiments. One involves the injection of large volumes or concentrations of barrier markers into fragile embryos or fetuses (21). The second, which is more directly relevant to toxicological studies, is the interpretation of differences in responses of the brains of adult and immature animals to drugs and toxins as due to barrier immaturity (22). The third type of experiment is measurement of entry of biologically important molecules such as amino acids into the developing brain. These show unequivocally that the level of transfer is generally greater in the developing rather than in the adult brain (reviewed below). Some have interpreted this as evidence of a "leaky" barrier (20,23). Most of those who did the actual experiments concluded that the greater transfer was a reflection of a developmental specialization during brain growth (24,25).

The term blood-brain barrier is unfortunate; for many people it implies a single mechanism that excludes certain classes of molecules from the brain. As will be described below, the term is now used to cover a wide range of mechanisms that determine and control the composition of the internal environment of the brain, the stability of which is essential for normal brain function. It is appropriate to be particularly cautious about the administration of drugs to pregnant women and children since all drugs have potential side effects and the immature nervous system may be more susceptible to these effects than the adult. Similar problems occur with respect to exposure to toxins. However, blanket assumptions that the blood-brain barrier in the fetus and newborn is immature may lead to potentially valuable drugs being withheld from treatment (26) although it is probable that the incidence of inappropriate use of drugs in pregnant women is much higher than under-use (27,28). Nevertheless, it would seem appropriate that clinical decisions about drug-use should be rational and based on sound evidence rather than supposition. In addition, if drugs and toxins have effects in the fetus or newborn, which are not apparent in the adult, it is important to know what mechanisms are involved.

Fundamental biological questions are: (*i*) What brain barrier mechanisms are present in the embryonic, fetal, and newborn brain? (*ii*) How do they contribute to brain development? (*iii*) When do adult mechanisms develop? (*iv*) Are there any brain barrier mechanisms that are specific or more prominent in the immature brain than in the adult? The key toxicological question is whether or not these mechanisms provide protection for the developing brain or render it more vulnerable.

BRAIN BARRIER MECHANISMS IN THE ADULT

The term "blood-brain barrier" was originally coined early in the 20th Century to explain the results of experiments showing that certain dyes and toxins, when injected into an animal, did not penetrate into the brain apart from a few limited areas, e.g., choroid plexus and area postrema. The explanation of this exclusion was shown by electron microscopy studies (29) to be due to the presence of specialized intercellular junctions ("tight junctions") between cells forming barrier interfaces between blood and brain. In physiological studies several authors (30,31) have shown that classical dyes used to demonstrate the blood-brain barrier bind to plasma proteins, particularly albumin, thus it was concluded that the barrier (tight junctions) was to protein rather than to the much smaller dye molecule (32). However, as a consequence of measurements of electrical resistance across different epithelia in vitro it was proposed that the tight junctions between cells in epithelia, such as the choroid plexus and between cerebral endothelial cells, are a low resistance pathway compared to the actual cell membranes through which ions and water are able to pass. From the point of view of understanding brain barrier mechanisms, the essential point is that the barrier interfaces are impermeable to the intercellular passage of proteins and they have low, if any, intrinsic permeability to small lipid insoluble molecules. The functional importance of this low permeability is that transport mechanisms in the cells of the barrier interfaces are able to generate gradients between the brain compartments [cerebrospinal fluid (CSF) and brain interstitial fluid] and the blood that are essential for normal brain functioning. Thus the overall effect of the brain barrier mechanisms is to maintain the internal environment of the brain with respect to a whole range of constituents (e.g., ions, glucose, amino acids, micronutrients). Some cellular transfer mechanisms result in a net influx of molecules into the brain [e.g., glucose (24,33), many amino acids (24,34), and efflux mechanisms that exclude certain classes of molecules (including many drugs) from entering the brain]. Among efflux mechanisms the multidrug resistance proteins MDR-1 (P-glycoprotein) and the MRPs are particularly important as their substrates are a wide range of drugs and chemicals; many of them neurotoxic (35–37). The developmental status of these efflux mechanisms is a key to understanding the potential for neurotoxicity in the fetus and newborn brain, although many textbooks do not mention these mechanisms (38–41).

The brain barrier interfaces are present at three main sites in the adult brain, illustrated in Figure 1: cerebral blood vessels (blood-brain barrier), choroid plexuses (blood-cerebrospinal fluid barrier), and pia-arachnoid barrier. In the fetus there is an additional barrier interface between the cerebrospinal fluid and the brain parenchyma (CSF-brain barrier) (Fig. 1 and section "Barrier Mechanisms in the Developing Brain"). Furthermore, the placenta provides an additional barrier interface between the maternal and fetal blood.

BARRIER MECHANISMS IN THE DEVELOPING BRAIN

Intracellular Tight Junctions in Barrier Interfaces

The fundamental structural barrier (tight junctions) present in both the blood-brain and blood-CSF interfaces is already functionally effective from very early in brain development. Molecules even as small as sucrose and inulin do not appear to permeate tight junctions either in the endothelial cells of the first blood vessels to grow into the brain or in the choroid plexus epithelial cells (42–44). This has been demonstrated at both the light and electron microscopical levels using biotin labelled molecules of different sizes that can be visualized microscopically and also have permeability properties similar to more familiar markers, such as sucrose mw 362 Da (cf. biotin ethylenediamine MW 286 Da) and inulin MW 5000 Da (cf. biotin dextran 3000 Da) as illustrated in Figure 2. Thus the fundamental structural barrier at the blood-brain and blood-CSF interfaces is present and functionally effective from very early in brain development. The assumption that the blood-brain barrier in the fetus is not present or only partly formed and therefore the developing brain is accessible to drugs and toxins (3,4,7,8,11,45) is incorrect. However, the tight junctions are only part of the system that provides the brain with its stable internal environment. In addition there is a range of cellular mechanisms that control entry and exit of molecules across the barrier interfaces. These are the essential determinants of the local environment in which the brain grows and develops. Much less is known about these mechanisms than about the development of the morphological components of brain barrier systems. Understanding these mechanisms is an essential prerequisite for interpreting data on toxicological effects in the fetus and newborn. Some of these mechanisms develop early and their functional effectiveness is, in itself, evidence of the presence of some aspects of barrier mechanisms in the immature brain. However, some mechanisms develop later. Whether the temporal differences in development of different brain barrier mechanisms can be interpreted as evidence of a degree of immaturity or

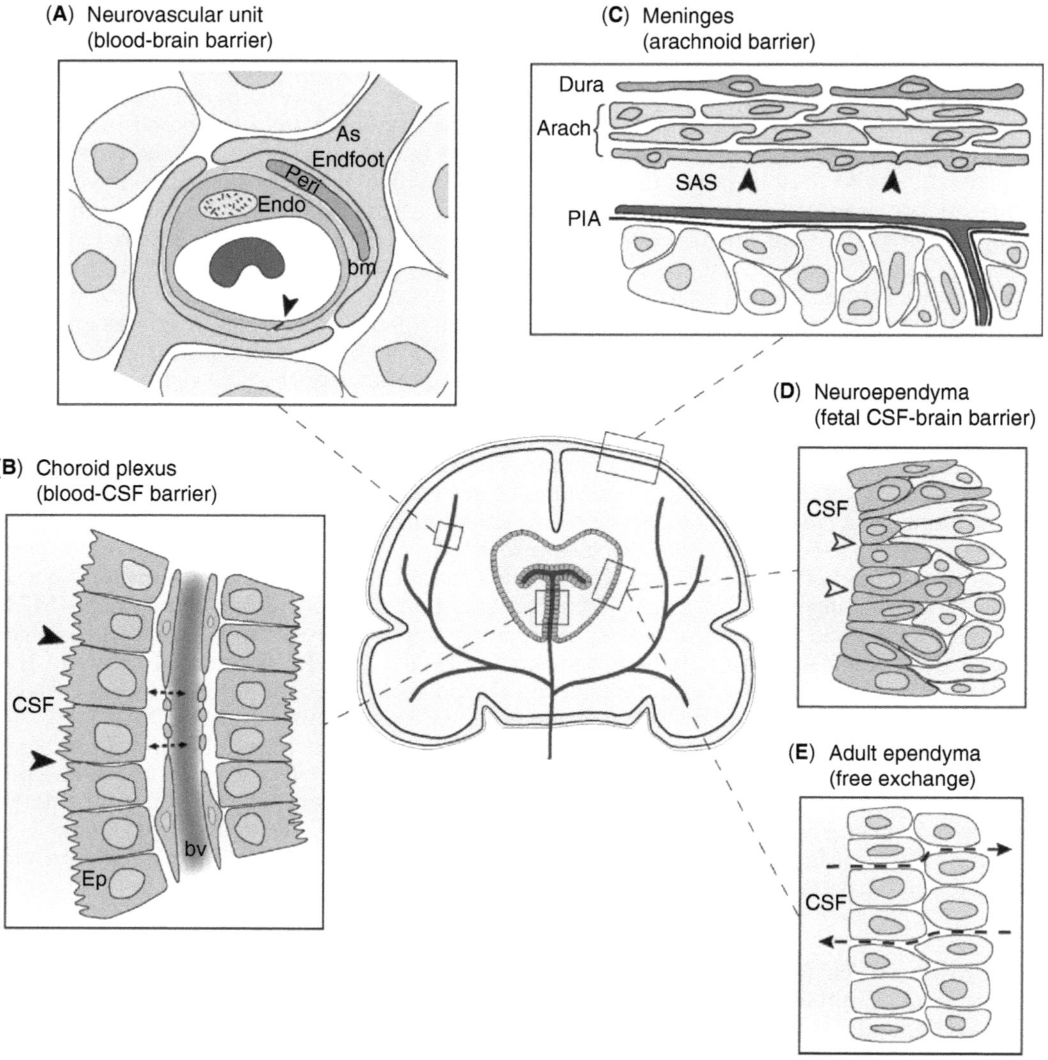

Figure 1 (*See color insert*) Schematics of the interfaces and barriers in the adult and developing brain. (**A**) The blood-brain barrier is a barrier between the lumen of cerebral blood vessels and brain parenchyma. The endothelial cells (Endo) have luminal tight junctions (arrowhead) forming the physical barrier of the interendothelial cleft. Outside the endothelial cell is a basement membrane (bm) which also surrounds the pericytes (Peri). Around all these structures are the astrocytic endfeet processes from nearby astrocytes (As Endfoot). All these structures together are often referred to as the neurovascular unit. (**B**) The blood-CSF barrier, a barrier between choroid plexus blood vessels (bv) and the CSF. The choroid plexus blood vessels are fenestrated and form a non-restrictive barrier (small arrows), however, the epithelial cells (Ep) have apical tight junctions (arrowheads) that restrict intercellular passage of molecules. (**C**) The meningeal barrier, is the least studied and structurally most complex of all the brain barriers. The blood vessels of the dura are fenestrated and provide little barrier function, however, the outer cells of the arachnoid membrane (Arach) have tight junctions (arrowheads) and this cell layer is believed to form the physical barrier between the CSF-filled subarachnoid space (SAS) and overlaying structures. The blood vessels in the arachnoid and on the pial surface (PIA) have tight junctions with similar barrier characteristics as cerebral blood vessels although lacking the surrounding pericytes and astrocytic end-feet. (**D**) The fetal CSF-brain barrier, a barrier between the CSF and brain parenchyma, and has only been shown to be a functional barrier in the early developing brain (245). In early development, the neuroependymal cells are connected to each other by strap-junctions (open arrowheads) that are believed to form the physical barrier restricting the passage of larger molecules such as proteins but not smaller molecules such as sucrose. (**E**) The adult ventricular ependyma. During development, the neuroependymal cells flatten and lose their strap-junctions. The mature ependyma does not restrict the exchange of large molecules at (e.g., proteins) between CSF and brain. *Source*: From Ref. 246.

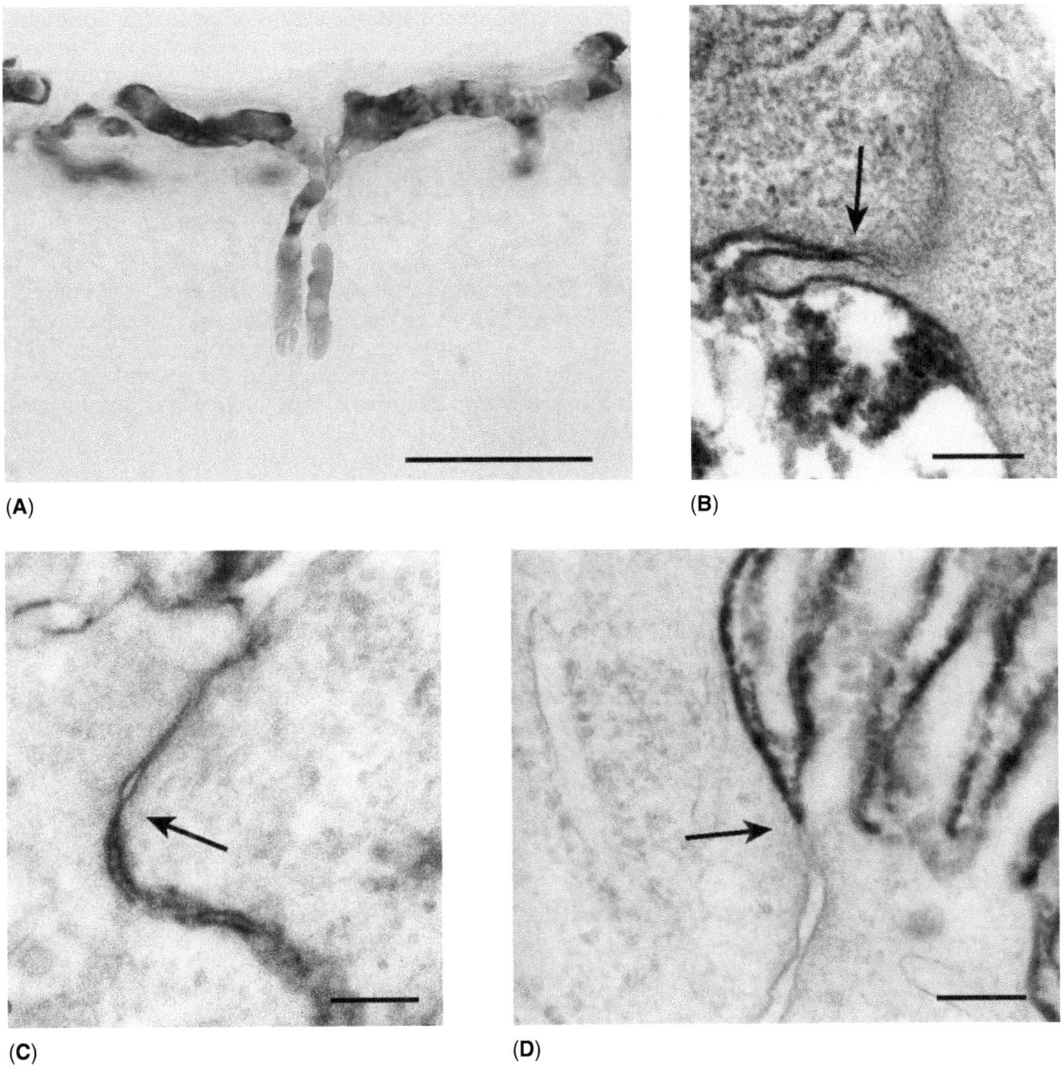

Figure 2 Micrographs of the blood-brain and blood-CSF barriers at early stages of brain development. (**A**) shows neocortical blood vessels on the brain surface of a newborn opossum (*Monodelphis domestica*) 30 minutes after an intraperitoneal injection of a small inert tracer (3 kDa biotin-dextranamine). The tracer, visualized as a dark reaction product, is clearly confined to the lumen of these blood-vessels, which are the first to grow into the neocortex of the developing opossum brain (scale bar is 50 μm). (**B**) Electron micrograph of an intercellular junction between two adjacent endothelial cells in the newborn opossum brain (blood-brain barrier). Note the biotin-dextran amine tracer (dark reaction product) which is present in the lumen of the blood vessel, but does not pass beyond the tight junction complex (arrow) located at the luminal end of the intercellular cleft (scale bar is 200 nm). (**C**) Electron micrograph of an intercellular junction between two adjacent choroid plexus epithelial cells (blood-CSF barrier) in a newborn opossum brain 30 minutes after the tracer was injected into the intraperitoneal cavity. Note that the tight junction (arrow) restricts the passage of this tracer along the intercellular cleft towards the cerebrospinal fluid in the ventricles (scale bar is 100 nm). (**D**) Electron micrograph of an intercellular junction between two adjacent choroid plexus epithelial cells in an embryonic rat brain at 15 days gestation. In this animal, the tracer was administered into the cerebrospinal fluid on the other side of the blood-CSF barrier rather than into the peritoneal cavity as in (**C**). Note that by 10 minutes after administration, the tracer is clearly visible in the intercellular cleft, but is unable to pass beyond the tight junction complex (arrow) located at the ventricular end of the intercellular cleft (scale bar is 200 nm). *Source*: From Ref. 245.

represent developmental specializations appropriate for specific stages of brain development is a matter of debate. It may be that mechanisms important for particular aspects of brain development become potentially damaging if toxic agents are able to use them to gain entry to the immature brain. One such mechanism is the transfer of proteins from blood to CSF, which results in a high concentration of proteins in fetal CSF (46). Since heavy metals and some drugs bind to plasma proteins, the higher level of these proteins in fetal CSF may represent a hazard to the developing brain, particularly since they will be entering the brain at periods of greater vulnerability because of the various growth processes that are occurring (3).

Blood-Brain and Blood-CSF Transport Mechanisms in the Developing Brain

It is the transport mechanisms in these barrier interfaces that determine the internal environment of the developing brain and supply essential nutrients and other molecules essential for the growth and differentiation of this complex integrated organ. Figure 3 summarizes what is known about inward (blood-brain) transport mechanisms that have been shown to be functional in the fetal and newborn brain.

Water

The key water channel, aquaporin-1, is present in the appropriate (adult) functional site of the choroid plexus epithelial cells as soon as the plexus begins to differentiate in all species studied, including the human fetus (47).

Ions

One of the most fundamental characteristics of the adult brain is stability of the ionic composition of CSF and brain extracellular fluid. This stability is essential for normal brain function in that the transmission of nerve impulses in axons and across synapses is dependent on a stable resting membrane potential and characteristic transient changes in ion permeability across axon membranes at postsynaptic terminals. An indication of the stability of ion gradients in

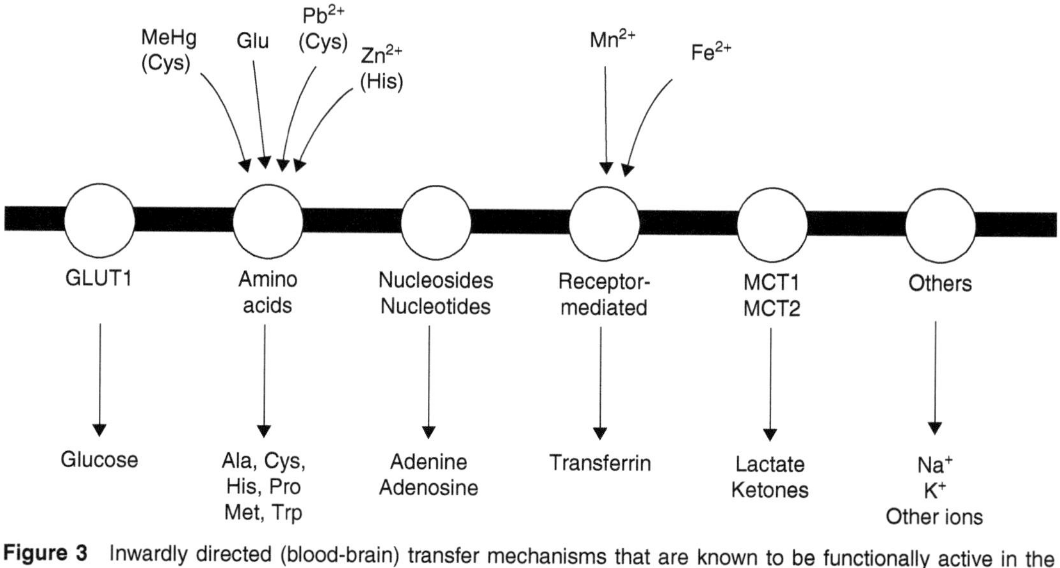

Figure 3 Inwardly directed (blood-brain) transfer mechanisms that are known to be functionally active in the barrier interfaces of the developing central nervous system and which could provide potential routes of entry for some neurotoxic compounds. Examples of the endogenous substrates for each transfer mechanism are shown below and neurotoxic compounds that can also enter via these mechanisms are shown above. GLUT1 is the main glucose transporter in cerebral endothelium and choroid plexus epithelium (33,51,53). There are several amino acid transporters (24,25,34,52,55). Receptor mediated transporters include the transferrin receptor (203,216). Other transporters identified in immature brain vessels are MCT1 and MCT2 (247).

brain extracellular fluid can be obtained from measurements of CSF and plasma ion concentrations. Gradients for some ions are present very early in brain development: E45 in fetal sheep (Mg^{2+}; 48), E50 in fetal monkeys (Mg^{2+}; 49), and E23 in fetal rabbits (Na^+, K^+, Cl^-; 50). The presence of even one ion gradient between blood and CSF is strong evidence that the tight junctions in the barrier interface are functionally effective.

Glucose

The first blood vessels to enter neural tissue acquire the Glut 1 that is characteristic of endothelial cells in brain barriers (51). Similarly in the choroid plexuses, although Glut 1 is absent from the blood vessels which lack barrier properties, it appears very early in the epithelial cells of the choroid plexuses (52,53). Glut 1 levels are actually highest in the rat choroid plexus epithelium during gestation and fall to 50% of that level by P1 and further decline during the first week of postnatal life (54) which is probably a reflection of the importance of the choroid plexuses for transfer of materials from blood to brain in early stages of brain development (46).

Amino Acids

Transport of amino acids and other important nutrients in newborn brains of rabbits and rats has been demonstrated (24,34,55). The uptake of these essential nutrients was greater in newborn than in older animals. Some metabolites (e.g., lysolecithin and cytosine) were excluded from entry into the developing brain and it was concluded that this was further evidence that "the concept of an 'immature' barrier at birth seems untenable and inconsistent with current understanding of the blood-brain barrier" (24). Expression of amino acid transporters in endothelial cells in developing brain does not appear to have been studied.

Choroid Plexus Transfer of Proteins from Plasma to CSF

The protein concentration in fetal CSF is high compared to the adult (56,57). There is now evidence from several species that this is produced by transcellular transfer of plasma proteins across the epithelial cells of the choroid plexuses (58,59) reinforced by the slow turnover of CSF in the developing brain which allows the proteins to accumulate in CSF early in brain development to a greater extent than in the adult (46).

Efflux Mechanisms in the Embryo and Fetus

An important barrier mechanism in the adult brain at both the blood-brain and blood-CSF barrier interfaces is the presence of efflux ATP-binding cassette (ABC) transporters. There are 48 known ABC proteins in humans (60) classified into seven subfamilies (A–G) of which three (B, C, G) have been shown to act as efflux transporters at the blood-brain and blood-CSF barriers in the adult (60,61). These transporters either intercept and export lipophilic compounds including xenobiotics as they pass across the cell membrane or export compounds from within the intracellular compartment in association with a transport moiety (Fig. 4). Because of the importance of these efflux mechanisms for protection of the brain from many drugs and toxic agents (62,63) knowledge of their status in the fetus and newborn is essential for understanding the potential neurotoxicity of a wide range of compounds.

P-Glycoprotein

The most studied of these ABC proteins is P-glycoprotein [also known as MDR1 or ABCB1 according to the Hugo Gene Nomenclature Committee]. P-glycoprotein is localized in the luminal membrane of cerebral endothelial cells of several species including humans (64) and has also been claimed to be located in the apical membrane of choroid plexus epithelial cells (65). The functional importance of MDR1 in choroid plexus, at least in mice, was demonstrated in triple knockout (KO) mice lacking *Mdr1a*, *Mdr1b*, and *Mrp1* in which the level of penetration of the anticancer drug etoposide from blood into CSF was about 10 times higher in the triple KO mice compared to the double KO *Mdr1a/Mdr1b* mice (66). P-glycoprotein has been shown to be present in human embryonic and fetal brain in three studies. P-glycoprotein expression was reported to be at adult levels in the fetal brain by the third trimester (67). Weak staining for

P-glycoprotein in some vessels in the mid and hindbrain of 30 mm CRL (8 weeks gestation) embryos, with strong staining by 123 mm CRL (15 weeks gestation) was identified by (68). Virgintino et al. (64) compared the immunoreactivity of four different antibodies to human P-glycoprotein and found that even as early as 12 weeks gestation there was strong staining for P-glycoprotein in the first vessels growing into the telencephalon (forerunner of the cerebral cortex) at a time when the characteristic layers of the cerebral hemispheres are beginning to form (69).

In mouse embryos, messenger RNA (mRNA) for *mdr1a* was already expressed at E10.5 in blood vessels associated with the neural tube (70). In a more detailed quantitative study of mRNA expression detected by reverse transcription-polymerase chain reaction and protein estimated by western blotting (71) measurable amounts of both at E16 (the earliest age studied) were detected. By the day of birth, there was no significant increase in the level of mRNA, but the protein was about four times higher. However, the main increases in both mRNA and protein did not occur until the postnatal period, the adult level of P-glycoprotein being reached at three weeks of age (although the mRNA level was only about a quarter of that in the adult). In the rat fetus expression of *mdr1a* was identified as early as E15 (9) whereas the protein was not detected until P7 (72). This apparent discrepancy compared to the mouse may have been due to lack of sensitivity of the antibodies used.

The developmental changes in expression of *mdr1a* in rodents reflect the pattern of vascularization of the developing rodent brain. Prior to birth most vascularization occurs in the more mature parts of the brain namely the mid and hindbrain. Although the characteristic neuronal layers of the neocortex are established in the period E15 to the time of birth (73) most vascularization of the cerebral hemispheres in the rodent occurs in the first three weeks of life (74). Thus much of the increase in P-glycoprotein in the postnatal period in rodents is probably accounted for by this expansion of the cerebral vasculature. However, in humans the equivalent stage of vascularization of the brain extends from early in the third trimester to the end of the second postnatal year (75). In addition to the stage of vascularization at birth, another key question from the point of view of the vulnerability of the developing brain is how active is this efflux transport mechanism during these critical stages of brain development. This seems not to have been studied. Postnatally, there appears to be no information available about the level of *MDRI* expression and the level of P-glycoprotein protein in the human infant brain; deductions from animal experiments depend critically on assumptions about appropriate age-related comparisons, given the different stages of brain development at which humans and rodents are born (see section "Species Differences").

MRPs in Barrier Interfaces in the Developing Brain

MRP1 to MRP6 have been identified in endothelial cells of cerebral blood vessels of a variety of species (76,77) and see (37) for brief review. Only those MRPs (and other efflux transporters) present in the apical/luminal membrane of endothelial cells (P-glycoprotein, MRP2, MRP4, and BCRP) would be expected to restrict entry of drugs and toxins into the brain. However, the evidence for localization of some of the other MRPs (1,3,5,6) to the basolateral membrane was determined in cell culture experiments (37); this may not reflect the situation in vivo. In the developing rat brain *mrp1* was expressed at E15 (earliest age studied) at a level that was 75% of that at birth and in the adult, although the difference was not significantly different (9).

Breast Cancer Resistance Protein in the Brain Barrier Interfaces

Another ABC transporter that has been identified in adult cerebral endothelial cells is breast cancer resistance protein (BCRP) (63,78). This product of the ABCG2 gene was first isolated from a multi-drug resistant human breast cancer cell line (79). The substrate specificity of BCRP overlaps with that of P-glycoprotein (62). The mouse equivalent, Bcrp1 shares 81% homology with human BCRP and is encoded by the *Abcg2* gene (80). *Brcp1* expression has been identified in embryonic rat brain as early as E12.5 (81), with expression said to be confined to blood vessels and not present in the rest of the brain (81). However, the evidence for this was based on a micrograph of immunocytochemical staining of Bcrp1 in E18.5 brain. Further studies will be required to define the developmental profile of this potentially important xenobiotic efflux transporter in the fetal brain. Figure 4 summarizes the three main ABC transporter types involved in xenobiotic exports at the blood-brain, blood-spinal cord, and placental barriers.

Detoxifying Enzyme Systems in Brain Barriers of the Developing Brain

As outlined in Figure 4, the MRPs primarily export compounds coupled with glutathione, glucuronic acid, or sulfates, either by co-transport with the compound or by biochemical conjugation onto the compound and subsequent efflux of the conjugate [reviewed by (82)]. There is some evidence that BCRP exports both conjugated (glucuronidate and sulfate) and unconjugated compounds (83,84). Conjugating enzymes form part of the natural Phase II metabolism machinery of cells for detoxifying and removing endogenous compounds and probably unwittingly contribute to the efflux of xenobiotics that manage to find their way into the cells in which these enzymes and MRPs are present. Glutathione is a tri-peptide that requires the presence of three main enzymes; firstly *Gamma-glutamylcysteine synthetase* to couple glutamate to cysteine, then *Glutathione synthetase* to add glycine and finally *Glutathione-S transferase* to attach the glutathione onto the drug to be exported. Glucuronidation requires the presence of the *UDP-glucuronyltransferase* enzymes to transfer the glucuronyl group from UDP (uridine-di-phosphoglucuronate) onto a polar functional group present on the compound to be exported. Sulfation requires the presence of *3-phosphoadenosine-5-phosphosulfate synthase* followed by the *sulfotransferase* enzymes to transfer the sulfonate group from 3-phosphoadenosine-5-phosphosulfate onto the compound to be exported. These conjugations result in a less toxic (less active) conjugate, due to the modification of the reactive polar group, making it less lipid soluble, thus restricting its distribution and ability to cross cell membrane, and allowing the compound to be excreted from the cell by MRP and BCRP.

Gamma-glutamylcysteine synthetase is present in many tissues in the adult, including the central nervous system (CNS) and mRNA from the two genes which encode for it has been shown to be present as early as gestational day 10 (E10) in fetal mouse brain and liver (85). There is no information available on the expression of *Glutathione synthetase* in the developing CNS.

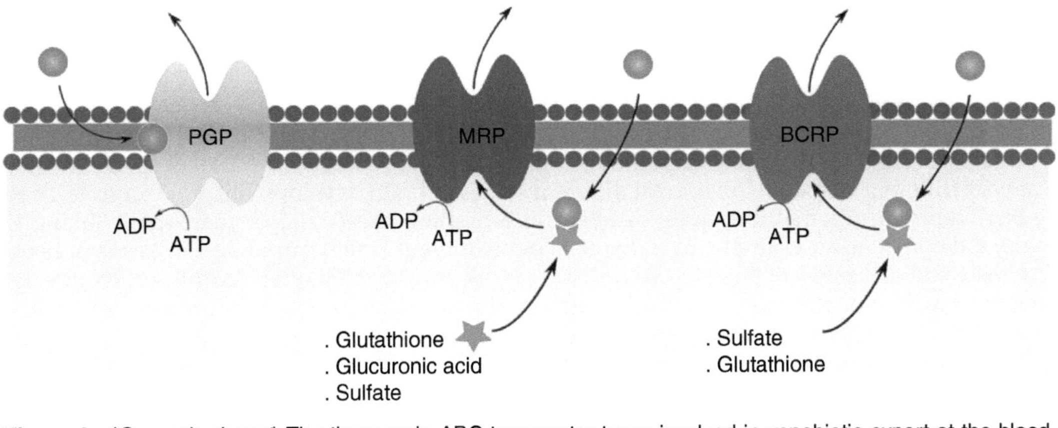

Figure 4 (*See color insert*) The three main ABC transporter types involved in xenobiotic export at the blood-brain, blood-spinal cord, and blood-placental barriers; PGP (p-glycoprotein, also known as MDR1, ABCB1), MRP (multidrug resistance related proteins, ABCC1-8) and BCRP (breast cancer related protein, ABCG2). PGP and the MRPs are arranged in two repeated halves, each half containing a nucleotide binding domain and six membrane spanning domains. BCRP is a half-transporter with a single nucleotide binding domain and six membrane spanning domains, but is thought to form a homodimer in order to be functionally active. All utilize energy from ATP hydrolysis to move substrates across the membrane. PGP's substrate binding sites are located within or close to the internal leaflet, thus PGP is thought to be able to intercept lipophilic compounds as they pass across the cell membrane. In contrast, the MRPs transport compounds from within the cell cytoplasm in conjunction with a transport moiety (glutathione, glucuronic acid, or sulfate). This is achieved either by co-transport or by conjugation of the transport moiety (shown as a star in the diagram) onto the substrate prior to export. The MRPs do not appear to be able to directly intercept compounds within the internal leaflet in the way that PGP does. BCRP also transports a wide range of unconjugated compounds, but it is not known if it is able to intercept these from within the internal leaflet. BCRP is also able to transport conjugated compounds, with a greater affinity for sulfate conjugates over glutathione conjugates (248), thus it may function in a similar manner to the MRPs. *Abbreviations*: ADP, adenosine diphosphate; ATP, adenosine triphosphate.

Glutathione S transferase is present in both cerebral endothelial cells and in choroid plexus epithelial cells (86). The Pi class isoenzyme of glutathione S-transferase has been shown to be present at least as early as 12 weeks gestation in human fetuses (87) and in rat choroid plexus from the day of its first differentiation (E15) and subsequently in the brain ventricular zone cells (E18 to birth) and subventricular zone and astroglial cells of grey and white matter (88). *UDP-glucuronyltransferases (UGTs)*: These are found in the CNS in barrier forming cerebral endothelial cells (89), but the highest concentration appears to be in choroid plexus epithelial cells (90). In fetal rats, *UGTs* are present in cerebral cortex at embryonic day 17 (the earliest age investigated) and increase to reach a peak between birth and postnatal day 5, before declining to 30% of the peak level by postnatal day 60 (91). Since *mrp1* is expressed as early as E15 in rat brain (9) its concurrent presence with the glutathione producing enzymes suggests that this efflux mechanism may be functionally active very early in brain development.

Placental Efflux and Detoxifying Mechanisms

Many of the efflux transporters and enzymes described above in the barrier interfaces of the developing brain are also present in the placenta (36,60,62,92,93). Studies of mice with deletions of either or both *mdr1a* or *mdr1b* genes have shown that there is a much greater transfer of P-glycoprotein substrates [e.g., digoxin (94), some pesticides (95), in *mdr1a* –/– or *mdr1a* –/– *mdr1b* –/– mice]. The phenotypes of these have been reported to be normal (96) but there is some evidence that if challenged with certain drugs the mutants may show congenital deformities (97).

In the embryo and fetus the placenta provides an important defence against the penetration of many drugs and toxins that would otherwise enter from the mother's circulation, should the mother be exposed. A consideration of whether P-glycoprotein is present and functionally active at the brain barrier interfaces in the embryo and fetus is important in relation to possible effects on the developing brain, should the level of such materials exceed the efflux capacity of the placenta. A related consideration is that there are genetic differences in levels of expression of P-glycoprotein genes (98); the fetuses of such mothers may be at greater risk that those of mothers with normal levels of P-glycoprotein gene expression.

SPECIES DIFFERENCES

In addition to the general need for caution when attempting to extrapolate from animals to humans, in the developmental field there is the additional problem that different species develop at very different rates and are born at different stages of brain development. For example, "newborn" represents something quite different in a human, sheep, rat, or marsupial. Taking the cerebral cortex as an example, in rodents its development is very rapid and in terms of neurogenesis and cell migration is largely complete by birth, although other features of its development such as synaptogenesis continue well into the postnatal period. In the sheep, which has a slowly growing brain and a long gestational period (150 days), the equivalent period for the end of cortical neurogenesis is mid gestation (99). In the human cortex the equivalent period is towards the end of gestation (100,101). This conclusion was based on studies in rhesus macaque monkey and extrapolated to humans (102). Correction for shrinkage and stereological errors in the extensive data of Conel on brains of children showed that after a decline in cortical neurons numbers at 6 to 18 months (depending on the region) there is a subsequent increase at least up to 72 months of about two fold (102). In late gestation and in the postnatal period there is also a substantial amount of organizational development, especially gliogenesis, and synaptogenesis (101). At the other extreme, in marsupials, the whole of neocortical development occurs after birth, postnatal day 15 (P15) being equivalent to the newborn (P0) rat neocortex. With respect to vascularization of the developing brain, most of this process in the neocortex of rodents occurs in the first three weeks of life (74) whereas in the human this occurs over a much longer period extending from the third trimester through the first two years of life (75). When transferring information from experimental animals to humans, particular caution is needed with respect to development; it is not appropriate to extrapolate findings from neonatal rodents to neonatal infants. CSF in developing brain has a high concentration of protein, which declines with age. Thus a simple comparison of total protein concentration in CSF could be a useful comparator between species. In the adult term human, CSF protein concentration is about two to four times the adult level (103) but is much higher in the fetus (57,104). On this basis the equivalent stage in the rat is about postnatal day (P) 10 to 20 (105); in fetal sheep it is late in gestation (106). In this and other respects, a newborn rat is quite different from a newborn human.

PHYSICOCHEMICAL PROPERTIES OF DRUGS AND TOXINS

The idea that it should be possible to predict brain barrier permeability of drugs and other compounds based on considerations of their chemical structure and physical properties has been around for a while (107). Thus lipid solubility and/or molecular size have been considered to be primary predictors of drug entry from blood into brain and CSF. It assumed that for drugs to permeate the brain barriers they would need to be small (<400 Da) if they are hydrophilic, or if they were larger, they would need to be lipid soluble. In real life this simplistic view is defeated by the presence of efflux mechanisms in the barrier interfaces (see section "Efflux Mechanisms in the Embryo and Fetus"). A good illustration of this point in neurotoxicology is the study of Watanabe et al. (108). These authors investigated placental and blood-brain barrier transfer of a series of drugs in pregnant rats. The drugs (propranalol, chlorpromazine, haloperidol, atropine, reserpine, dopamine, epinephrine, and norepinephrine) were chosen to be of very different lipid solubility in the expectation that this would be the main determinant of their entry into the fetus and its brain. However, with respect to placental transfer, there was no significant correlation between log P_{corr} [P_{corr} is the corrected partition coefficient, chloroform/water. Thus log $P_{corr} = \log[P_{app}/(1 - \alpha)$ where $\alpha = 1/[1 + \text{antilog}({}_pH - {}_pK_a)]$ and the fetal plasma/maternal plasma concentration ratios. Thus the entry of the highly lipid soluble drugs propranalol, chlorpromazine, and haloperidol was much less than predicted from their lipid solubility. The explanation for this is likely to be that these drugs are P-glycoprotein substrates (109,110) and the presence of this efflux protein in the placenta would have reduced their entry into the fetus. However, at the time their paper was published it had only just been shown that P-glycoprotein was present in the placenta (111) and cerebral endothelial cells (112,113). Watanabe et al. (108) did find that there was a significant correlation for log P_{corr} and brain/plasma concentration ratios for the drugs in the mothers, their fetuses (E19) and the postnatal animals at P2, P7, and P14, but brain/plasma concentration ratios for the three most lipid soluble drugs (propranalol, chlorpromazine, and haloperidol) were appreciably less in the fetus and neonatal animals compared to the mother and by two weeks postnatal the ratio was approaching maternal levels. For the less lipid soluble drugs, some reached higher brain/plasma concentration ratios in the fetus (atropine, epinephrine, norepinephrine) whereas others achieved lower or similar ratios in the fetus compared to the mother (dopamine, reserpine). Possibly, age-related differences in brain receptor binding, metabolism, or efflux proteins in developing cerebral endothelial cells account for these differences. In a detailed compilation of the physicochemical characteristics and developmental toxicology of 50 compounds, neither P-glycoprotein nor other efflux mechanisms are mentioned (41). Their presence in the placenta, cerebral blood vessels, and other sites in the fetus and newborn will affect the distribution of compounds that are substrates, thus limiting the predictive value of the physicochemical data. Results such as those in (108) make it clear that for both the adult and developing brain the question of whether a particular compound will enter it and to what extent and whether it has neurodevelopmental effects, can only be answered by direct experimentation.

VULNERABILITY OF THE DEVELOPING BRAIN TO DRUGS AND TOXINS

The evidence described above strongly supports the notion that the fetal and newborn brain has well developed mechanisms in its brain barrier interfaces that determine the composition and control of the internal environment in which the brain grows. In spite of the wealth of morphological and functional evidence, a belief in the "immaturity" of the blood-brain barrier persists (1,4–13,15,16). This perhaps explains why experimental evidence about barrier permeability and brain vulnerability in the fetus and newborn is so fragmentary. The findings that key barrier mechanisms, particularly the efflux mechanisms, are already present early in brain development as well as in the placenta, does not necessarily mean that the immature brain will be unaffected by potentially toxic agents ingested by the mother or young child. This can only be determined by experimentation, in most cases in suitable animal systems. The rest of this review will outline what is known about the penetration of potential neurotoxins into the developing brain and attempt to assess the extent to which the very real neurotoxic effects may be a property of the barrier interfaces or to some other feature of brain development. However, the design of many experiments is such that it is often difficult to distinguish between protective (or imperfections in protection) provided by placental mechanisms compared to brain barrier mechanisms.

The number of potentially neurotoxic agents is considerable. A recent review (12) listed 201 industrial chemicals known to be neurotoxic in humans and >1000 known to be neurotoxic from experimental studies. The "chemical universe" was assessed as exceeding 80,000 compounds. The review did not consider drugs, food additives, microbial toxins, snake venoms, or other biogenic toxins. Many of these potential neurotoxins are substrates for the various efflux transporters described above. Several reviews have published lists of substrates for the efflux transporters (35,114–116) but these are not comprehensive. A systematic review of the literature on effects of prenatal toxin exposure on mental health in children and adolescents has been published (116). The number of drugs that have been administered to pregnant women is very large. A risk assessment for nearly 1200 drugs used in pregnant and lactating mothers has been published (117). Most of these have never been evaluated formally for neurotoxic or damaging effects and testing is not included in the regulatory approval of most of these drugs. There is little serious evidence (as opposed to anecdote and belief) on the specific problem of the extent to which different drugs enter the developing brain and by what mechanism. As indicated in the introduction to this chapter, there is a general tendency to assume that the presence of a drug (or other toxin) in the developing brain is due to "immaturity" of the blood-brain barrier, rather than to specific developmental mechanisms.

Medically Important Drugs

Anticonvulsants

Various abnormalities have been reported in offspring of mothers treated with anticonvulsant drugs during pregnancy [reviewed in (118)]. The incidence appears to vary with different antiepileptic drugs and drug dose, with valproate having the highest incidence (over 15% in some studies) (119) and lamotrigine possibly the lowest (around 3%) (120) although this might reflect the fact that it is a much newer treatment. Use of polytherapies (98,121), and large doses, appear to be associated with a higher incidence of congenital malformations; but the reported incidences are very variable. The main disorders affecting the CNS that have been reported are neural tube defects for exposure early in pregnancy and behavioral (including autism) and cognitive effects that may be severe enough to cause mental retardation (118). Greater blood-brain barrier transfer of the protein-bound anticonvulsant drugs phenytoin and phenobarbitone has been reported in immature animals (122). This was attributed to greater extraction from more slowly perfusing cerebral circulation, which was estimated in newborn rabbits. It was suggested that this was due to uptake of protein-bound drug in the newborn, but not adult brain. However, proteins in plasma are not taken up into developing brain via the cerebral blood vessels (21) but via the choroid plexuses (46). This would also be expected to transfer protein-bound drugs into CSF, which might account for some of the greater extraction of protein-bound drugs reported by (122). Only recently has it been appreciated that many of the antiepileptic drugs are substrates for P-glycoprotein. The likely importance of fetal genotype for efflux transporters in the placenta in determining fetal sensitivity to these drugs has been discussed (98). This will also be important for the level of protection efflux transporters provide against the nervous system damaging effects of these drugs. Evidence for potentially damaging effects of the GABA-acting antiepileptic drugs, carbamazepine, valproate, vigabatrine, on specific features of development of the hippocampus and cortex in fetal rats whose mothers had been given antiepileptic drugs from E14 to E18 have been described (123). Other authors have suggested that genetic differences in detoxifying mechanisms may play a role in the predisposition of some infants to induction of congenital malformations by antiepileptics (124). A further complication is that in prolonged, high dose therapy, there maybe up-regulation of efflux transporter activity (98).

Opiates

For many years morphine has been considered to have a greater effect in newborn infants than in adults. Decades ago this was attributed to immaturity of the blood-brain barrier (125) and also in more recent studies in which fetal brain morphine levels were five times higher than in adult brain following administration of morphine to pregnant rats (125). In support of the interpretation of barrier immaturity the authors (126) cited (127) a fetal rat study, in which horseradish peroxidase (HRP) was used as a marker for blood-brain barrier permeability. However, this is not relevant because HRP tests protein (tight junction) permeability. Morphine is a much smaller and more lipid soluble compound and therefore enters the brain by a different mechanism, although its entry is reduced by P-glycoprotein. Fetal brain levels of morphine could only be interpreted if fetal blood levels were available, but these were not measured (126). Studies of blood levels of morphine in neonates (128) raised an important issue, which was ignored for a long period by neonatal pediatricians and led to a number of disasters (perhaps most notably the potentially lethal "grey syndrome" in neonates from the toxic effect of chloramphenicol). The blood-levels of a drug may be very different in the neonate because of differences in metabolism. Often liver enzymes that metabolize drugs are less active in the newborn, leading to higher blood levels of the drug. Also, the metabolic pathways used may be different, resulting in more active forms of the drug than occurs in adults. Metabolic and pharmacokinetic differences in preterm infants, term infants and adults have been reviewed in (129). The supposed greater susceptibility of the immature brain to morphine in causing respiratory depression could relate to a number of different factors: (*i*) possible differences in metabolism in neonates, (*ii*) the level of morphine receptors in a number of key brain areas may be greater in the newborn than later in life (130), (*iii*) morphine is protein bound (131) and the concentration of proteins in plasma of immature animals and humans is substantially less than in the adult (57,59,105,106), or (*iv*) morphine is a substrate for P-glycoprotein (132–134); this would result in a brain level of morphine that is less than would be predicted for its physicochemical characteristics.

Theophylline and Related Xanthines

Studies of metabolism and distribution of theophylline in the pregnant rat have been published (135). Pregnant rats were injected with labelled theophylline; fetal brain-blood ratios of around 1.0 compared with 0.41 in adult, were reported. It was proposed that this indicated the lack of a blood-brain barrier in the fetus. The authors do not seem to have realized that theophylline is protein bound and that since plasma protein concentration is much lower in the fetus and newborn, this could increase availability of theophylline for exchange in younger animals. Accumulation of theophylline, theobromine, and paraxanthine in the fetal rat brain has been measured following a single oral dose of caffeine (136). Pregnant rats at 20 days gestation (E20) were given a single dose of caffeine (5 or 25 mg/kg). Fetal and maternal concentrations of caffeine were estimated both in the blood and the brain. Brain/blood ratios of around three were obtained for theophylline for both doses of caffeine in fetuses, but adult results depended on the dose administered. In adults, the brain/plasma ratio for theophylline at a caffeine dose of 5 mg/kg was about 3, whereas at a caffeine dose of 25 mg/kg, the ratio was around 0.1. This was a consequence both of *less* theophylline getting into the brain *and* a higher concentration in plasma. The authors acknowledged that theophylline is lipid soluble (although less so than caffeine) and has a higher binding affinity than caffeine for proteins in plasma (50–60% and 24–35%, respectively). Nevertheless they still concluded that the greater brain/blood ratios for theophylline found in fetuses were due to differences in blood-brain barrier permeability. Concern has been expressed about possible teratogenic effects of caffeine ingested by pregnant women (137) but as indicated above, the developmental state of blood-brain barrier mechanisms is not likely to be a significant contributor, although transfer of protein-bound drug across the choroid plexuses and greater accumulation of drug due to low turnover of CSF, may be important.

Theophylline and L-glucose entry from blood into different brain regions and CSF have been studied in postnatal rats (138). Although theophylline and L-glucose have similar molecular weights, theophylline is much more lipid soluble, and therefore unlikely to be hindered by a blood-brain barrier, as theophylline appears not to be a P-glycoprotein substrate (139). L-glucose is a non-transported low molecular weight probe of brain barrier permeability. Hypercapnia increased penetration of L-glucose, but not theophylline, into CSF and brain. This supports the contention that the difference in lipid solubility means that the route of entry is likely to be different and that an increase in protein binding of theophylline accounts for the decrease of its brain/plasma ratios in older animals (138). In addition, the transfer of theophylline bound to proteins that are transported across choroid plexus epithelial cells (140,141) combined with the lower turnover of CSF in the developing brain (46) would be expected to lead to higher amounts of theophylline in fetal CSF and brain, which is a quite different mechanism from "immaturity" of the barrier in cerebral blood vessels.

Psychoactive Drugs

Drugs such as selective serotonin uptake inhibitors (SSRIs) (e.g., fluoxetine) presumably enter the brain in adults (otherwise they would be functionally ineffective). Many do so because they are lipid soluble and are not substrates of efflux proteins such as P-glycoprotein or if they are substrates, this activity is insufficient to prevent the entry of therapeutically effective concentrations of the drug. Such drugs would be expected to cross the placenta and enter the fetal brain, whatever the state of development of brain barrier mechanisms. The amount reaching the fetal brain might be affected by the level of fetal metabolism of the drug or degree of protein binding, but is unlikely to be related to brain barrier mechanisms. However, if psychoactive drugs are even low-level substrates for P-glycoprotein (as appears to be the case for some SSRIs) (110) this may influence the level of entry into fetal brain because of the presence of P-glycoprotein in human cerebral blood vessels (64,68). Because the activity of this P-glycoprotein in the fetal brain is unknown, it is not clear how important this effect would be. These uncertainties emphasize the need for evaluation of entry of individual drugs into the fetal or neonatal brain, rather than relying on predictions from physicochemical properties of drugs.

Steroids and the Developing Brain

In view of the widespread use of steroids in prematurely born infants (e.g., as a means of mitigating the effects of respiratory distress syndrome and intraventricular hemorrhage) (142), it is important to know whether steroids have any deleterious effects on brain development or the properties of blood-brain barrier mechanisms. Permeability to AIB (α-aminoisobutyric acid, a low molecular weight, hydrophilic, inert compound) was increased by dexamethasone, and the effect was slightly greater in the youngest fetuses (143). This effect might make the brain more vulnerable to exposure to low molecular weight drugs or toxins. An extensive series of studies concerning repeated doses of betamethasone in pregnant sheep has shown deleterious effects on brain growth and myelination in the fetuses (144,145).

Environmental Agents

Metals

Metals such as mercury and lead have no known biological function but are present in the environment in amounts that have increased as a result of human activity. If they enter the body (e.g., by ingestion or inhalation) they are extremely toxic in small amounts and have deleterious effects on the nervous system especially when immature. Other metals such as, cobalt, copper, and manganese are incorporated into various biological systems, but have toxic effects, including neurotoxicity, when present in excess amounts. Although some authors have suggested that immaturity of the blood-brain barrier contributes to the greater toxicity of several metals (3,8,146,147) it may actually be the presence of specific barrier transfer properties at a time of brain vulnerability due to developmental growth processes that actually contributes to brain damage.

Mercury

Mercury is present in the environment in a variety of forms, the most abundant and neurotoxic of which is methylmercury (MeHg). Elemental mercury (Hg^0), which may be inhaled from mercury used in manufacturing processes or in instruments, is also toxic. MeHg is a particular hazard because other less toxic forms of mercury are converted to MeHg by a variety of organisms, particularly aquatic and MeHg thus enters the food chain and becomes concentrated in fish. There have been some notable epidemics of mercury poisoning due to contamination of fish and crops by MeHg (148,149). The neurotoxic effect of mercury have been described in detail and it is clear that they are greater and more diverse in the immature brain (148,150–152) but there is some disagreement about the mechanisms by which mercury enters the brain. MeHg is said by many authors to be highly lipid soluble (153,154) but actual estimates of its solubility are hard to find. However, in the body MeHg is water soluble rather than lipid soluble (155). The presence of cerebral endothelial amino acid transporters in immature brain blood vessels (24,156) is more likely to be of significance. Their transfer capacity has been estimated to be larger than in the adult (55) and this probably accounts for the greater accumulation of mercury in fetal than in the adult brain (157) as MeHg appears to form a complex with cysteine and cysteine administration to pregnant rats has been shown to increase brain uptake of MeHg in their fetuses (146). However, these authors have suggested that immaturity of the blood-brain barrier also contributes to the greater damaging effect of MeHg in the developing brain (146,158). Given the evidence summarized earlier about the effectiveness of brain barrier mechanisms in the embryo and fetus, this seems unlikely to play a role in mercury toxicity in the developing brain. In adult rats, administration of either MeHg or $HgCl_2$ has been shown to produce acute damage to cerebral blood vessels (159,160). It is not known if this occurs in the developing brain or if such blood-brain barrier breakdown is part of the neuropathology in human exposure to mercury. However, the observation that ^{75}Se-selenomethionine uptake into brain was *less* in rats exposed to MeHg (159) is additional evidence of the importance of amino acid transport across the blood-brain barrier for entry of MeHg into the brain.

Lead

The developing nervous system appears to be particularly susceptible to lead toxicity. Above 80–100 μg/100 mL children are more likely to develop signs of lead encephalopathy (including fits). At levels that do no appear to produce discernible problems in adults (e.g., 10–20 μg/100 mL) there is evidence of effects on neurocognitive development, including reduction in IQ (161). Extensive experimental studies in animals (mainly in rats, but also in guinea pigs and monkeys) have shown that key features of the neuro- and developmental pathology can be reproduced (162–166) although some early studies used such high levels of lead exposure that there were confounding effects due to malnutrition (167). Two main features of lead neurotoxicity have been established: a direct damage of the cerebrovascular endothelium (168–170) and disruption of a number of biochemical processes within the brain (171,172). The blood vessels in the cerebellum appear to be more susceptible than those, for example, in the cerebral hemispheres (162,173–175). This has been suggested to be due to the later development of the cerebellum (174,175). The relative contributions of damage to the blood-brain barrier and the effects on biochemical processes to the pathogenesis of lead intoxication is not clear due to the wide range of experimental protocols for lead administration and different analytical techniques used. At the blood-CSF interface the choroid plexus cells concentrate lead taken up from the circulation (20,165). The level of lead in CSF is much lower than in brain (176,177) probably because of the turnover of CSF (sink effect of CSF secretion and flow) (178). Cerebral endothelial cells also concentrate lead to a high degree (179,180). However, unlike lead crossing the choroid plexuses into CSF, with subsequent return to the circulation by the flow of CSF, lead crossing the cerebral endothelial interface is taken up by brain cells, particularly end feet of astrocytes (179).

Fundamental questions include how lead enters the brain and if the process quantitatively or qualitatively different in the developing brain. Brain uptake of ^{203}Pb using steady state in vivo and short pass arterial techniques has been studied in anesthetized adult rats (177,181). It was concluded that lead enters the brain as $PbOH^+$. Lead is bound to albumin and to L-cysteine in plasma (182). In a limited number of intra-arterial perfusions in P16 to P17 and P26 postnatal rats, the brain uptake of lead was substantially greater than in the adult (177). An alternative mechanism for lead entry into brain that does not seem to have been considered is that lead bound to L-cysteine may be transported across cerebral endothelial cells, analogous to the entry into brain of MeHg coupled to L-cysteine (see above). Since amino acid transfer has been reported to be greater in the developing brain (24,156) this could account for the greater lead uptake. Lead bound to albumin might also reach the immature brain via the CSF in greater amounts than in the adult.

There is general agreement that the blood-brain barrier in immature animals is susceptible to disruption by lead administration, detected by vascular markers such as HRP, trypan blue, or labelled albumin, up to the age of about two weeks in rats (162,167,168). However, in rats administered different levels of lead, administration from P5 up to P21 at a daily IP dose of 10 μg/g body weight, which achieved a blood lead level of 3.4 μg/mL (16.4 μM) there was no effect of this level of lead intoxication on blood-brain barrier transfer of several nutrients

(D-glucose, phenylalanine, lysine, proline, pyruvate, and uridine) (183). In other experiments with similar blood levels of lead also using IP injections, but from birth to P15, there was a blood-brain barrier leak to Evans blue-albumin and to plasma proteins at P15 but not P20 or P30 (167). This may reflect greater resistance of the blood-brain barrier to lead toxicity in older animals as suggested by several authors (162,167). It may explain the lack of effect of lead on glucose, amino acid and pyruvate transport in the immature brain (183) and also on tyrosine and choline blood-brain barrier permeability from birth to 70 days of age in rata (184). In addition, there is evidence that in prolonged exposure to lead from the neonatal period, the blood-brain barrier shows recovery from the earlier pathological changes (180). It is unclear how much of the neuropathology in the immature brain (169,171,172) is due disruption of a susceptible blood-brain barrier and how much is directly due to primary neurotoxicity.

Cadmium

Cadmium has been implicated in a number of neurological disturbances in children, although exposure to cadmium was also often combined with exposure to other neurotoxic metals such as lead. Problems reported included deficits in psychomotor development (185) and cognitive skills (186). Much less information is available about the effects of cadmium on the fetus via maternal exposure (187,188) although it appears that the placenta does limit its entry into the fetus (188). Most of the available information on possible mechanisms of cadmium neurotoxicity in development comes from animal studies. Differences in experimental protocols and the lack of blood measurements of cadmium in many experiments make it difficult to relate the results of animal studies to reports on toxicity in infants and children. The acute effect of cadmium appears to be on the vascular endothelium in the brain (189–191) and some other vascular beds (192). Blood vessels in the immature brain appear to be appreciably more susceptible to lower levels of injected cadmium (190). Pathological changes consisting of extravasated red blood cells and occasional severely damaged capillaries were observed within two to four hours after cadmium chloride ($CdCl_2$) injection, with increasing signs of hemorrhage and vacuolation of endothelial cells by four to six hours, which increased further by six to eight hours, by which time neuronal and glial cells showed pathological changes. Inter-endothelial cell tight junctions were widened, but the junctions appeared intact (190). Increased permeability of cerebral vessels in cadmium toxicity reported by (193) appears to be due to damage to the endothelial cells, which were reported to have definite intracellular gaps, rather than to disruption of the intercellular junctions (190). Detailed descriptions of neuropathological lesions in immature rats and rabbits exposed to different regimes of cadmium intoxication have been published (190,192,194,195). Cadmium has been reported to accumulate in immature rodent brain to a much greater extent than in the adult (194) this was attributed to immaturity of the blood-brain barrier by some (196). However, others found no age-related difference in brain cadmium uptake (195). Given that the damaging effects of cadmium on cerebral vascular endothelial cells do not appear to include disruption of tight junctional complexes (190) it seems more likely that if there is an effect on brain uptake, it is direct damage to endothelial cells that is involved. The relevance of this to acute cadmium toxicity in children is illustrated by a case report on a child (2 years 10 months) who died of cadmium poisoning and at post-mortem had signs of blood-brain barrier disruption and cerebral edema (197).

In addition, loading the diet of rats with cadmium resulted in much greater accumulation of cadmium in brains of adult animals, than in controls, but this effect was barely apparent at the P15 (195). These results suggest that there is a significant barrier to cadmium entry into immature brain at least as early as P15 in rats. The immature brain, if anything, appears to be better protected from exposure to cadmium via the circulation than the adult, but the mechanisms are unclear. One factor that does not appear to have been considered is that the choroid plexuses have been shown to concentrate cadmium in adults (198) but this property has been little studied in the immature brain [but see (199)]. The choroid plexuses have been suggested to be substantially more important than the cerebral vasculature for blood to brain exchange (46) in early stages of brain development, thus it may be that choroid plexuses also play a larger role in protecting the brain from toxic effects of cadmium and other xenobiotics.

Iron

The fetus has been suggested not to be at risk from iron overdose in the mother because placental transport of iron is a saturable process (200). However, recent evidence suggests that high maternal iron stores correlate with reduced IQ scores in children (201). Iron is the commonest cause of poisoning in children, due to gastrointestinal and liver toxicity and there is evidence from animal studies that iron administration may have adverse neurodevelopmental effects (201). Chronic low-level ingestion of iron by children might lead to its accumulation in the brain, as transport of iron from blood into brain appears to be much higher during development (202,203). Another aspect of iron transport across the blood-brain and blood-CSF interfaces that may have neurotoxicological relevance lies in the utilization and interactions of transport mechanisms by other metals such as manganese. The main mechanism transporting iron into brain involves the transferrin receptor, which is only present on cerebral vessel endothelial cells (204) and possibly other proteins such as divalent metal transporter-1 (DMT-1; 205) and ferroportin-1 (206) although ferroportin-1 does not yet seem to have been localized to cerebral endothelial cells. Studies of neurotoxic metal entry into brain of rats from P15 to adult showed that there is a substantially effective barrier to entry of metals (copper, lead, and cadmium) into the brain at least as early as P15 (195,207,208).

Manganese

Manganese is an essential metal found in various tissues including the brain, but in excess it is toxic. The biological functions of manganese have been summarized in (209); it is a constituent of enzymes involved in metabolic pathways for synthesis of amino acids, lipids, proteins, and carbohydrates. Manganese enters the brain via both the cerebral endothelial cells and the choroid plexuses, but the relative importance of these interfaces appears to depend on its plasma concentration. Thus at high plasma manganese levels, transport across the choroid plexuses predominates (210) as may also be the case for iron (211). At least six mechanisms for manganese entry into brain have been described (209) the major ones are via DMT-1 and transferrin-dependent transport, as well as entry of unbound Mn^{2+}. In adults, the main source of manganese toxicity is from industrial exposure, resulting in neurological disturbances similar to Parkinson's disease (147). There have been reports of adverse neurodevelopmental effects in children associated with manganese-contaminated water (147). In infants, deficiency and liver disease predispose to manganese toxicity. In neonates, the main source of toxicity appears to be associated with parenteral feeding. Several studies have shown that the uptake of manganese in developing brain is higher than in adults (198,212). This has been suggested to reflect both an increased requirement for manganese during development and an "incomplete" blood-brain barrier (147). Administration by inhalation of manganese to pregnant rats did not cause increased levels of manganese in the fetal brain (213) from which the authors concluded that the placenta partially sequesters inhaled manganese, thus limiting exposure of the fetus (214). In contrast, increased manganese in the diet of pregnant rats from E7 into the neonatal and postnatal period, resulted in higher brain levels of manganese, zinc, and chromium, but decreased iron when the young were examined at three weeks of age (147). It is not clear whether this increase began in the fetal period, because only postnatal animals were examined. DMT-1 and transferrin receptor levels were increased, as was the level of the inhibitory transmitter γ-amino butyric acid. $MgCl_2$ added to the drinking water only in the postnatal period resulted in substantially increased brain levels of manganese that were much greater at P5 and P10 than in older animals, a finding that the authors attributed to "closure" of the blood-brain barrier, development of mechanisms of ionic homeostasis, CSF flow, and secretion. However, as reviewed above, ionic mechanisms controlling CSF composition are already becoming functional at this early stage and the blood-brain barrier in terms of functionally impermeable intercellular junctions develops early in embryonic life (44,48–50). A combination of lead and manganese exposure during gestation in rats resulted in a 10-fold increase in brain uptake of lead by three weeks of age (215); this is an example of several of interactions between different metals in their transport into brain. Part of the explanation for the higher levels of brain manganese uptake in the fetus and neonate is probably that the brain is exposed to higher levels in the blood, because renal secretion of manganese appears to be limited in the first 17 to 18 days of life in the rat (212). The transport mechanisms that are likely to contribute to the greater accumulation of manganese in immature brain are the early presence of transferrin receptors on cerebral endothelial cells (216) and greater activity of transferrin-mediated metal transfer, as described for (203). In addition, DMT-1 in cerebral endothelial cells of the developing brain (147) may contribute to increased manganese transfer into the immature brain. Another mechanism likely to be of significance in the developing brain is transfer of manganese bound to transferrin and albumin that are transferred across the choroid plexus epithelial cells from blood into CSF. The lower turnover of CSF (46) would also be expected to contribute to the greater accumulation of manganese in early postnatal life.

Aluminum

Mice fed a diet containing excess aluminum had the highest brain levels of aluminum in the youngest animals (P6); in addition, aluminum binds to plasma transferrin (217) thus a possible route of entry and explanation for the higher levels in the younger animals is transfer of transferrin across the epithelial cells of the choroid plexuses as may be the case with a number of metals that are bound to proteins in plasma (e.g., iron and manganese).

Arsenic

Two types of developmental neurotoxic effects have been ascribed to exposure to arsenic. Numerous papers have associated early maternal exposure both in animals and humans with neural tube defects. The literature has been reviewed in detail (218) and it was concluded that deficiencies in the reported studies were such that it could not be accepted that there is a risk of neural tube defects in the offspring of pregnant women exposed to arsenic. The other neurotoxic effect reported is that of neurological and intellectual impairment resulting from environmental exposure (219). In studies of the levels of different arsenic species in newborn mice following exposure of pregnant mothers to inorganic arsenic in the drinking water, it was found that in the newborn, brain levels of all forms of arsenic measured were 50% to 70% higher than in the maternal brains (220). This was mainly in the form of dimethylarsinic acid. The levels of this form of arsenic were higher in the newborn brain than in the newborn liver, which may indicate that dimethylarsinic acid is synthesized from administered inorganic arsenic. The mechanism by which arsenic enters the brain (whether immature or not) appears not to be known. There is some evidence for binding of arsenic by proteins in plasma (221) so that transfer across the choroid plexus of the developing brain is a possible explanation (cf. 46). In adult animals there is evidence that arsenic compounds are concentrated in the CSF (20). Also of possible relevance is the report (222) that arsenic is transported by MRP1 although this was based on in vitro studies and does not seem to have been investigated in the fetus or newborn.

Other Metals

Tellurium has been reported to produce hydrocephalus in fetus when administered to pregnant rats (223). The mechanism by which tellurium enters the fetal brain is not known, but in the adult it concentrates in the choroid plexuses (20). Tellurium binds strongly to serum proteins (224). Thus, as suggested for some other metals (see above) it may be that the mechanism in the developing brain is via the choroid plexus with subsequent uptake into brain cells from the CSF. Most of the literature on neurotoxicity of copper and zinc deals with effects of nutritional deficiency (201). These authors point out that descriptions of excess copper producing developmental neurobehavioral effects are rare and effects of excess zinc appear not to have been described. Nevertheless, some caution is warranted, as both are known to have neurotoxic effects in the adult. Both copper and zinc are bound to plasma proteins and would be expected to enter the fetal and neonatal brain via the choroid plexuses and CSF (see above). In addition, the main mechanism of transport of zinc into adult brain is as a zinc-histidine complex (225) that is transported across cerebral endothelial cells. As discussed earlier the greater transport of amino acids in the developing brain could lead to transport of metals that bind to amino acids in excessive amounts if there were environmental or dietary exposure.

Naturally Occurring Toxins: Excitotoxic Amino Acids

Glutamate and aspartate are excitatory neurotransmitters; these and several other amino acids (e.g., cystine) have been shown to produce brain damage particularly during development, as has also been found for the food additive, monosodium-L-glutamate (226,227). Much of the damage is in areas outside the blood-brain barrier such as the hypothalamus and area postrema. However, there are reports of neuronal damage in brain regions that might be expected to be protected from damage by brain barrier mechanisms (228,229). Many amino acids including excitotoxic ones are transported into the developing brain to a greater extent than in the adult (Fig. 3; see above). This mechanism might contribute to the neurotoxicity of excitotoxic amino acids in the developing brain.

Acrylamide

Ikeda et al. (230) conducted maternal-fetal distribution studies of [^{14}C] acrylamide in tissues of beagle dogs and miniature pigs in late pregnancy. Acrylamide was described as moderately lipid soluble. Radioactivity in brains of both mothers and fetuses indicated that it penetrated easily, which would be expected simply on the basis of its lipid solubility, although the authors explained its presence in fetal brain as due to an absence of a blood-brain barrier. As only 30% to 40% of the labelled acrylamide was protein bound, there should still have been plenty of free lipid soluble acrylamide to distribute into the brain. In addition, some would probably have penetrated into CSF across the choroid plexus epithelial cells bound to plasma proteins.

Pesticides

The neurotoxicology of pesticides, including developmental aspects, has been reviewed in (13,231). A large number are relatively lipid soluble (231). This might lead to a conclusion that brain barrier mechanisms would be an unlikely impediment to entry into the brain in either the adult or developing brain. However, many pesticides are to some degree substrates for P-glycoprotein (232,233). As indicated above, P-glycoprotein is present in both the placenta and fetal cerebral blood vessels, but the degree to which it is functionally effective is unknown. The distribution of organochlorines in stranded pilot whales (*Globicephala melaena*) from the coast of Massachusetts has been reported (234). The study included some whole fetuses and young whales. The absolute concentration of all organochlorines measured in whale fetuses was considerably less than in mothers, males, or non-pregnant females, which suggests that placental P-glycoprotein or other efflux mechanisms were effective in limiting entry of these toxins into the fetus. Only one fetal brain was examined. Levels were lower than in adults, but as no blood levels were available, this finding is difficult to interpret. The low fetal brain level suggests that the placental and possibly the brain efflux mechanisms were effective in limiting brain entry of these toxins.

Pipecolic Acid

Transport of pipecolic acid has been examined in adult and developing mouse brain (235,236). Pipecolic acid is a plant amino acid found, for example, in some types of bean. Marked hyper-pipecolinemia has been described in a group of human genetic disorders connected with peroxisomal defects such as in the Zellweger syndrome; these syndromes are associated with brain damage (237). It was found in mice that the net uptake at 10 minutes of injected pipecolic acid was more than two times greater at P1 than in adults (236). This was interpreted as evidence of postnatal development of the blood-brain barrier. However, these findings are more likely to reflect a more active carrier mechanism, related to early (growing) brain development, rather than to immaturity (cf. 24). Alternatively, or in addition to more active transport in the developing brain, the finding that in the immature brain small molecules such as L-glucose and sucrose (42,138) reached higher brain/plasma and CSF plasma ratios than in the adult, due to low turnover of CSF (46) may also contribute to higher levels of water soluble pipecolic acid (MW = 129) in the developing brain.

Phytoestrogens

These are plant compounds that produce estrogen-like activity in mammals. The isoflavanoids diadzin and genestein, included in soy-based foods, are the most commonly consumed (238). These and coumestrol, present in red clover in high concentrations, appear to be substrates for Brcp but not for Mdr1a (238). These authors also found that fetal brain levels of these phytoestrogens were higher in Brcp$^{-/-}$ mice than in controls. However, the increase in brain levels of adult Brcp$^{-/-}$ mice compared to adult controls was about seven times greater, which suggests that although some barrier function of Bcrp was present in the fetal brain, it was much less than in the adult.

Herbicides

In experiments in which the herbicide dichlorophenoxyacetic acid labelled with ^{14}C ([^{14}C]-2,4-D) was injected intra-peritoneal (IP) into pregnant mice on E17, following (IP) pre-treatment with saline or 40 or 80 mg/kg [^{14}C]-2,4-Dichlorophenoxyacetic Acid ([^{14}C]-2,4-D) no [^{14}C]-2,4-D was detectable by autoradiography in the brains of mothers or fetuses (239). For zero pre-treatment, maternal brain/plasma ratios were around 3% to 4%. The fetal level was about 8% of *maternal* plasma concentration, which might have given an overestimate of true brain/blood ratio in the fetus. This finding was interpreted as indicating the existence of a substantial barrier to 2,4-D in both adult and fetus. Pre-treatment of mothers with 2,4-D over two days resulted in substantial increases in brain levels for both fetuses and mothers. In similar experiments, adult rabbits were studied, with the addition that the CSF and choroid plexus were sampled. At the lowest dose of [^{14}C]-2,4-D, the CSF concentration was only about 10% of brain concentration. This increased to 50% at highest dose of [^{14}C]-2,4-D. The level in the choroid plexus was about seven times the brain concentration at the lowest dose, but only 50% higher at the highest dose. Permeability to the small organic solute, 2-deoxyglucose (2-DG) was unaffected by 2,4-D administration. This was interpreted as indicating that blood-brain barrier permeability was unaffected by the pre-treatment with 2,4-D. However, 2-DG is a non-metabolized glucose analogue which is transported into the brain by the glucose transporter, it is therefore an inappropriate test of barrier function for 2,4-D. In vitro experiments with choroid plexus demonstrated specific inhibition of 2,4-D uptake by increasing concentrations of 2,4-D in the incubation medium. It was concluded that the concentration of 2,4-D in choroid plexus and lower values in CSF indicated that 2,4-D is cleared via the choroid plexus and the elevated brain levels found at higher exposure levels of 2,4-D were probably due to reduced clearance rather than increased barrier permeability. In terms of fetal/neonatal brain toxicology, the important conclusion is that in the rat, a significant barrier mechanism to 2,4-D exists at least as early as late gestation. Since the brain barrier mechanism appears to involve active exclusion of the compound across the choroid plexus, this presumably develops at an earlier stage of brain development. This could be an MDR, as up regulation of *MDR2* by dichlorophenoxyacetic acid in mouse liver has been reported (240). How much earlier is not known, nor is it clear when this happens in the human fetus. A physiologically based pharmacokinetic model for 2,4-D dosimetry in the developing rabbit brain (241) is probably the only detailed pharmacokinetic model that describes the entry of a compound into fetal brain. Modelling results gave a reasonably good fit with the experimental data (242).

Paraquat is a low molecular weight (186 Da) highly toxic herbicide. In 10 day old and adult rats, there was similar limited penetration into brain at 0.5 hour after subcutaneous injection (243). The penetration at 24 hours was significantly greater in the immature brains, but mainly in regions outside the blood-brain barrier. The restricted entry after 0.5 hour exposure suggests strongly that there was an effective barrier to paraquat. The later accumulation could have been due to the low turnover of CSF in postnatal rats. In addition, there is a possibility that paraquat is a substrate for P-glycoprotein (244) that would contribute its exclusion from the brain.

SUMMARY AND CONCLUSIONS

There is clearly substantial evidence that brain barrier mechanisms show significant functional capacity from early in brain development. Tight junctions in barrier interfaces are functionally impermeable to proteins and molecules smaller than sucrose from as early as blood vessels and the choroid plexuses begin to appear in the embryonic brain. At least one of the efflux mechanisms (MDR1, p-glycoprotein) is present in early embryonic cerebral blood vessels of human and rodent brains. Furthermore, inward transport mechanisms for nutrients such as amino acids, glucose, and trace metals are present and functional early in brain development and most appear to be more active than in the adult brain. In addition, systemic injections of classical markers of barrier function (dyes and HRP) show that they do not enter the developing brain (except in regions without a barrier as in the adult) providing only moderate amounts of marker are injected.

The vulnerability of the developing brain lies mainly in the susceptibility of developmental processes (cell division, differentiation, migration, and synaptogenesis) to the presence of drugs and toxins. This vulnerability may be enhanced by the activity of some barrier transport mechanisms in the developing brain: Influx carrier mechanisms may facilitate the entry of neurotoxins such as mercury (usually in the form of MeHg) and manganese. The lower concentration of proteins in fetal and neonatal plasma may result in the binding capacity for drugs and toxins to be exceeded more easily in the developing than in the mature brain; since many drug and toxins are lipid soluble, they easily penetrate into the brain when unbound unless they are substrates of the efflux proteins (MDRs, MRPs, BCRP). However, these although present may not be fully functional. In addition, the presence of transport mechanisms that are unique to the developing brain may exacerbate the neurotoxic effects of drugs and toxins by increasing their entry. A well-documented example of such a mechanism is the developmentally regulated transfer of plasma proteins across the fetal and neonatal choroid plexus. Proteins such as albumin and transferrin, in addition to acting as carriers for endogenous molecules such as hormones and trace metals, also bind many drugs and toxins such as unconjugated bilirubin and heavy metals.

The widespread notion that the blood-brain barrier is absent or immature is long overdue to be laid to rest. However, the developing brain is undoubtedly vulnerable to drugs and toxins, thus caution in administering drugs to pregnant women remains important. What is really required is systematic testing of drugs that may be essential for medical conditions in pregnant women or for diseases in the newborn. At present the most vulnerable members of society, fetuses and neonates, are the least protected by the regulatory authorities. Most of the drugs used in pregnancy and for neonates and children have never been fully evaluated for effectiveness and risks in these populations. Most such drugs are used for treatment not covered by regulatory approval and problems such as the "grey syndrome" with chloramphenicol and phocomelia and other deformities with thalidomide, only became clear after extensive clinical use. Effects of drugs and toxins on brain development and behavior may be more subtle and difficult to detect, but the consequences could be lifelong.

It is essential that proper mechanisms are developed for preclinical and clinical surveillance of drugs used in pregnancy and in neonates. These should include not only appropriate toxicological studies but also detailed studies of the mechanisms involved; making use of this information it may be possible to modify potentially useful drugs to decrease their unwanted effects.

REFERENCES

1. Schettler T. Toxic threats to neurologic development of children. Environ Health Perspect 2001; 109(Suppl 6): 813–16.
2. Scheuplein R, Charnley G, Dourson M. Differential sensitivity of children and adults to chemical toxicity. I. Biological basis. Regul Toxicol Pharmacol 2002; 35: 429–47.
3. Costa LG, Aschner M, Vitalone A, et al. Developmental neuropathology of environmental agents. Annu Rev Pharmacol Toxicol 2004; 44: 87–110.
4. Rodier PM. Developing brain as a target of toxicity. Environ Health Perspect 1995; 103(Suppl 6): 73–6.
5. Harry GJ. Introduction to develomental neurotoxicology. In: Harry GJ, ed. Developmental Neurotoxicology. Boca Raton, Florida: CRC, 1994: 1–7.
6. Andersen HR, Nielsen JB, Grandjean P. Toxicologic evidence of developmental neurotoxicity of environmental chemicals. Toxicology 2000; 144: 121–7.
7. Latasch L, Frey E. Pain and opioids in preterm and newborns. Anaesthesist 2002; 51: 272–84.
8. Järup L. Hazards of heavy metal contamination. Br Med Bull 2003; 68: 167–82.
9. Rosati A, Maniori S, Decorti G, et al. Physiological regulation of P-glycoprotein, MRP1, MRP2 and cytochrome P450 3A2 during rat ontogeny. Dev Growth Differ 2003; 45: 377–87.
10. Levy Y, Zadok D. Systemic side effects of opthalmic drops. Clin Pediatr (Phila) 2004; 43: 99–101.
11. Dietrich KN, Eskenazi B, Schantz S, et al. Principles and practice of neurodevelopmental assessment in children: lessons learned from the Centers for Children's Environmental Health and Disease Prevention Research. Environ Health Perspec 2005; 113: 1437–46.
12. Grandjean P, Landrigan PJ. Developmental neurotoxicity of industrial chemicals. Lancet 2006; 368: 2167–78.
13. Costa LG, Giordano G, Guizzetti M, et al. Neurotoxicity of pesticides: a brief review. Front Biosci 2008; 13: 1240–9.
14. Risau W, Hallman R, Albrecht U. Differentiation-dependent expression of proteins in brain endothelium during development of the blood-brain barrier. Dev Biol 1996; 117: 537–45.

15. Ohsugi M, Sato H, Yamamura H. Transfer of 125I-albumin from blood to brain in newborn rats and the effect of hyperbilirubinemia on the transfer. Biol Neonate 1992; 62: 47–54.
16. Ohsugi M, Sato H, Yamamura H. Transfer of bilirubin covalently bound to 125I-albumin from blood to brain in the Gunn rat newborn. Biol Neonate 1992; 62: 416–23.
17. Kniesel U, Risau W, Wolburg H. Development of blood-brain barrier tight junctions in the rat cortex. Brain Res Dev Brain Res 1996; 96: 229–40.
18. Engelhardt B. Development of the blood-brain barrier. Cell Tissue Res 2003; 314: 119–29.
19. Barcroft J. The Brain and Its Environment. New Haven: Yale University Press, 1938.
20. Zheng W. Blood-brain barrier and blood-CSF barrier in metal-induced neurotoxicities. In: Massaro EJ, ed. Handbook of Neurotoxicology, Vol. 1. Totawa, NJ: Humana Press, 2002: 161–93.
21. Saunders NR. Ontogenetic development of brain barrier mechanisms. In: Bradbury MWB, ed. Handbook of Experimental Physiology and Pharmacology, Vol. 103, 1992: 327–69.
22. Saunders NR. Function of the blood-brain barrier in infants, children and the unborn. (Chapter 5). In: U.S. EPA. Exploration of Perinatal Pharmacokinetic Issues. Washington, DC: U.S. Environmental Protection Agency, Risk Assessment Forum, EPA/630/R-01/004, 2001: 5.1–5.20. [Available from: http://cfpub.epa.gov/ncea/cfm/recordisplay.cfm?deid=29420].
23. Johanson CE. Ontogeny and phylogeny of the blood-brain barrier. In: Neuwelt EA, ed. Implications of the blood-brain barrier and its manipulation. New York: Plenum, 1989: 157–98.
24. Cornford EM, Braun LD, Oldendorf WH. Developmental modulations of blood-brain barrier permeability as an indicator of changing nutritional requirements in the brain. Pediatr Res 1982; 16: 324–8.
25. Lefauconnier J-M, Trouvé R. Developmental changes in the pattern of amino acid transport at the blood-brain barrier in rats. Dev Brain Res 1983; 6: 175–82.
26. Wooltorton E. Oseltamavir (Tamiflu) unsafe in infants under 1 year old. CMAJ 2004; 170: 336.
27. Lacroix I, Damase-Michel C, Lepeyre-Mestre M, et al. Prescription of drugs during pregnancy in France. Lancet 2000; 356: 1735–6.
28. Beyens MN, Guy C, Ratrema M, et al. Prescription of drugs to pregnant women in France: the HIM-AGE study. Therapie 2003; 58: 505–11.
29. Brightman M, Reese T. Junctions between intimately apposed cell membranes in the vertebrate brain. J Cell Biol 1969; 40: 648–77.
30. Rawson RA. The binding of T-1824 and structurally related diazo dyes by the plasma proteins. Am J Physiol 1943; 138: 708–17.
31. Freedman FB, Johnson JA. Equilibrium and kinetic properties of the Evans blue-albumin system. Am J Physiol 1969; 216: 675–81.
32. Tschirgi RD. Protein complxes and the impermeability of the blood-brain barrier to dyes. Am J Physiol 1950; 163: 756P.
33. Bauer H. Glucose transporters in mammalian brain development. In: Pardridge WM, ed. An Introduction to the Blood-Brain Barrier Methodology and Biology. Cambridge, UK: Cambridge University Press, 1998: 175–87.
34. Braun LD, Cornford EM, Oldendorf WH. Newborn rabbit blood-brain barrier is selectively permeable and differs substantially from the adult. J Neurochem 1980; 34: 147–52.
35. Abu-Qare AW, Elmasry E, Abou-Donia MB. A role for P-glycoprotein in environmental toxicology. J Toxicol Environ Health B Crit Rev 2003; 6: 279–88.
36. Fromm MF. Importance of P-glycoprotein at blood-tissue barriers. Trends Pharmacol Sci 2004; 25: 423–9.
37. Löscher W, Potschka H. Blood-brain barrier efflux transporters: ATP-binding cassette gene family. NeuroRx 2005; 2: 86–98.
38. Ford MD, Delaney KA, Ling LJ, et al., eds. Clinical Toxicology. Philadelphia: Saunders, 2001.
39. Massaro EJ, ed. Handbook of Neurotoxicology, Vol. 1. Totowa, NJ: Humana Press, 2002.
40. Bellinger DC, ed. Human Developmental Neurotoxicology. New York: Taylor & Francis, 2006.
41. Schardein JL, Macina OT, eds. Human Developmental Toxicants. Boca Raton, FL: CRC Taylor & Francis, 2007: 1–425.
42. Ek CJ, Habgood MD, Dziegielewska KM, et al. Permeability and route of entry for small lipid insoluble molecules across brain barriers in developing opossums (*Monodelphis domestica*). J Physiol 2001; 536: 841–53.
43. Ek CJ, Habgood MD, Dziegielewska KM, et al. Structural characteristics and barrier properties of the choroid plexuses in developing brain of the opossum (*Monodelphis domestica*). J Comp Neurol 2003; 460: 451–64.
44. Ek CJ, Dziegielewska KM, Saunders NR. Functional effectiveness of the blood-brain barrier to small water soluble molecules in developing and adult opossum (*Monodelphis domestica*). J Comp Neurol 2006; 496: 13–26.

45. Watson RE, DeSasso JM, Hurtt ME, et al. Postnatal growth and morphological development of the brain: a species comparison. Birth Defects Res B Dev Reprod Toxicol 2006; 77: 471–84.
46. Johansson PA, Dziegielewska KM, Liddelow SA, et al. The blood-CSF barrier explained: when development is not immaturity. Bioessays 2008; 30: 237–48.
47. Johansson PA, Dziegielewska KM, Ek CJ, et al. Aquaporin-1 in the choroid plexuses of developing mammalian brain. Cell Tissue Res 2005; 322: 353–64.
48. Bradbury MWB, Crowder J, Desai S, et al. Electrolytes and water in the brain and cerebrospinal fluid of the foetal sheep and guinea pig. J Physiol 1972; 227: 591–610.
49. Bito LZ, Myers RE. The ontogenesis of haematoencephalic cation transport processes in the rhesus monkey. J Physiol 1970; 208: 153–70.
50. Amtorp O, Sørensen SC. The ontogenetic development of concentration differences for protein and ions between plasma and cerebrospinal fluid in rabbits and rats. J Physiol 1974; 243: 387–400.
51. Bauer H, Sonnleitner U, Lametschwandtner A, et al. Ontogenetic expression of the erythroid-type glucose transporter (Glut 1) in the telencephalon of the mouse: correlation to the tightening of the blood-brain barrier. Brain Res Dev Brain Res 1995; 86: 317–25.
52. Dwyer KJ, Pardridge WM. Developmental modulation of blood-brain barrier and choroid plexus GLUT1 glucose transporter messenger ribonucleic acid and immunoreactive protein in rabbits. Endocrinology 1993; 132: 558–65.
53. Vannucci SJ, Seaman LB, Brucklacher RM, et al. Glucose transport in developing rat brain: glucose transporter proteins, rate constants and cerebral glucose utilisation. Mol Cell Biochem 1994; 140: 177–84.
54. Kohle SJ, Vannucci RC. Glycogen metabolism in fetal and postnatal rat brain: influence of birth. J Neurochem 1977; 28: 441–3.
55. Pardridge WM, Mietus LJ. Kinetics of neutral amino acid transport through the blood-brain barrier of the newborn rabbit. J Neurochem 1982; 38: 955–62.
56. Dziegielewska KM, Saunders NR. The development of the blood-brain barrier: proteins in fetal and neonatal CSF, their nature and origins. In: Meisami E, Timiras PS, eds. CRC Press Handbook of Human Growth and Developmental Biology. Neural, Sensory, Motor and Integrative Development, Vol. 1. Boca Raton, FL: CRC Press, 1988, 169–91.
57. Bell JE, Fryer AA, Collins M, et al. Developmental profile of plasma proteins in human fetal cerebrospinal fluid and blood. Neuropathol Appl Neurobiol 1991; 17: 441–56.
58. Habgood MD, Knott GW, Dziegielewska KM, et al. The nature of the blood-CSF barrier permeability decrease during postnatal brain development in the rat. J Physiol 1993; 468: 73–83.
59. Knott G.W, Dziegielewska KM, Habgood MD, et al. Albumin transfer across the choroid plexus of South American opossum (Monodelphis domestica). J Physiol 1997; 499: 179–94.
60. Scherrmann J-M. Expression and function of multidrug resistance transporters at the blood-brain barriers. Expert Opin Drug Metab Toxicol 2005; 1: 233–46.
61. Kim WS, Shannon C, Garnet B. Role of ATP-binding cassette transporters in brain lipid transport and neurological disease. J Neurochem 2008; 104: 1145–66.
62. Leslie EM, Deeley RG, Cole SPC. Multidrug resistance proteins: role of P-glycoprotein, MRP1, MRP2 and BCRP (ABCG2) in tissue defence. Toxicol Appl Pharmacol 2005; 204: 216–37.
63. Graff CL, Pollack GM. Drug transport at the blood-brain barrier and the choroid plexus. Curr Drug Metab 2004; 5: 95–108.
64. Virgintino D, Errede M, Girolamo F, et al. Fetal blood-brain barrier P-glycoprotein contributes to brain protection during human development. J Neuropathol Exp Neurol 2007; 67: 50–61.
65. Rao VV, Dahlheimer JL, Bardgett ME, et al. Choroid plexus epithelial expression of MDR1 P glycoprotein and multidrug resistance-associated protein contribute to the blood-cerebrospinal-fluid drug-permeability barrier. Proc Natl Acad Sci U S A 1999; 96: 3900–5.
66. Wijnholds J, deLange EC, Scheffer GL, et al. Multidrug resistance protein 1 protects the choroid plexus epithelium and contributes to the blood-cerebrospinal fluid barrier. J Clin Invest 2000; 105: 279–85.
67. van Kalken CK, Giacone G, van der Valk P, et al. Multidrug resistance gene (P-glycoprotein) expression in the human fetus. Am J Pathol 1992; 141: 1063–72.
68. Schumacher U, Møllgård K. The multidrug-resistant P-glycoprotein (Pgp. MDR1) is an early marker of blood-brain barrier development in the microvessels of the developing human brain. Histochem Cell Biol 1997; 108: 179–82.
69. Bystron I, Blakemore C, Rakic P. Development of the human cerebral cortex: Boulder Committee revisited. Nat Rev Neurosci 2008; 9: 110–22.
70. Qin Y, Sato TN. Mouse multidrug resistance 1a/3 gene is the earliest known endothelial cell differentiation marker during blood-brain barrier development. Dev Dyn 1995; 202: 172–80.

71. Tsai CE, Daood MJ, Lane RH, et al. P-glycoprotein expression in mouse brain increases with maturation. Biol Neonate 2002; 81: 58–64.
72. Matsuoka Y, Okazaki M, Kitamura Y, et al. Developmental expression of P-glycoprotein (multidrug resistance gene product) in the rat brain. J Neurobiol 1999; 39: 383–92.
73. Bayer SA, Altman J. Neocortical Development. New York: Raven Press, 1991.
74. Caley DW, Maxwell DS. Development of the blood vessels and extracellular spaces during postnatal maturation of rat cerebral cortex. J Comp Neurol 1970; 138: 31–47.
75. Nelson MD, Gonzales-Gomez I, Gilles FH. The search for human telencephalic ventriculofugal arteries. Am J Neuroradiol 1991; 12: 215–22.
76. Nies AT, Jeditshcky G, Konig J, et al. Expression and immunolocalisation of the multidrug resistance proteins, MRP1-MRP6 in human brain. Neuroscience 2004; 129: 349–60.
77. Zhang Y, Schuetz JD, Elmquist WF, et al. Plasma membrane localization of multidrug resistance-associated protein homologs in brain capillary endothelial cells. J Pharmacol Exp Ther 2004; 311: 449–55.
78. Cooray HC, Blackmore CG, Maskell L, et al. Localisation of breast cancer resistance protein in microvessel endothelium of human brain. Neuroreport 2002; 13: 2059–63.
79. Doyle LA, Yang W, Abruzzo LV, et al. A multidrug resistance transporter from human MCF-7 breast cancer cells. Proc Natl Acad Sci U S A 1998; 95: 15665–70.
80. Allen JD, Brinkhuis RF, Wijnholds J, et al. The mouse Bcrp1/Mxr/Abcp gene: amplification and overexpression in cell lines selected for resistance to topotecan, mitoxantrone, or doxorubicin. Cancer Res 1999; 59: 4237–41.
81. Kalabis GM, Petropoulos S, Gibb W, et al. Breast cancer resistance protein (Bcrp1/Abcg2) in mouse placenta and yolk sac: ontogeny and its regulation by progesterone. Placenta 2007; 28: 1073–81.
82. Chang XB. A molecular understanding of ATP-dependent solute transport by multidrug resistance-associated protein MRP1. Cancer Metastasis Rev 2007; 26: 15–37.
83. Imai Y, Asada S, Tsukahara S, et al. Breast cancer resistance protein exports sulfated estrogens but not free estrogens. Mol Pharmacol 2003; 64: 610–18.
84. Nakatomi K, Yoshikawa M, Oka M, et al. Transport of 7-ethyl-10-hydroxycamptothecin (SN-38) by breast cancer resistance protein ABCG2 in human lung cancer cells. Biochem Biophys Res Commun 2001; 288: 827–32.
85. Díaz D, Krejsa CM, Kavanagh TJ. Expression of glutamate-cysteine ligase during mouse development. Mol Reprod Dev 2002; 62: 83–91.
86. Ghersi-Egea JF, Strazielle N, Murat A, et al. Brain protection at the blood-cerebrospinal fluid interface involves a glutathione-dependent metabolic barrier mechanism. J Cereb Blood Flow Metab 2006; 26: 1165–75.
87. Carder PJ, Hume R, Fryer AA, et al. Glutathione S-transferase in human brain. Neuropathol Appl Neurobiol 1990; 16: 293–303.
88. Senjo M, Ishibashi T, Terashima T, et al. Successive appearance of glutathione S-transferase-positive cells in developing rat brain: choroid plexus, pia mater, ventricular zone and astrocytes. Neurosci Lett 1986; 66: 131–4.
89. Minn A, Ghersi-Egea JF, Perrin R, et al. Drug metabolizing enzymes in the brain and cerebral microvessels. Brain Res Brain Res Rev 1991; 16: 65–82.
90. Ghersi-Egea JF, Leininger-Muller B, Cecchelli R, et al. Blood-brain interfaces: relevance to cerebral drug metabolism. Toxicol Lett 1995; 82–83: 645–53.
91. Chou DK, Flores S, Jungalwala FB. Expression and regulation of UDP-glucuronate: neolactotetraosylceramide glucuronyltransferase in the nervous system. J Biol Chem 1991; 266: 17941–7.
92. Molsa MT, Heikkinen TJ, Hakkola JK, et al. Drug disposition. Functional role of P-glycoprotein in the human blood-placental barrier. Clin Pharmacol Ther 2005; 78: 123–31.
93. Sun M, Kingdom J, Baczyk D, et al. Expression of the multidrug resistance P-glycoprotein (ABCB1 glycoprotein) in the human placenta decreases with advancing gestation. Placenta 2006; 27: 602–9.
94. Ito K, Olsen SL, Qiu W, et al. Mutation of a single conserved tryptophan in multidrug resistance protein 1 (MRP1/ABCC1) results in loss of drug resistance and selective loss of organic anion transport. J Biol Chem 2001; 276: 15616–24.
95. Smit JW, Huisman MT, van Tellingen O, et al. Absence or pharmacological blocking of placental P-glycoprotein profoundly increases fetal drug exposure. J Clin Invest 1999; 104: 1441–7.
96. Schinkel AH, Mayer U, Wagenaar E, et al. Normal viability and alteed pharmacokinetics in mice lacking mdr1-type (drug transporting) P-glycoproteins. Proc Natl Acad Sci U S A 1997; 94: 4028–33.
97. Lankas GR, Wise LD, Cartwright ME, et al. Placental P-glycoprotein deficiency enhances susceptibility to chemically induced birth defects in mice. Reprod Toxicol 1998; 12: 457–63.

98. Atkinson DE, Brice-Bennett S, D'Souza SW. Antiepileptic medication during pregnancy: does fetal genotype affect outcome? Pediatr Res 2007; 62: 120–7.
99. Åström KE. On the early development of the isocortex in fetal sheep. Prog Brain Res 1967; 26: 1–59.
100. Jacobson M. Developmental Neurobiology. New York: Plenum, 1991.
101. Volpe JJ. Neurology of the Newborn. Philadelphia: Saunders, 2008.
102. Shankle WR, Landing BH, Rafii MS, et al. Evidence for postnatal doubling of neuron number in the developing human cerebral cortex between 15 months and 6 years. J Theor Biol 1998; 191: 115–40.
103. Saunders NR. Ontogeny of the blood-brain barrier. Exp Eye Res 1977; 25(Suppl): 523–50.
104. Adinolfi M, Haddad S. Levels of plasma proteins in human and rat fetal CSF and development of the blood-CSF barrier. Neuropadiatrie 1977; 8: 345–53.
105. Dziegielewska KM, Evans CAN, Lai PCW, et al. Proteins in cerebrospinal fluid and plasma of fetal rats during development. Dev Biol 1981; 83: 192–200.
106. Dziegielewska KM, Evans CAN, Fossan G, et al. Proteins in cerebrospinal fluid and plasma of fetal sheep during development. J Physiol 1980; 300: 441–55.
107. Levin VA. Relationship of octanol/water partition coefficient and molecular weight to rat brain capillary permeability. J Med Chem 1980; 23: 682–4.
108. Watanabe T, Matsuhashi K, Takayama S. Placental and blood-brain barrier transfer following prenatal and postnatal exposures to neuroactive drugs: relationship with partition coefficient and behavioural teratogenesis. Toxicol Appl Pharmacol 1990; 105: 66–77.
109. Bachmakov I, Wernet U, Endress B, et al. Charactrization of beta-adrenoceptor antagonists as substrates and inhibitors of the drug transporter P-glycoprotein. Fundam Clin Pharmacol 2006; 20: 273–82.
110. Wang JS, Zhu HJ, Markowitz JS, et al. Evaluation of antipsychotic drugs as inhibitors of multidrug resistance transporter P-glycoprotein. Psychopharmacology (Berl) 2006; 187: 415–23.
111. Croop JM, Raymond M, Haber D, et al. The three mouse multidrug resistance (mdr) genes are expressed in a tissue-specific manner in normal mouse tissues. Mol Cell Biol 1989; 9: 1346–50.
112. Cordon-Cardo C, O'Brien JP, Casals D, et al. Multidrug resistance gene (P-glycoprotein) is expressed by endothelial cells at blood-brain barrier sites. Proc Natl Acad Sci U S A 1989; 86: 695–8.
113. Thiebault F, Tsuruo T, Hamada H, et al. Immunohistochemical localisation in normal tissues of different epitopes of in the multidrug transport protein P170: evidence for localization in brain capillaries and cross reactivity of one antibody with a muscle protein. J Histochem Cytochem 1989; 37: 159–64.
114. Kusuhara H, Sugiyama Y. Efflux transport systems for drugs at the blood-brain barrier and blood-cerebrospinal fluid barrier (Part 1). Drug Discov Today 2001; 6: 150–6.
115. Terasaki T, Ohtsuki S. Brain-to-blood transporters for endogenous substrates and xenobiotics at the blood-brain barrier: an overview of biology and methodology. NeuroRx 2005; 2: 63–72.
116. Williams JHG, Ross R. Consequences of prenatal toxin exposure for mental health in children and adolescents. Eur Child Adolesc Psychiatry 2007; 16: 243–53.
117. Briggs GG, Freeman RK, Yaffe SJ. Drugs in Pregancy and Lactation. Philadelphia: Walters Kluwer, 2008.
118. Ornoy A. Neuroteratogens in man: an overview with special emphasis on the teratogenicity of antiepileptic drugs in pregnancy. Reprod Toxicol 2006; 22: 214–26.
119. Vajda Z, Pedersen M, Doczi T, et al. Studies of MDX mice. Neuroscience 2004; 129: 993–8.
120. Morrow J, Russell A, Guthrie E, et al. Malformation risks of antiepileptic drugs in pregnancy: a prospective study from the UK Epilepsy and Pregnancy register. J Neurol Neurosurg Psychiatry 2006; 77: 193–8.
121. Oguni M, Osawa M. Epilepsy and pregnancy. Epilepsia 2004; 45(Suppl 8): 37–41.
122. Cornford EM, Pardridge WM, Braun LD, et al. Increased blood-brain barrier transport of protein bound anticonvulsant drugs in the newborn. J Cereb Blood Flow Metab 1983; 3: 280–6.
123. Manent J-B, Jorquera I, Mazzucchelli A, et al. Fetal exposure to GABA-acting antiepileptic drugs generates hippocampal and cortical dysplasias. Epilepsia 2007; 48: 684–93.
124. Lajoie J, Moshé SL. Effects of seizures and their treatment on fetal brain. Epilepsia 2004; 45(Suppl 8): 48–52.
125. Kupferberg HJ, Way EL. Pharmacologic basis for the increased sensitivity of the newborn rat to morphine. J Pharmacol Exp Ther 1963; 141: 105–12.
126. DeVane CL, Simpkins JW, Boulton DW, et al. Disposition of morphine in tissues of the pregnant rat and foetus following single and continuous intraperitoneal administration to the mother. J Pharm Pharmacol 1999; 51: 1283–7.
127. Stewart PA, Hayakawa EM. Interendothelial junctional changes underlie the developmental "tightening" of the blood-brain barrier. Brain Res 1987; 32: 271–81.

128. Olkkola KT, Maunuksela EL, Korpela R, et al. Kinetics and dynamics of postoperative intravenous morphine in children. Clin Pharmacol Ther 1988; 44: 128–36.
129. Ward RM. Drug disposition in the late preterm ("near-term") newborn. Semin Perinatol 2006; 30: 48–51.
130. Barg J, Simantov R. Transient expression of opioid receptors in defined regions of developing brain: are embryonic receptors selective? J Neurochem 1991; 57: 1978–84.
131. Lynn AM, McRorie TI, Slattery JT, et al. Age-dependent morphine partitioning between plasma and cerebrospinal fluid in monkeys. Dev Pharmacol Ther 1991; 17: 200–4.
132. Letrent SP, Polli JW, Humphreys JE, et al. P-glycoproteinmediated transport of morphine in brain capillary endothelial cells. Biochem Pharmacol 1999; 58: 951–7.
133. Xie R, Hammarlund-Udenaes M, de Boer AG, et al. The role of P-glycoprotein in blood-brain-barrier transport of morphine: transcortical microdialysis studies in mdr1a (−/−) and (+/+) mice. Br J Pharmacol 1999; 128: 563–8.
134. Thompson SJ, Koszdin K, Bernards CM. Opiate-induced analgesia is increased and prolonged in mice lacking P-glycoprotein. Anesthesiology 2000; 92: 1392–9.
135. Arnaud MJ, Bracco I, Welsch C. Metabolism and distribution of labelled theophylline in the pregnant rat. Impairment of theophylline metabolism by pregnancy and absence of a blood-brain barrier in the fetus. Pediatr Res 1982; 16: 167–71.
136. Wilkinson JM, Pollard I. Accumulation of theophylline, theobromine and paraxanthine in the fetal brain following a single dose of caffeine. Brain Res Dev Brain Res 1993; 75: 193–99.
137. Eteng MU, Eyong EO, Akpanyung EO, et al. Recent advances in caffeine and theobromine toxicities: a review. Plant Foods Hum Nutr 1997; 51: 231–43.
138. Habgood MD, Knott GW, Dziegielewska, et al. Permeability of the developing and mature blood-brain barriers to theophylline in rats. Clin Exp Pharmacol Physiol 1998; 25: 361–8.
139. Tsai TH, Liu MC. Determination of unbound theophylline in rat blood and brain by microdialysis and liquid chromatography. J Chromatogr A 2004; 1032: 97–101.
140. Dziegielewska KM, Habgood MD, Mollgard K, et al. Species-specific transfer of plasma albumin from blood into different cerebrospinal fluid compartments in the fetal sheep. J Physiol 1991; 439: 215–37.
141. Habgood MD, Sedgwick JEC, Dziegielewska KM, et al. A developmentally regulated blood-cerebrospinal fluid transfer mechanism for albumin in immature rats. J Physiol 1992; 456: 181–92.
142. Avery GB, Fletcher MA, MacDonald MG. Neonatology. Philadelphia: Lippincott Williams and Wilkins, 1998.
143. Stonestreet BS, Sadowska GB, McKnight AJ, et al. Exogenous and endogenous corticosteroids modulate blood-brain barrier development in the ovine fetus. Am J Physiol Regul Integr Comp Physiol 2000; 279: R468–77.
144. Dunlop SA, Archer MA, Quinlivan JA, et al. Repeated prenatal corticosteroids delay myelination in the ovine central nervous system. J Matern Fetal Med 1997; 6: 309–13.
145. Quinlivan JA, Archer MA, Evans SF, et al. Fetal sciatic nerve growth is delayed following repeated maternal injections of corticosteroids in sheep. J Perinat Med 2000; 28: 26–33.
146. Aschner M, Clarkson TW. Distribution of mercury 203 in pregnant rats and their fetuses following systemic infusions with thiol-containing amino acids and glutothione during late gestation. Teratology 1988; 38: 145–55.
147. Garcia SJ, Gellein K, Syversen T, et al. A manganese-enhanced diet alters brain metals and transporters in the developing rat. Toxicol Sci 2006; 92: 516–25.
148. Weiss B. Methylmercury: a model neurotoxicant and risk assessment. In: Bellinger DC, ed. Human Developmental Toxicology. New York: Taylor & Francis, 2006: 1–23.
149. Myers GJ, Davidson PW, Strain JJ. Nutrient and methyl mercury exposure from consuming fish. J Nutr 2007; 137: 2805–8.
150. Chang LW, Reuhl KR, Lee GW. Degenerative changes in the developing nervous system as a result of in utero exposure to methylmercury. Environ Res 1977; 14: 414–23.
151. Geelen JAG, Dormans JAMA, Vehoef A. The early effects of methylmercury on the developing rat brain. Acta Neuropathol 1990; 80: 423–38.
152. Myers GJ, Davidson PW. Prenatal methylmercury exposure and childen: neurologic, developmental and behavioural research. Environ Health Perspect 1998; 106(Suppl 3): 841–7.
153. Eisler R. Mercury hazards to fish, wildlife and invertebrates: a synoptic review. In: US Fish and Wildlife Service Biological Report 85(1.10), Contaminant Hazard Reviews, Report No. 10. Laurel, MD: Patuxent Wildlife Research Center, 1987.
154. Goldman LR, Shannon MW; American Academy of Pediatrics: Committee on Environmental Health. Technical report: mercury in the environment: implications for pediatricians. Pediatrics 2001; 108: 197–205.

155. Aschner M, Aschner JL, Kimelbeg HK. Methylmercury neurotoxicityand its uptake across the blood-brain barrier. In: Isaacson RL, Jensen KF, eds. The Vulnerable Brain and Environmental risks: Toxins and Foods, Vol. 2. New York: Plenum, 1992: 3–17.
156. Cornford EM, Cornford ME. Nutrient transport and the blood-brain barrier in developing animals. Fed Proc 1986; 45: 2065–72.
157. Inoueye M, Kajiwara Y, Hirayama K. Dose-and sex-dependent alterations in mercury distribution in fetal mice following methylmercury exposure. J Toxicol Environ Health 1986; 19: 425–35.
158. Aschner M, Syversen T. Methylmercury. Recent advances in understanding its neurotoxicity. Ther Drug Monit 2005; 27: 278–83.
159. Steinwall O, Olsson Y. Impairment of the blood-brain barrier in mercury poisoning. Acta Neurol Scand 1969; 45: 351–61.
160. Chang LW, Hartmann HA. Blood-brain barrier dysfunction in experimental mercury intoxication. Acta Neuropathol (Berl) 1972; 21: 179–84.
161. Kosnett MJ. Lead. In: Ford MD, Delaney KA, Ling LJ, et al., eds. Clinical Toxicology. Philadelphia: Saunders, 2001: 723–36.
162. Pentschew A, Garro F. Lead encephalo-myopathy of the suckling rat and its implications on the porphyrinopathic nervous diseases. Acta Neuropathol 1966; 6: 266–78.
163. Krigman MR, Hogan EL. Effect of lead intoxication on the postnatal growth of the rat nervous system. Environ Health Perspect 1974; 7: 187–99.
164. Michaelson A, Sauerhoff MW. Animal models of human disease: severe and mild lead encephalopathy in the neonatal rat. Environ Health Perspect 1974; 7: 201–25.
165. O'Tuama LA, Kim CS, Gatzy JT, et al. The distribution of inorganic lead in guinea pig brain and neural barrier tissues in control and lead-poisoned animals. Toxicol Appl Pharmacol 1976; 36: 1–9.
166. Clasen RA, Hartman JF, Coogan PS, et al. Experimental acute lead encephalopathy in the juvenile Rhesus monkey. Environ Health Perspect 1974; 7: 175–85.
167. Sundström R, Muntzing K, Kalimo H, et al. Changes in the integrity of the blood-brain barrier in suckling rats with low dose lead encephalopathy. Acta Neuropathol (Berl) 1985; 68: 1–9.
168. Clasen RA, Hartman JF, Starr AJ, et al. Electron microscopic and chemical studies of the vascular changes and edema of lead encephalopathy. Am•J Pathol 1974; 74: 215–40.
169. Hertz MM, Bolweg TG, Grandjean P, et al. Lead poisoning and the blood-brain barrier. Acta Neurol Scand 1981; 63: 286–96.
170. Moorehouse SR, Carden S, Drewitt PN, et al. The effect of chronic low level lead exposure on blood-brain barrier function in the developing rat. Biochem Pharmacol 1988; 37: 4539–47.
171. Holtzmann D, Herman MM, Hsu JS, et al. The pathogenesis of lead encephalopathy. Virchows Arch A Path Anat Histol 1980; 387: 147–64.
172. Bressler JP, Goldstein GW. Mechanisms of lead toxicity. Biochem Pharmacol 1991; 41: 479–84.
173. Goldstein GW, Astbury AK, Diamond I. Pathogenesis of lead encephalopathy. Arch Neurol 1974; 31: 382–9.
174. Press MF. Lead encephalopthy in neonatal Long-Evans rats: morphological studies. J Neuropathol Exp Neurol 1977; 36: 169–95.
175. Lefauconnier J-M, Hauw JJ, Bernard G. Regressive or lethal lead encephalopathy in the suckling rat. J Neuropathol Exp Neurol 1983; 42: 177–90.
176. Bradbury MWD, Deane R. Rate of uptake of lead-203 into brain and other soft tissues of the rat at constant radiotracer levels in plasma. Ann N Y Acad Sci 1986; 481: 142–60.
177. Bradbury MWB, Deane R. Permeability of the blood-brain barrier to lead. Neurotoxicology 1993; 14: 131–6.
178. Davson H, Segal MB. Physiology of the CSF and Blood-Brain Barriers. Boca Raton: CRC, 1996.
179. Thomas JA, Dallenbach FD, Thomas M. The distribution of radioactive lead (210Pb) in the cerebellum of developing rats. J Pathol 1973; 109: 45–50.
180. Toews AD, Kolber A, Hayward J, et al. Experimental lead encephalopathy in the suckling rat: concentration of lead in cellular fractions enriched in brain capillaries. Brain Res 1978; 147: 131–8.
181. Deane R, Bradbury MWB. Transport of lead-203 at the blood-brain barrier during short cerebrovascular perfusion with saline in the rat. J Neurochem 1990; 54: 905–14.
182. Simons TJB, Al-Modhefer A, Bradbury MWB. The lead binding characteristics of human serum. Neurotoxicology 1992; 13: 489.
183. Lefauconnier J-M, Lavielle E, Terrien G, et al. Effect of lead doses on some cerebral capillary functions in the suckling rat. Toxicol Appl Pharmacol 1980; 55: 467–76.
184. Michaelson A, Bradbury MWB. Effect of early inorganic lead exposure on rat blood-brain barrier permeability to tyrosine or choline. Biochem Pharmacol 1982; 31: 1881–5.
185. Bonithon-Kopp C, Huel G, Moreau T, et al. Prenatal exposure to lead and cadmium and psychomotor development of the child at 6 years. Neurobehav Toxicol Teratol 1986; 8: 307–10.

186. Lewis M, Worobey J, Ramsey DS, et al. Prenatal exposure to heavy metals: effect on childhood cognitive skills and health status. Pediatrics 1992; 89: 1010–15.
187. Fréry N, Nessmann C, Girard F, et al. Environmental exposure to cadmium and human birthweight. Toxicology 1993; 79: 109–18.
188. Schoeters G, Den Hond E, Zuurbier M, et al. Cadmium and children: exposure and health effects. Acta Paediatr Suppl 2006; 95: 50–4.
189. Rohrer SR, Shaw SM, Lamar CH. Cadmium induced endothelial cell alterations in fetal brain from prenatal exposure. Acta Neuropathol (Berl) 1978; 44: 147–9.
190. Webster WS, Valois AA. The toxic effects of cadmium on the neonatal mouse CNS. J Neuropathol Exp Neurol 1981; 40: 247–57.
191. Nolan CV, Shaikh ZA. The vascular endothelium as a target tissue in acute cadmium toxicity. Life Sci 1986; 39: 1403–9.
192. Gabbiani G, Baddonnel M-C, Mathewson S, et al. Acute cadmium intoxication. Early selective lesions of endothelial clefts. Lab Invest 1974; 30: 686–95.
193. Shukla A, Shukla GS, Srimal RC. Cadmium-induced alterations in blood-brain barrier permeability and its possible correlation with decreased microvesel antioxidant potential in rat. Hum Exp Toxicol 1996; 15: 400–5.
194. Wong K-L, Klaassen CD. Neurotoxic effects of cadmium in young rats. Toxicol Appl Pharmacol 1982; 63: 330–7.
195. Crowe A, Morgan EH. Effect of dietary cadmium on iron metabolism in growing rats. Toxicol Appl Pharmacol 1997; 145: 136–46.
196. Klassen CD, Wong KL. Cadmium toxicity in the newborn rat. Can J Physiol Pharmacol 1982; 60: 1027–36.
197. Provias JP, Ackerley CA, Smith C, et al. Cadmium encephalopathy: a report with elemental analysis and pathological findings. Acta Neuropathol 1994; 88: 583–6.
198. Takeda A, Takefuta S, Ijiro H, et al. 109Cd transport in rat brain. Brain Res Bull 1999; 49: 453–7.
199. Valois AA, Webster WS. The choroid plexus and cerebral vasculature as target sites for cadmium following acute exposure in neonatal and adult mice: an autoradiographic and gamma counting study. Toxicology 1987; 46: 43–55.
200. Tenenbein M. Iron. In: Ford MD, Delaney KA, Ling Lj, et al., eds. Clinical Toxicology. Philadelphia: Saunders, 2001: 305–9.
201. Wright RO, Baccarelli A. Metals and neurotoxicology. J Nutr 2007; 137: 2800–13.
202. Taylor EM, Morgan EH. Developmental changes in transferrin and iron uptake by the brain in the rat. Brain Res Dev Brain Res 1990; 55: 35–42.
203. Morgan EH, Moos T. Mechanism and developmental changes in iron transport across the blood-brain barrier. Dev Neurosci 2002; 24: 106–13.
204. Jeffries WA, Brandon MR, Hunt SV, et al. Transferrin on endothelium of brain capillaries. Nature 1984; 132: 162–3.
205. Siddappa AJ, Rao RB, Wobken JD, et al. Developmental changes in the expression of iron regulatory proteins and iron transport proteins in the perinatal brain. J Neurosci Res 2002; 68: 761–75.
206. Jiang DH, Cheng YZ, Ho KP, et al. Distribution of ferroportin-1 in different regions of developing rat brain. Dev Neurosci 2002; 24: 94–8.
207. Crowe A, Morgan EH. Interactions between tissue uptake of lead and iron in normal and iron-deficient rats during development. Biol Trace Elem Res 1996a; 52: 249–61.
208. Crowe A, Morgan EH. Iron and copper interact during their uptake and deposition in the brain and other organs of developing rats exposed to dietary excess of the two metals. J Nutr 1996b; 126: 183–94.
209. Aschner M, Guilarte TR, Schneider JS, et al. Manganese: recent advances in understanding its transport and neurotoxicity. Toxicol Appl Pharmacol 2007; 221: 131–47.
210. Murphy VA, Wadhwani KC, Smith QR, et al. Saturable transport of manganese(II) across the rat blood-brain barrier. J Neurochem 2001; 57: 948–54.
211. Moos T, Skjoerrinje T, Gosk S, et al. Brain capillary endothelial cells mediate iron transport into the brain by segregating iron from transferrin without the involvement of divalent metal transporter 1. J Neurochem 2006; 98: 1946–58.
212. Miller ST, Cotzias GC, Evert HA. Control of tissue manganese: initial absence and sudden emergence of excretion in the neonatal mouse. Am J Physiol 1975; 229: 1080–4.
213. Dorman DC, McElveen AM, Marshall MW, et al. Maternal-fetal distribution of manganese in the rat following inhalation exposure to manganese sulfate. Neurotoxicology 2005; 26: 625–32.
214. Fechter LD. Distribution of manganese in development. Neurotoxicology 1999; 20: 197–202.
215. Chandra SV, Murthy RC, Saxena DK, et al. Effects of pre- and postnatal combined exposure to Pb and Mn on brain development in rats. Ind Health 1983; 21: 273–9.

216. Moos T, Morgan EH. A morphological study of the developmentally regulated transport of iron into the brain. Dev Neurosci 2002; 24: 99–105.
217. Golub MS, Han B, Keen CL, et al. Developmental patterns of aluminum in mouse brain and effects of dietary aluminum excess on manganese deficiency. Toxicology 1993; 81: 33–47.
218. DeSesso JM, Jacobson CF, Scialli A, et al. An assessment of the developmental toxicity of inorganic arsenic. Reprod Toxicol 1998; 12: 385–433.
219. Tsuji JS, Benson R, Schoof RA, et al. Health effect levels for risk assessment of childhood exposure to arsenic. Regul Toxicol Pharmacol 2004; 39: 99–110.
220. Jin Y, Xi S, Li X, et al. Arsenic speciation transported through the placenta from mother mice to their newborn pups. Environ Res 2006; 101: 349–55.
221. Naramandura H, Suzuki KT. Identification of the major arsenic-binding protein in rat plasma as the ternary dimethylarsinous-hemoglobin-haptoglobin complex. Chem Res Toxicol 2008; 21: 678–85.
222. Leslie EM, Haimura A, Waalkes MP. Arsenic transport by the human multidrug resistance protein 1 (MRP1/ABCC1). J Biol Chem 2004; 279: 32700–8.
223. Johnson EM, Christian MS, Hoberman AM, et al. Developmental toxicology investigation of tellurium. Fundam Appl Toxicol 1988; 11: 691–702.
224. Agnew WF, Cheng JT. Protein binding of tellurium-127m by maternal and fetal tissues of the rat. Toxicol Appl Pharmacol 1971; 20: 346–56.
225. Aiken SP, Horn NM, Saunders NR. Effects of histidine on tissue zinc distribution in rats. Biol Met 1992; 5: 235–43.
226. Olney JW, Ho OL, Rhee V. Cytotoxic effects of acidic and sulphur containing amino acids on the infant mouse central nervous system. Exp Brain Res 1971; 14: 61–76.
227. Toth L, Karcsu S, Feledi J, et al. Neurotoxicity of monosodium-L-glutamate in pregnant and fetal rats. Acta Neuropathol (Berl) 1987; 75: 16–22.
228. Lemkey-Johnston N, Reynolds WA. Nature and extent of brain lesions in mice related to ingestion of monosodium glutamate. J Neuropathol Exp Neurol 1974; 33: 74–97.
229. Seress L, Lázár G, Kosaras B, et al. Regional effect of monosodium-L-glutamate on the superficial layers of superior colliculus in rat. Cell Tissue Res 1984; 235: 453–7.
230. Ikeda GJ, Miller E, Sapienza PP, et al. Maternal-foetal distribution studies in late pregnancy. II. Distribution of [1-14C]acrylamide in tissues of beagle dogs and miniature pigs. Food Chem Toxicol 1985; 23: 757–61.
231. Barr DB, Bradman A, Freeman N, et al. Studying the relation between pesticide exposure and human development. In: Bellinger DC, ed. Human Developmental Neurotoxicology. New York: Taylor & Francis, 2006: 252–85.
232. Bain LJ, LeBlanc GA. Interaction of structurally diverse pesticides with the human MDR1 gene Product P-glycoprotein. Toxicol Appl Pharmacol 1996; 141: 288–98.
233. Shabbir A, DiStasio S, Zhao J, et al. Differential effects of the organochlorine pesticide DDT and it smetabolite p,p′-DDE on p-glycoprotein activity and expression. Toxicol Appl Pharmacol 2005; 203: 91–8.
234. Tilbury KL, Adams NG, Krone CA, et al. Organochlorines in stranded pilot whales (Globicephala melaena) from the coast of Massachusetts. Arch Environ Contam Toxicol 1999; 37: 125–34.
235. Nishio H, Giacobini E, Oriz J, et al. Accumulation and metabolism of pipecolic acid in the developing brain of the mouse. Brain Res 1983; 282: 107–12.
236. Kim JS, Giacobini E. Transport of pipecolic acid in adult and developing mouse brain. Neurochem Res 1985; 10: 1405–15.
237. Steinberg RJ, Dodt G, Raymond GV, et al. Peroxisome biogenesis disorders. Biochim Biophys Acta 2006; 1763: 1733–48.
238. Enokizono J, Kusuhara H, Sugiama Y. Effect of breast cancer resistance protein (Bcrp/Abcg2) on the disposition of phytoestrogens. Mol Pharmacol 2007; 72: 967–75.
239. Kim CS, Keizer RF, Pritchard JB. 2,4-Dichlorophenoxyacetic acid intoxication increases its accumulation within the brain. Brain Res 1988; 440: 216–26.
240. Miranda S, Vollrath V, Wielandt AM, et al. Overexpression of mdr2 gene by peroxisome proliferators in mouse liver. J Hepatol 1997; 26: 1331–9.
241. Kim CS, Binienda Z, Sandberg JA. Construction of a physiologically-based pharmacokinetic model for 2,4-dichlorophenoxyacetic acid dosimetry in the developing rabbit brain. Toxicol Appl Pharmacol 1996; 136: 250–9.
242. Sandberg JA, Duhart HA, Lipe G, Slikker W Jr, Kim CS. Distribution of 2,4- dichlorophenoxyacetic acid (2,4-D) in maternal and fetal rabbits. J Toxicol Environ Health 1995; 49: 497–509.
243. Widdowson PS, Farnworth MJ, Simpson MG, et al. Influence of age on the passage of paraquat through the blood-brain barrier in rats: A distribution and pathological examination. Hum Exp Toxicol 1996; 15: 231–6.

244. Dinis-Oliveira RJ, Remlao F, Durte JA, et al. P-glycoprotein induction: an antidotal pathway for paraquat-induced lung toxicity. Free Radic Biol Med 2006; 41: 1213–24.
245. Saunders NR, Knott GW, Dziegielewska KM. Barriers in the immature brain. Cell Mol Neurobiol 2000; 20: 29–40.
246. Saunders NR, Ek CJ, Habgood MD, et al. Barriers in the brain: a renaissance? Trends Neurosci 2008; 31: 279–86.
247. Baud O, Faytol L, Gressens P, et al. Perinatal and early postnatal changes in the expression of monocarboxylate transporters MCT1 and MCT2 in the rat forebrain. J Comp Neurol 2003; 465: 445–54.
248. Chen Z-S, Robey RW, Belinsky MG, et al. Transport of methotrexate, methotrexate polyglutamates, and 17β-Estradiol17-(β-D-glucuronide) by ABCG2: effects of acquired mutations at R482 on methotrexate transport. Cancer Res 2993; 63: 4048–54.

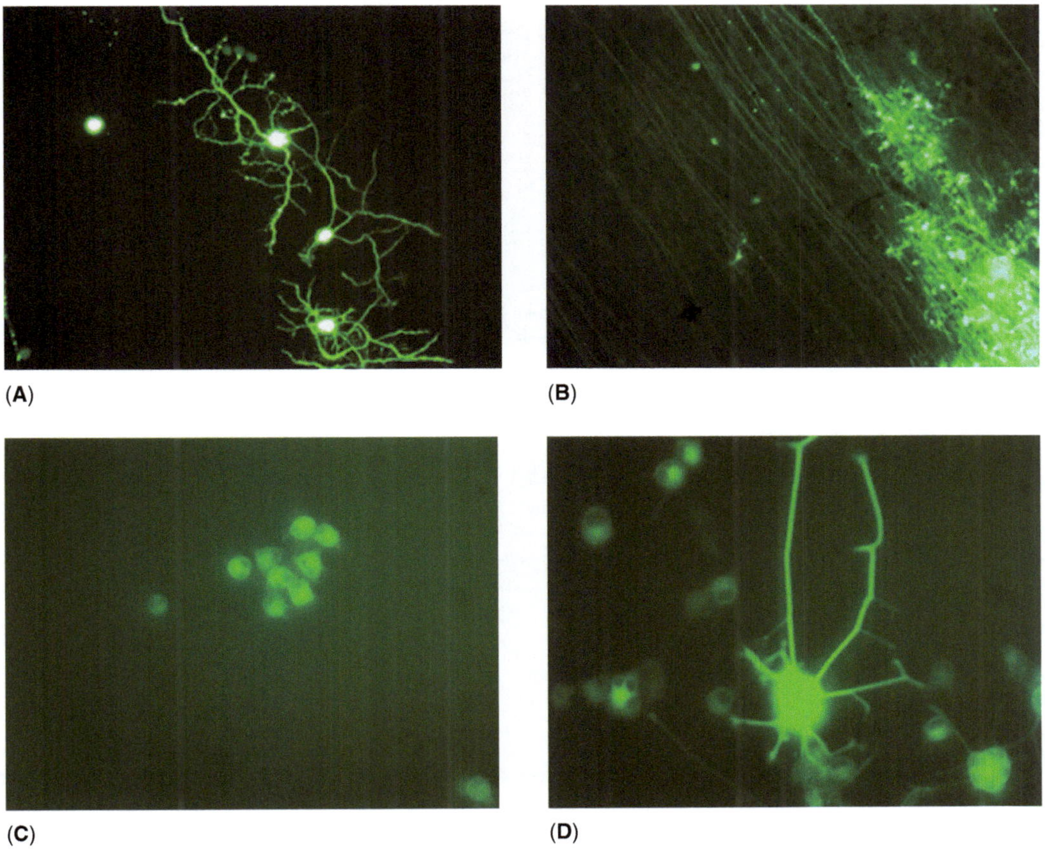

Figure 1.1 Lentiviral transduction of neuronal cells. Lentiviral vectors expressing green fluorescent protein (GFP) under the constitutive cytomegalovirus or elongation factor alpha promoters were used to transduce cells and explants. Panel (**A**) isolated rat retinal ganglion cells in culture, three days after transduction; (**B**) porcine retinal ganglion cells in explant one week after transduction; (**C**) PC12 cells stably transduced and expressing GFP over several passages; (**D**) neuronal differentiation of cells in panel C, exposed to nerve growth factor. *Source*: Panel B reproduced from Ref. 93. *See page 9.*

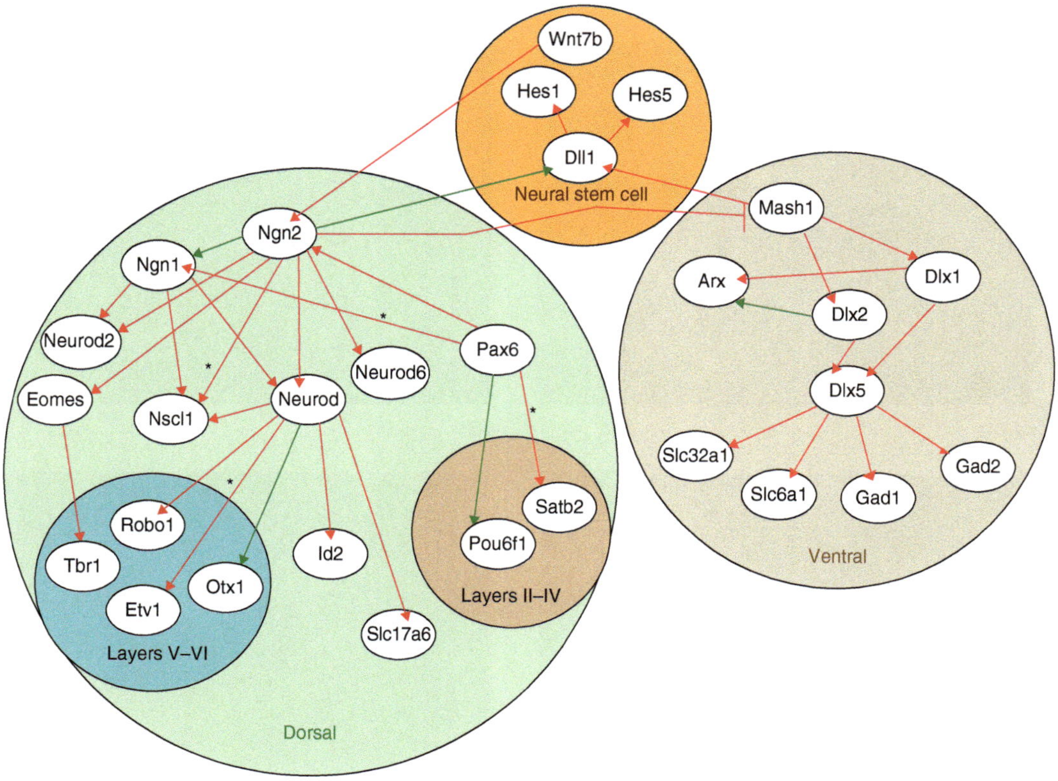

Figure 2.1 Literature-based gene regulatory network describing proneural bHLH regulation of telencephalon neurogenesis. Activations are identified with an arrow and repressions are identified with a barred line. Those connections that are non-significant based on the current microarray dataset are represented as dashed line. Significant relationships were determined through analysis of the distribution of the strength of linkage parameter (β) after 500,000 MCMC simulations. If >95% of the simulations have values above zero they are considered significant. *Connections which were significant, but as inhibition. *Abbreviations*: bHLH, basic-helix-loop-helix; MCMC, Markov Chain Monte Carlo. *Source*: From Ref. 106 (originally published by Biomed Central). *See page 37.*

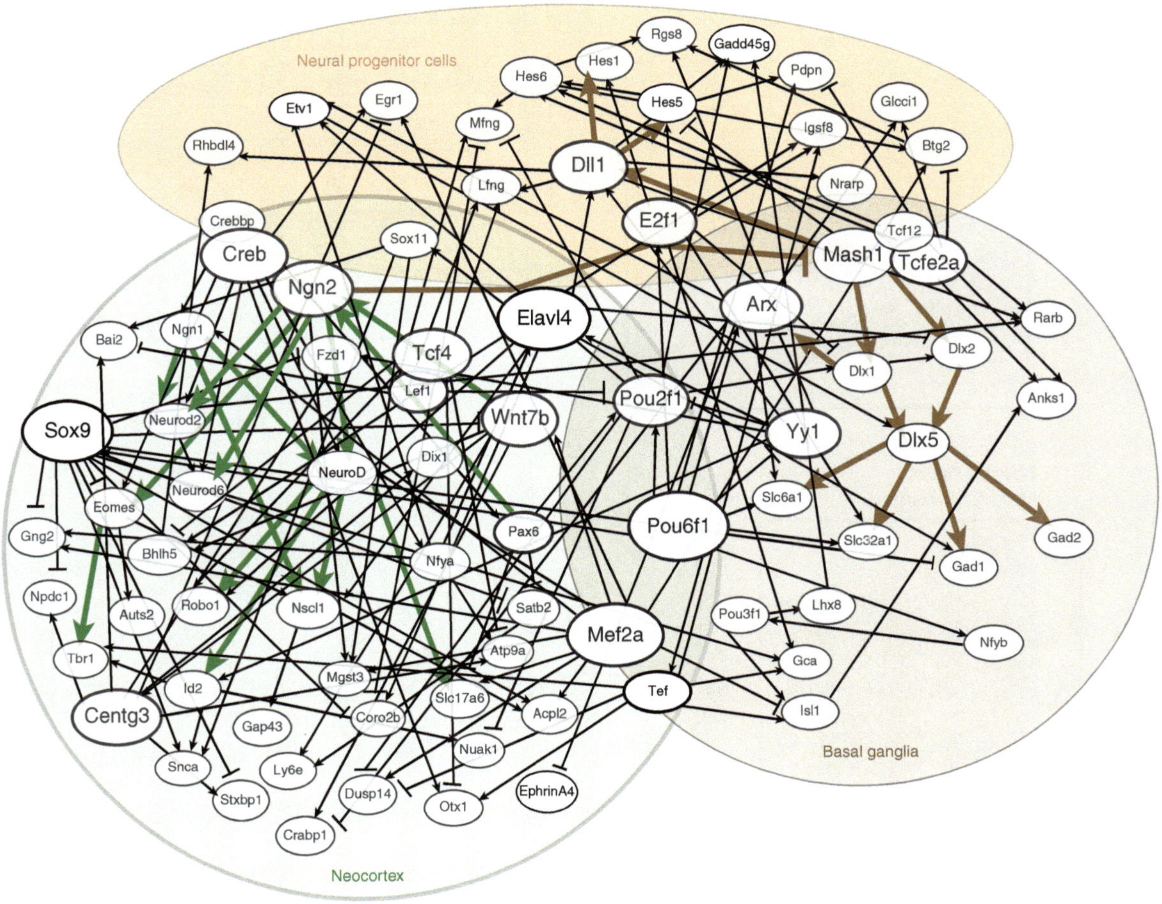

Figure 2.2 Algorithm-based gene regulatory network structure for dorsal and ventral telencephalon development. The Bayesian-based TAO-Gen algorithm (139) was implemented with an informative prior structure to predict the optimal network structure based on the LOF and GOF microarray datasets, evolutionarily conserved transcription factor binding site data, prior literature-based knowledge, and spatial and time specific expression patterns. To highlight the key regulators, the nodes representing genes predicted to be the parent of at least nine other genes are largest in size (Sox9, Mef2a, Elavl4, and Pou6f1), whereas those that are predicted to regulate at least five other genes are medium in size (Ngn2, Centg3, Tef, Tcf4, Wnt7b, Pou2f1, Yy1, Dll1, E2f1, Tcfe2a, Arx, and Creb). Arrows indicate activation and barred lines indicate inhibition. *Abbreviations*: GOF, gain of function; LOF, loss of function. *Source*: From Ref. 106 (originally published by Biomed Central). *See page 42.*

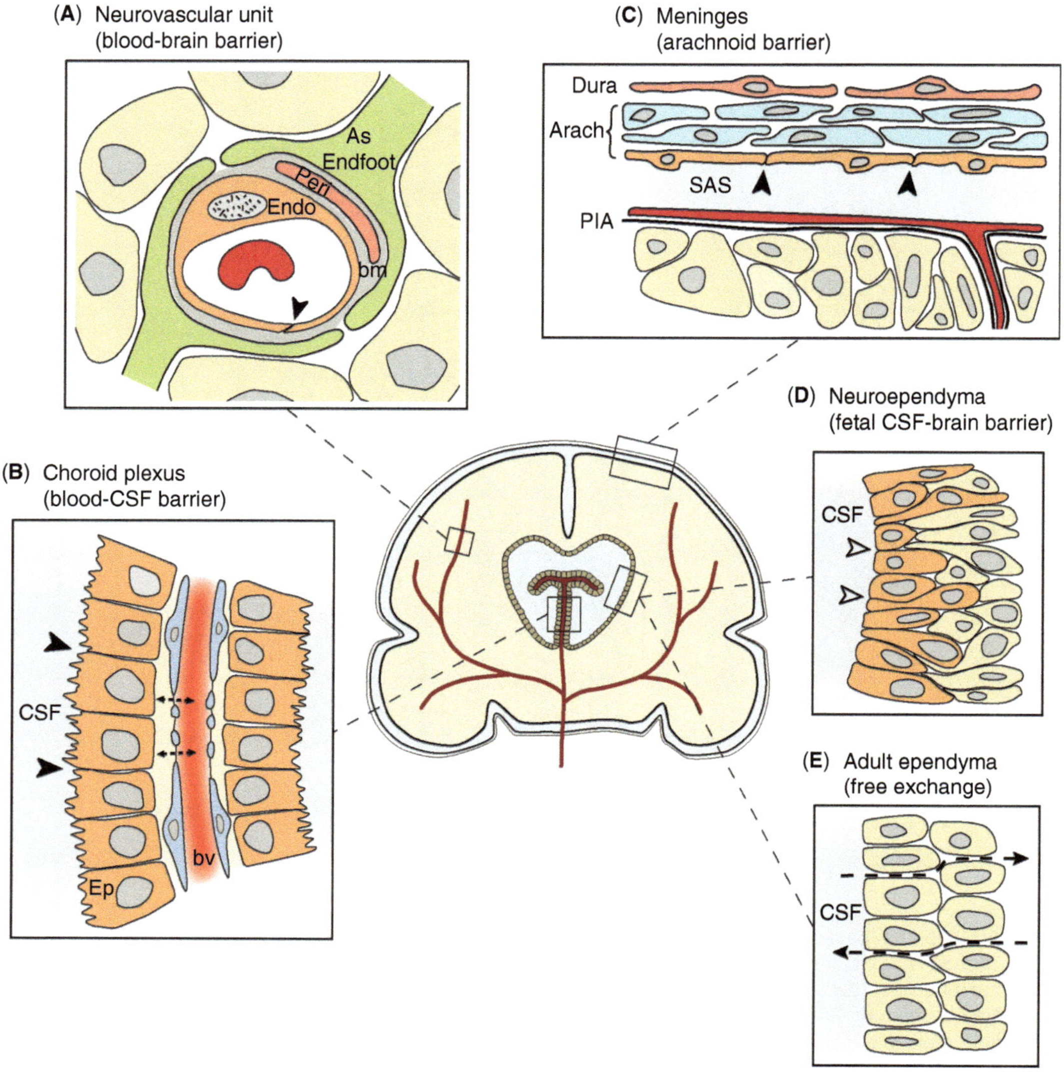

Figure 3.1 Schematics of the interfaces and barriers in the adult and developing brain. (**A**) The blood-brain barrier is a barrier between the lumen of cerebral blood vessels and brain parenchyma. The endothelial cells (Endo) have luminal tight junctions (arrowhead) forming the physical barrier of the interendothelial cleft. Outside the endothelial cell is a basement membrane (bm) which also surrounds the pericytes (Peri). Around all these structures are the astrocytic endfeet processes from nearby astrocytes (As Endfoot). All these structures together are often referred to as the neurovascular unit. (**B**) The blood-CSF barrier, a barrier between choroid plexus blood vessels (bv) and the CSF. The choroid plexus blood vessels are fenestrated and form a non-restrictive barrier (small arrows), however, the epithelial cells (Ep) have apical tight junctions (arrowheads) that restrict intercellular passage of molecules. (**C**) The meningeal barrier, is the least studied and structurally most complex of all the brain barriers. The blood vessels of the dura are fenestrated and provide little barrier function, however, the outer cells of the arachnoid membrane (Arach) have tight junctions (arrowheads) and this cell layer is believed to form the physical barrier between the CSF-filled subarachnoid space (SAS) and overlaying structures. The blood vessels in the arachnoid and on the pial surface (PIA) have tight junctions with similar barrier characteristics as cerebral blood vessels although lacking the surrounding pericytes and astrocytic end-feet. (**D**) The fetal CSF-brain barrier, a barrier between the CSF and brain parenchyma, and has only been shown to be a functional barrier in the early developing brain (245). In early development, the neuroependymal cells are connected to each other by strap-junctions (open arrowheads) that are believed to form the physical barrier restricting the passage of larger molecules such as proteins but not smaller molecules such as sucrose. (**E**) The adult ventricular ependyma. During development, the neuroependymal cells flatten and lose their strap-junctions. The mature ependyma does not restrict the exchange of large molecules at (e.g., proteins) between CSF and brain. *Source*: From Ref. 246. *See page 52.*

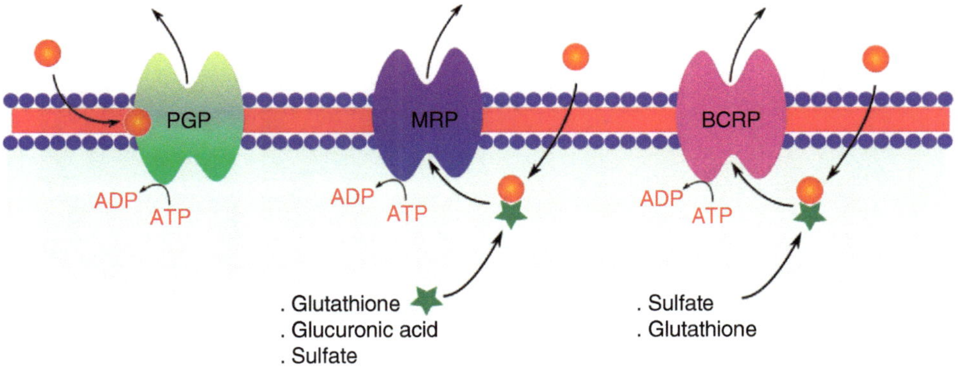

Figure 3.4 The three main ABC transporter types involved in xenobiotic export at the blood-brain, blood-spinal cord, and blood-placental barriers; PGP (p-glycoprotein, also known as MDR1, ABCB1), MRP (multidrug resistance related proteins, ABCC1-8) and BCRP (breast cancer related protein, ABCG2). PGP and the MRPs are arranged in two repeated halves, each half containing a nucleotide binding domain and six membrane spanning domains. BCRP is a half-transporter with a single nucleotide binding domain and six membrane spanning domains, but is thought to form a homodimer in order to be functionally active. All utilize energy from ATP hydrolysis to move substrates across the membrane. PGP's substrate binding sites are located within or close to the internal leaflet, thus PGP is thought to be able to intercept lipophilic compounds as they pass across the cell membrane. In contrast, the MRPs transport compounds from within the cell cytoplasm in conjunction with a transport moiety (glutathione, glucuronic acid, or sulfate). This is achieved either by co-transport or by conjugation of the transport moiety (shown as a star in the diagram) onto the substrate prior to export. The MRPs do not appear to be able to directly intercept compounds within the internal leaflet in the way that PGP does. BCRP also transports a wide range of unconjugated compounds, but it is not known if it is able to intercept these from within the internal leaflet. BCRP is also able to transport conjugated compounds, with a greater affinity for sulfate conjugates over glutathione conjugates (248), thus it may function in a similar manner to the MRPs. *Abbreviations*: ADP, adenosine diphosphate; ATP, adenosine triphosphate. *See page 57.*

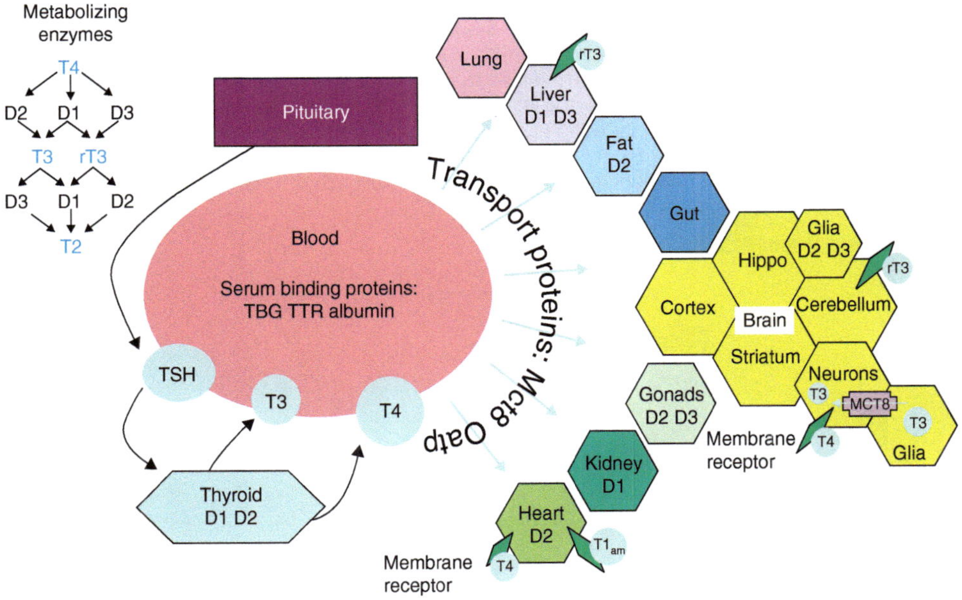

Figure 4.3 Recent information has required significant refinement of the original model of the HPT axis presented in Figure 1. A number of factors contribute to the differential regulation of thyroid hormone supply and action in different target tissues. In addition to the classic molecular action of thyroid hormones on gene expression, nongenomic effects mediated by membrane bound receptors for T3, T4, and T1 amines can mediate thyroid hormone action. Active transport proteins (e.g., MCT8, OATP) and metabolizing enzymes are differentially expressed and regulate the presentation of hormone to different target tissues. The impact of environmental contaminants on these different regulatory mechanisms, especially as they pertain to brain development has not been addressed. *Abbreviations*: HPT, hypothalamus-pituitary-thyroid; MCT8, monocarboxylate transporter-8; OATP, organic anion transporting polypeptide; TBG, thyroxine-binding globulin; TSH, thyroid-stimulating hormone; TTR, transthyretin. *See page 84.*

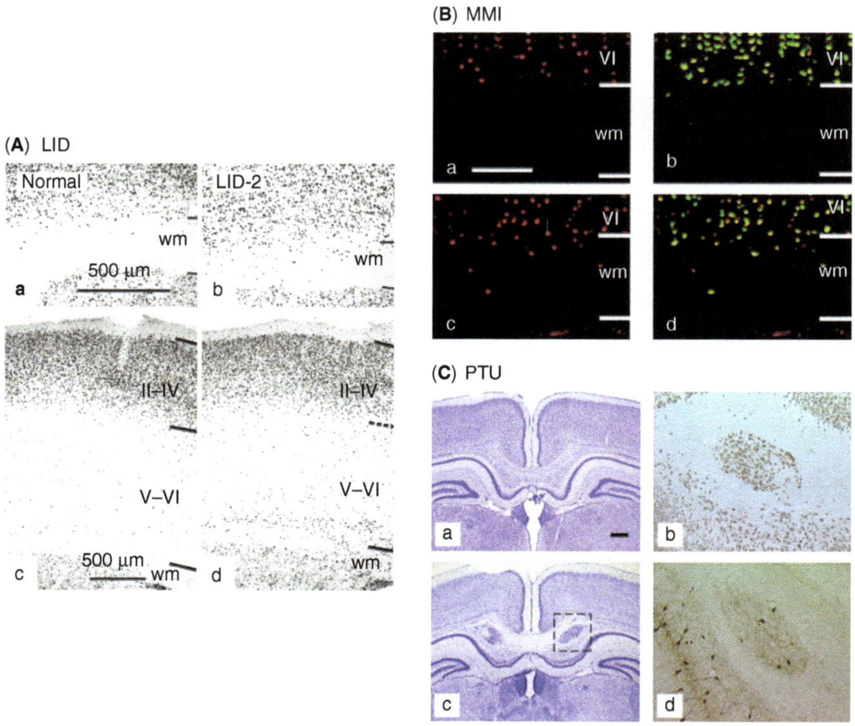

Figure 4.8 Cortical histogenesis and neuronal migration are disrupted with low level thyroid hormone (TH) disruption. (**A**) In the cortex of offspring born to iodine deficient rat dams (LID-2), cells are evident in the white matter (wm) of the corpus callosum (top) and the distinction between layers within the cortex are blurred (bottom). *Source*: From Ref. 118. (**B**) TH reductions induced over three days during gestation with methimazole (MMI) also impaired neuronal migration. Left is bromodeoxyruracil (BrdU)-positive cells, right is double labelling of BrdU and NeuN, indicating that cortical neurons persist in the white matter (wm) at PN40. *Source*: From Ref. 114. (**C**) Moderate degrees of TH insufficiency limited to the prenatal period induced by propylthiouracil (PTU) produced an abnormal cluster of cells, a heterotopia, in the corpus callosum (top left is control, bottom left is treated). Cells within the heterotopia were neurons as they stained positively with NeuN (top right), and some were inhibitory, staining positively for parvalbumin (bottom right). *Source*: From Ref. 157. Collectively, these data indicate that permanent structural changes are induced by mild or transient reductions in TH in the absence of other overt toxicity. *See page 93.*

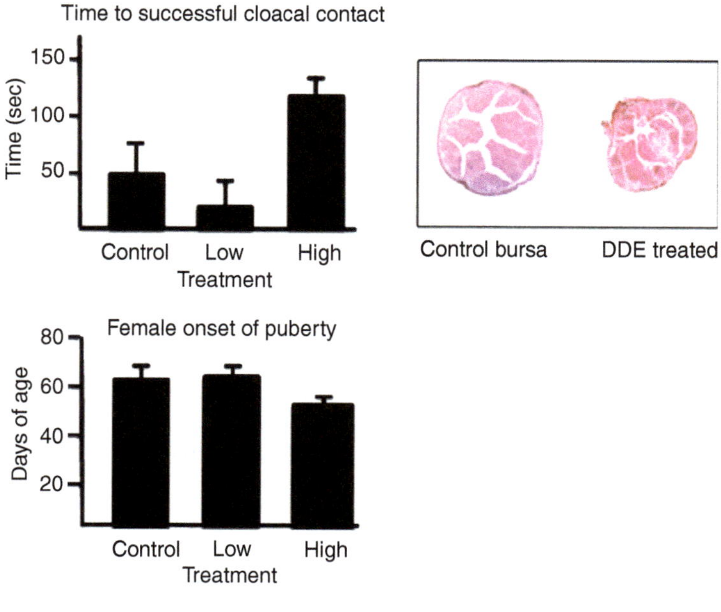

Figure 6.1 Representative bursal sections from control and DDE treated birds. Onset of puberty in females (first egg) and mating behavior in males were both impaired in high DDE (40 µg/kg) treatment groups. *Abbreviation*: DDE, dichlorophenyldichloroethylene. *Source*: From Ref. 28. *See page 141.*

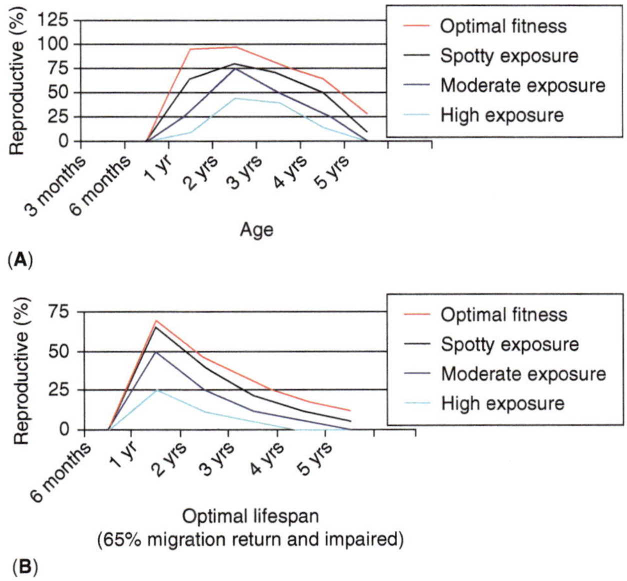

Figure 6.3 Diagrammatic reproduction addressing: (**A**) whether reproductive function is altered by EDC exposure at varied degrees due to variable timing of exposure or at different concentrations; or (**B**) considering a consistent rate of return from migration in combination with impaired reproduction due to EDC impacts. *See page 149.*

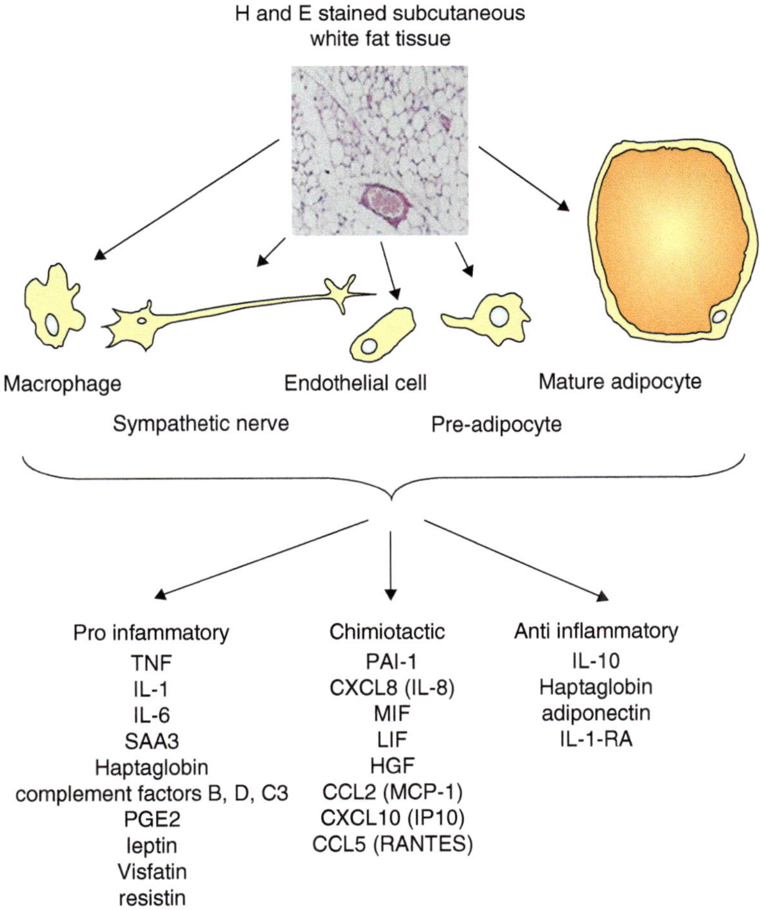

Figure 7.3 Inflammatory factors produced by fat tissue cells. These factors could be classified in pro-inflammatory or anti inflammatory based on literature (160–165). Chemokines produced by white adipose tissue may play a role in the monocyte/macrohage recruitment. *Source*: Adapted from Refs. 2, 7, and 20. *See page 159.*

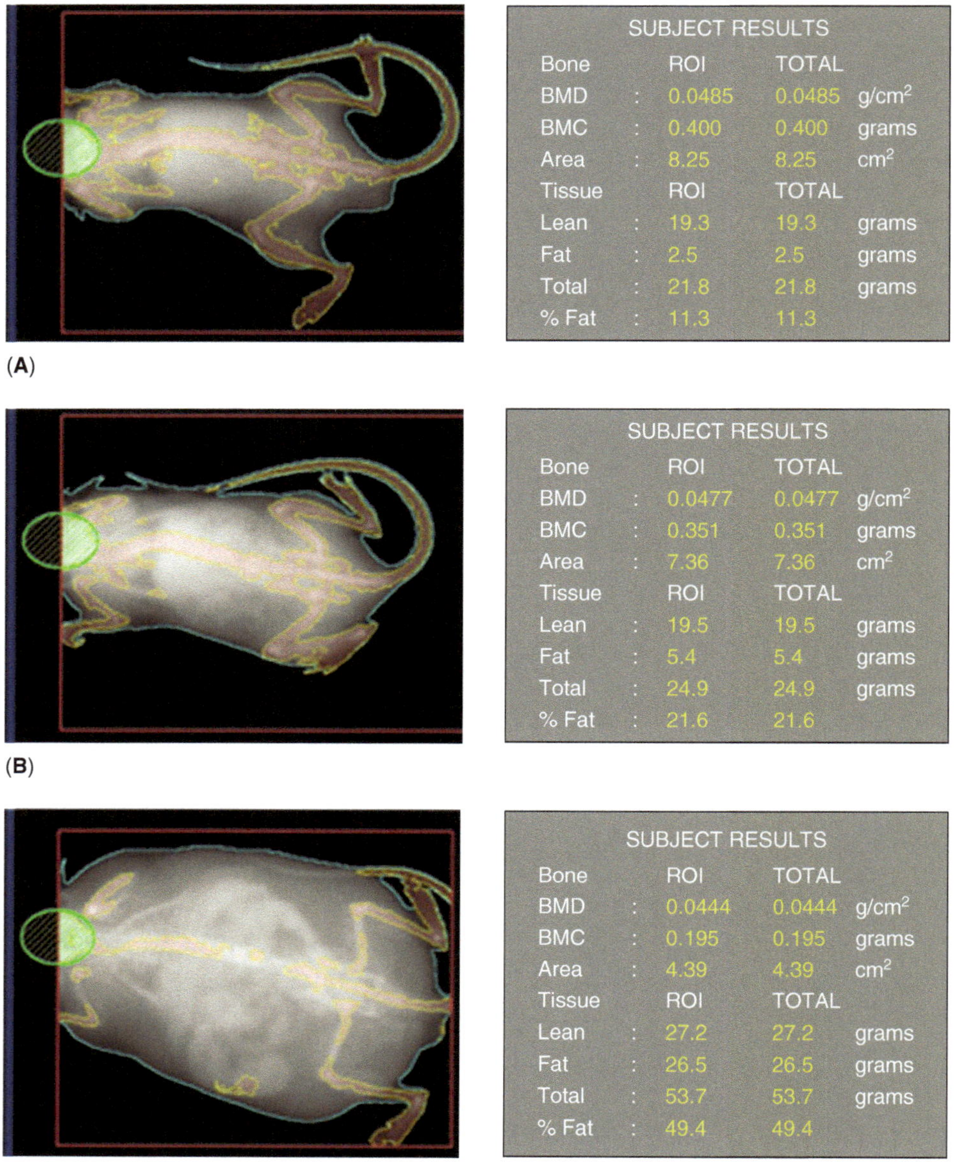

Figure 7.5 Fat measurement. Estimation of fat content by dual energy X-ray absorptiometry using a LUNAR PIXImus bone densitometer (GE Healthcare, Fairield, CT) as described previously (166) in 10 week-old male mouse (Jackson Laboratory, Bar Harbor, Maine, U.S.A.). (**A**) C57BL/6J mouse, (**B**) agouti (B6.Cg-A^y/J) mouse, (**C**) Ob/ob mouse (B6.V-*Lep*^{ob}/J). *Abbreviations*: BMD, bone mineral density; BMC, bone mineral content; ROI, region of interest. *See page 164.*

4 Thyroid Hormones—Impact on the Developing Brain: Possible Mechanisms of Neurotoxicity

Mary E. Gilbert

Neurotoxicology Division, U.S. Environmental Protection Agency, Research Triangle Park, and Department of Psychology, University of North Carolina, Chapel Hill, North Carolina, U.S.A.

R. T. Zoeller

Department of Biology, University of Massachusetts, Amherst, Massachusetts, U.S.A.

INTRODUCTION

In the adult human, thyroid hormone (TH) deficiency or excess can lead to a wide array of neurological and psychiatric symptoms, which are largely reversible with adequate treatment (1–3). In contrast, TH deficiency or excess during nervous system development leads to neurological and psychiatric symptoms that are not reversible, and are associated with permanent alterations in brain structure and function (4). Recently, the potential for environmental contaminants that disrupt the thyroid axis to induce neurodevelopmental impairments has been a topic of considerable interest (5–10). A large number of environmental contaminants with diverse structures have been shown to decrease circulating levels of TH (6,11–13). Animal studies have indicated that this action of xenobiotics on the thyroid system may contribute to alterations in nervous system development and function (5,6,9,11–21). Some of these environmental contaminants have also been evaluated in humans and been found to lower serum TH levels (22–24). Moreover, human exposures to some of these same contaminants are associated with neurodevelopmental impairments (25–29). Thus, an important mechanism by which some environmental contaminants may produce neurotoxic effects on the developing or adult human is by interfering with thyroid function, or with TH action.

Considering the importance of TH for normal brain development and for adult physiology, it is of significant concern that a large number of chemicals to which the human population is incidentally exposed can potentially impact thyroid function through a number of different mechanisms. Biomonitoring studies conducted by the center for disease control (CDC) and others (24,30) document that the general population is contaminated with large variety of industrial chemicals. Of particular concern are the numbers and concentrations of chemicals found in human amniotic fluid, fetal blood, and breast milk, rendering it unlikely for a child to be born without some exposure to xenobiotics. Many of these chemicals have been shown to interfere with TH signaling in experimental systems. However, there are significant gaps in our understanding of the clinical and fundamental elements of thyroid endocrinology and toxicology that require resolution before a complete understanding the functional implications of these observations can be derived. These knowledge gaps also confound our ability to identify some types of thyroid toxicants or to test their potential to produce adverse effects during development or in the adult.

The effects of thyroid disruption from environmental sources are likely to be mild and may be difficult to detect in the individual using standard clinical tests of thyroid hormone levels in blood. Until very recently, experimental studies almost uniformly modeled severe TH depletions rather than mild to subtle TH insufficiency. The lack of adequate information on low level thyroid dysfunction has hampered efforts to determine the potential impact of this action of environmental contaminants on brain development. In addition, the possibility exists that specific chemicals may bind to the nuclear TH receptor, alter the expression of transport proteins, or interfere with brain deiodinases and could, in principle, affect TH action directly in the absence of effects on serum hormones. As such, they would remain undetected by simple assessments of serum hormone change. The impact of acute actions of TH through non-traditional, nongenomic mechanisms must also be appreciated in the context of developmental and adult neurotoxicity of thyroid disrupting agents. This chapter will detail the function of the hypothalamic-pituitary-thyroid (HPT) axis, the role of TH in brain development and in adult physiology, and will summarize recent findings on the effects of modest perturbations of the thyroid axis on neurodevelopment.

THE HPT AXIS

Overview

The thyroid system is organized as a classic neuroendocrine axis, involving the hypothalamus, pituitary gland, and the thyroid gland. As depicted in Figure 1, TH synthesis begins with the active uptake of iodine into the thyroid follicles via the sodium-iodide symporter. Thyroglobulin (Tg) is a large glycoprotein stored within the colloid of the thyroid gland and is the substrate upon which TH are synthesized. Iodine must be oxidized, under the control of the thyroid oxidase enzymes (31–34), and bound to Tg by the synthesis enzyme thyroperoxidase (TPO). Tg contains four major tyrosyl residue sites where iodine becomes covalently bound when TPO is activated. In the adult rat thyroid gland, Tg is stored at concentrations sufficient to support hormone release for several days. Human thyroid glands contain a supply of Tg to last several months (35). Chemicals that act primarily by inhibiting TH synthesis may have no effect on hormone release until these stores of Tg are depleted. However, in contrast to the adult, the fetal and neonatal human thyroid gland is in short supply of iodinated Tg. At birth, neonates contain Tg stores to support only a single day of TH, making the developing organism particularly vulnerable to hormone deficiencies induced through a variety of means (36,37). Parameters of the adult and infant thyroid system that have implications for tolerance to perturbation of the thyroid axis are summarized in Table 1.

The classic view is that the thyroid gland produces and secretes the thyronine hormones thyroxine (T4) and triiodothyronine (T3), and T3 represents the "active form" of the hormone (Fig. 1).As depicted in Figure 1, synthesis and release of TH into the circulation are

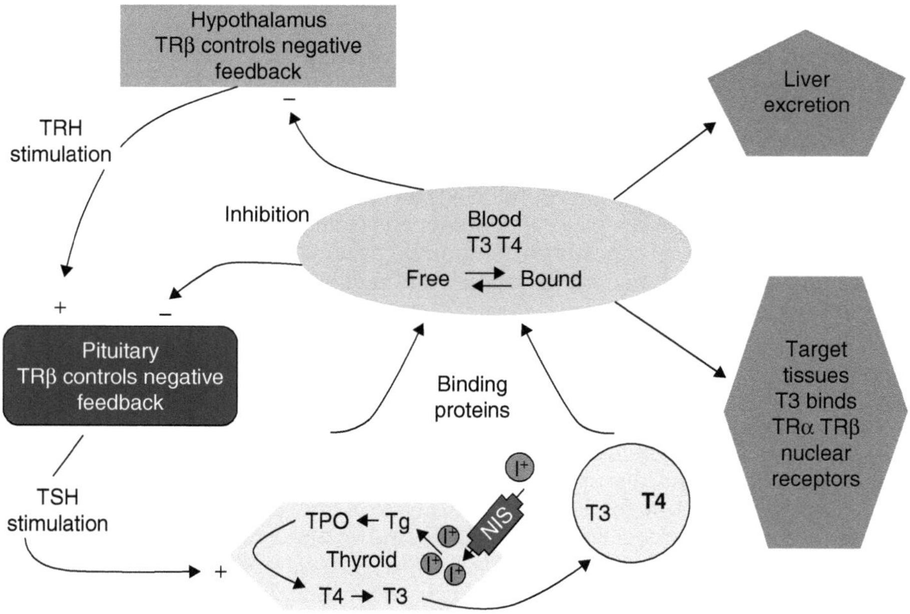

Figure 1 Classic model of negative feedback of hypothalamus-pituitary-thyroid axis. Iodine is an essential anion actively transported into the thyroid gland for synthesis of TH (T3 and T4). Upon release from the gland, TH are bound to proteins in the serum and delivered to a number of target tissues, including brain. TH are metabolized in the liver and excreted in the bile. Serum TH is regulated by a classic negative feedback loop to the hypothalamus and pituitary gland. Reductions in serum hormones are corrected by hypothalamic release of TRH to activate pituitary release of TSH. TSH increases synthesis and release of TH from the thyroid gland. *Abbreviations*: NIS, sodium-iodide symporter; Tg, thyroglobulin; TH, thyroid hormone; TPO, thyroperoxidase; TRH, thyrotropin-releasing hormone; TSH, thyroid stimulating hormone.

Table 1 The Adult and Infant Thyroid System Differ from Each Other in a Number of Parameters that Have Implications for Tolerance to Perturbations of the Thyroid Axis. The Thyroid System of the Neonate Is Less Able to Tolerate Fluctuations in THs due to its Limited Reserve Capacity Relative to the Adult

	Adult	Neonate/fetus
Serum half-life T4	5–7 days	3 days
Thyroglobulin stores (Tg)	Months	<1 day, so minimal with functional reserve
Iodine stores	Significant iodine stores in thyroid gland	Adequate placental and lactational supply needed on daily basis to maintain circulating TH
Synthesis of TH	Reduced hormone levels appear after a delay in reduced synthesis	High rate of synthesis necessary to meet daily demand—placental transfer of maternal T4 in early gestation, fetal T4 production later in gestation
Maturity of adaptive mechanisms	TSH, serum binding proteins, deiodinases, iodothyronine transporter proteins	Immaturity of these sources of compensation in face of dropping serum hormone concentrations
Reversibility of insult	All effects completely reversible with pharmacological treatment	Vulnerability of brain to even small insufficiencies leads to permanent effects

Abbreviations: TH, thyroid hormone; TSH, thyroid-stimulating hormone.

under the control of negative feedback loops of the HPT axis. A drop in circulating levels of T4 induces the secretion of thyrotropin releasing hormone (TRH) from the hypothalamus. TRH stimulates thyroid-stimulating hormone (TSH) secretion from the anterior pituitary. TSH then initiates TH synthesis and release from the thyroid gland. Although opposing TRH and TH inputs regulate the HPT axis, TH negative feedback at the level of the hypothalamus has always been believed to be the primary regulator. Until recently, however, this assertion had not been directly confirmed in vivo. Using a combination of TRH and TRβ knock-out mice, Nikrodhanond et al. (39) demonstrated the dominant role of the hypothalamus in the control of serum TH. Hypothalamic control of serum hormones through negative feedback is mediated through activation of the TH nuclear receptor, TRβ2 (40). These observations are of toxicological significance as they indicate that the negative feedback control of serum hormone concentrations might be more sensitive to TH insufficiency than some other endpoints of hormone action. In addition, environmental contaminants that interfere selectively with the TRβ2 may be particularly potent at affecting serum TH levels. In contrast, TRH also controls the set-point around which TH exert a negative feedback on TSH (41). This action may account for the paradoxical decreases in both TSH and TH that accompany food deprivation and caloric restriction (42,43). Although the neural pathways controlling this function are not completely understood, toxicants may interact with these systems in a manner that mimicks the effects of caloric restriction.

Furthermore, the iodothyronines, T3 and T4, may not be the only hormones produced by the thyroid gland. New information indicates that decarboxylated and deiodinated metabolites of TH, thyronamines, are also produced in the thyroid gland and when released into the circulation can interact with membrane-bound G protein-coupled receptors (38). In vivo administration of thyronamines (e.g., $T1_{am}$) produce a rapid decline in cardiac output and hypothermia, physiological functions that are also modulated by TH themselves. In isolated cell and synaptosomal preparations, thyronamines act as specific dopamine and norepinephrine reuptake inhibitors and also block the transport of monoamines into synaptic vesicles. These actions may underlie the pharmacological effects of thyronamines in cardiac and thermoregulatory functions. Toxicants that interfere with TH synthesis or metabolism may also interfere with thyronamine production as well. The contribution of altered thyroamine levels to the constellation of symptoms that comprise hypothyroidism has yet to be addressed.

Deiodination of THs

In addition to negative feedback mechanisms to control serum concentrations of TH, metabolism by peripheral and central deiodinases and sulfotransferases serve to further modulate local thyroid action (see inset Fig. 3) (44,45). The only source of circulating T4 is the thyroid gland, whereas circulating T3 is derived both from the thyroid gland and from conversion of T4 by iodothyronine deiodinases in tissues. Each deiodinase possesses unique characteristics and the complex tissue-specific and developmental pattern of their expression is indicative of their ability to impact diverse physiological systems (46). Type 1 (D1) and Type 2 deiodinase (D2) catalyze the outer ring deiodination of T4, converting T4 to T3. In brain, D2 is the active enzyme, residing in astrocytes or the tanocytes that line the third ventricle (44,47). Type 3 deiodinase (D3) has the reciprocal action of D2, deactivating T3 and is predominantly localized in neurons (48–51). D1 is a dual purpose enzyme and can either activate or deactivate TH.

Among the many TH-responsive tissues, there exists a differential reliance on deiodination mechanisms to modulate the action of TH. In brain, deiodinases represent an important point of physiological control as 80% of brain T3 is produced locally through activation of these enzymes (44,46,52). Because T3 has an affinity for the nuclear receptor that is 10-fold greater than that of T4 (53), it is T3 availability inside the cell that is important for hormone action. As such, TH signaling in individual tissues can be affected by these tissue-specific processes even as serum hormone concentrations remain "normal." Conversely, chemically-induced reductions or increases in serum hormone levels may be compensated for, within limits, at the tissue level by upregulation of D2 or down regulation of D3. However, the degree to which these cellular adaptive mechanisms can compensate for low T4 or for chemically-induced alterations in TH receptor function has not been widely studied.

Developmentally programmed temporally and spatially specific changes in deiodinase expression regulate the intracellular concentrations of T3 essential for normal development of the brain, cochlea, and retina (44,46,54,55). Studies in early human gestation and embryonic cells studied in vitro, however, have failed to demonstrate appropriate regulation of T3 by deiodinases, suggesting immaturity of this regulatory system (49,50,56). The inability of this system to regulate T3 in early brain development may contribute to the particular sensitivity of the developing central nervous system (CNS) to even mild maternal hormonal deficiencies (49). The potential for environmental contaminants to interfere with the ontogeny and function of central and peripheral deiodinase pathways can have significant toxicological consequence in the developing and the adult nervous system.

Transporter Proteins

In the classic model of HPT function, circulating levels of TH are directly related to hormone action in tissues (57). Recent work indicates that active transport of TH through plasma membrane plays a significant role in the regulation of local T3 action. Monocarboxylate transporter-8 (MCT8) is a specific and powerful transporter of both T3 and T4. Mutations of the MCT8 gene in humans produce mental retardation and global neurological dysfunction in the absence of severe physical characteristics of hypothyroidism (45,58). These observations confirm that the tissue-specific selective uptake of TH is an important regulatory step in the pathway of TH action (59). Expression of MCT8 in human fetal cortex begins early in gestation, specifically implicating this transport protein in the TH-mediated control of fetal brain development (49). In addition, recent studies show that the severity of the neurological deficits are directly related to the degree to which specific mutations abrogate T3 uptake (60).

Selective regulation of transport proteins may also contribute to the regional sensitivity of some brain areas to hormone insufficiency. Hypothyroidism reduces the expression of neurogranin (RC3), a thyroid-dependent signaling protein, in the dentate gyrus but not the CA1 region of the developing hippocampus (54,61,62). TH receptors and MCT8 are expressed in both pyramidal and granule cell neurons of the hippocampus. However, under conditions of moderate TH insufficiency a selective upregulation of the expression of MCT8 is seen in pyramidal cells (63). Consistent with previous work in a model of severe hypothyroidism, this altered expression pattern of MCT8 is accompanied by a differential expression of neurogranin in these two neuronal populations. These observations indicate a degree of protection may be conferred upon CA1 pyramidal neurons as a function of transporter expression and suggest a potential mechanism for selective vulnerability of neuronal subtypes within the hippocampus (Fig. 2).

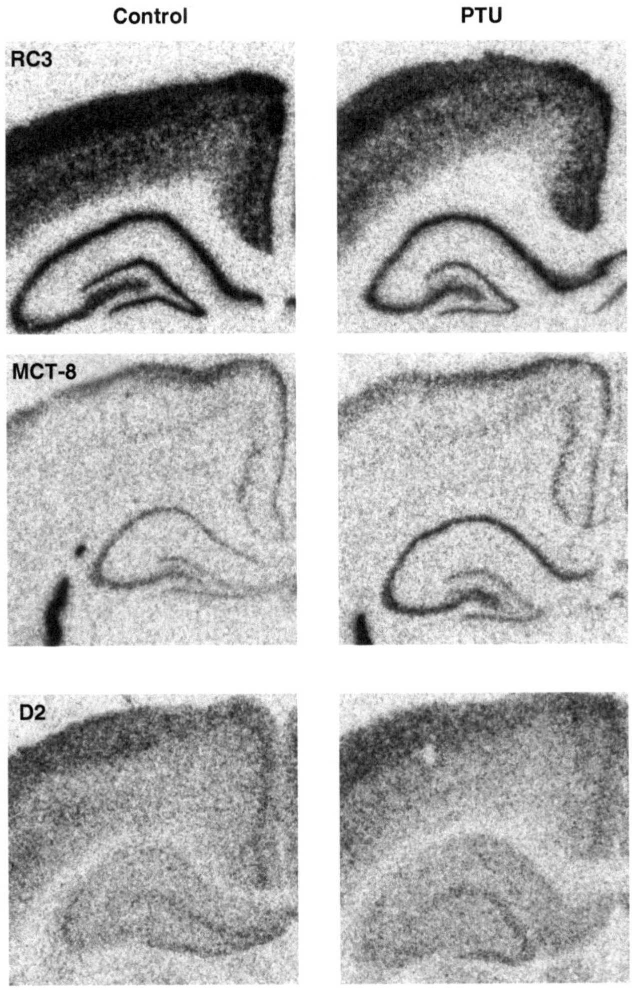

Figure 2 Active transport of TH through plasma membrane plays a significant role in the regulation of local T3 action. Developmental exposure to PTU lowers circulating levels of TH, reduces TH-dependent expression of neurogranin (RC3) mRNA in the dentate gyrus of the hippocampus, but not area CA1. Conversely, deiodinase-2 is upregulated in the dentate gyrus but not area CA1. These data reflect regional variation in compensatory responses to disruption of the thyroid axis. Differential capacity in these compensatory mechanisms may contribute to vulnerability of some brain regions and sparing of others under conditions of TH insufficiency. *Source*: From Ref. 63. *Abbreviations*: D2, deiodinase-2; MCT-8, monocarboxylate transporter-8; PTU, propylthiouracil; TH, thyroid hormone.

Environmental contaminants that interfere with TH transporters like MCT8 and others (e.g., OATP) may produce unique patterns of thyroid disruption.

In summary, a number of recent developments in endocrinology of the thyroid system indicate that the HPT axis is controlled in ways not fully appreciated previously, but that are important for thyroid toxicology. As outlined above, new information accumulating over the past decade has significantly increased our understanding of the regulation of TH in the brain and other tissues. New insights into feedback control mechanisms, plasma membrane transporters, and peripheral and central deiodinases demonstrate a modulation of TH action with a degree of precision and regional specificity not previously recognized. Incorporating these findings requires modifications to the idealized view of the HPT axis depicted in Figure 1 to that shown in Figure 3. Refining the model of TH function will also necessarily influence the course of research in neurotoxicology of anti-thyroid agents, as well as the specific kinds of endpoints evaluated in a screening and testing program designed to identify thyroid toxicants.

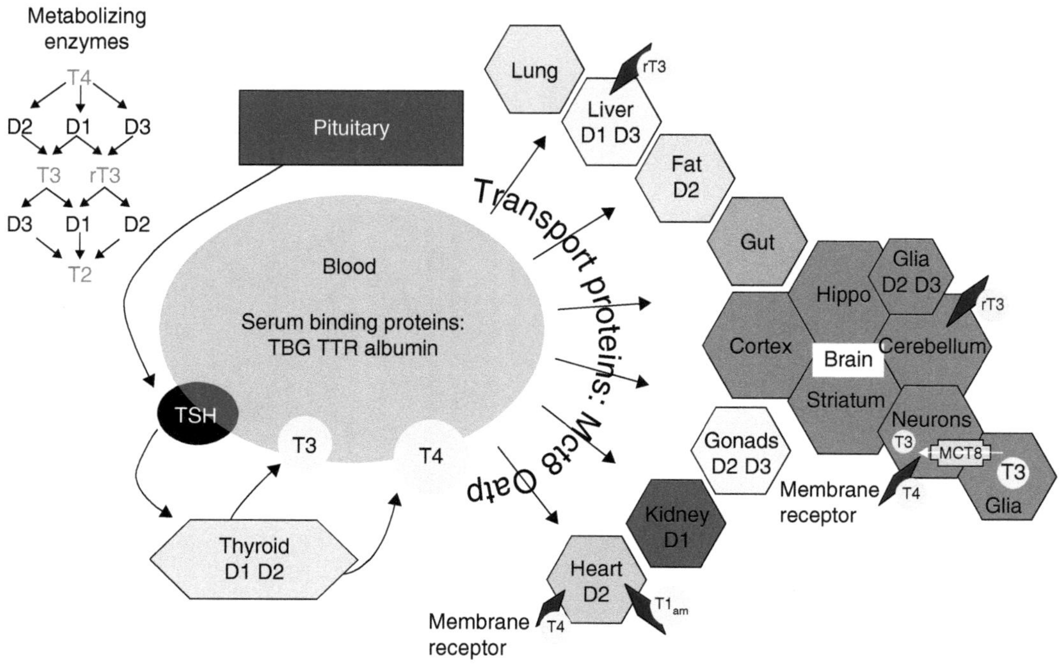

Figure 3 (*See color insert*) Recent information has required significant refinement of the original model of the HPT axis presented in Figure 1. A number of factors contribute to the differential regulation of thyroid hormone supply and action in different target tissues. In addition to the classic molecular action of thyroid hormones on gene expression, nongenomic effects mediated by membrane bound receptors for T3, T4, and T1 amines can mediate thyroid hormone action. Active transport proteins (e.g., MCT8, OATP) and metabolizing enzymes are differentially expressed and regulate the presentation of hormone to different target tissues. The impact of environmental contaminants on these different regulatory mechanisms, especially as they pertain to brain development has not been addressed. *Abbreviations*: HPT, hypothalamus-pituitary-thyroid; MCT8, monocarboxylate transporter-8; OATP, organic anion transporting polypeptide; TBG, thyroxine-binding globulin; TSH, thyroid-stimulating hormone; TTR, transthyretin.

THs REGULATE GENE TRANSCRIPTION

During brain development, TH transiently regulates the expression of many genes involved in a variety of developmental processes in a temporally and spatially specific manner. The cellular actions of TH are mediated by nuclear receptors encoded by two genes, TRα and TRβ (48,64). TRα1 is widely expressed in developing brain, whereas TRβ is more restrictive, found primarily in sensory organs and extensively studied in the cochlea and retina (51). As previously described, TRβ in hypothalamic neurons has recently been implicated in regulation of serum hormone concentrations (39). Thyroid receptors (TRs) bind to DNA in the unliganded state (i.e., the "aporeceptor"), and can enhance or repress gene transcription depending on a number of factors including the cellular mileau of co-factors available to the receptor as well as the particular cis-regulatory element to which the TR is bound (Fig. 4). The hormone-receptor complex binds to DNA sequences, TH response elements (TREs), that directly regulate specific genes. Few genes have been identified that are directly activated by TH, and of them, most are transcription factors (e.g., hairless, *hr*) that modulate the expression of other genes (65). TH regulation of these downstream genes is therefore indirect as these genes do not themselves possess a TRE and do not bind the T3-receptor complex. In this manner, TH can modulate the expression of a multitude of downstream genes, an action that dramatically increases the sphere of their influence on brain development, and that has increased the complexity of identifying developmentally important TH-responsive genes (66–68).

One concept of TH action is to accelerate developmental changes in gene expression. In the absence of TH, these changes still take place but at a slower rate, disrupting the exquisite choreography of timing of events that underlie normal brain development (48,69,70). As such,

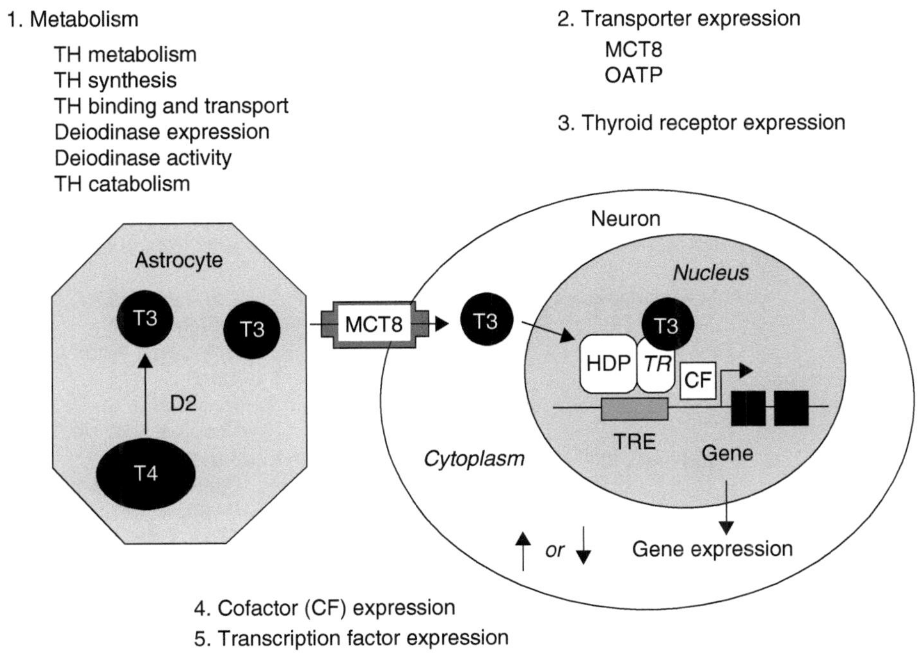

Figure 4 Mechanisms controlling molecular action of thyroid hormone (TH). The molecular action of TH in the brain can be modulated in a variety of ways. This figure depicts metabolism of THs in the astrocyte, active transport of T3 into the neuron were it binds to TR in the nucleus. The TR, often paired with a HDP, is bound to a TRE on the chromatin of a thyroid-responsive gene. TH action can be modulated at several sites: (1) alterations in TH metabolism, (2) expression of transport proteins, (3) TR, (4) cofactors, and (5) transcription factors. *Abbreviations*: HDP, heterodimeric partner; MCT8, monocarboxylate transporter-8; OATP, organic anion transporting polypeptide; TH, thyroid hormone; TR, thyroid receptors; TRE, thyroid response element; D2, deiodinase-2.

TH are necessary but not sufficient for normal development and TH must interact with a number of factors to exert their full effect. Many comprehensive reviews are available on the role of TH receptor activation and gene expression in brain development (4,48,64–66,70). Table 2 was derived from these reviews and represents a summary of genes whose expression is altered in the brains of hypothyroid animals. Recently, using "functional genomics" (e.g., microarrays), a number of these genes were also demonstrated to be differentially expressed in animals experiencing varying degrees of TH insufficiency, from very modest to severe (71,72). It is not at all clear, however, which genes are directly regulated by TH, which genes are downstream and only indirectly regulated by TH, and which genes are indirectly modified by a change in developmental trajectory induced by TH insufficiency (48,65,68).

This complex interplay of TH, gene expression, and developmental timeline is exemplified by the role of TH in white matter development. TH act on a common precursor to oligodendrocytes and astrocytes to favor differentiation of the former at the expense of the latter (73,74). As recently demonstrated in vivo, the ratio of oligodendrocytes to astrocytes is altered by thyroid status (19). Cataloguing changes in messenger RNA (mRNA) levels that occur at varying degrees of TH insufficiency will identify those genes directly regulated by TH uniquely in oligodendrocytes, in astrocytes, and in their common precursors. As TH levels decline, the number of oligodendrocytes declines and the number of astrocytes increases; thus, astrocyte-specific genes will appear to be up-regulated while oligodendrocyte-specific genes will appear to be down-regulated. These interrelationships may be further clarified through examination of dose-dependency at different developmental timelines and through the use of chromatin immunoprecipitation (ChIP) followed by whole-genome approaches.

Adding to the complexity of the molecular basis of neural development, the regulation of TH-responsive genes in brain is subtle in nature, standing in marked contrast to the gene expression in the liver and pituitary. Most known neural genes exhibit transient responsiveness to TH and undergo changes in expression of only two- to three-fold in response to TH (75).

Table 2 Thyroid Hormone-Sensitive Genes Categorized by Developmental Process and Functional Domain[a]

Developmental process	Functional description	Genes previously identified	Genes with related functions
Myelination	Myelin genes	***Mbp, Plp, Mag,*** CPNase	***Mal, Cldn11, MoBP,*** **Mog, Fndc5**
Differentiation	Neurotrophins	BDNF NT-3, trkA, p75ntr	***Ntf3,*** **Edg2, Bmp3, Nrn1**
Migration	Extracellular matrix	***Tnc*** Laminin, L1, reelin	***Slit1, Dcx, Sox4,*** **Itih3**
	Adhesion molecules	**Ncam**	***Col11a2, Hapln1, Dpysl,*** **Amigo2, Itih3**
	Cytoskeletal components	Tubulin $\alpha1$ $\alpha2$ $\beta4$ tau	***Arc, Csrpl, Tmod1, Pdlim,*** **Larp1**
Apoptosis			***Hop/Hod*** **Casp3, Epha7, Aaft**
Synaptic function	Intracellular signaling	**CaMK-IV, Krox-24/Egr-1, Rhes/Rasd2**, RC3, Pgd2	***Ntsr2,*** **CaMK-II, Homer1, Calb2**
	Synaptic transmission	Synaptotagmin-related gene-1	***Pvalb, Kcna1, Slc1a2,*** **Grik4, Grin3a, Gabrd, Tac2, Vamp1, Elfin, Gprc5b, Lynx1**
Transcriptional regulation	Transcription factors	***Hr*** RORα, Bten	***Junb, Pias3, Mllti, Agpat7,*** **Jundp2, Csen**

[a]Genes previously identified in the literature are summarized in column 3. Bolded transcript names in this column represent literature-reported genes, identified under conditions of severe hypothyroidism, that were also detected by microarray at low doses of propylthiouracil. The last column lists novel gene transcripts implicated in similar functional domains. Italicized names denote those whose expression level was significantly altered at the lowest dose of propylthiouracil (1 ppm) examined that induced a state of hypothyroxinemia in dams and pups (71).

Consequently, it has proven challenging to link the changes in expression of a particular gene or family of genes to the well known effects of TH on brain development (48,64,65). Difficulty linking alterations in gene expression and brain function may also derive from the paradigm often used in these studies. Specifically, models of severe hypothyroidism may lead to a plethora of effects on somatic development of many organ systems that may obscure the more direct actions of TH on brain development, inducing alterations in brain development that are secondary to TH insufficiencies.

Work with TH receptor knock-out mouse models has not aided in the search for thyroid-responsive genes. Few obvious abnormalities in brain development appear in these mutants and certainly TR knock-out mice do not exhibit the neural phenotype characteristic of chemically-induced hypothyroidism (4,67,76). Although originally somewhat enigmatic, recent work has demonstrated clearly that the severe neural phenotype associated with chemically-induced developmental hypothyroidism is caused by the presence of the unliganded TR. This was clearly demonstrated by Hashimoto et al. (77) who developed a mouse model that expressed a mutant TRβ gene that was incapable of binding TH, but retained its ability to bind to DNA. This mouse exhibits many structural and behavioral abnormalities that are distinct from TR knock-out mice, and are similar to hypothyroid wild-type mice. Thus, these and other studies demonstrate that the aporeceptor is responsible for the severe neurological phenotype associated with hypothyroidism, and implicates co-repressors in the mechanism whereby hypothyroidism causes brain damage (78).

NON-GENOMIC ACTIONS OF THs

It is also becoming increasingly clear that a subset of important actions of TH—and of products of the thyroid gland—are mediated by membrane receptors (see reviews by Davis and colleagues (79). These actions of TH have been characterized as "nongenomic" as they have traditionally been demonstrated using in vitro preparations lacking TREs transiently exposed to TH. They are also described as "acute," with rapid onset relative to the traditional genomic mechanisms that require gene transcription and consequent translation of mRNA. However, as noted by Davis et al. (79) this characterization is inaccurate in the context of the intact organism where endogenous levels of TH can mediate acute membrane and cytosolic effects or modulate background levels of phosphorylation of specific signaling molecules that do impact gene transcription, albeit through a more indirect route. The nongenomic actions of THs clearly represent an important array of signaling pathways that deserve increased study and awareness. However, no studies to date have investigated the ability of environmental chemicals to interfere with these important pathways. Thus, our goal will be only to provide a brief overview of these signaling pathways and to provide references sufficient for the interested reader to pursue this topic.

A number of specific cellular events are targets of TH action mediated by nongenomic mechanisms. TH initiation of rapid biological responses include the regulation of intracellular pH in myoblasts (80), changes in the rates of protein trafficking (81), changes in phosphorylation of specific nuclear proteins (82), and regulation of actin polymerization and cell motility (83,84), and calcium-mediated signaling events (79). Occupancy of T4 of the integrin receptor, a cell surface structural protein of the plasma membrane phosphorylates the mitogen activated protein kinase (MAPK or erk1/2) (81,85,86). In the developing brain this class of protein kinases regulates cell proliferation, differentiation, survival, apoptosis, and plasticity (87,88). In the adult brain, phosphorylation of erk1/2 is critical for some forms of synaptic plasticity and learning (89,90) through its interaction with protein kinases and neurotrophins and activation of transcription factors including cyclic AMP response-element binding protein (CREB) (91).

In astrocytes in culture, interaction of TH with laminin, a member of the integrin receptor family regulates the dynamics of actin fiber remodeling to promote cell adhesion and guide cell migration. As described below, migration of granule cells within the cerebellum is significantly delayed in hypothyroid animals, possibly due to absence of this action of TH (84,92). However the recent report of nuclear TH receptors in astrocytes indicates a direct nuclear genomic action may also contribute to TH modulation of astrocyte maturation (93).

THs also activate the phosphoinositol 3-kinase/Akt pathway, an important regulator of gene expression, cell cycle progression, cell growth, differentiation, metabolism, and survival (94–98). Direct administration of T3 into hippocampus activates phosphorylation of Akt (99). Clearly important developmentally, recent data also implicates this signaling pathway and its downstream targets in the maintenance phase of long-term potentiation (LTP), the leading experimental model of synaptic changes that underlie learning and memory (100–102).

It is of particular significance that both T4 and reverse T3 may be even more potent than T3 to induce some of these non-genomic actions. As some hydroxylated environmental chemicals [e.g., polychlorinated biphenyls (PCBs), and polybrominated diphenyl ethers] interact with the T4-specific transport protein transthyretin, these actions of TH may have toxicological relevance (103–105). Thus, toxicants may well interfere with the nongenomic actions of TH without altering the more visible effects of the classically defined genomic actions of TH. Collectively, these data indicate that actions of TH can occur at non-nuclear sites to alter developmental processes and by this action may also contribute to effects of TH insufficiency on adult nervous system and mediate some of the actions of environmental contaminants on brain function.

In addition to the nongenomic mechanisms of the iodothyronines, TH can be progressively deiodinated to lower iodine states, and undergo decarboxylation to form monothyronamine, $T1_{am}$ (38,106). First produced synthetically, $T1_{am}$ has since been identified in vivo and when bound to plasma membrane receptors, exerts potent effects on cardiovascular function, body temperature regulation, the balance between glucose and lipid utilization, and can inhibit specific monoamine transporters in the brain (38). A great deal remains to be learned about this novel signaling pathway, but it is clear that the physiological impacts of thyronamines may be quite important. The potential for environmental chemicals to interfere with the synthesis, transport, or action of thyroamines has yet to be examined.

THs AND BRAIN DEVELOPMENT IN HUMANS

In humans, severe deficiencies in TH during development are associated with irreversible damage to virtually all organ systems, a condition termed cretinism (107,108). The primary causes of severe developmental hypothyroidism in humans are mainly iodine deficiency and congenital hypothyroidism (109–112). Each condition can produce a different spectrum of symptoms, the specific nature of which is dependent on the timing, duration, and severity of the deficiency. TH were once believed to be blocked from reaching the fetus by the placenta; however, direct evidence that maternal T4 reached the fetus in substantial amounts changed this perception (113). Moreover, recent epidemiological evidence also indicates that even mild reductions in maternal THs during early and/or late pregnancy has consequences on brain development (87,97,114–117). These observations are supported by work in experimental animals in which mild and transient reductions in TH in maternal serum early in pregnancy can produce specific deficits in cortical and hippocampal histogenesis (114,118,119).

Congenital Hypothyroidism

Congenital hypothyroidism refers to a condition whereby children are born with very low levels of serum TH (120). In such cases, children appear normal at birth and their mothers have normal thyroid function. In fact, it is the lack of overt clinical symptoms in children with congenital hypothyroidism that initially supported the concepts that TH were not important in fetal development and that the placental barrier restricted access of maternal TH to the fetus (113). The most common causes of congenital hypothyroidism are ectopic thyroid gland, thyroid agenesia and hypoplasia, and inborn errors of TH biosynthesis (121). Because of the lack of specific symptoms in the early neonatal period, the diagnosis of congenital hypothyroidism was often delayed and the prognosis deteriorated with the passage of time (120). The full clinical picture prior to the era of systematic screening included growth retardation, puffy features and myxedema, and mental retardation. Severely affected children exhibited neurological signs including spasticity, incoordination, cerebellar ataxia, strabismus, speech problems, and hearing loss (70,107,110,122). With the introduction of routine neonatal screening, severe mental retardation associated with this condition has been eliminated. However, late diagnosis and delayed or inadequate treatment is associated with poorer intellectual outcome. Lower global IQ scores, language delays and poor verbal skills, motor weakness, attentional deficits, and learning impairments are evident in children with delayed or inadequate treatment (123). Even in cases where the condition is diagnosed early and treated effectively, subtle impairments in mental function remain. Standard tests of IQ function in congenitally hypothyroid children are approximately six points below expected values and selective deficits on visuo-spatial, motor, language, memory, and attention tests are observed (124–126).

Endemic Iodine Deficiency

Iodine is an essential element for the biosynthesis of TH (127,128). Some of the most serious neurologic impairments associated with thyroid dysfunction from an environmental cause have been documented in children born in regions of the world where dietary iodine deficiency is prevalent (108,109,111,122). These children are characterized by stunted growth and a high incidence of mental retardation. Motor impairments involve both gross and fine motor control, spasticity, gait disturbances, and inability to stand. Anatomical and physiological alterations of the ear produce deaf-mutism in a large portion of affected children. Beyond the sensory organ, deficits in perceptual hearing, learning deficits, and speech problems are also evident. Neurological deficiencies seen in endemic iodine deficiency are more severe than those resulting from congenital hypothyroidism, suggesting that early fetal thyroid function is necessary for normal development. Supplementation of iodine by mid-gestation improves the outcome but does not completely prevent damage. These observations suggested that development of the fetal brain, before the onset of fetal thyroid function, is dependent on maternal THs derived from placental transfer of T4 (113,129–131).

Maternal Hypothyroidism/Hypothyroxinemia

Because the symptoms of congenital hypothyroidism were not apparent at birth, it was once thought that THs were only necessary for fetal brain development after the onset of fetal thyroid function (122). TH of maternal origin were not believed to cross the placenta and therefore could not contribute to fetal brain development (132). However, Vulsma et al. (113) were the first to provide clinical evidence of concentrations of circulating THs in the congentially athyroid fetus. TH has also been shown to cross the placental in rats and as much at 17.5% of fetal TH at birth are maternal in origin (129,130,133,134).

A number of studies have now demonstrated that subclinical hypothyroidism in pregnant women can result in neuropsychological deficits in their offspring, despite normal thyroid status of the child at birth. Low circulating levels of TH during the second trimester of pregnancy were associated with poorer scores on the Bayley Scale of infant development and lower IQ scores at four to seven years of age (135). Pop et al. (136,137) supported these findings and extended them to an earlier gestational time window. Low T4 levels in maternal serum and the presence of circulating antibodies for thyroid peroxidase at 12 weeks of gestation were strong predictors of infant mental development and IQ. Haddow et al. (115) reported 4–7 point IQ deficits in children born to women with low T4 levels in early gestation (<12 weeks), serum hormone levels that were still within the subclinical range. None of these children were hypothyroid at birth yet at seven to nine years of age had lower neuropsychological tests scores in attention, language, reading, and visuo-motor performance. Recently, suboptimal neurological development as assessed by the Neurobehavioral Assessment Scale was identified as early as three weeks of age in children born to women with low serum T4 (138). In this study, low maternal T4 and not TSH or T4 later in gestation was a significant predictor of outcome. These observations of neurological impairments following subclinical reductions in maternal T4 have raised the level of concern over the influence of environmental contaminants on thyroid function and brain development (23,24,139,140).

ANIMAL MODELS OF DEVELOPMENTAL HYPOTHYROIDISM

The role of THs in brain development has been extensively studied in rodent models. Hypothyroidism is induced by thyroidectomy, radioactive iodine, or administration of TPO inhibitors such as propylthiouracil (PTU) and methimazole (MMI). Hypothyroidism induced by maternal exposure to these TH synthesis inhibitors is the most common model (46). Because TH were not believed to be necessary for fetal brain development until late in gestation, many animal models have incorporated only the late gestational and early postnatal period to evaluate the impact of altered TH on brain development (47,108). In addition to dismissing the role of TH in early fetal development, these models are of severe TH deprivation (48,64,65,108). In the majority of these studies, T4 is typically below the level of detection of most assays, T3 is reduced, and TSH is increased several fold. Such treatments are accompanied by severe growth retardation, decreased brain size, delayed developmental milestones, and abnormal gait reminiscent of the cases of neurological cretinism in humans. As such, the clinical symptoms of severe hypothyroidism in animals came to be viewed as the clinical signs of developmental TH insufficiency. In the absence of this phenotype, an erroneous but not uncommon conclusion was that no effects on brain development were produced.

It is now clear that TR activation in fetal brain occurs early in gestation through the placental transfer of T4 from the mother to influence brain development (49,134,141). In humans, fetal thyroid function begins in the second trimester, in rodent not until late gestation (approximately gestational day 17). A gradual decline in dependence on maternal sources occurs from the onset of fetal thyroid function until birth (Fig. 5). Until recently few experimental studies were designed to examine mild or subclinical forms of maternal or neonatal TH disruption.

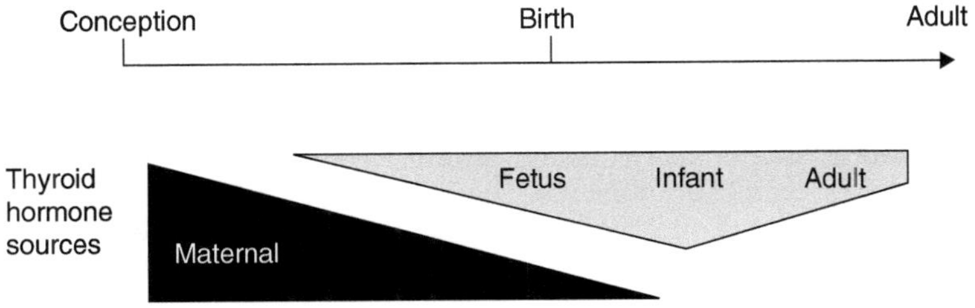

Figure 5 Thyroid hormone sources during fetal and postnatal life. Development of the fetal brain requires sufficient levels of thyroid hormone. Thyroid hormones are active long before the fetal thyroid is functional. In early gestation, the fetus is fully dependent upon the mother for its supply of thyroid hormone. In humans, fetal thyroid function begins in the second trimester, in rodents on GD17, but the status of brain development is comparable at these times between the two species. *Source*: From Ref. 76.

Milder forms of hypothyroidism producing graded levels of hormone reduction have been produced through manipulations of dietary iodine or administration of lower doses or shorter duration exposures to the model TH synthesis inhibitors. These studies are the most informative in defining the untoward effects of mild perturbations thyroid function of concern today. Developmental processes disturbed by overt hypothyroidism in severe models of hormone deficiency will be described below, followed by a summary of more recent findings with newer models that more closely approximate conditions of subclinical hypothyroidism.

TH Modulates Brain Development

THs are generally regarded as regulating cell proliferation, synthesis of microtubule associated proteins, neuronal outgrowth, and synapse formation. However, it is important to recognize that TH effects are quite pleiotropic and it is likely that the specific role of TH will depend on the cell type and time of development. Pre- and postnatal models of hypothyroidism have revealed reductions in cell number, synaptogenesis and dendritic arborization, altered patterns of cell migration, and reductions in axonal myelination. Different time windows of hormone deficits dictate the regionality and the nature of the observed effect (Fig. 6). Abnormal development has been well documented in the cerebellum, neocortex, hippocampus, and the heavily myelinated fiber tracts of the corpus callosum. Normal development of sensory system organs including the cochlea and the retina is also dependent upon TH.

TH and Development of the Cerebellum

The cerebellum is a largely postnatally developing structure in rodent and has been extensively studied for its developmental dependence on TH (65,72,142,143). Postnatal TH deficiency during the critical period of cerebellar development leads to a multitude of irreversible morphological abnormalities including reduced neurogenesis, defects in granule migration, increased granule death, and blunted dendritic arborization of Purkinje cells. THs promote granule cell proliferation in the cerebellum in the early postnatal period. Between PN2 and PN12, with a peak on PN8, programmed cell death removes the supernumerary neurons within the internal granule cell layer (GCL). Severe hypothyroidism decreases cell proliferation and dramatically increases the incidence and extends the duration of apoptotic cell death in the internal GCL (142–146).

One of the hallmarks of TH deficiency on cerebellar development is the persistence of an external GCL resulting from delays in neuronal migration. On PN15, the cerebellum from a normal animal has two distinct GCLs, an internal and an external layer. By PN21, migration of granule cells from the external to the internal layer is complete, the external layer has disappeared leaving a single layer. Hypothyroidism delays this temporal pattern of neuronal migration such that a distinct external GCL persists on PN21 (Fig. 7).

The mechanisms responsible for altered migration patterns in cerebellum have been elucidated using cultures of cerebellar granule cells (83,92). Migrating neurons find their path through interactions with extracellular matrix proteins and adhesion cell molecules (e.g., NCAM, laminin, integrins). The organization of actin filaments within astrocytes is regulated by TH and dictates the secretion, deposition, and patterning of adhesion molecules. Extracellular matrix proteins and cell adhesion molecules are important because they serve as guidance cues for cell migration and axonal growth. Actin polymerization is disrupted by TH deficiencies, laminin is not deposited, and migrating neurons do not recognize the appropriate guidance molecules (83). In vivo studies support these observations and reveal a marked suppression in the expression of astrocytic proteins in the cerebellum of hypothyroid animals. The appearance of laminin is delayed and is reduced in quantity. These effects are accompanied by the persistence of external GCL in the developing cerebellum as described above and evident in Figure 7 (144,147,148). Despite the well-characterized effects of postnatal hypothyroidism on cerebellar development, it is surprising that this system has not been examined in models of low level TH disruption.

TH and Development of the Cortex

THs are also critical for neuronal migration within the neocortex. Cortical neuron migration occurs earlier in development than in the cerebellum, beginning in the mid to late gestational period in rodents. The mammalian neocortex is highly ordered, the deepest layer cells are born

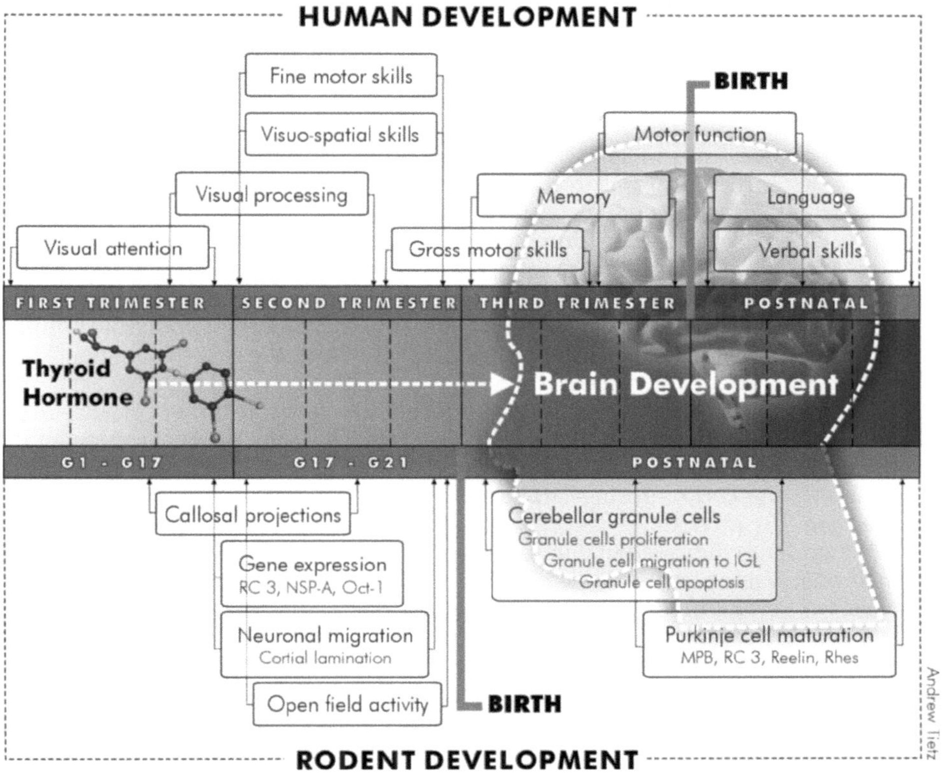

Figure 6 Timeline of development and functional brain maturation in humans and rodents. Events critical for normal development occur in different brain regions along a distinct ontogenic timeline. Thyroid hormones are necessary for the normal pattern and timing of critical developmental events. Distinct patterns of functional deficits result under different windows of hormone insufficiency. The relative state of brain maturity is different between humans and rodents, but the pattern and sequence of developmental events is common across the two species. *Source*: From Ref. 126.

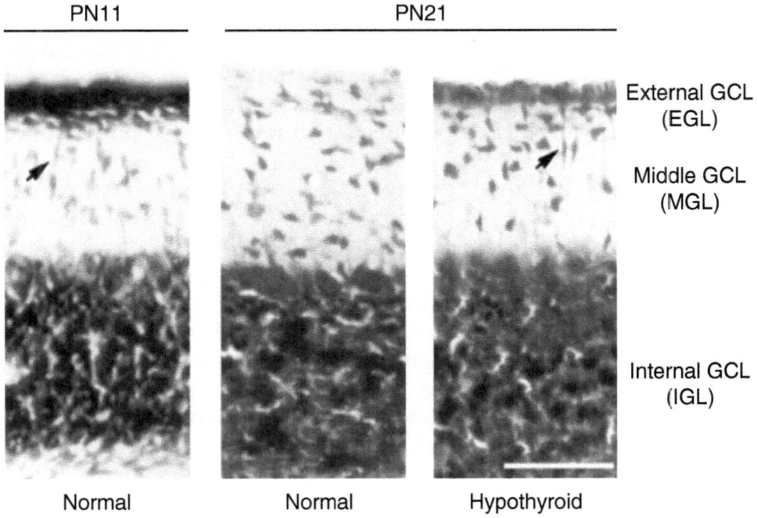

Figure 7 Cerebellar development is thyroid hormone dependent. On postnatal day 11 (P11), the EGL is still visible. Cells migrate from the EGL to the MGL over the next 10 days and in the euthyroid mouse, the EGL has disappeared by P21. In the absence of thyroid hormone, cell migration is slowed and the EGL remains at P21 and persists for several more days before cells either migrate or are apoptotically removed. Arrows identify spindle-shaped migrating cells. Bar = 50 μm. *Abbreviation*: GCL, granule cell layer. *Source*: From Ref. 144.

first followed by the cells in middle, then the upper layers. Each successively generated post-mitotic neuron must bypass predecessors, which have migrated along the same glial fibers before settling in the outmost level of the cortical plate. This migratory pattern is known as the inside-out gradient of cortical plate neurons is exhibited by excitatory glutamatergic neurons of the cortex (149).

Within the cortex, radial glial cells, an immature form of astrocyte, form a physical scaffold for the migratory cortical neurons to track to their final destination. Radial glial cells initially express nestin and vimentin in an early stage, then undergo a morphological transformation to mature astrocytes to express GFAP during the late gestational, early postnatal period. TH may affect the balance of production of neurons and glia in the early cortex (150). Immunohistochemical analysis of nestin and GFAP reveals a disruption in temporal pattern of expression of these markers of astrocyte maturity in the hippocampus and neocortex of hypothyroid animals (151–153). There are no data to indicate a direct action of TH on radial glial cells. Rather, it is posited that TH indirectly mediate the release of signaling molecules from neurons that act to modulate glial cell differentiation (48). One candidate signaling molecule is reelin, a substance released by Cajal-Retzial cells in layer 1 of the cortex. Reelin is critical for laminar organization and maintenance of the inside-out gradient of the neocortex (154,155). Hypothyroidism reduces the expression of reelin in the cortex during late gestation altering neuronal migration patterns and cytoarchitecture (151,152,156).

Recent work has indicated that induction of a severe hypothyroid state is not necessary to disrupt neuronal migration and lamellar cytoarchitecture of the hippocampus and neocortex (Fig. 8A and B). Perturbation of the laminar pattern of cortical neurons has been observed in response to very brief and transient exposure of the dam to MMI between gestational day 12 and 15, before the onset of fetal thyroid function (114). In a model of pre- and post-natal iodine deficiency, cytoarchitectural abnormalities were also apparent in the hippocampus and somatosensory cortex (118). In both models, neurons were present at locations that were inappropriate for their birth date and detected in areas of white matter that do not typically have a neuronal constituent. Importantly, these aberrations were found in cases of moderate degrees of iodine deficiency that did not produce the constellation of effects typically characteristic of severe hypothyroidism. Similarly cytoarchitectural abnormalities following a three-day exposure MMI were produced by only a modest drop in TH levels in dams to 70% of normal. These findings are particularly significant as they indicate that a brief episode of maternal hypothyroxinemia at a critical time is sufficient to impair brain development.

Another demonstration of aberrant neuronal migration was recently reported by Goodman and Gilbert (157). In this model low doses of PTU were administered to pregnant dams to produce graded degrees of TH insufficiency. An aberrant cluster of neurons, named a heterotopia, was detected in the corpus callosum of prenatally exposed animals (Fig. 8C). Cell dating studies using bromodeoxyruracil delivered to the dam indicated these cells were born in the late gestational period. A heterotopia was evident in the brains of offspring of dams with ~35% reduction in circulating T4 and in pups exhibiting no reduction in body weight, delay in eye opening, or other overt physical signs of hypothyroidism.

In contrast to excitatory neuronal cell migration, GABAergic inhibitory neurons originate in the medial ganglionic eminence (MGE), transgress the corticostriatal boundary and migrate tangentially into the developing neocortex. These interneurons form the cortical inhibitory circuits in the cortex and tightly regulate the activity of excitatory glutamatergic neurons. Migration of inhibitory neurons from the MGE is more difficult to track in vivo. So too is identification of perturbations in positioning as these inhibitory neurons do not display the lamellar profile of cortical glutamatergic neurons. Some but not all neurons within the heterotopia of the corpus callosum reported by Goodman and Gilbert (157) described above expressed proteins specific to inhibitory cortical neurons, suggesting that this migratory pathway is also disturbed by TH insufficiency.

Migration of inhibitory neurons was assessed more directly by Cuevas and colleagues (158) using transgenic mice expressing green fluorescent protein (GFP). Wildtype and GFP-expressing transgenic mice were exposed prenatally to MMI from GD10 to GD13. Embryos were taken on GD13 and flat cortical mounts prepared. Explants of GFP-MGE cells were transfected on control or hypothyroxinemic cortices and the migratory pattern observed over two days in culture. Cells from MMI-treated embryos implanted into control cortical mounts migrated in a normal, preferentially medial direction. However, neurons from control embryos placed within a hypothyroxinemic cortex, migrated long distances but displayed a concentric

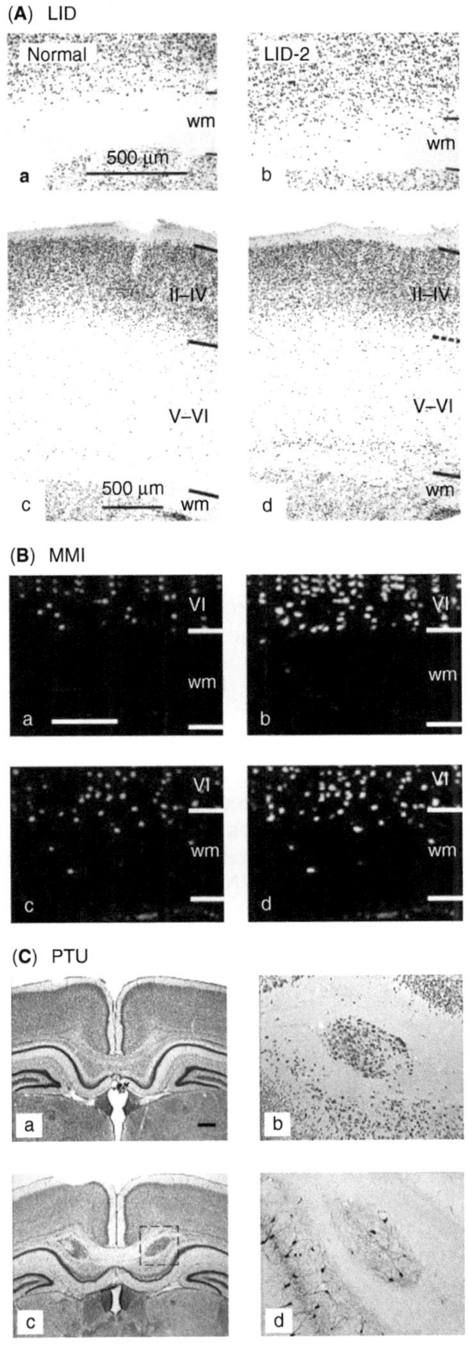

Figure 8 (*See color insert*) Cortical histogenesis and neuronal migration are disrupted with low level thyroid hormone (TH) disruption. (**A**) In the cortex of offspring born to iodine deficient rat dams (LID-2), cells are evident in the white matter (wm) of the corpus callosum (top) and the distinction between layers within the cortex are blurred (bottom). *Source*: From Ref. 118. (**B**) TH reductions induced over three days during gestation with methimazole (MMI) also impaired neuronal migration. Left is bromodeoxyruracil (BrdU)-positive cells, right is double labelling of BrdU and NeuN, indicating that cortical neurons persist in the white matter (wm) at PN40. *Source*: From Ref. 114. (**C**) Moderate degrees of TH insufficiency limited to the prenatal period induced by propylthiouracil (PTU) produced an abnormal cluster of cells, a heterotopia, in the corpus callosum (top left is control, bottom left is treated). Cells within the heterotopia were neurons as they stained positively with NeuN (top right), and some were inhibitory, staining positively for parvalbumin (bottom right). *Source*: From Ref. 157. Collectively, these data indicate that permanent structural changes are induced by mild or transient reductions in TH in the absence of other overt toxicity.

rather than preferential medial direction of travel. These observations indicate that the migratory potential of the neurons was not changed by TH insufficiency. Rather the transient hypothyroxinemia in the dam altered the cortical milieu upon which appropriate migration occurs.

Collectively, these findings indicate that severe hormone deficiencies and hypothyroidism are not necessary to disrupt normal brain development. Modest reductions in hormones are sufficient to induce a delay in the temporal development of the radial glial scaffold, prevent the normal pattern of neuronal migration, and promote a disorganization of cortical layering. THs modulate both proximal and distal cues responsible for guiding the migrational map of both excitatory and inhibitory neurons. Alterations in temporal patterning of cortical neuronal migration and subtle changes in cytoarchitectonic organization indicate that the normal trajectory of brain development has been disrupted.

In the presence of altered cortical development, it is not unlikely that the establishment of normal brain functions is also compromised and has significant consequences for the organism. Impaired migration of inhibitory neurons may contribute to increased susceptibility to seizures observed by Auso et al. (114) following brief and transient reductions in maternal THs. In our laboratory, we have also observed an enhanced responsiveness to pentylenetetrazol-induced seizures in PTU-exposed animals (Gilbert, unpublished observations). Expression of parvalbumin, a calcium-binding protein in inhibitory neurons is suppressed in cortex and hippocampus of hypothyroid animals and may alter physiological properties on these local circuit neurons (159,160). Reductions in synaptic inhibition and excitatory synaptic transmission in the hippocampus of adult rats developmentally exposed to PTU have also been reported (161–166). Perinatal exposure to environmental contaminants that reduce circulating levels of T4 (e.g., PCBs, diphenyl ethers, perchlorate) also alter synaptic function in the hippocampus and cortex (14,15,20,167–172).

TH and Myelination

Myelination is a predominantly postnatal process that extends in rodents from ~PN10 until beyond PN30 and from before birth to three years of age in humans (173). The myelin coating of neuronal axons in the CNS provides the electrical insulation for rapid and efficient conduction of electrical impulses that forms the basis of intercellular synaptic transmission and communication in the brain. Oligodendrocytes are a specialized type of glial cell responsible for synthesis, assembly, and maintenance of CNS myelin. The formation of myelin by oligodendrocytes requires the coordination and synthesis of large quantities of specific proteins and lipids and the integration of these into a highly organized multilamellar structure (174).

The role of TH in myelination has been well documented. Oligodendrocytes express TH receptors, and in cell culture systems, the addition of TH promotes the differentiation and proliferation of oligodendrocyte precursors as well as the amount of myelin each mature oligodendrocyte produces (175,176). TH accelerate the rate of accumulation but not typically the final concentration of specific oligodendrocyte mRNAs and proteins, and regulate the temporal expression of a number of key enzymes involved in the synthesis of complex lipids and protein constituents of the myelin membrane. The wave of myelination that starts in caudal brain regions and progresses rostrally is delayed in hypothyroid animals (48,66,174,177).

A number of classic myelin markers are reduced in the brains of hypothyroid animals. CNPase (2′3′-cyclic nucleotide 3′-phosphodiesterase) is a myelin protein expressed in early stages of oligodendrocyte development, whereas proteolipid protein (PLP), myelin basic protein (MBP), and myelin associated glycoprotein (MAG) appear later during the construction of the myelin sheath. The expression of myelin associated oligodendrocyte basic protein (MOBP) is associated with myelin compaction. CNPase, PLP, MAG, and MBP are down regulated in the cerebellum, cortex, striatum, and corpus callosum as a function of MMI- or PTU-induced hypothyroidism (48,66,152,174,177–182). MBP is reduced on PN10 in cerebellum and cortex of hypothyroid animals but recovery occurs despite continued reductions in T3 (183,184). In contrast, permanent downregulation of MOBP and PLP expression are evident in corpus callosum and cerebellum (178). Iodine deficiency is also associated with reductions in MBP-immunoreactivity in the internal capsule on PN10 and PN15, which were no longer evident by PN20 (180). The pattern was similar to that induced by MMI but transient and less severe, consistent with the more modest effects of iodine deficiency relative to MMI on brain T3. Recently, microarrays have been used to identify genes responsive to mild disruptions of the thyroid axis (71). In the hippocampus of PN14 pups, a number of myelin genes were significantly downregulated at very low doses of PTU producing modest reductions in circulating levels of T4 (Table 2). These data suggest that myelin genes may provide a very sensitive biomarker of low level TH disruption.

The corpus callosum and anterior commissure have long been a focus for the study of THs on the myelination process. These extensive fiber pathways represent the primary means of communication between the two cerebral hemispheres. They are present in the embryo, undergo substantial postnatal development, and contain a large number of heavily myelinated fibers. The number of axons in corpus callosum and anterior commissure of hypothyroid rats does not differ, but the density of fibers is increased, the cross-sectional area is smaller, and the ratio of myelinated to unmyelinated fibers is substantially reduced (185–187).

Myelination has not been well studied in animals with other than severe TH deficiencies. In a recent paper by Sharlin et al. (182) the number of oligodendrocytes and astrocytes was examined in rat pups exposed perinatally to a series of low doses of PTU that produced graded levels of TH insufficiency (Fig. 9). PTU induced a dose-dependent reduction in the

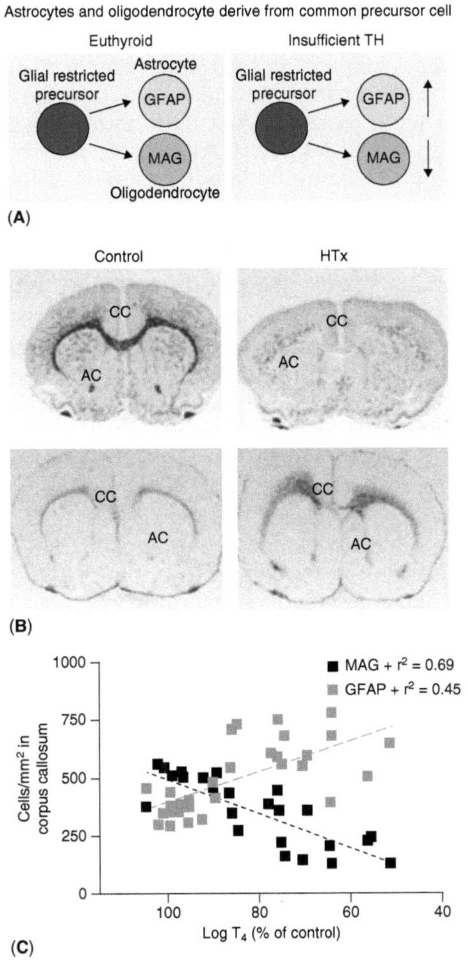

Figure 9 Oligodendrocytes and astrocytes derive from a common glial precursor cell. Under conditions of severe thyroid hormone insufficiency, there is a reduction in oligodendrocytes and myelination of axons in the corpus callosum and anterior commissure is impaired. Oligodendrocyte expression is also reduced following low level thyroid hormone disruption. Offspring (PN15) of dams exposed to 0, 1, 2, or 3 ppm PTU from GD6 until pups were sacrificed and brains prepared for in situ hybridization of mRNA for myelin associated glycoprotein (MAG) to mark oligodendrocytes, or glial fibrillary acidic protein (GFAP) to identify astrocytes. The reduction in MAG-positive cells was matched by a parallel increase in astrocytes staining for GFAP. A positive correlation between number of GFAP-positive cells and serum T4 and a negative correlation between MAG-positive cell number and serum T4 suggests that TH insufficiency altered cell fate specificity in the developing brain. *Source*: From Ref. 182.

expression of MAG, as well as the number of oligodendrocytes in corpus callosum and anterior commissure. This alteration in MAG expression was accompanied by a commensurate increase in the density of astrocytes expressing GFAP. The total number of cells, MAG and GFAP-expressing, did not differ between control and PTU-treated animals at any dose level, but the ratio of oligodendrocytes to astrocytes was significantly reduced at all but the lowest dose. Because oligodendrocytes and astrocytes derive from a common glial precursor cell, these observations suggest that adequate TH is essential in directing cells into the oligodendrocyte over the astrocytic cell lineage (73,74,188). Importantly, these findings are the first to demonstrate that shifts in fate specificity can occur at moderate levels of hormone insufficiency.

TH and Development of the Hippocampus

In a series of papers, Madeira and colleagues (189–191) examined the impact of neonatal hypothyroidism on development of the hippocampal formation in rats. No change in volume of the CA1 pyramidal cell layer was seen, but the neuronal volume and cell number were reduced in hypothyroid animals. In the neonatal brain, the pyramidal cells of Ammon's horn show a gradation of progressive differentiation over time from area CA1 to CA4. Pyramidal neurons of hypothyroid animals have fewer synapses and an impoverished dendritic arbor, the extent of impairment dependent upon the position of the cells within the layer. Area CA3-4 pyramidal neurons are born later and were more affected than neurons within area CA1 (190,192). The apparent greater vulnerability of CA3-4 over CA1 neurons may be a function of timing of PTU initiation, exposure in this model beginning late in gestation when pyramidal cells of CA1 had already differentiated.

In contrast to pyramidal cells, granule cells (the principal cell type of the dentate gyrus region of the hippocampal formation) are born postnatally, and are more affected by hypothyroidism initiated just before birth. PTU decreased the volume of the GCL, the density of cells within the layer, and estimates of total granule cell number (189). Migration of granule cells from the proliferative zone to the GCL is retarded by thyroid deficiency as is dendritic arborization and synaptogenesis assessed by immunohistochemistry for the synaptic protein, synaptophysin (145,192,193). Studies of this nature have not been performed in models other than severe neonatal hypothyroidism so it is not known if such changes will be evident in hippocampus with more modest hormone insufficiency. Some evidence of a blurring of cell layers was reported in area CA1 by Auso et al. (114) in response to brief episodes of maternal hypothyroxinemia in the mid to late gestational period.

Functional impairments in hippocampal synaptic transmission have been reported following developmental TH deficiencies. In models of severe hypothyroidism, beginning either in the early or late gestation period and continuing to weaning, excitatory synaptic function is disrupted in area CA1 of slices taken from hypothyroid animals in the preweaning period (162,164–166). These changes in synaptic transmission are permanent as deficiencies are also evident in slices from adult animals exposed developmentally but tested following return to euthyroidism (161,194). In a much more restricted dosing paradigm, similar to that described by Auso et al. (114), Opazo and colleagues (119) also reported deficits in hippocampal synapatic plasticity in adult offspring following a brief and transient reduction in TH in utero. These electrophysiological observations were associated with impaired performance in a spatial learning task. In vivo recordings from the dentate gyrus in adult offspring of hypothyroid dams also exhibit permanent suppression of excitatory and inhibitory synaptic function (20,159,163,195). The latter observations by Gilbert and Sui (163) and Opazo et al. (119) are noteworthy as they were evident at modest levels of TH insufficiency (Fig. 10). Alterations in synaptic function have also been demonstrated in adult offspring of PCB- and perchlorate-exposed dams, environmental contaminants that produce a state of hypothyroxinemia in the dam (20).

TH and Development of the Sensory Organs

The Auditory System

THs are essential for normal development of the auditory system (196). Hearing impairment is the most frequent sensorineural defect in humans and is commonly associated with

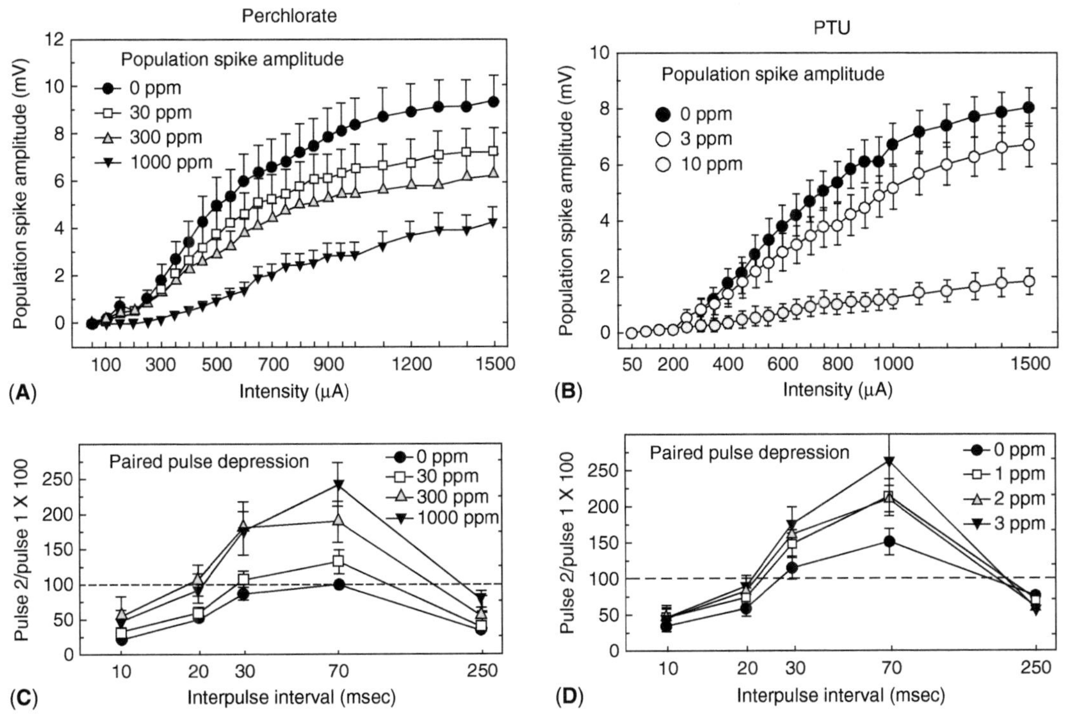

Figure 10 Permanent impairments of synaptic function in hippocampus following low level thyroid hormone (TH) disruption during development. Developmental exposure to propylthioruacil (PTU) or perchlorate reduce excitatory (**A, B**) and inhibitory (**C, D**) synaptic transmission in the dentate gyrus of adult offspring. These treatments produced moderate reductions in serum TH with no effects on growth or development, yet permanently altered function within these neural circuits. *Source*: From Refs. 20,159,163.

thyroid disorders, including congenital hypothyroidism (197). In rodents, changes in the relative availability of TH in late fetal and early neonatal life results affect the expression of functionally important proteins involved in the structural development and physiology of the inner ear (76). In addition to structural defects at the sensory organ level, hearing impairments may also derive from hypomyelination of the auditory nerve and delayed maturation of central auditory system (198). Aberrant synaptic connections in the auditory cortex, reductions in synaptic spine densities and hypomyelination of the interhemispheric connections, and loss of topographic specificity have been described in hypothyroid animals (186,187).

Structural defects in the auditory system are reflected in behavioral and electrophysiological indices of hearing impairment (76,198). Graded levels of hormone insufficiency induced by PTU in the late gestational/early postnatal period are also associated with varying degrees of hearing loss measured behaviorally (199). These impairments were evident in adults following transient exposure to PTU and were produced at doses and concentrations of serum hormones that were without effects on other physical and behavioral measures of hypothyroidism. These data are significant as they were the first to demonstrate dose-dependency of the hearing deficits at moderate levels of hormone insufficiency. Using brainstem auditory evoked potentials, Knipper et al. (198) also demonstrated irreversible elevation of hearing thresholds with very brief periods of exposure to MMI (GD17 to PN3), in the absence of cochlear structural defects, implicating impairment in central auditory processing systems. Low frequency hearing loss and structural damage to hair cells has been observed in animals perinatally exposed to PCBs, an environmental contaminant with thyroidal action, an effect that could be blocked by concomitant treatment with T4 (16,199,200). Crofton (5), summarizing a number of studies, demonstrated a strong correlation between degree of T4 reduction and the magnitude of the induced hearing loss.

The Visual System

Visual system development, particularly retinal development, is regulated by TH. Frank hypothyroidism results in fewer dividing retinal progenitor cells, reduced retinal thickness, and impaired photoreceptors with improperly formed outer segments (201,202). Structural alterations in development of the retina have not been examined in models of modest reductions in TH.

In rodents, retinal progenitor cells generate cones from E10 to E16, and rod cells are then generated from this pool of cones. TH activated transcription factors are necessary for instruction of photoreceptor cell fate decisions (76). Rodents, like most mammals have two opsin proteins that respond maximally at different wavelengths: S-opsin to short and M-opsin to longer wavelengths. In early retinal development, THs determine the "developmental choice" of which opsin gene to express (203). The default for cone development is to express S-opsins (blue) and this begins between E10 and E16. TH regulates the switch from S-opsin to M-opsin (red-green) cone expression around PN10. There is a dorsal ventral gradient of opsin expression on the mouse retina that also displays an ontogenetic profile. This gradient and profile is mirrored by TH expression in the retina.

Data to date have revealed that experimental elevation of T3 at the time of S-opsin onset (mid-late gestation and just after birth) inhibits S-opsin expression. Treatment with T3 at the time of M-opsin activation (~PN10) increases M-opsin expression. TH receptor knock-out mice also show disruption of the S- and M-opsin gradient in the retina and an overexpression of S-opsins at the expense of M-opsin mRNA (203). Although not studied directly, these data suggest that minor hormone insufficiencies may produce subtle alterations in the expression of S- and M-opsin in the developing retina, and that the nature of the effects will be dependent upon developmental window over which it occurs. Animal studies of visual function following modest perturbations of the thyroid axis deserve further investigation, especially in light of evidence of color vision and contrast sensitivity deficits in children suffering congenital hypothyroidism (204). Preliminary findings in rats treated perinatally with PTU reveal irreversible impairments in visual contrast sensitivity. Whether these deficits are dependent upon disruption at the retinal or central visual processing level has not been determined (Boyes and Gilbert, personal communication).

Behavioral Deficits Following Developmental Hypothyroidism

As outlined above, recent reports indicate that children born to women experiencing modest subclinical perturbations of the thyroid axis during pregnancy have reduced IQ scores and subtle deficits in cognition, memory, and visuo-spatial ability (115,124,125,131,137). In experimental models of developmental hypothyroidism, increases in locomotor activity, hearing loss, and reduced seizure thresholds accompany severe hormonal deprivation beginning in the late gestational or early postnatal period (114,199,205). Others have reported deficits in passive avoidance, spatial learning, and operant conditioning (21,206–212), but these observations are often limited to animals suffering fairly severe hormonal deprivation. Given the critical observations in human infants that very modest perturbations of the thyroid axis can alter cognition, animal research efforts have begun to focus on the timing and degree of hormone insufficiency as well as the nature and persistence of functional deficits. These include increased seizure responsiveness, reductions in synaptic transmission, alterations in hippocampal LTP, and spatial learning deficits (20,114,119,166,195,213). These recent findings demonstrate synaptic and behavioral deficits that persist in the adult despite return to the euthyroid state and are induced at relatively modest perturbations in the thyroid axis.

Knock-out models have also been assessed for subtle alterations in behavior, including tests of anxiety and learning. As described above, TH receptor knock-outs and other mutants that exhibit perturbations of the thyroid axis exhibit a phenotype that is much milder than that apparent in conditions of severe chemical-induced hypothyroidism. TR α accounts for ~75% of all TH receptors in brain and as such the effects of TH are most likely to be mediated via this receptor (214). TR β has a more circumscribed localization in purkinje cells of the cerebellum, the hypothalamus, the retina, and the cochlea (40,76). Hearing loss, visual deficits, and cerebellar ataxia are evident in these mice, but no alterations in open field behavior, spatial learning in a Morris water maze, or in context fear conditioning were observed (215). TR α knockouts display increased anxiety in locomotor tests in an open field, spending a greater percentage of time in the periphery than in the central regions of the open field. Acquisition of fear conditioning was similar in knockout and wildtype animals, but delays in extinction were evident in the knockouts (216). These behavioral deficits were associated with reductions in parvalbumin immunoreacitivity in GABAergic interneurons in the hippocampus and decreases in GABA-mediated synaptic inhibition as described above in chemically-induced hypothyroidism (20,160). In another model, a mutation introduced to the TR α that greatly reduces the affinity of the receptor for T3 resulted in a reduction in the number of parvalbumin immunoreactive neurons. A similar behavioral phenotype of locomotor dysfunction, increased anxiety in open field and elevated plus maze, and memory deficits in the delayed object recognition task was also produced (217).

NEUROLOGICAL IMPAIRMENTS ACCOMPANYING ADULT ONSET HYPOTHYROIDISM

Hypothyroidism is a prevalent condition in humans with an incidence of 8% in the adult population (218). The nongenomic actions of TH described previously may be one mechanism whereby TH insufficiencies beginning in adulthood lead to psychiatric disturbances, dementias and major depression (219,220). In young and middle-aged adults, thyroid dysfunction is associated with both neurological and behavioral abnormalities, including mood disturbances and cognitive dysfunction (221). Cognitive decline in the elderly and in cases of geriatric dementia and Alzheimer's disease are often associated with decreases in TH that can be partially alleviated by treatment with thyroxine (222–226).

Animal models of adult onset hypothyroidism have demonstrated increased reactivity to stress in the forced swim model of depression (227). Spatial learning and hippocampal plasticity in the form of LTP are also impaired in adult rats subjected to thyroidectomy, effects that are reversed by treatment with T3 or T4 (228–231). Adult onset hypothyroidism induced by PTU also reduces activity-dependent synaptic plasticity in the hippocampal-prefrontal cortex circuit (232). In contrast to electrophysiological assessments in adult offspring following hypothyroidism limited to the developmental period, reductions in synaptic plasticity in adult onset hypothyroidism exist in the absence of alterations in baseline synaptic responsiveness. This distinction suggests different life stages over which hormone insufficiency is induced may result in different patterns of disruption of synaptic function, both of which, however, may ultimately present as behavioral impairment (163,230). Another primary distinction between the impact of thyroid dysfunction in the developing and the adult organism is the reversibility of deficits. Hypothyroidism beginning in adulthood can be effectively treated with hormone therapy, in contrast to impairments produced by developmental TH insufficiencies that persist despite return to euthyroidism in adulthood (119,159,163).

Disruption of the thyroid axis by physical, chemical or genetic manipulations is associated with a reduction in the generation and survival of new of neurons in the dentate gyrus of the adult hippocampus and olfactory bulb (233–236). This process, known as adult neurogenesis, has been implicated in learning, memory and affective disorders such as depression (237–241). Initial reports using PTU or MMI demonstrated a specific decline in the survival of newly born granule cells in the hippocampus in thyroid-deficient rats (233,234). Disruption of the thyroid axis by thyroidectomy, however, produced deficits in cell proliferation and maturation without impacting survival (236). Consistent with these reports, administration of T3 into the dorsal hippocampus of adult rats improves performance on an emotional memory task (242), whereas hypothyroidism initiated in adulthood impairs spatial learning, hippocampal LTP, and increases stress, anxiety, and depression in a variety of animal models (227,230,231,236,243). Collectively, these findings suggest that in addition to the established role of TH in early neurogenesis, TH are essential components of the endocrine environment that modulates neural stem cell progenitor growth, migration, and apoptosis in the mature nervous system. TH modulation of adult neurogenesis may represent one mechanism underlying cognitive impairments and mood disorders in adults with TH imbalance.

IMPLICATIONS FOR NEUROTOXICITY OF TH DISRUPTORS

A number of recent developments indicate that the HPT axis is controlled in ways not fully appreciated previously, but that are important for thyroid toxicology. New information accumulating over the past decade has significantly increased our understanding of the regulation of TH in the brain and other tissues and requires a refinement to the traditional model of the HPT axis. New insights into feedback control mechanisms, plasma membrane transporters, and metabolizing enzymes demonstrate modulation of thyroid action to a degree of precision and regional specificity not previously recognized. Incorporation of these findings into our conceptual framework requires modifications to the classic model of the HPT depicted in Figure 1. Refining the model of TH function will also necessarily influence the course of research in neurotoxicology of anti-thyroid agents, as well as the specific kinds of endpoints evaluated in a screening and testing program designed to identify thyroid toxicants (244). We feel that there are three particular areas of uncertainty about the thyroid system that will continue to require attention.

1. Determining the relationship between alterations in circulating levels of TH and TH action in the production of adverse effects.
2. Determining the relationship between life stage, vulnerability to thyroid toxicants, and specific adverse consequences.
3. Identification of biomarkers of thyroid toxicity that reflect adverse "downstream" effects.

Determining the Relationship Between Circulating Levels of TH and TH Action

The general dogma of thyroid endocrinology is that serum total and free THs (T4 and/or T3) are linearly related to TH receptor occupancy and define TH action in tissues (57). Modest reductions in serum T3 lead to activation of the thyroid axis and increases in TSH. However, recent findings in thyroid toxicology indicate that this mutual balance with synchronous control of T4 and TSH is not always observed (10,245,246). PCBs, for example, are well known to cause a reduction in circulating total and free TH (10,245), yet administration of PCB mixtures such as Aroclor 1254 do not also increase serum TSH (10). Furthermore, animal studies have not typically evaluated the dose-dependent relationship between circulating levels of TH and cellular levels of genes regulated by TH. Nor has there been an adequate examination of the relationship of circulating TH and developmental events. In light of recent human findings of neurodevelopmental impairments following subclinical hypothyroxemia (115), the animal research community has been challenged to develop models of moderate degrees of TH disruption that more directly parallel the human conditions.

As data from low-level TH disruption begins to accumulate, how best can these models be characterized? Serum hormones are used diagnostically and provide a valuable metric to compare human and animal findings, and to contrast animal studies among themselves. Figure 11 presents a scheme of TH insufficiency ranging from mild to severe, based on relationships of T3, T4, and TSH derived from classic TH disruptors, MMI, and PTU. The graphic presents a gradation of symptomotology that parallels increasing degrees of TH disruption. There is a wealth of information on mechanisms of thyroid action at the molecular, cellular, structural and behavioral levels based on animal models and human conditions of severe hypothyroidism. There is now accumulating a research base that has begun to characterize the impacts and mechanisms of much more subtle disturbances to the thyroid axis and to define gradations in the degree of TH insufficiency that lead to graded levels of neurological dysfunction. However, not all TH disruptors are created equal. It is becoming increasingly clear that physiological responses to toxicants that interfere with the thyroid axis are not well represented in models derived from PTU or MMI. Is it wise then to base decisions of public health standards on concepts derived from these models?

Determining the Relationship Between Life Stage, Vulnerability to Thyroid Toxicants, and Specific Adverse Consequences

In the adult, the range of fluctuations in hormone levels is large and the consequences of even a relatively protracted time of hormonal imbalance, although undesirable, is not irrecoverable.

In the fetus and neonate, however, the range over which hormones can change is small as the timing of critical developmental events dependent on TH signaling must be exquisitely maintained. Neither is the fetus or the neonate equipped with the same compensatory responses as the adult (as previously summarized in Table 1). Even modest and transient alterations in the thyroid axis have the potential to produce irreversible changes in brain development that ripple well beyond the initiating event.

Given genetic variation among individuals (52,247), the narrow range of variation of TH on an individual level (248), and the number of checks and balances incorporated into the mammalian physiology to maintain optimal hormone levels, how do we determine if the system has been perturbed, and if perturbed, perturbed to a sufficient degree to have an adverse impact on brain development? In the evaluation of environmental contaminants that are thyroid disruptors, the issue is more complicated. Environmental contaminants disrupt the thyroid axis through a variety of mechanisms and exhibit significant deviations from the profile of hormone disruption depicted in Figure 11.

Identification of Biomarkers of Thyroid Toxicity That Reflect Adverse "Downstream" Effects

Clinical measures of thyroid function, most often total T3 and T4, have guided the measurement and interpretation of altered thyroid status in rodent models. Contaminants, such as PCBs and pesticides, reduce circulating levels of T4 without changing T3 or TSH (7,11,249). At what level does a measurable change in circulating T4 become of toxicological significance? When does an increase in TSH represent a simple adaptive response to a minor fluctuation within the normal physiological range or an indication of hormone disruption with physiological consequences? How reflective are circulating levels of total TH in the blood of critical tissue concentrations of free TH in a specific brain region requiring that signal at this particular time? Is there a critical magnitude or duration of hormone change that can be identified where

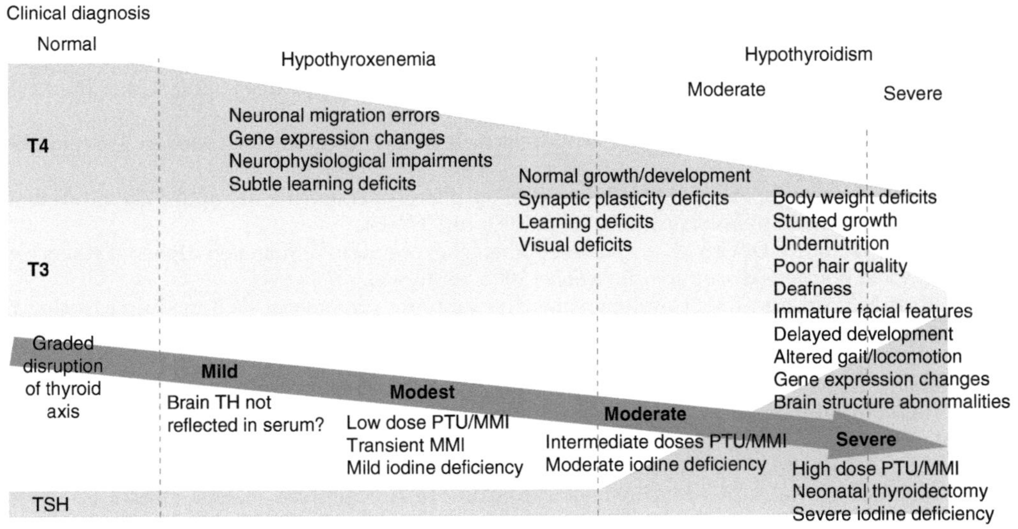

Figure 11 Gradient of thyroid hormone insufficiency. Both human and animal studies reveal that graded degrees of TH insufficiency can lead to different phenotypes. Initial models of TH function as depicted in Figure 1 were based on conditions of severe hypothyroidism (dotted line to far right). This depiction of graded levels of TH disruption is based on animal studies using MMI, PTU, ID, treatments that primarily act to reduce the synthesis of TH at the level of the thyroid gland. Environmental contaminants that perturb the thyroid axis through different mechanisms do not necessarily maintain these same relationships between T3, T4, and TSH, yet impact brain development. Defining these relationships across structural and mechanistically distinct chemicals represents a significant challenge to the field of thyroid toxicology. The degree to which serum TH can be disturbed without adverse consequences on brain development has not been adequately defined. Abbreviations: MMI, methimazole; PTU, propylthiouracil; TH, thyroid hormone; TSH, thyroid-stimulating hormone; ID, iodine deficiency.

"adaptive" responses are indicative of disturbances that harbor significant physiological consequences? As described above, it is now clear that other targets exist within the thyroid system that may be prey to environmental toxicants and that will not be reflected in changes in serum hormone (e.g., TH receptors, local tissue specific deiodinases, transport proteins). The use of traditional molecular techniques and more recently genomic microarrays have identified numerous genes whose expression is altered by low level TH disruption, the functional consequences of these changes have yet to be determined (71,182). To adequately address these issues considerably more information on dose-response relationships at low levels of thyroid disruption of a variety of measures, hormonal, genomic, cellular, physiological, and functional will be required.

REFERENCES

1. Burmeister LA, Ganguli M, Dodge H, et al. Hypothyroidism and cognition: preliminary evidence for a specific defect in memory. Thyroid 2001; 11: 1177–85.
2. Haggerty JJ Jr, Garbutt JC, Evans DL, et al. Subclinical hypothyroidism: a review of neuropsychiatric aspects. Int J Psychiatry Med 1990; 20: 193–208.
3. Monzani F, DelGuerra P, Caraccio N, et al. Subclinical hypothyroidism: neurobehavioral features and beneficial effect of L-thyroxine treatment. Clin Investig 1993; 71: 367–71.
4. Bernal J. Thyroid hormone receptors in brain development and function. Nat Clin Pract Endocrinol Metab 2007; 3: 249–59.
5. Crofton KM. Developmental disruption of thyroid hormone: correlations with hearing dysfunction in rats. Risk Anal 2004; 24: 1665–71.
6. Crofton KM, Zoeller RT. Mode of action: neurotoxicity induced by thyroid hormone disruption during development—hearing loss resulting from exposure to PHAHs. Crit Rev Toxicol 2005; 35: 757–69.
7. Howdeshell KL. A model of the development of the brain as a construct of the thyroid system. Environ Health Perspect 2002; 110(Suppl 3): 337–48.
8. Porterfield SP. Vulnerability of the developing brain to thyroid abnormalities: environmental insults to the thyroid system. Environ Health Perspect 1994; 102(Suppl 2): 125–30.
9. Zoeller RT, Dowling AL, Vas AA. Developmental exposure to polychlorinated biphenyls exerts thyroid hormone-like effects on the expression of RC3/neurogranin and myelin basic protein messenger ribonucleic acids in the developing rat brain. Endocrinology 2000; 141: 181–9.
10. Zoeller RT, Dowling AL, Herzig CT, et al. Thyroid hormone, brain development, and the environment. Environ Health Perspect 2002; 110(Suppl 3): 355–61.
11. Brucker-Davis F. Effects of environmental synthetic chemicals on thyroid function. Thyroid 1998; 8: 827–56.
12. Colborn T, vom Saal FS, Soto AM. Developmental effects of endocrine-disrupting chemicals in wildlife and humans. Environ Health Perspect 1993; 101: 378–84.
13. Zhou T, Taylor M, DeVito M, et al. Developmental exposure to brominated diphenyl ethers results in thyroid hormone disruption. Toxicol Sci 2002; 66: 105–16.
14. Gilbert ME, Crofton KM. Developmental exposure to a commercial PCB mixture (Aroclor 1254) produces a persistent impairment in long-term potentiation in the rat dentate gyrus in vivo. Brain Res 1999; 850: 87–95.
15. Gilbert ME, Mundy WR, Crofton KM. Spatial learning and long-term potentiation in the dentate gyrus of the hippocampus in animals developmentally exposed to Aroclor 1254. Toxicol Sci 2000; 57: 102–11.
16. Goldey ES, Crofton KM. Thyroxine replacement attenuates hypothyroxinemia, hearing loss, and motor deficits following developmental exposure to Aroclor 1254 in rats. Toxicol Sci 1998; 45: 94–105.
17. Goldey ES, Kehn L, Lau C, et al. Developmental exposure to polychlorinated biphenyls (Aroclor 1254) reduces circulating thyroid hormone concentrations and causes hearing deficits in rats. Toxicol Appl Pharmacol 1995; 135: 77–88.
18. Nguon K, Baxter MG, Sajdel-Sulkowska EM. Perinatal exposure to polychlorinated biphenyls differentially affects cerebellar development and motor functions in male and female rat neonates. Cerebellum 2005; 4: 112–22.
19. Sharlin DS, Bansal R, Zoeller RT. Polychlorinated biphenyls exert selective effects on cellular composition of white matter in a manner inconsistent with thyroid hormone insufficiency. Endocrinology 2006; 147: 846–58.

20. Gilbert ME, Sui L. Developmental exposure to perchlorate alters synaptic transmission in hippocampus of the adult rat. Environ Health Perspect 2008; 116: 752–60.
21. van Wijk N, Rijntjes E, van de Heijning BJ. Perinatal and chronic hypothyroidism impair behavioural development in male and female rats. Exp Physiol 2008 (Epub 20 June).
22. Turyk ME, Anderson HA, Persky VW. Relationships of thyroid hormones with polychlorinated biphenyls, dioxins, furans, and DDE in adults. Environ Health Perspect 2007; 115: 1197–203.
23. Chevrier J, Eskenazi B, Bradman A, et al. Associations between prenatal exposure to polychlorinated biphenyls and neonatal thyroid-stimulating hormone levels in a Mexican-American population, Salinas Valley, California. Environ Health Perspect 2007; 115: 1490–6.
24. Blount BC, Pirkle JL, Osterloh JD, et al. Urinary perchlorate and thyroid hormone levels in adolescent and adult men and women living in the United States. Environ Health Perspect 2006; 114: 1865–71.
25. Jacobson JL, Jacobson SW. Intellectual impairment in children exposed to polychlorinated biphenyls in utero. N Engl J Med 1996; 335: 783–9.
26. Koopman-Esseboom C, Morse DC, Weiglas-Kuperus N, et al. Effects of dioxins and polychlorinated biphenyls on thyroid hormone status of pregnant women and their infants. Pediatr Res 1994; 36: 468–73.
27. Koopman-Esseboom C, Weiglas-Kuperus N, deRidder MA, et al. Effects of polychlorinated biphenyl/dioxin exposure and feeding type on infants' mental and psychomotor development. Pediatrics 1996; 97: 700–6.
28. Lonky E, Reihman J, Darvill T, et al. Neonatal behavioral assessment scale performance in humans influenced by maternal consumption of enviornmentally contaminated Lake Ontario fish. J Great Lakes Res 1996; 22: 198–212.
29. Patandin S, Lanting C, Mulder PG, et al. Effects of environmental exposure to polychlorinated biphenyls and dioxins on cognitive abilities in Dutch children at 42 months of age. J Pediatr 1999; 134: 33–41.
30. Needham LL, Calafat AM, Barr DB. Uses and issues of biomonitoring. Int J Hyg Environ Health 2007; 210: 229–38.
31. Vigone MC, Fugazzola L, Zamproni J, et al. Persistent mild hypothyroidism associated with novel sequence variants of the DUOX2 gene in two siblings. Hum Mutat 2005; 26: 395.
32. Pachucki J, Wang D, Christophe R, et al. Structural and functional characterization of the two human ThOX/Duox genes and their 5'-flanking regions. Mol Cell Endocrinol 2004; 214: 53–62.
33. Dupuy C, Pomerance M, Ohayan R, et al. Thyroid oxidase (THOX2) gene expression in the rat thyroid cell line FRTL-5. Biochem Biophys Res Commun 2000; 277: 287–92.
34. Wang D, deDeken X, Milenkovic M, et al. Identification of a novel partner of duox: EFP1, a thioredoxin-related protein. J Biol Chem 2005; 280: 3096–103.
35. Greer MA, Goodman G, Pleus R, et al. Health effects assessment for environmental perchlorate contamination: the dose response for inhibition of thyroidal radioiodine uptake in humans. Environ Health Perspect 2002; 110: 927–37.
36. Savin S, Cvejic O, Nedic O, et al. Thyroid hormone synthesis and storage in the thyroid gland of human neonates. J Pediatr Endocrinol Metab 2003; 16: 521–8.
37. van den Hove MF, Beckers C, Devielger H, et al. Hormone synthesis and storage in the thyroid of human preterm and term newborns: effect of thyroxine treatment. Biochimie 1999; 81: 563–70.
38. Snead AN, Santos M, Seal R, et al. Thyronamines inhibit plasma membrane and vesicular monoamine transport. ACS Chem Biol 2007; 2: 390–8.
39. Nikrodhanond AA, Ortiga-Carvalho T, Shibusawa N, et al. Dominant role of thyrotropin-releasing hormone in the hypothalamic-pituitary-thyroid axis. J Biol Chem 2006; 281: 5000–7.
40. Wan W, Farboud B, Privalsky ML. Pituitary resistance to thyroid hormone syndrome is associated with T3 receptor mutants that selectively impair beta2 isoform function. Mol Endocrinol 2005; 19: 1529–42.
41. Greer MA, Sato M, Wang X, et al. Evidence that the major physiological role of TRH in the hypothalamic paraventricular nuclei may be to regulate the set-point for thyroid hormone negative feedback on the pituitary thyrotroph. Neuroendocrinology 1993; 57: 569–75.
42. Guo F, Bakal K, Minokoshi Y, et al. Leptin signaling targets the thyrotropin-releasing hormone gene promoter in vivo. Endocrinology 2004; 145: 2221–7.
43. Rondeel JM, Heide R, deGreef WJ, et al. Effect of starvation and subsequent refeeding on thyroid function and release of hypothalamic thyrotropin-releasing hormone. Neuroendocrinology 1992; 56: 348–53.
44. Guadano-Ferraz A, Escamez M, Rausell E, et al. Expression of type 2 iodothyronine deiodinase in hypothyroid rat brain indicates an important role of thyroid hormone in the development of specific primary sensory systems. J Neurosci 1999; 19: 3430–9.

45. Visser WE, Friessma E, Jansen J, et al. Thyroid hormone transport by monocarboxylate transporters. Best Pract Res Clin Endocrinol Metab 2007; 21: 223–36.
46. St Germain DL, Hernandez A, Schneider M, et al. Insights into the role of deiodinases from studies of genetically modified animals. Thyroid 2005; 15: 905–16.
47. Cooper DS, Kieffer J, Halpern R, et al. Propylthiouracil (PTU) pharmacology in the rat. II. Effects of PTU on thyroid function. Endocrinology 1983; 113: 921–8.
48. Bernal J. Action of thyroid hormone in brain. J Endocrinol Invest 2002; 25: 268–88.
49. Chan S, Franklyn JA, Kilby MD. Maternal thyroid hormones and fetal brain development. Curr Opin Endocrinol Diabetes Obes 2005; 12: 23–30.
50. Chan S, Kachilele S, McCabe C, et al. Early expression of thyroid hormone deiodinases and receptors in human fetal cerebral cortex. Brain Res Dev Brain Res 2002; 138: 109–16.
51. Forrest D, Vennstrom B. Functions of thyroid hormone receptors in mice. Thyroid 2000; 10: 41–52.
52. Peeters RP, VandenBeld A, Attalke H, et al. A new polymorphism in the type II deiodinase gene is associated with circulating thyroid hormone parameters. Am J Physiol Endocrinol Metab 2005; 289: E75–81.
53. Gauger KJ, Kato Y, Haraguchi K, et al. Polychlorinated biphenyls (PCBs) exert thyroid hormone-like effects in the fetal rat brain but do not bind to thyroid hormone receptors. Environ Health Perspect 2004; 112: 516–23.
54. Guadano-Ferraz A, Escamez M, Morte B, et al. Transcriptional induction of RC3/neurogranin by thyroid hormone: differential neuronal sensitivity is not correlated with thyroid hormone receptor distribution in the brain. Brain Res Mol Brain Res 1997; 49: 37–44.
55. Schneider MJ, Fiering S, Pallud S, et al. Targeted disruption of the type 2 selenodeiodinase gene (DIO2) results in a phenotype of pituitary resistance to T4. Mol Endocrinol 2001; 15: 2137–48.
56. Karmarkar MG, Prabarkaran D, Godbole MM. 5′-Monodeiodinase activity in developing human cerebral cortex. Am J Clin Nutr 1993; 57(2 Suppl): 291S–4S.
57. Oppenheimer J. The nuclear-receptor-triiodothyronine complex: relationship to thyroid hormone distribution, metabolism, and biological action. In: Samuels HH, ed. Molecular Basis of Thyroid Hormone Action. New York: Academic Press, 1983: 1–34.
58. Gruters A. Thyroid hormone transporter defects. Endocr Dev 2007; 10: 118–26.
59. Jansen J, Fiesema E, Kester M, et al. Functional analysis of monocarboxylate transporter 8 mutations identified in patients with X-linked psychomotor retardation and elevated serum triiodothyronine. J Clin Endocrinol Metab 2007; 92: 2378–81.
60. Jansen J, Kester FE, Schwartz MH, et al. Genotype-phenotype relationship in patients with mutations in thyroid hormone transporter MCT8. Endocrinology 2008; 149: 2184–90.
61. Iniguez MA, deLecea L, Guadano-Ferraz A, et al. Cell-specific effects of thyroid hormone on RC3/neurogranin expression in rat brain. Endocrinology 1996; 137: 1032–41.
62. Iniguez MA, Rodriguez-Pena A, Ibarrola N, et al. Thyroid hormone regulation of RC3, a brain-specific gene encoding a protein kinase-C substrate. Endocrinology 1993; 133: 467–73.
63. Sharlin DS, Gilbert ME, Taylor M, et al. The nature of compensatory response to low level thyroid hormone in the developing brain. J Neuroendocrinology 2009; in press.
64. Oppenheimer JH, Schwartz HL. Molecular basis of thyroid hormone-dependent brain development. Endocr Rev 1997; 18: 462–75.
65. Thompson CC, Potter GB. Thyroid hormone action in neural development. Cereb Cortex 2000; 10: 939–45.
66. Anderson GW, Schoonover CM, Jones SA. Control of thyroid hormone action in the developing rat brain. Thyroid 2003; 13: 1039–56.
67. Flamant F, Samarut J. Thyroid hormone receptors: lessons from knockout and knock-in mutant mice. Trends Endocrinol Metab 2003; 14: 85–90.
68. Quignodon L, LeGrand C, Allioli N, et al. Thyroid hormone signaling is highly heterogeneous during pre- and postnatal brain development. J Mol Endocrinol 2004; 33: 467–76.
69. Bernal J, Guadano-Ferraz A, Morte B. Perspectives in the study of thyroid hormone action on brain development and function. Thyroid 2003; 13: 1005–12.
70. Chan S, Kilby MD. Thyroid hormone and central nervous system development. J Endocrinol 2000; 165: 1–8.
71. Royland JE, Parker JS, Gilbert ME. A genomic analysis of subclinical hypothyroidism in hippocampus and neocortex of the developing rat brain. J Neuroendocrinol 2008; 20: 1319–38.
72. Dong H, Wade M, Williams A, et al. Molecular insight into the effects of hypothyroidism on the developing cerebellum. Biochem Biophys Res Commun 2005; 330: 1182–93.
73. Billon N, Jolicoeur C, Tokumoto Y, et al. Normal timing of oligodendrocyte development depends on thyroid hormone receptor alpha 1 (TRalpha1). EMBO J 2002; 21: 6452–60.

74. Billon N, Tokumoto Y, Forrest D, et al. Role of thyroid hormone receptors in timing oligodendrocyte differentiation. Dev Biol 2001; 235: 110–20.
75. Poguet AL, LeGrand C, Feng X, et al. Microarray analysis of knockout mice identifies cyclin D2 as a possible mediator for the action of thyroid hormone during the postnatal development of the cerebellum. Dev Biol 2003; 254: 188–99.
76. Forrest D, Reh TA, Rusch A. Neurodevelopmental control by thyroid hormone receptors. Curr Opin Neurobiol 2002; 12: 49–56.
77. Hashimoto K, Curty F, Borges P, et al. An unliganded thyroid hormone receptor causes severe neurological dysfunction. Proc Natl Acad Sci U S A 2001; 98: 3998–4003.
78. Wondisford FE. Thyroid hormone action: insight from transgenic mouse models. J Investig Med 2003; 51: 215–20.
79. Davis PJ. Leonard JL, Davis FB. Mechanisms of nongenomic actions of thyroid hormone. Front Neuroendocrinol 2008; 29: 211–18.
80. D'Arezzo S, Incerpi S, Davis F, et al. Rapid nongenomic effects of 3,5,3′-triiodo-L-thyronine on the intracellular pH of L-6 myoblasts are mediated by intracellular calcium mobilization and kinase pathways. Endocrinology 2004; 145: 5694–703.
81. Davis PJ, Davis FB, Cody V. Membrane receptors mediating thyroid hormone action. Trends Endocrinol Metab 2005; 16: 429–35.
82. Tang HY, Lin H, Zheng S, et al. Thyroid hormone causes mitogen-activated protein kinase-dependent phosphorylation of the nuclear estrogen receptor. Endocrinology 2004; 145: 3265–72.
83. Farwell AP, Dubord-Tomasetti SA. Thyroid hormone regulates the expression of laminin in the developing rat cerebellum. Endocrinology 1999; 140: 4221–7.
84. Farwell AP, DuBord-Tomasetti S, Pretrzykowski A, et al. Dynamic nongenomic actions of thyroid hormone in the developing rat brain. Endocrinology 2006; 147: 2567–74.
85. Davis PJ, Davis FB. Nongenomic actions of thyroid hormone on the heart. Thyroid 2002; 12: 459–66.
86. Lin HY, Davis F, Gardinier J, et al. Thyroid hormone induces activation of mitogen-activated protein kinase in cultured cells. Am J Physiol 1999; 276(5 Pt 1): C1014–24.
87. Bonni A, Brunet A, West A, et al. Cell survival promoted by the Ras-MAPK signaling pathway by transcription-dependent and -independent mechanisms. Science 1999; 286: 1358–62.
88. Kaplan DR, Miller FD. Neurotrophin signal transduction in the nervous system. Curr Opin Neurobiol 2000; 10: 381–91.
89. Atkins CM, Selcher J, Petraitis J, et al. The MAPK cascade is required for mammalian associative learning. Nat Neurosci 1998; 1: 602–9.
90. Chen SJ, Sweatt JD, Klann E. Enhanced phosphorylation of the postsynaptic protein kinase C substrate RC3/neurogranin during long-term potentiation. Brain Res 1997; 749: 181–7.
91. Roberson ED, English JD, Sweatt JD. A biochemist's view of long-term potentiation. Learn Mem 1996; 3: 1–24.
92. Farwell AP, DuBord-Tomasetti S, Pretryzkowski A, et al. Regulation of cerebellar neuronal migration and neurite outgrowth by thyroxine and 3,3′,5′-triiodothyronine. Brain Res Dev Brain Res 2005; 154: 121–35.
93. Manzano J, Cuadrado M, Morte B, et al. Influence of thyroid hormone and thyroid hormone receptors in the generation of cerebellar gamma-aminobutyric acid-ergic interneurons from precursor cells. Endocrinology 2007; 148: 5746–51.
94. Cao X, Kambe F, Moeller L, et al. Thyroid hormone induces rapid activation of Akt/protein kinase B-mammalian target of rapamycin-p70S6K cascade through phosphatidylinositol 3-kinase in human fibroblasts. Mol Endocrinol 2005; 19: 102–112.
95. Coffer PJ, Jin J, Woodgett JR. Protein kinase B (c-Akt): a multifunctional mediator of phosphatidylinositol 3-kinase activation. Biochem J 1998; 335(Pt 1): 1–13.
96. Dygas A, Baranska J. Lipids and signal transduction in the nucleus. Acta Biochim Pol 2001; 48: 541–9.
97. Franke TF, Hornik C, Segev L, et al. PI3K/Akt and apoptosis: size matters. Oncogene 2003; 22: 8983–98.
98. Moeller LC, Cao X, Dumitrescu A, et al. Thyroid hormone mediated changes in gene expression can be initiated by cytosolic action of the thyroid hormone receptor beta through the phosphatidylinositol 3-kinase pathway. Nucl Recept Signal 2006; 4: e020.
99. Sui L, Wang J, Li BM. Administration of triiodo-L-thyronine into dorsal hippocampus alters phosphorylation of Akt, mammalian target of rapamycin, p70S6 kinase and 4E-BP1 in rats. Neurochem Res 2008; 33: 1065–76.
100. Cammalleri M, Lutjens R, Berton F, et al. Time-restricted role for dendritic activation of the mTOR-p70S6K pathway in the induction of late-phase long-term potentiation in the CA1. Proc Natl Acad Sci U S A 2003; 100: 14368–73.

101. Sanna PP, Cammalleri M, Berton F, et al. Phosphatidylinositol 3-kinase is required for the expression but not for the induction or the maintenance of long-term potentiation in the hippocampal CA1 region. J Neurosci 2002; 22: 3359–65.
102. Tang SJ, Reis G, Kang H, et al. A rapamycin-sensitive signaling pathway contributes to long-term synaptic plasticity in the hippocampus. Proc Natl Acad Sci U S A 2002; 99: 467–72.
103. Hallgren S, Darnerud PO. Polybrominated diphenyl ethers (PBDEs), polychlorinated biphenyls (PCBs) and chlorinated paraffins (CPs) in rats-testing interactions and mechanisms for thyroid hormone effects. Toxicology 2002; 177: 227–43.
104. Lans MC, Spiertz C, Brouwer A, et al. Different competition of thyroxine binding to transthyretin and thyroxine-binding globulin by hydroxy-PCBs, PCDDs and PCDFs. Eur J Pharmacol 1994; 270: 129–36.
105. Purkey HE, Palaninathan S, Kent K, et al. Hydroxylated polychlorinated biphenyls selectively bind transthyretin in blood and inhibit amyloidogenesis: rationalizing rodent PCB toxicity. Chem Biol 2004; 11: 1719–28.
106. Scanlan TS, Suchland K, Hart M, et al. 3-Iodothyronamine is an endogenous and rapid-acting derivative of thyroid hormone. Nat Med 2004; 10: 638–42.
107. Delange F. The disorders induced by iodine deficiency. Thyroid 1994; 4: 107–28.
108. Schwartz HL. Effect of thyroid hormone on growth and development. In: Oppenheimer JH, Samuels HH, eds. Molecular Basis of Thyroid Hormone Action. New York: Academic Press, 1983: 413–44.
109. Glinoer D. The regulation of thyroid function in pregnancy: pathways of endocrine adaptation from physiology to pathology. Endocr Rev 1997; 18: 404–33.
110. Glinoer D. Potential consequences of maternal hypothyroidism on the offspring: evidence and implications. Horm Res 2001; 55: 109–14.
111. Glinoer D. Clinical and biological consequences of iodine deficiency during pregnancy. Endocr Dev 2007; 10: 62–85.
112. Glinoer D, Delange F. The potential repercussions of maternal, fetal, and neonatal hypothyroxinemia on the progeny. Thyroid 2000; 10: 871–87.
113. Vulsma T, Gons MH, de Vijlder JJ. Maternal-fetal transfer of thyroxine in congenital hypothyroidism due to a total organification defect or thyroid agenesis. N Engl J Med 1989; 321: 13–16.
114. Auso E, Lavado-Autric R, Cuevas E, et al. A moderate and transient deficiency of maternal thyroid function at the beginning of fetal neocorticogenesis alters neuronal migration. Endocrinology 2004; 145: 4037–47.
115. Haddow JE, Palamaki G, Allan W, et al. Maternal thyroid deficiency during pregnancy and subsequent neuropsychological development of the child. N Engl J Med 1999; 341: 549–55.
116. Man EB, Holden RH, Jones WS. Thyroid function in human pregnancy. VII. Development and retardation of 4-year-old progeny of euthyroid and of hypothyroxinemic women. Am J Obstet Gynecol 1971; 109: 12–19.
117. Kooistra L, Crawford S, vanBaar A, et al. Neonatal effects of maternal hypothyroxinemia during early pregnancy. Pediatrics 2006; 117: 161–7.
118. Lavado-Autric R, Auso E, Garcia-Velasca J, et al. Early maternal hypothyroxinemia alters histogenesis and cerebral cortex cytoarchitecture of the progeny. J Clin Invest 2003; 111: 1073–82.
119. Opazo MC, Gianini A, Pancetti F, et al. Maternal hypothyroxinemia impairs spatial learning and synaptic nature and function in the offspring. Endocrinology 2008; 149: 5097–106.
120. Klein RZ. History of congenital hypothyroidism. In: Burrow GN, Dussault JH, eds. Neonatal Thyroid Screening. New York: Raven Press, 1980: 51–9.
121. Hanukoglu A, Perlman K, Shamis I, et al. Relationship of etiology to treatment in congenital hypothyroidism. J Clin Endocrinol Metab 2001; 86: 186–91.
122. Dussault JH, Ruel J. Thyroid hormones and brain development. Annu Rev Physiol 1987; 49: 321–34.
123. Derksen-Lubsen G, Verkerk PH. Neuropsychologic development in early treated congenital hypothyroidism: analysis of literature data. Pediatr Res 1996; 39: 561–6.
124. Mirabella G, Westall C, Asztalos E, et al. The effect of abnormal intrauterine thyroid hormone economies on infant cognitive abilities. J Pediatr Endocrinol Metab 2000; 13: 191–4.
125. Rovet JF. Congenital hypothyroidism: an analysis of persisting deficits and associated factors. Child Neuropsychol 2002; 8: 150–62.
126. Zoeller RT, Rovet J. Timing of thyroid hormone action in the developing brain: clinical observations and experimental findings. J Neuroendocrinol 2004; 16: 809–18.
127. Pedraza PE, Obregon M, Escobar-Morreale H, et al. Mechanisms of adaptation to iodine deficiency in rats: thyroid status is tissue specific. Its relevance for man. Endocrinology 2006; 147: 2098–108.
128. Spitzweg C, Heufelder AE, Morris JC. Thyroid iodine transport. Thyroid 2000; 10: 321–30.

129. Morreale de Escobar G, Calvo R, Obregon M, et al. Contribution of maternal thyroxine to fetal thyroxine pools in normal rats near term. Endocrinology 1990; 126: 2765–7.
130. Morreale de Escobar G, Obregon MJ, Escobar del Rey F. Fetal and maternal thyroid hormones. Horm Res 1987; 26: 12–27.
131. Morreale de Escobar G, Obregon MJ, Escobar del Rey F. Is neuropsychological development related to maternal hypothyroidism or to maternal hypothyroxinemia? J Clin Endocrinol Metab 2000; 85: 3975–87.
132. Hamburgh M, Lynn E, Weiss EP. Analysis of the influence of thyroid hormone on prenatal and postnatal maturation of the rat. Anat Rec 1964; 150: 147–61.
133. Porterfield SP. Prenatal exposure of the fetal rat to excessive L-thyroxine or 3,5-dimethyl-3′-isopropyl-thyronine produces persistent changes in the thyroid control system. Horm Metab Res 1985; 17: 655–9.
134. Porterfield SP, Hendrich CE. The role of thyroid hormones in prenatal and neonatal neurological development—current perspectives. Endocr Rev 1993; 14: 94–106.
135. Man EB, Brown JF, Serunian SA. Maternal hypothyroxinemia: psychoneurological deficits of progeny. Ann Clin Lab Sci 1991; 21: 227–39.
136. Pop VJ, Brouwers E, Vader H, et al. Maternal hypothyroxinaemia during early pregnancy and subsequent child development: a 3-year follow-up study. Clin Endocrinol (Oxf) 2003; 59: 282–8.
137. Pop VJ, Kuijpens J, vanBaar A, et al. Low maternal free thyroxine concentrations during early pregnancy are associated with impaired psychomotor development in infancy. Clin Endocrinol (Oxf) 1999; 50: 149–55.
138. Vermiglio F, LoPresti V, Moleti M, et al. Attention deficit and hyperactivity disorders in the offspring of mothers exposed to mild-moderate iodine deficiency: a possible novel iodine deficiency disorder in developed countries. J Clin Endocrinol Metab 2004; 89: 6054–60.
139. Ginsberg GL, Hattis D, Zoeller RT, et al. Evaluation of the U.S. EPA/OSWER preliminary remediation goal for perchlorate in groundwater: focus on exposure to nursing infants. Environ Health Perspect 2007; 115: 361–9.
140. Porterfield SP. Thyroidal dysfunction and environmental chemicals—potential impact on brain development. Environ Health Perspect 2000; 108(Suppl 3): 433–8.
141. Iannacone EA, Yan A, Gauger K, et al. Thyroid hormone exerts site-specific effects on SRC-1 and NCoR expression selectively in the neonatal rat brain. Mol Cell Endocrinol 2002; 186: 49–59.
142. Koibuchi N, Chin WW. Thyroid hormone action and brain development. Trends Endocrinol Metab 2000; 11: 123–8.
143. Rabie A, Favre C, Calvel M, et al. Effects of thyroid dysfunction on the development of the rat cerebellum, with special reference to cell death within the internal granular layer. Brain Res 1977; 120: 521–31.
144. Morte B, Manzano J, Scanlan T, et al. Deletion of the thyroid hormone receptor α1 prevents the structural alterations of the cerebellum induced by hypothyroidism. Proceedings National Academy of Science 2002; 99: 3985–9.
145. Rabie A, Clavel MC, Legrand J. Analysis of the mechanisms underlying increased histogenetic cell death in developing cerebellum of the hypothyroid rat: determination of the time required for granule cell death. Brain Res 1980; 190: 409–14.
146. Xiao Q, Nikodem VM. Apoptosis in the developing cerebellum of the thyroid hormone deficient rat. Front Biosci 1998; 3: A52–7.
147. Farwell AP, Dubord-Tomasetti SA. Thyroid hormone regulates the extracellular organization of laminin on astrocytes. Endocrinology 1999; 140: 5014–21.
148. Li GH, Post J, Koibuchi N, et al. Impact of thyroid hormone deficiency on the developing CNS: cerebellar glial and neuronal protein expression in rat neonates exposed to antithyroid drug propylthiouracil. Cerebellum 2004; 3: 100–6.
149. Jacobson M. Developmental Neurobiology, 3rd edn. New York: Plenum Press, 1991.
150. Bansal R, You S, Herzig C, et al. Maternal thyroid hormone increases HES expression in the fetal rat brain: an effect mimicked by exposure to a mixture of polychlorinated biphenyls (PCBs). Brain Res Dev Brain Res 2005; 156: 13–22.
151. Martinez-Galan JR, Escobar-del Rey F, Morreale de Escobar G, et al. Hypothyroidism alters the development of radial glial cells in the term fetal and postnatal neocortex of the rat. Brain Res Dev Brain Res 2004; 153: 109–14.
152. Martinez-Galan JR, Pedraza P, Santacana M, et al. Early effects of iodine deficiency on radial glial cells of the hippocampus of the rat fetus. A model of neurological cretinism. J Clin Invest 1997; 99: 2701–9.
153. Rami A, Rabie A. Effect of thyroid deficiency on the development of glia in the hippocampal formation of the rat: an immunocytochemical study. Glia 1988; 1: 337–45.

154. Gilmore EC, Herrup K. Cortical development: receiving reelin. Curr Biol 2000; 10: R162–6.
155. Gleeson JG, Walsh CA. Neuronal migration disorders: from genetic diseases to developmental mechanisms. Trends Neurosci 2000; 23: 352–9.
156. Alvarez-Dolado M, Ruiz M, del Rio J, et al. Thyroid hormone regulates reelin and dab1 expression during brain development. J Neurosci 1999; 19: 6979–93.
157. Goodman JH, Gilbert ME. Modest thyroid hormone insufficiency during development induces a cellular malformation in the corpus callosum: a model of cortical dysplasia. Endocrinology 2007; 148: 2593–7.
158. Cuevas E, Auso E, Telefont M, et al. Transient maternal hypothyroxinemia at onset of corticogenesis alters tangential migration of medial ganglionic eminence-derived neurons. Eur J Neurosci 2005; 22: 541–51.
159. Gilbert ME, Sui L, Walker M, et al. Thyroid hormone insufficiency during brain development reduces parvalbumin immunoreactivity and inhibitory function in the hippocampus. Endocrinology 2007; 148: 92–102.
160. Berbel P, Marco P, Cerezo J, et al. Distribution of parvalbumin immunoreactivity in the neocortex of hypothyroid adult rats. Neurosci Lett 1996; 204: 65–8.
161. Sui L, Anderson WL, Gilbert ME. Impairment in short-term but enhanced long-term synaptic potentiation and ERK activation in adult hippocampal area CA1 following developmental thyroid hormone insufficiency. Toxicol Sci 2005; 85: 647–56.
162. Gilbert ME. Alterations in synaptic transmission and plasticity in area CA1 of adult hippocampus following developmental hypothyroidism. Brain Res Dev Brain Res 2004; 148: 11–18.
163. Gilbert ME, Sui L. Dose-dependent reductions in spatial learning and synaptic function in the dentate gyrus of adult rats following developmental thyroid hormone insufficiency. Brain Res 2006; 1069: 10–22.
164. Dong J, Yin H, Liu W, et al. Congenital iodine deficiency and hypothyroidism impair LTP and decrease C-fos and C-jun expression in rat hippocampus. Neurotoxicology 2005; 26: 417–26.
165. Vara H, Martinez B, Santos A, et al. Thyroid hormone regulates neurotransmitter release in neonatal rat hippocampus. Neuroscience 2002; 110: 19–28.
166. Sui L, Gilbert ME. Pre- and postnatal propylthiouracil-induced hypothyroidism impairs synaptic transmission and plasticity in area CA1 of the neonatal rat hippocampus. Endocrinology 2003; 144: 4195–203.
167. Altmann L, Lilienthal H, Hany J, et al. Inhibition of long-term potentiation in developing rat visual cortex but not hippocampus by in utero exposure to polychlorinated biphenyls. Brain Res Dev Brain Res 1998; 110: 257–60.
168. Altmann L, Mundy WR, Ward TR, et al. Developmental exposure of rats to a reconstituted PCB mixture or Aroclor 1254: effects on long-term potentiation and [3H]MK-801 binding in occipital cortex and hippocampus. Toxicol Sci 2001; 61: 321–30.
169. Carpenter DO, Hussain R, Berger D, et al. Electrophysiologic and behavioral effects of perinatal and acute exposure of rats to lead and polychlorinated biphenyls. Environ Health Perspect 2002; 110(Suppl 3): 377–86.
170. Dingemans MM, Ramakers G, Gardoni F, et al. Neonatal exposure to brominated flame retardant BDE-47 reduces long-term potentiation and postsynaptic protein levels in mouse hippocampus. Environ Health Perspect 2007; 115: 865–70.
171. Kitamura S, Jinno N, Ohta S, et al. Thyroid hormonal activity of the flame retardants tetrabromobisphenol A and tetrachlorobisphenol A. Biochem Biophys Res Commun 2002; 293: 554–9.
172. Morse DC, Groen D, Veerman M, et al. Interference of polychlorinated biphenyls in hepatic and brain thyroid hormone metabolism in fetal and neonatal rats. Toxicol Appl Pharmacol 1993; 122: 27–33.
173. Wei Q, Miskimins WK, Miskimins R. Cloning and characterization of the rat myelin basic protein gene promoter. Gene 2003; 313: 161–7.
174. Ferreira AA, Pereira M, Manhaes A, et al. Effects of experimental hypothyroidism on myelin sheath structural organization. J Neurocytol 2004; 33: 225–31.
175. Almazan G, Honegger P, Matthieu JM. Triiodothyronine stimulation of oligodendroglial differentiation and myelination. A developmental study. Dev Neurosci 1985; 7: 45–54.
176. Schoonover CM, Seibel M, Jolson D, et al. Thyroid hormone regulates oligodendrocyte accumulation in developing rat brain white matter tracts. Endocrinology 2004; 145: 5013–20.
177. Ferreira AA, Periera M, Manhaes A, et al. Ultrastructural identification of oligodendrocyte/myelin proteins in corpus callosum of hypothyroid animals. Int J Dev Neurosci 2007; 25: 87–94.
178. Barradas PC, Vieira RS, De Freitas MS. Selective effect of hypothyroidism on expression of myelin markers during development. J Neurosci Res 2001; 66: 254–61.

179. Ibarrola N, Rodriguez-Pena A. Hypothyroidism coordinately and transiently affects myelin protein gene expression in most rat brain regions during postnatal development. Brain Res 1997; 752: 285–93.
180. Martinez-Galan JR, Pedraza P, Santacana M, et al. Myelin basic protein immunoreactivity in the internal capsule of neonates from rats on a low iodine intake or on methylmercaptoimidazole (MMI). Brain Res Dev Brain Res 1997; 101: 249–56.
181. Rodriguez-Pena A. Oligodendrocyte development and thyroid hormone. J Neurobiol 1999; 40: 497–512.
182. Sharlin DS, Tighe D, Gilbert ME, et al. The balance between oligodendrocyte and astrocyte production in major white matter tracts is linearly related to serum total thyroxine. Endocrinology 2008; 149: 2527–36.
183. Farsetti A, Mitsuhashi T, Desvergne B, et al. Molecular basis of thyroid hormone regulation of myelin basic protein gene expression in rodent brain. J Biol Chem 1991; 266: 23226–32.
184. Figueiredo BC, Almazon G, Ma Y, et al. Gene expression in the developing cerebellum during perinatal hypo- and hyperthyroidism. Brain Res Mol Brain Res 1993; 17: 258–68.
185. Berbel P, Gaudano-Ferraz A, Angulo A, et al. Role of thyroid hormones in the maturation of interhemispheric connections in rats. Behav Brain Res 1994; 64: 9–14.
186. Berbel P, Guandano-Ferraz A, Martinez M, et al. Organization of auditory callosal connections in hypothyroid adult rats. Eur J Neurosci 1993; 5: 1465–78.
187. Gravel C, Hawkes R. Maturation of the corpus callosum of the rat: I. Influence of thyroid hormones on the topography of callosal projections. J Comp Neurol 1990; 291: 128–46.
188. Raff MC, Miller RH, Noble M. A glial progenitor cell that develops in vitro into an astrocyte or an oligodendrocyte depending on culture medium. Nature 1983; 303: 390–6.
189. Madeira MD, Cadete-Leite A, Andrade J, et al. Effects of hypothyroidism upon the granular layer of the dentate gyrus in male and female adult rats: a morphometric study. J Comp Neurol 1991; 314: 171–86.
190. Madeira MD, Paula-Barbosa MM. Reorganization of mossy fiber synapses in male and female hypothyroid rats: a stereological study. J Comp Neurol 1993; 337: 334–52.
191. Madeira MD, Sousa N, Lima-Andrade M, et al. Selective vulnerability of the hippocampal pyramidal neurons to hypothyroidism in male and female rats. J Comp Neurol 1992; 322: 501–18.
192. Rami A, Patel AJ, Rabie A. Thyroid hormone and development of the rat hippocampus: morphological alterations in granule and pyramidal cells. Neuroscience 1986; 19: 1217–26.
193. Rami A, Rabie A, Patel AJ. Thyroid hormone and development of the rat hippocampus: cell acquisition in the dentate gyrus. Neuroscience 1986; 19: 1207–16.
194. Gilbert ME. Perinatal exposure to polychlorinated biphenyls alters excitatory synaptic transmission and short-term plasticity in the hippocampus of the adult rat. Neurotoxicology 2003; 24: 851–60.
195. Gilbert ME, Paczkowski C. Propylthiouracil (PTU)-induced hypothyroidism in the developing rat impairs synaptic transmission and plasticity in the dentate gyrus of the adult hippocampus. Brain Res Dev Brain Res 2003; 145: 19–29.
196. Meyerhoff WL. Hypothyroidism and the ear: electrophysiological, morphological, and chemical considerations. Laryngoscope 1979; 89(10 Pt 2 Suppl 19): 1–25.
197. Uziel A, Legrand C, Ohresser M, et al. Maturational and degenerative processes in the organ of Corti after neonatal hypothyroidism. Hear Res 1983; 11: 203–18.
198. Knipper M, Zinn C, Maier H, et al. Thyroid hormone deficiency before the onset of hearing causes irreversible damage to peripheral and central auditory systems. J Neurophysiol 2000; 83: 3101–12.
199. Goldey ES, Kehn L, Rehnberg G, et al. Effects of developmental hypothyroidism on auditory and motor function in the rat. Toxicol Appl Pharmacol 1995; 135: 67–76.
200. Herr DW, Goldey ES, Crofton KM. Developmental exposure to Aroclor 1254 produces low-frequency alterations in adult rat brainstem auditory evoked responses. Fundam Appl Toxicol 1996; 33: 120–8.
201. Harpavat S, Cepko CL. Thyroid hormone and retinal development: an emerging field. Thyroid 2003; 13: 1013–19.
202. Sevilla-Romero E, Munoz A, Pinazo-Duran MD. Low thyroid hormone levels impair the perinatal development of the rat retina. Ophthalmic Res 2002; 34: 181–91.
203. Roberts MR, Srinivas M, Forrest D, et al. Making the gradient: thyroid hormone regulates cone opsin expression in the developing mouse retina. Proc Natl Acad Sci U S A 2006; 103: 6218–23.
204. Mirabella G, Westall C, Asztalos E, et al. Development of contrast sensitivity in infants with prenatal and neonatal thyroid hormone insufficiencies. Pediatr Res 2005; 57: 902–7.
205. Pacheco-Rosado J, Hernandez-Garcia A, Ortiz-Butron R. Perinatal hypothyroidism increases the susceptibility to lidocaine-kindling in adult rats. Neurosci Lett 2004; 367: 186–8.

206. Akaike M, Kato N, Ohna H, et al. Hyperactivity and spatial maze learning impairment of adult rats with temporary neonatal hypothyroidism. Neurotoxicol Teratol 1991; 13: 317–22.
207. Brosvic GM, Taylor JN, Dihoff RE. Influences of early thyroid hormone manipulations: delays in pup motor and exploratory behavior are evident in adult operant performance. Physiol Behav 2002; 75: 697–715.
208. Darbra S, Balado F, Marti-Carbonell M, et al. Perinatal hypothyroidism effects on step-through passive avoidance task in rats. Physiol Behav 2004; 82: 497–501.
209. Darbra S, Garou A, Balada F, et al. Perinatal hypothyroidism effects on neuromotor competence, novelty-directed exploratory and anxiety-related behaviour and learning in rats. Behav Brain Res 2003; 143: 209–15.
210. Davenport JW, Dorcey TP. Hypothyroidism: learning deficit induced in rats by early exposure to thiouracil. Horm Behav 1972; 3: 97–112.
211. Schalock RL, Brown WJ, Smith RL. Long-term effects of propylthiouracil-induced neonatal hypothyroidism. Dev Psychobiol 1979; 12: 187–99.
212. Tamasy V, Meisami E, Vallerga A, et al. Rehabilitation from neonatal hypothyroidism: spontaneous motor activity, exploratory behavior, avoidance learning and responses of pituitary—thyroid axis to stress in male rats. Psychoneuroendocrinology 1986; 11: 91–103.
213. Axelstad M, Hansen PR, Boberg J, et al. Developmental neurotoxicity of Propylthiouracil (PTU) in rats: relationship between transient hypothyroxinemia during development and long-lasting behavioural and functional changes. Toxicol Appl Pharmacol 2008; 232: 1–13.
214. Ercan-Fang S, Schwartz HL, Oppenheimer JH. Isoform-specific 3,5,3′-triiodothyronine receptor binding capacity and messenger ribonucleic acid content in rat adenohypophysis: effect of thyroidal state and comparison with extrapituitary tissues. Endocrinology 1996; 137: 3228–33.
215. Forrest D, Hanebuth E, Smeyne R, et al. Recessive resistance to thyroid hormone in mice lacking thyroid hormone receptor beta: evidence for tissue-specific modulation of receptor function. EMBO J 1996; 15: 3006–15.
216. Guadano-Ferraz A, Benavides-Piccione R, Venero C, et al. Lack of thyroid hormone receptor alpha1 is associated with selective alterations in behavior and hippocampal circuits. Mol Psychiatry 2003; 8: 30–8.
217. Venero C, Guadano-Ferraz A, Herrero A, et al. Anxiety, memory impairment, and locomotor dysfunction caused by a mutant thyroid hormone receptor alpha1 can be ameliorated by T3 treatment. Genes Dev 2005; 19: 2152–63.
218. Vanderpump MP, Tunbridge WM. Epidemiology and prevention of clinical and subclinical hypothyroidism. Thyroid 2002; 12: 839–47.
219. Bauer M, London E, Rasgon N, et al. Supraphysiological doses of levothyroxine alter regional cerebral metabolism and improve mood in bipolar depression. Mol Psychiatry 2005; 10: 456–69.
220. Baumgartner A. Thyroxine and the treatment of affective disorders: an overview of the results of basic and clinical research. Int J Neuropsychopharmacol 2000; 3: 149–65.
221. Miller KJ, Parsons T, Whybrow P, et al. Verbal memory retrieval deficits associated with untreated hypothyroidism. J Neuropsychiatry Clin Neurosci 2007; 19: 132–6.
222. Faldt R, Passant U, Nilsson K, et al. Prevalence of thyroid hormone abnormalities in elderly patients with symptoms of organic brain disease. Aging (Milano) 1996; 8: 347–53.
223. Hendrick, V, Altshuler L, Whybrow P. Psychoneuroendocrinology of mood disorders. The hypothalamic-pituitary-thyroid axis. Psychiatr Clin North Amer 1998; 21: 277–92.
224. Heyman A, Wilkinson W, Hurwitz B, et al. Alzheimer's disease: genetic aspects and associated clinical disorders. Ann Neurol 1983; 14: 507–15.
225. Levitt AJ, Joffe RT. Folate, B12 and thyroid function in depression. Biol Psychiatry 1993; 33: 52–3.
226. Loosen PT. Effects of thyroid hormones on central nervous system in aging. Psychoneuroendocrinology 1992; 17: 355–74.
227. Kulikov A, Torresani J, Jeanningros R. Experimental hypothyroidism increases immobility in rats in the forced swim paradigm. Neurosci Lett 1997; 234: 111–14.
228. Alzoubi KH, Aleisa A, Gerges N, et al. Nicotine reverses adult-onset hypothyroidism-induced impairment of learning and memory: behavioral and electrophysiological studies. J Neurosci Res 2006; 84: 944–53.
229. Alzoubi KH, Gerges NZ, Alkadhi KA. Levothyroxin restores hypothyroidism-induced impairment of LTP of hippocampal CA1: electrophysiological and molecular studies. Exp Neurol 2005; 195: 330–41.
230. Gerges NZ, Alkadhi KA. Hypothyroidism impairs late LTP in CA1 region but not in dentate gyrus of the intact rat hippocampus: MAPK involvement. Hippocampus 2004; 14: 40–5.

231. Tong H, Chen G, Liu R, et al. Age-related learning and memory impairments in adult-onset hypothyroidism in Kunming mice. Physiol Behav 2007; 91: 290–8.
232. Sui L, Wang F, Li BM. Adult-onset hypothyroidism impairs paired-pulse facilitation and long-term potentiation of the rat dorsal hippocampo-medial prefrontal cortex pathway in vivo. Brain Res 2006; 1096: 53–60.
233. Ambrogini P, Cuppini R, Ferri P, et al. Thyroid hormones affect neurogenesis in the dentate gyrus of adult rat. Neuroendocrinol 2005; 81: 244–53.
234. Desouza LA, Ladawala U, Daniel S, et al. Thyroid hormone regulates hippocampal neurogenesis in the adult rat brain. Mol Cell Neurosci 2005; 29: 414–26.
235. Lemkine GF, Raj A, Afama G, et al. Adult neural stem cell cycling in vivo requires thyroid hormone and its alpha receptor. FASEB J 2005; 19: 863–5.
236. Montero-Pedrazuela A, Verero C, Lavado-Autric R, et al. Modulation of adult hippocampal neurogenesis by thyroid hormones: implications in depressive-like behavior. Mol Psychiatry 2006; 11: 361–71.
237. Malberg JE, Schechter LE. Increasing hippocampal neurogenesis: a novel mechanism for antidepressant drugs. Curr Pharm Des 2005; 11: 145–55.
238. Santarelli L, Saxe M, Gross C, et al. Requirement of hippocampal neurogenesis for the behavioral effects of antidepressants. Science 2003; 301: 805–9.
239. Shors TJ, Miesegaes G, Beylin A, et al. Neurogenesis in the adult is involved in the formation of trace memories. Nature 2001; 410: 372–6.
240. Snyder JS, Hong N, McDonald R, et al. A role for adult neurogenesis in spatial long-term memory. Neuroscience 2005; 130: 843–52.
241. van Praag H, Schinder A, Christie B, et al. Functional neurogenesis in the adult hippocampus. Nature 2002; 415: 1030–4.
242. Sui L, Wang F, Liu F, et al. Dorsal hippocampal administration of triiodothyronine enhances long-term memory for trace cued and delay contextual fear conditioning in rats. J Neuroendocrinol 2006; 18: 811–19.
243. Redei EE, Solberg L, Kluczynski J, et al. Paradoxical hormonal and behavioral responses to hypothyroid and hyperthyroid states in the Wistar-Kyoto rat. Neuropsychopharmacology 2001; 24: 632–9.
244. Kohn M, Lazar J, McMaster S, et al. Screening methods for thyroid hormone disruptors. Environ Health Perspect 1999; 107: 407–15.
245. Hood A, Klaassen CD. Effects of microsomal enzyme inducers on outer-ring deiodinase activity toward thyroid hormones in various rat tissues. Toxicol Appl Pharmacol 2000; 163: 240–8.
246. Thibodeaux JR, Hanson RG, Rogers JM, et al. Exposure to perfluorooctane sulfonate during pregnancy in rat and mouse. I: maternal and prenatal evaluations. Toxicol Sci 2003; 74: 369–81.
247. Peeters RP, van der Deure WM, Visser TJ. Genetic variation in thyroid hormone pathway genes; polymorphisms in the TSH receptor and the iodothyronine deiodinases. Eur J Endocrinol 2006; 155: 655–62.
248. Andersen S, Pedersen K, Bruun N, et al. Narrow individual variations in serum T(4) and T(3) in normal subjects: a clue to the understanding of subclinical thyroid disease. J Clin Endocrinol Metab 2002; 87: 1068–72.
249. DeVito M, Biegel L, Brouwer A, et al. Screening methods for thyroid hormone disruptors. Environ Health Perspect 1999; 107: 407–15.

5 Dopaminergic Nerves as Targets for Neurotoxins

Richard M. Kostrzewa
Department of Pharmacology, Quillen College of Medicine, East Tennessee State University, Johnson City, Tennessee, U.S.A.

Lucyna Antkiewicz-Michaluk
Department of Neurochemistry, Institute of Pharmacology, Polish Academy of Sciences, Krakow, Poland

Francisco Fornai
Department of Human Morphology and Applied Biology, University of Pisa, Pisa, and Laboratory of Neurobiology of Movement Disorders IRCCS, INM Neuromed Pozzilli, Isernia, Italy

INTRODUCTION

The dopaminergic nervous system in brain has been the most-intensively studied of all the known neuronal phenotypes, owing in large part to the availability of neurotoxins able to produce selective destruction and thereby model Parkinson disease (PD) (1–3), attention-deficit hyperactivity disorder (ADHD) (4–8), and Lesch-Nyhan syndrome (9–11). The first available dopaminergic neurotoxin was 6-hydroxydopamine (6-OHDA), discovered before 1970 and used in more than 9000 published studies (3,11,12). Substituted AMPHs were similarly found to be neurotoxic to dopaminergic as well as serotoninergic nerves (13). However, it was the neurotoxin MPTP (1-methyl-4-phenyl-1,2,3,6-tetrahydropyridine) that reinvigorated studies of the dopaminergic system, owing to its inadvertent use in the early 1980s by humans abusing drugs and its rapid (days) production of an irreversible parkinsonian state (14–16). MPTP is widely used to produce animal modeling of PD. Over the past 15 years the cytotoxin rotenone, a mitochondrial complex I inhibitor of oxidative phosphorylation, has been widely used chronically in low amount to model PD because this treatment results in the production of protein aggregates in brain similar to that accompanying human PD (17). Most recently, MPTP administrated chronically in low amount has been shown to also result in the presence of protein aggregates in the brain of animal models of PD (18).

However, even before the discovery of selective neurotoxins, there was already a debate as to the potential of dopamine metabolites perhaps producing neurotoxic effects, since many metabolites are oxidizing agents and some can resonate into a quinone counterpart. In this paper we will discuss the liability of the dopaminergic system to neurotoxins from several distinct vantage points. Initially, the focus will be on traditional selective neurotoxins, principally 6-OHDA, substituted AMPH, and MPTP. Next, the use of low dosage rotenone and MPTP will be discussed in relation to their utility in modeling PD. Focus will then move to non-classical neurotoxins, to introduce the concept of neurotoxicity occurring in the absence of cell injury. Finally, endogenously produced species endowed with neurotoxic or neuroprotective properties will be presented and discussed.

CLASSICAL NEUROTOXINS FOR DOPAMINERGIC NERVES

Although many neurotoxins are able to directly damage dopaminergic nerves, this paper will focus on 6-OHDA, MPTP—the two most-selective and most-frequently used dopaminergic neurotoxins—and substituted AMPHs.

6-OHDA

6-OHDA was originally proposed to be a metabolite of dopamine (DA) and/or norepinephrine (NE) (19–21). However, in 1967, 6-OHDA was shown to be overtly neurotoxic to sympathetic noradrenergic neurons (22,23). Soon, 6-OHDA was also shown to be toxic to dopaminergic nerves in brain (24). 6-OHDA selectivity for dopaminergic neurons owes to the high affinity of 6-OHDA to the DA transporter (DAT), providing the means for selective

Table 1 Changes in mRNA expression of Iron Genes Involved in Iron Homeostasis in the Substantia Nigra and Cortex of PD Patients Compared with Controls in Post Mortem Tissue

Descriptor	Gene	SnM	Sni	Cortex
			p-value	
Upregulated				
IRP-binding protein 1	IRP1	ns	ns	ns
IRP-binding protein 2	IRP2	0.025	ns	ns
Transferrin	Tf	0.0001	0.0030	ns
Transferrin receptor 2	TfR2	ns	0.0017	ns
Transferrin receptor 2	TfR2	ns	0.0184	ns
Ferritin H	FTH1	0.0019	ns	ns
Ferritin H pseudogene 1	FTHP1	0.0010	0.0348	ns
Ferritin L	FTL	0.0291	ns	ns
Ferritin L	FTL	0.0335	ns	ns
Ferritin L	FTL	ns	0.006	ns
Caeruloplasmin	Cp	0.0276	0.0276	ns
Caeruloplasmin	Cp	ns	0.0343	ns
Caeruloplasmin	Cp	ns	0.0336	ns
Hephastin	HEPH	ns	0.009	ns
Haemochromatosis	HFE	0.0416	0.0005	ns
Haemochromatosis	HFE	ns	0.0111	ns
Haemochromatosis	HFE	ns	0.0295	ns
Haemochromatosis	HFE	ns	0.0039	ns
Ferroportin	FPN1	0.0192	ns	ns
Ferroportin	FPN1	0.0353	ns	ns
Solutecarrier family11	SLC11A2	ns	ns	0.0291
Downregulated				
Ferrochetalase		0.006	0.0223	ns
Sideroflexin 1		0.006	0.0314	ns
Friedreich ataxia		0.031	ns	ns

accumulation of 6-OHDA in DA nerves. Inside the nerve 6-OHDA autooxidizes to both *ortho*- and *para*-quinones which rearrange to form a multitude of reactive intermediaries, each of which is a potent oxidant (12,25). Also, both the *ortho*- and *para*-quinones are overtly neurotoxic (26), as well as generated malonyldialdehyde (27,28). Further, 6-OHDA generates superoxides, which then promote intraneuronal formation of hydrogen peroxide (H_2O_2) and hydroxyl radical ($HO^\bullet$) (29–32). In light of the abundance of iron in the substantia nigra pars compacta (SNpc) (Table 1) (33), the Fenton reaction [$Fe^{2+} + H_2O_2 \rightarrow HO^\bullet$] conceivably can take place in SNpc to aid in PD progression (34,35). The iron chelator desferoxamine reduces 6-OHDA neurotoxicity, although other neuroprotective mechanisms may be involved (36–39). Ultimately, these reactive oxygen species (ROS) uncouple oxidative phosphorylation, leading to energy depletion (37,38,40), oxidation of unsaturated lipids, disruption of proteins, and inactivation of enzymes (12).

Substituted AMPHs

In the 1960s *para*-chloroamphetamine (PCA) was shown to induce a long-lasting reduction in brain 5-HT levels (41), and a few years later a neurotoxic effect was confirmed (42). Soon afterwards, other AMPH derivatives, methamphetamine (METH), and methylene-dioxy-methamphetamine (MDMA) were found to be neurotoxic, and these were investigated as well, particularly in light of their explosive abuse among youngsters in Western countries. Repeated METH induces a marked reduction in striatal DA, L-3,4-dihydroxyphenylacetic acid (DOPAC), and homovanillic acid (HVA) (43), consequent to destruction of nigrostriatal dopaminergic neurons (44), while sparing of DA levels in the nucleus accumbens (i.e., "mesolimbic" pathway) (45–48). However, in rats the

METH effect is not selective for DA terminals, since marked serotonin (5-HT) loss in several brain areas is observed as well (46–48). Despite inducing marked 5-HT loss in rats and primates, MDMA is considered to be a relatively selective DA neurotoxin in mice (49–55). In selected mouse strains (e.g., C57Bl) repeated low-dose METH (5 mg/kg × 2, i.p.) induces selective nigrostriatal DA degeneration in the absence of significant 5-HT loss (56–58). High-dose METH (up to 10 mg/kg × 4) damages both DA and 5-HT neurons (59).

METH acts at several molecular targets in monoaminergic neurons. METH competitively and selectively inhibits monoamine oxidase A (MAO-A) fully at 10^{-3} M (60), the level achieved when mice are treated with a METH i.p. dose of 2 mg/kg (61). METH (and AMPHs in general) is a substrate for DAT, through which it is transported into DA terminals (62) "displacing" and releasing DA from cytoplasmic sites. At higher concentrations METH enters DA neurons by passive diffusion (63). Finally, AMPHs bind to the VMAT (vesicular monoamine transporter) and are transported into vesicles or, for higher i.c. concentrations, diffuse into the vesicles through their membrane (64,65); DA would be released from the vesicles into the cytosol due to the alkalinization of vesicle contents (with an impairment of the electrochemical drive usually present into these organelles).

In the last decades several investigators explored the mechanisms by which AMPHs (and METH in particular) induce nigrostriatal toxicity. It is now widely accepted that DA release plays a pivotal role in this toxicity (66,67). This is confirmed by the fact that the tyrosine hydroxylase inhibitor α-methyl-para-tyrosine (AMPT) prevents DA toxicity (68). For decades free radical formation and excitotoxicity have been recognized as the main contributors to toxicity. In fact, DA per se can induce free radical formation either by auto-oxidation (69,70) or by H_2O_2 formation during MAO metabolism. Cadet et al. (51) reported that METH-induced neurotoxicity was attenuated in copper/zinc superoxide dismutase transgenic mice. In the early 1980s Seiden and Vosmer (71), reporting high levels of 6-OHDA in the striatum of METH-treated rats, hypothesized that there was non-synaptic formation of 6-OHDA from DA. However, the finding has not been replicated.

Excitotoxicity has been proposed as one of the mechanisms involved in METH toxicity. Nash and Yamamoto (72) demonstrated that METH-induced DA release was associated with massive glutamate release in the striatum, in line with the observation by Sonsalla et al. (73). Moreover, dizolcipine (MK-801) prevented METH-induced DA toxicity. METH-induced 5-HT neuronal lesioning is likely related to the massive release of DA, since DA uptake inhibitors or DAT knock out prevents damage (74,75). Again, DA synthesis inhibitors or nigrostriatal pre-lesioning prevents METH-induced 5-HT loss (74,76).

MPTP

The discovery of MPTP neurotoxicity by J.W. Langston, a young Santa Barbara neurologist, was serendipitous. He was intrigued by the fact that in only a few months he was visited by a number of young parkinsonian outpatients. This contrasts with the epidemiologic characteristic of PD as a disease of aging (14). After diagnosing each one as a heroin addict, and analyzing for heroin, he was intrigued by the presence of MPTP, a contaminant by-product of the synthetic heroin derivative 1-methyl-4-phenyl-propion-oxypiperidine. In light of an isolated report by Davis et al. (77) a few years before, he ultimately confirmed in animal experiments that MPTP was indeed a neurotoxin with selectivity for dopaminergic neurons (for an extensive review, see 78).

MPTP is a meperidine derivative, highly lipophilic, and able to cross the blood-brain barrier. Once in the brain MPTP is converted by glial MAO-B to the toxic metabolite 1-methyl-4-phenyl-1,2,3,6-tetrahydropyridinium ion (MPP^+) via the enzymatic action of MAO-B. MPP^+ has high affinity for the DAT and is transported into DA neurons (79), after which MPP^+ selectively inhibits complex I of the mitochondrial respiratory chain, producing (*i*) energetic failure of the cell and (*ii*) formation of ROS due to leakage of electrons from the impaired respiratory chain. Furthermore, MPTP induces an increase of free cytosolic DA, as well as DA release, events which are likely indirectly related to the energy depletion per se.

MODELING PARKINSON'S DISEASE

Introduction

PD is a neurological disorder with a prevalence in the general population of 50–200/100,000 (depending on the different areas/studies). Most cases are idiopathic (i.e., no observed external cause), even though in the last few decades many familial cases have been described as well, and in some of them the underlying genetic defect has been identified. In the last decades it has been established from autopsy specimens that in PD, apart from the DA nigrostriatal system, several other pathways/areas undergo neurodegenerative changes as well, in particular, noradrenergic neurons of the nucleus locus coeruleus (LC) (80,81). Since the very early pathological description of PD brain, the existence of eosinophilic neuronal inclusions, so-called Lewy Bodies (LBs), have also been noted in the nigrostriatal system. These LBs are localized in DA neurons of the SNpc, noradrenergic neurons of the LC, catecholamine cells of the medulla oblongata, serotonergic neurons of the raphe nuclei, and specific cholinergic neurons (82,83). Furthermore, atypical LBs (i.e., smaller and with a more heterogeneous ultrastructure as compared to classical LBs) are observed in telencephalic structures such as cingulate cortex, insular cortex, amygdale, and frontotemporal cortex (82).

Apart from the spatial distribution of LB in PD, Braak et al. (84) have proposed a scenario for time-related recruitment of lesioned regions. According to this schema of disease progression, there is a caudal-rostral diffusion of neurodegenerative features, starting from the medulla, through a recruitment of mesencephalic structures (and, at this stage showing the onset of motor disturbances due to involvement of SNpc), and up to telencephalic structures. This hypothesis concerning the progression of PD pathology diverts from the simplistic description of PD as a neurodegenerative disorder of nigrostriatal DA neurons (or additional involvement of LC NE-pathways), and relates PD to a diffuse "multisystemic" disorder (i.e., multiple neuronal phenotypes). Since the early 1960s several attempts have been made to produce experimental animal models that could help to elucidate the pathophysiology of PD and thus gain insight into therapeutic strategies. As a general rule, the ideal animal model of a human disorder should have the following features: (*i*) pathology analogous to that of the human disorder; (*ii*) a neurochemical pattern of derangement resembling that of the human disorder; (*iii*) symptomatic features that mimic the human disorder; and (*iv*) a reversal of behavioral deficits by treatments effective in the human disorder. Below, we will show the current state of the art in the modelistic approaches for PD in experimental animals, with a special emphasis on rodents, since they are more easily adaptable to laboratory studies. Additionally, LB involvement in PD will be addressed in relation to their role in the neurodegenerative disease progression.

6-OHDA and Neonatal Lesioning of Rodents to Model PD

When administered systemically in perinatal rats, 6-OHDA tends to more-selectively damage noradrenergic versus dopaminergic nerves in brain (12,85,86). However, when administered into the cerebrospinal fluid (CSF) of perinatal rodents, following treatment with the NE transporter inhibitor desipramine, 6-OHDA produces near-total destruction of nigrostriatal dopaminergic neurons in the relative absence of noradrenergic fiber damage (87–89). Prominent DA D_1 receptor supersensitivity (RSS; DARSS) develops in the lesioned rats. Not overt at first, and unable to be observed after the first or second treatment with a D_1 agonist, the third and subsequent D_1 agonist treatments "unmask" the DARSS—and this is termed a priming process. That is, the third and subsequent D_1 agonist treatments produce a spectrum of exaggerated stereotypies and enhanced locomotor activity. Such rats, marked by greater than 99% dopaminergic denervation of striatum, have been proposed as a rodent model for severe PD (90).

6-OHDA and Adulthood Lesioning of Rodents to Model PD

In adult rats 6-OHDA must be administered directly into the brain because of its inability to cross the blood-brain barrier (86). Ungerstedt (1) was the first to model PD with 6-OHDA. When infused into the medial forebrain bundle (MFB) of adult rats, 6-OHDA produces as much as 97% loss of nigrostriatal neurons ipsilateral to the lesion, and similar loss of noradrenergic fibers traversing the MFB (91). Alternatively, if infused into the striatum, 6-OHDA induces a slow-progressing neuronal death to less than 50% of SNpc neurons. Finally, 6-OHDA infusion into the lateral ventricles produces degeneration of both SNpc dopaminerg neurons as well as LC noradrenergic neurons, since 6-OHDA has high affinity for both DATs and noradrenaline transporters (NETs). 6-OHDA damage to NE neurons can be prevented by pretreatment of rodents with the NET inhibitor desmethylimipramine (DMI) (92), although the desired effect still may not be achieved (93).

The effectiveness of 6-OHDA in producing ipsilateral destruction of DA traversing the MFB is gauged by the number of contralateral rotations induced in rats, about two weeks later, by the predominate DA D_2 agonist apomorphine; or number of ipsilateral rotations induced by AMPH, which reverses the DAT to release DA onto D_2 receptors that have become supersensitized. Apomorphine is considered to be a better indicator than AMPH (94). Rats lesioned unilaterally, as adults, do not demonstrate the akinesis, adipsia, and aphagia accompanying bilateral lesioning of the same pathway (1,91).

Although 6-OHDA was reported to be a metabolite in caudate nucleus following METH treatment of rats (71), that finding is unable to be replicated. Nevertheless, 6-OHDA even recently has been proposed as an endogenous neurotoxin in PD brain (95,96). The 6-OHDA model of PD has been of value in reference to postsynaptic (i.e., in the intrinsic GABAergic or cholinergic striatal neurons) adaptations to DA denervation, including DA receptor subtype expression (e.g., 97), expression of second messengers (98) or immediate early genes (99) or neuropeptides (100) in subpopulations of medium sized spiny neurons. A compensatory increase in DA turnover in the 6-OHDA model of PD is now established (101) and considered to be critical in the therapeutic management of PD (3,102). Finally, the 6-OHDA hemiparkinsonian rat model of PD has been valuable in assessing the dyskinetic potential of different therapeutics including chronic L-3,4-dihydroxyphenylalanine (l-DOPA) (103). A caveat, however, relates to the assumption that motor complications following DOPA therapy of PD are strictly related to the post-synaptic modifications. In conclusion, the 6-OHDA hemiparkinsonian rat model is considered to be important in addressing issues related to PD progression, therapeutic complications and treatment-related pharmacodynamic issues. Obviously, any model can address only appropriate questions: in the case of the 6-OHDA model, a surprisingly high number of questions can still be answered to by its appropriate use.

Substituted-AMPH Lesioning to Model PD

AMPH derivatives constitute another class of compounds that have been used in the last decade to induce experimental parkinsonism. In primates and humans there are conflicting results concerning the neurotoxic effects of AMPHs. Since the early 1970s it has been known that AMPHs produce long-lasting nigrostriatal DA loss in Rhesus monkeys (104) and other monkey strains (105–107). Up to now there is no evidence of METH-induced parkinsonism in people abusing METH. Positron emission tomography (PET) or single-photon-emission-tomography (SPECT) has been used to show that there is a reduction in striatal DAT in METH abusers, and despite early data on persistence of the effect, there is recovery of DAT number after months of METH withdrawal (108–110). This interpretation, however, is not unequivocal. For instance, (*i*) it is not clear to what extent the reduction in DAT striatal levels is related to down-regulation of the DAT molecule, versus true DA terminal loss and (*ii*) recovery in striatal DAT levels after METH withdrawal might represent either a re-innervation rendered by the sprouting from surviving terminals, or simply recovery in the expression of DAT in the original number of terminals. A recent detailed description of the monoaminergic neurotoxic effects of AMPH derivatives in different species can be found in Mc Cann and Ricaurte (111). Now there is evidence from post-mortem analysis that previous METH abusers bear neuronal inclusions in midbrain monoaminergic neurons (112).

The Rotenone Model of PD

Heikkila and colleagues (113) were the first to demonstrate that the pesticide and rodenticide rotenone, infused by stereotaxic means into brain, was neurotoxic to catecholamine neurons. Rotenone, a known inhibitor of complex I in the mitochondrial respiratory chain (114,115), was engaged since it had been discovered that another complex I inhibitor, namely MPTP, produced parkinsonism in humans (see section "The MPTP model of PD" on p. 117). Cellular energy failure and production of superoxide were considered to be the causes of neuronal cell death (116). Generally administered systemically, rotenone induces a varied pattern of neuronal degeneration, according to its dosage, route of administration, and the animal species under study. Surprisingly, when administered to rats at a daily dose of 7–18 mg/day, rotenone failed to produce destruction of DA terminals or DA perikarya (117). Rather, there was selective damage within the globus pallidus and caudate putamen, confined to the rostral medio-lateral part of striatum and lateral pallidus, contiguous to the lesioned part of striatum (117). In another study performed several years later in mice (118), rotenone again failed to induce nigrostriatal degeneration, despite the presence of marked acute toxicity/lethality. Subsequent studies in rats gave conflicting results. In particular, Greenamyre's group (17), by administering rotenone as a constant daily i.v. dose via an Alzet osmotic minipump, demonstrated, for the first time, rotenone-induced nigrostriatal cell loss. By treating groups of rats in this way for as long as five weeks, with different daily doses of rotenone, they determined that was the most effective in inducing nigrostriatal lesioning. Lewis rats, more sensitive than Sprague Dawleys, had SNpc neuronal loss along with a 55% loss of striatal DA innervation, but no cell damage at striatal level—confirming the utility of this method to model PD. Remarkably, in the nigral dopaminergic neurons there was accumulation of fibrillary intracellular inclusions immunopositive for ubiquitin and α-synuclein. This was the first demonstration of LB-like nigral inclusions in an animal model of PD. Several years later, the same group reproduced all of the morphological/neurochemical changes in Lewis rats treated 28 or 56 days with rotenone administered s.c. via Alzet minipumps releasing 2–3 mg/kg/day. Finally, with either approach, many rats with severe DA loss developed bradykinesia/rigidity and hunched posture, thus reproducing some cardinal signs of PD.

Despite the positive results obtained with these methods, the authors highlighted limitations of the procedure. First, there is large inter-animal variability for each dose and, at least for the method used by Betardet et al. (17), and associated acute toxicity/lethality. Again, striatal DA denervation involved mainly the central part of striatum, sparing the dorsolateral part of it [except in a few, severely lesioned animal, as described in the original paper by Betarbet et al. (17)], which is invariably affected in PD. Second, the selective DA nigrostriatal toxicity described in the two studies was not replicated in subsequent studies in which a similar protocol and dose regimen was used (119–121). In the latter studies a constant degeneration of striatal intrinsic neurons was observed, closely resembling what was already shown by Ferrante et al. (117) with the already mentioned "sub-chronic" pattern of rotenone administration. In conclusion, probably the use of rotenone in rodents does not produce models bearing enough reproducibility in different laboratories/species to be used routinely to replicate PD in rodents. By the fact that rotenone (2–3 mg/kg/day) induces neuronal degeneration after a low steady aministration, it is apparent that the low steady levels attained in brain are far lower than the levels needed to inhibit complex I of the mitochondrial respiratory chain (122). Selective DA lesioning by rotenone is now associated with intraneuronal ROS formation related to enzymatic or DA auto-oxidation, with generation of peroxide and $HO^{\bullet}$. The prominent role of oxidative stress, rather than bioenergy deficit, has been demonstrated by Sherer et al. (123) in organotypic midbrain culture, in human neuroblastoma cell culture (SK-N-MC), and in vivo in his rat chronic s.c. model (122). Obviously, if one also takes into account the reports in which a striatal neuronal lesion occurs, in this case the toxic ROS-mediated effects cannot be demonstrated as easily as for the DA-rich neurons/terminals. Again, a features arising from the Betarbet in vivo model, in which there was lesioning of NE and 5-HT neurons in selected areas (as constantly observed in PD as well), could not obviously be explained easily by oxidative stress induced by DA metabolism.

Despite a lack of chronic damaging effects of rotenone in mice, Thiffault et al. (118) showed that systemic rotenone acutely increases DA turnover, similarly to what was observed for toxic doses of MPTP—suggesting that this might be a common aspect of toxins acting on complex I of respiratory chain. However, the authors did not explore the mechanisms of increased DA turnover, but discussed this finding in light of previous data, suggesting that in hypoxic conditions DA release and turnover might be increased by a failure of the energy-dependent re-uptake mechanisms (124–127).

The MPTP Model of PD

MPTP has been used most frequently in mice to model PD because of their size, cost, and ease of handling. However, it is in primates that systemic MPTP most closely reproduces the signs and symptoms of PD. Even a very low dose of MPTP (vs. the dose used in rodents) induces a marked loss of nigrostriatal DA, accompanied by marked bradykinesia and rigidity (128); and the effects are reversed by L-DOPA. Further, it is possible to induce a hemiparkinsonian syndrome by injecting MPTP into the carotid artery. Animals so-treated also have a longer survival time (129).

MPTP is typically administered by an i.p. route, either as a single dose or by repeated administrations every few hours (for a review on technical issues, see 78). A single 30 mg/kg dose of MPTP induces a moderate nigrostriatal DA loss of several months duration (130); a higher single dose increases acute toxicity and lethality. Conversely, repeated low-doses of MPTP induce severe nigrostriatal DA loss (78,131).

In mice MPTP produces additional LC neuronal lesioning (132), similar in extent to the DA loss, and this has been considered as a further positive element for confirming the usefulness of the mouse MPTP model in reproducing PD (132). This duality of effect is a constant feature also in MPTP-treated primates (83,133). Behavioral counterparts of PD are present in MPTP-treated primates, but are absent in MPTP-treated mice. Obviously, there are huge phylogenetic differences in neuronal complexity and connectivity of the motor circuit and motor behavior of primates versus rodents, and this can account for differences in the MPTP behavioral effect in each species.

The role of excitatory amino acids (EAAs) in MPTP toxicity has been explored in several animal species (131). MK-801, a selective non-competitive blocker of NMDA receptors, prevents intranigral MPP^+ nigrostriatal toxicity in rats (134), although this effect has not been confirmed in MPTP-treated mice (135). MK-801, as well as CPP, is neuroprotective in primates (136,137). Thus, EAAS are considered to be important in the pathogenesis of PD, and the several NMDA blockers are under development.

The role of the noradrenergic nucleus LC in modulating/determining the neurotoxic effects of MPTP has been explored: (*i*) a prior 6-OHDA lesion of LC transforms a sub-threshold MPTP dose into one that is frankly toxic to SNpc DA neurons (138,139) and (*ii*) a prior DSP-4 [*N*-(2-chloroethyl)-*N*-ethyl-2-bromobenzylamine] lesion of LC NE terminals increases MPTP nigrostriatal damage in mice, an effect that is likely due to increased MPTP toxicity (131). Also, NE neuronal damage does not alter the distribution of MPP^+ in mice (131,140). Enhancement of MPTP toxicity by LC lesioning has been shown with METH (58) and 6-OHA in different species (e.g., monkey) (140), lending credence to a role of NE in PD pathogenesis.

As already indicated, the most crucial issue relating to MPTP modeling of PD is the presence or absence of LBs. With the sole exception—but not confirmed by others' studies—of findings in primate brains described by Forno et al. (141), of the occurrence of LBs in the brain of patients, primates, or rodents treated with the single or few injections of MPTP. In 2002 Meredith et al. reported the description of lysosomal inclusion formation into the SN of mice treated with repeated $MPTP^+$ probenecid (142). More recently, chronic s.c. infusion of a small daily amount of MPTP was found to induce formation of inclusion bodies strictly resembling LBs (18). Surprisingly, i.c.v. continuous infusion (143) of a low daily dose of MPP^+ in rats has been shown to be able to overcome this species resistance to MPTP toxicity, inducing a selective nigrostriatal loss and, at least at ultrastructural level, cellular alterations which have been interpreted as an early stage of LB formation. These two latter studies will be discussed more in detail in a following paragraph.

The Occurrence of Neuronal Inclusions in PD Models

As already mentioned above, the pathological hallmark of PD is the presence of LBs in the SNpc, LC and other subcortical nuclei and cortical areas. LBs are defined as pale inclusions inside neuronal processes or cell bodies, in specimens stained with routine staining techniques. More recently, it has been demonstrated that such inclusions can be selectively stained with antibodies anti-α-synuclein protein (144). In the recent years it has been proposed a staging of PD according to the distribution of LBs in the brain (84). According to such staging scale, LBs appear first in the lower brain stem, and afterward in the midbrain and cortex, in a caudal-rostral progression fashion. LBs are not specific for PD since they have been found in other neurodegenerative diseases, even though less abundantly and consistently. Furthermore, a specific form of Dementia, Dementia with LBs (also called Diffuse Lewy Body disease), has been named specifically after this kind of inclusion because of the widespread distribution of LBs as the main pathological marker.

As already mentioned in the specific paragraph above, recently some authors have been able to induce the occurrence of inclusions reminiscent of LBs in rodents. This was considered crucial for defining a PD model as a true pathogenetic model for PD. A first report of the occurrence of α-synuclein-immunopositive deposits into the SNpc and cortex of mice was observed by Meredith and colleagues (142) who treated mice with repeated daily MPTP injections, combined with probendecid. The latter compound was administered in order to slow the clearance of MPTP from the brain. As already described above, in 2000 Greenamyre and colleagues had induced the occurrence of inclusions in the SNpc by continuous s.c. administration of rotenone to rats (17); specifically, such intracellular bodies were electron dense cytoplasmic inclusions with a homogeneous dense core surrounded by fibrillary elements. Since these inclusions were immunopositive for ubiquitin and α-synuclein they were considered by these authors as equivalent to LB (17). This led Fornai et al. (18) to reconsider results obtained with the neurotoxin MPTP. In fact, it is well known that MPTP shares the inhibition of the complex I of the mitochondrial respiratory chain mechanism with rotenone. Until that time, MPTP was, as rotenone, usually administered acutely or in a schedule consisting of a few injections few hours apart from each other. Therefore, the lack of LB formation could have been due to the fact that by this approach DA neurons were injured too fast. In other words, it could be possible that the speed of intoxication might influence subsequent neuropathology. In line with this hypothesis, Fornai et al. (18) demonstrated that chronic s.c. administration of MPTP in mice induces the occurrence of LB-like inclusions as well.

In the three paradigms of administration described above (17,18,142), the crucial factor in determining the occurrence of inclusion bodies was likely to be due to the chronic paradigm of administration used. In particular, recent studies have demonstrated that some molecular/cellular steps play a key role in the formation of LB, and that these are likely to be fully recruited when low but prolonged, and appropriate, neurotoxic insults are repeated for long enough. Conversely, if acute, strong neurotoxic insults occur, they are strong enough to induce the activation of those molecular/cellular steps leading directly to severe cell impairment, and eventually up to cell death. In particular, in the formation of LB in nigral DA neurons in PD, a key role appears to be played by the so-called Ubiquitin-Proetasome-System (UPS) and its relation to cytoplasmic DA and α-synuclein. These aspects deserve, in our opinion, specific paragraphs, which follow below.

Role of the UPS in PD

The UPS acts as an intracellular protein-clearing system (145,146) which consists of a multienzymatic pathway placed mainly in the endoplasmic reticulum, which becomes activated during oxidative stress to process misfolded protein. The UP system becomes activated either in the physiological degradation of misfolded proteins occurring during oxidative stress, or in the degradation of abnormal proteins. In these conditions of excessive oxidative stress, or excess of altered proteins, the UPS may not be able to clear damaged proteins, which ultimately would accumulate and impair neuronal function, to lead to death of the cell (146). Furthermore, a similar effect can occur when there is a primary reduction in the efficacy of the UP system. This is indirectly confirmed by the occurrence of familiar forms of PD (bearing LB and pathological features similar to those observed in idiopathic PD) associated with mutations of UP components such as UchL-1 and parkin (147,148). Furthermore, impairment of the proteasome has been described also in idiopathic PD (149,150). Recent experimental findings confirm this hypothesis. In particular, direct infusion of a UPS inhibitor into the striatum induces selective degeneration of nigrostriatal terminals, and retrograde cell loss in the SNpc, while sparing both striatal serotonin levels and GABA cells; in spared nigral neurons, the presence of α-synuclein, parkin, E1 and ubiquitin immunostained inclusions was observed (151).

Remarkably, a selective nigrostriatal lesion has been observed also after *systemic* administration of proteasome inhibitors (152). In our study we observed after intrastriatal infusion of proteasome inhibitor that inclusion bodies occur in surviving nigral DA neurons. Such inclusions are formed by multilamellar envelops delimitating an electron-dense core ("whorls"). The inclusions appeared ultrastructurally very close to what is observed in vivo following chronic MPTP (153), METH (150), or MDMA (53,54). What we found as membrane whorls in transmission electron microscopy correspond to the autophagic granules first described by light microscopy in vitro after METH (154,155), and they are likely to represent an early step during the slow formation of authentic inclusions (not membrane-limited), and, presumably, up to "true" LBs. In fact, similar to LB, lactacystin-induced inclusions observed in vivo are exclusively cytoplasmic, never affect glial cells and their filamentous component is strongly immunopositive for α-synuclein.

Role of DA in Selective Neurotoxicity

Catecholamine-containing structures are the most significantly affected ones in PD, and contain the highest amount of inclusions (84). Among them, nigrostriatal neurons are constantly affected. In light of this observation DA itself could play a critical role in such toxicity. In other words, DA-containing neurons are at higher risk of damage under stressful conditions when other classes of neurons are not affected. This might explain why there is selective DA toxicity even in circumstances of ubiquitous impairment of the UPS, such as in selected forms of familiar PD or in animals after systemic administration of UPS inhibitors (152). The neurotoxic effects of DA itself might be related to its auto oxidation into DA-quinones. Toxic adducts between DA-quinones and α-synuclein could initiate critical biochemical steps towards PD pathogenesis (see section "Merging DA and α-Synuclein" on this page) (156–158).

In vitro, cell lines bearing the synthesis and storing apparati for DA display dramatic sensitivity to the effects of neurotoxins which increase cytosolic DA content. This is the case for METH (74,159), malonate (73,160), and MPTP (161). Thus, DA-containing neurons represent by themselves a system in delicate metabolic equilibrium consequent to the presence of high cytoplasmic levels of DA (156,159,162). α-synuclein could play a role in the deleterious effects of the oxidative stress produced by DA auto oxidation.

Role of α-Synuclein

α-Synuclein is a protein considered to play a role in membrane-associated processes at the presynaptic level (163). Because α-synuclein selectively accumulates in LB (143), and its mutation is present in selected familial PD forms (164–166), α-synuclein may be a key player in PD pathogenesis. In conformity with this, the injection of either human wild-type or mutant α-synuclein-expressing viral vectors into nigrostriatal pathways of animal models was shown to cause dopaminergic neurodegeneration and formation of inclusions that contain α-synuclein (167). In vitro, over-expression of either wild-type or mutant α-synuclein produces the formation of cytoplasmic aggregates (168,169). α-Synuclein is a substrate for the UPS, and it has been suggested that under specific conditions that UPS might not be able to cope with the rate of formation of damaged and/or mutant α-synuclein: this might lead to α-synuclein-aggregate formation (146).

Merging DA and α-Synuclein

A report of Conway et al. (156) suggests that α-synuclein and DA might interact to stabilize the α-synuclein protofibrils and to inhibit their conversion to fibrils (156,170). An increased amount of free cytosolic DA can induce the formation of DA-derived quinone (DAQ), a highly reactive metabolite which can covalently bind to proteins. In particular, cytoplasmic DAQ may be a crucial factor in determining the accumulation of pathogenic protofibrils (156). In humans, there is indirect evidence for the presumed crucial role of free cytosolic DA in neuronal degeneration occurring in PD. For instance, there is an inverse correlation between the expression of $VMAT_2$ and degenerative phenomena. Neurons of the ventral tegmental area (VTA), which are more resistant to PD cell death, express high levels of $VMAT_2$, which sequesters cytoplasmic DA into vesicles. High expression of $VMAT_2$, coupled with low levels of DAT produce very low levels of free cytosolic DA in VTA compared with SNpc DA neurons. Furthermore, monoaminergic regions more susceptible to degenerative phenomena in PD (i.e., SNpc and LC) are rich in neuromelanin, and the latter is a by-product of DAQ deriving from cytosolic DA (156).

As already indicated, an abnormal increase of cytoplasmic DA correlates with oxidation of cytoplasmic α-synuclein protofibrils, attributal to DA derivatives, which might be the final target to cluster cytoplasmic inclusions within DA neurons. Thus, DA cells are critically bound to an efficient UPS to clear damaged proteins formed by ROS interaction. Interestingly, in PD LBs are exclusively cytoplasmic and this might be due to the specific composition and dynamics of protein aggregates. Furthermore, when inclusion bodies arise in PC12 cells subsequent to treatment with METH, (1) they localize exclusively within cytoplasm (54) and (2) the occurrence of inclusions within PC12 cells is DA-dependent, since the presence of inclusions can be changed by modulating the synthesis and the metabolism of DA (53,151).

α-Synuclein and Parkinson's Disease

α-Synuclein is considered to be the best immunological marker of LBs. An excessive increase of free DA promotes conversion of α-synuclein oligomers to protofibrils stabilized by DAQ, and eventually inclusion formation and degeneration (171,172). In particular, the conversion of α-synuclein oligomers to protofibrils and their accumulation within neurons seems to play a fundamental role in the neurodegenerative process, since protofibrils may be neurotoxic intermediates while fibrils may be indifferent or even protective. These findings suggest that fibrillation and LB formation may be the last attempt by the cell to survive, by removing protofibrils. The central role of α-synuclein is confirmed by the occurrence of two mutations in α-synuclein (A53T and A30T) causing an early-onset autosomal dominant form of familial PD, suggesting that this protein plays a central role in PD. Furthermore, even normal, non-mutated α-synuclein produces neurotoxicity and LBs, when in excess, as in inherited Parkinsonism due to three alleles coding for α-synuclein.

NON-CLASSICAL NEUROTOXINS FOR DOPAMINERGIC NERVES

Agents acting directly at DA receptors are able to replicate neurological or psychiatric disorders without producing damage directly to dopaminergic nerves. Two agents, quinpirole and haloperidol, respective DA D_2 receptor agonist and antagonist, are discussed for this "neurotoxic effect," because of their relevance to human disorders.

Dopaminergic Neurotoxicity in the Absence of a Lesion

Quinpirole, a DA D_2 Agonist

Szechtman and colleagues (173,174) initially found that repeated treatments with the DA D_2 agonist quinpirole resulted in enhanced responses to a later quinpirole challenge. By treating rats during early postnatal ontogeny, for short periods (i.e., P0-P11 or P12-P22 or P23-P33 or P0-P28) and even with low-dose (50 μg/kg i.p.) quinpirole (175), it was possible to produce long-lived D_2 RSS, as manifested by enhanced quinpirole-induced effects in adulthood.

When administered once-a-day from birth, quinpirole (1 mg/kg) had seemingly little effect until rat pups were about two weeks old. At that age, quinpirole acutely produced intense gnawing and/or eating (176). At 19 days of age there was more locomotor activity, stereotyped behaviors (176), and the appearance of a darting-like behavior. At 21 days of age, if rats were place in chambers without lids, quinpirole acutely produced vertical jumping over a period of hours, and resulting in several hundred jumps. Simultaneously, there was prominent paw treading. The same quinpirole effect could be observed over the span of the following week, but by 35 days of age, jumping was no longer observed after quinpirole treatment (175). In rats placed in single chambers, quinpirole produced an enhanced yawning response (vs. control) (177,178) accompanied by penile erection (179,180). In a hot-plate test, quinpirole-primed rats displayed an antinociceptive effect that added to opiate antinociceptive action (179). At a later age, rats that had been treated with quinpirole daily during postnatal development displayed spacial memory deficits with enhanced skilled reaching (180). In some ways these rats model schizophrenia, and it is possible that quinpirole-priming is an alternative to the phencyclidine model. The induced D_2 RSS was life-long (181,182), and it was not induced by the predominately D_3 receptor agonist 7-hydroxy-*N*-*N*-di-n-propyl-2-aminotetralin (7-OH-DPAT) (183). If not acutely challenged with drugs, rats displaying D_2 RSS were indistinguishable from control. Enhanced behavioral responses to AMPH in these rats were associated with an approximate five-fold enhancement of AMPH-induced release of DA in the striatum, as demonstrated by in vivo microdialysis (184). When 6-OHDA was administered to neonatal rats to selectively destroy nigrostriatal dopaminergic innervation, repeated quinpirole treatments failed to produce an enhanced yawning response (185,186), while the ability of quinpirole to induce vertical jumping was preserved (unpublished). When the 5-HT neurotoxin, 5,7-dihydroxytryptamine, was administered to neonatal rats, D_2 RSS was observed in adulthood (187). Despite the nom de plume, "supersensitivity," it is quite possible that "D_2RSS" reflects D_2 autoreceptor subsensitivity (8,184,188–190), which conceivably would result in enhancement of D_2 agonist-induced behaviors. The description of behavioral alterations in quinpirole-primed rats is provided as an example of a neurotoxic effect produced by a substance (e.g., quinpirole) that seemingly does not produce overt destruction of a nerve but yet produces permanent behavioral alterations that persist for the life of the species (191).

Haloperidol, a DA D_2 Antagonist

Waddington and colleagues noted the appearance of spontaneous orofacial movements, resembling tardive dyskinesia in humans, in rats treated daily, for months, with the DA D_2 receptor antagonist haloperidol. The orofacial dyskinesia persisted for as long as haloperidol continued to be administered, but when haloperidol was discontinued as a treatment the orofacial dyskinesia disappeared within a month (192,193). When haloperidol was administered to 6-OHDA-lesioned rats, daily for one year, there was an approximate two-fold increase in the number of spontaneous orofacial dyskinesias (vs. non-lesioned rats receiving haloperidol), and the effect persisted for at least eight months after haloperidol had been withdrawn as a treatment (194). These findings represent other examples by which a neurotoxic *effect* is produced in the apparent absence of overt neuronal destruction (193); and the production of a permanent abnormality by a non-toxic substance (i.e., haloperidol) when a prior neurotoxic effect (i.e., by 6-OHDA) was produced. One might wonder if environmental exposure to substances, or pharmaceutical treatments by humans, might produce undesired "neurotoxic" effects if there is even minute damage in brain.

ENDOGENOUS NEUROTOXINS AND NEUROPROTECTANTS FOR DOPAMINERGIC NERVES

Introduction

Many features of PD (tremor, muscle rigidity, akinesia) are associated with DA deficiency in neostriatum, consequent to neurodegeneration and cell death of nigrostriatal dopaminergic neurons. Although the causes of idiopathic PD are still unknown, both environmental and endogenous toxins have been proposed as etiologic factors. The neurotoxin hypothesis of PD, originally proposed after discovery of 6-OHDA, gathered insurmountal momentum with the discovery of MPTP, the contaminant in "synthetic heroin" (i.e., China white) which produced irreversible Parkinson syndrome in substance abusers in California (14). On the assumption that MPTP-like neurotoxins may trigger PD, CSF as well as extracts of postmortem brains of PD patients have been taken and analyzed for putative endogenous MPTP-like compounds. Two groups of amine related compounds, which appeared chemically similar to MPTP, were detected in both human brain and CSF: β-carbolines (BCs-synonym norharmanes) and 1,2,3,4-tetrahydroisoquinolines (TIQs) (195–197). These heterocyclic compounds are formed endogenously from phenylalanine/tyrosine (TIQs) and tryptophan or indoloamines (BCs), respectively, and exert a wide spectrum of psychopharmacological and behavioral effects (198–200). The observation of various psychopharmacological effects of TIQs and BCs suggest an important role of the trace amines as neuromodulators. The TIQs and BCs may bind to their own high affinity sites on neuronal membranes associated with or located close to the receptors of neurotransmitters, and some of them such as 1,2,3,4-TIQ or its derivatives (salsolinol, 1-methyl-1,2,3,4-TIQ) bind directly to catecholamine brain receptors: dopaminergic $D_{2/3}$ and noradrenergic α_2 (201–203). Research on TIQs and BCs has intensified, because of their possible role(s) in pathological conditions, notably PD and alcoholism. Recently, from a clinical perspective, a promising role has been explored for 1-methyl-1,2,3,4-TIQ (1MeTIQ) as a neuroprotectant, anticonvulsant and anti-abuse substance (204–209).

β-Carbolines

Due to its structural similarity to MPTP the BCs have also been suggested as possible endogenous toxins leading to parkinsonism. BCs may be formed in vivo from condensation between tryptophan derivatives and aldehydes. Mass spectroscopy validated the presence of 1,2,3,4-tetrahydro-β-carboline (THBC; Fig. 1) in rat brain (210), and the aldehyde chloral as well as trichloroethylene (a solvent widely used in industry) may rapidly react with endogenous tryptamine to form the new 1-trichloromethyl-1,2,3,4-THBQ (TaClo) (211). BCs were detected in the human brain, CSF, and blood (195,212). Interestingly, *N*-methylated aromatic BC ions can be superimposed on MPP^+, the neurotoxic metabolite of MPTP (196). The unsubstituted BC easily crosses the blood-brain barrier. In vivo microdialysis studies in mice revealed that 2-me-BC^+ but not 2,9-dimethyl-β-carbolinium ion (2,9-dime-BC^+) penetrates the blood-brain barrier, possibly owing to transformation of 2-me-BC^+ to a neutral base (after a proton has been split off at position 9) under physiological conditions (213). The conversion of the unsubstituted

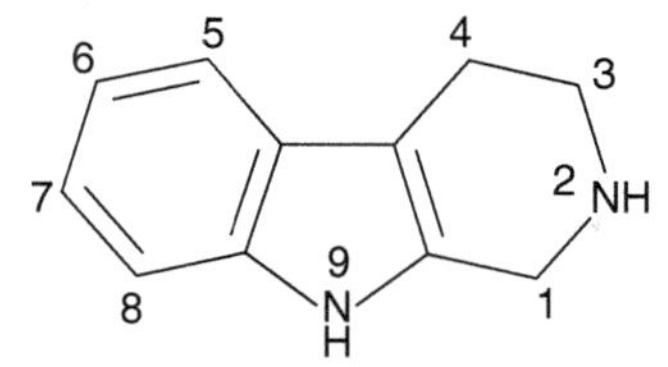

Figure 1 Chemical structure of tetrahydro-β-carboline.

BC norharman (NH) to the BC cations occurs endogenously in particulate fractions of the mammalian brain via *S*-adenosyl-methionine-dependent *N*-methylations, sequentially forming 2-mono-*N*-methylated (2-me-BC^+) and neurotoxic 2,9-di-*N*,*N*-methylated (2,9-dime-BC^+) β-carbolinium cations (214,215).

Neurotoxic Effects of BCs

Intracerebral injections of some analogs of BC such as *N*-methylated β-carboline 2-methyl-norharman (structural resemblance to MPP^+) into SN or MFB of rats results in a depletion of striatal DA and its metabolites. Indeed, three weeks after intranigral β-carboline (2-methyl-norharman) striatal DA and its metabolite DOPAC and HVA are reduced 40–60% (216). However, the lesion produced by 2-methyl-norharman appeared to be non-specific, affecting both dopaminergic and non-dopaminergic cells and fibers. NH, the precursor of 2,9-dime-BC^+ is present in human brain, and was measured in the parietal association cortex (0.58 pmol/g tissue) and even in higher concentration in SN (16 pmol/g tissue) (217). The endogenous neurotoxin found in human brain, 2,9-dime-BC^+, when injected into rat SN, potently induces apoptosis, and a reduction in striatal DA at three weeks (218). These findings strongly implicate BCs as endogenous neurotoxins. Clinical studies have shown elevated levels of BCs in the CSF and plasma of medication-free Parkinsonian patients (212). Also, subchronic TaClo had long-term neurotoxic effects in rats, reducing spontaneous locomotor activity and reducing apomorphine-induced activity at 12 weeks (219). Accordingly, it is hypothesized that TaClo may exert a progressive neurotoxic effect on the dopaminergic nigrostriatal system. Other studies have shown that the *N*-methylated TaClo is more toxic than TaClo (220), and Bonnet et al. (221) showed that dimethylation of the amino moiety at positions 2 and 9 leads to even greater neurotoxicity. Recently, the methylated β-carbolinium ion (2,9-DIME-BC^+) was shown to produce a dose-dependent degeneration of nigrostriatal neurons, leading to deficits in dopaminergic neurotransmission and an increase of muscle resistance and electromyographic activity—a syndrome resembling muscle rigidity in PD (222). Although neurotoxic effects of BCs are well documented, the mechanisms by which the BCs induce their neurotoxic effects are still unclear.

Mechanisms of BC Neurotoxicity

Analogues of MPTP and MPP^+ lacking an *N*-methyl group are virtually devoid of toxicity (223). Therefore activation of neurotoxicity of BCs requires an *N*-methylation. Numerous endogenous compounds undergo methylation by *N*-methyltransferases (phenylethanolamine, histamine, indolethylamine), as well as selected BCs (224,225). The resultant oxidation products, β-carboliniums, inhibit NAD^+ linked O_2 consumption in rat liver mitochondria. Janetzky and coworkers (226) have demonstrated that TaClo, and its *N*-methylated derivative are strong inhibitors of complex I (total inhibition at 400 μM and 250 μM, respectively). These results thus indicate that respiratory inhibition may underlie the neurotoxicity of BCs, as observed in primary cell cultures of mouse mesencephalon containing dopaminergic neurons (227) and in vivo (216,220). While *N*-methylation of BCs has been explored, the possible oxidation (aromatization) of *N*-methyl-THBQs to *N*-methyl-β-carbolinium cations resembling MPP^+ has not been investigated. As with MPTP oxidation, MAO and cytochrome P450 could be key enzymes involved in BC oxidation. As shown by Herraiz et al. (228) the *N*-methyl-THBQ analogs 2-methyl-THBQ (2-ME-BC) and 2,9-dimethyl-THBQ (2,9-DIME-BC) are oxidized to neurotoxic β-carbolinium cations by heme peroxidases; but unlike MPTP, not by MAO or P450. This suggests a novel bioactivation to derivative pyridinium-like cations resembling the aggressive neurotoxin MPP^+. Conceivably, production of THBQs, continuously and/or for prolonged period (perhaps years), could result in their accumulation, with *N*-methylation and oxidation by heme peroxidases to β-carbolinium toxins that progressively destroy SN neurons. β-carbolinium toxins have been detected in normal and Parkinsonian human brains (195,217), and if heme peroxidases are involved in toxicity of BCs, inhibitors of these enzymes could act as neuroprotectants.

Endogenous Ligands of Benzodiazephine Receptors (BDZ)

BCs have affinity equal to that of the most potent BDZs at the $GABA_A$ receptor, and thus BCs potently displace radiolabeled BDZs from this site. Some ligands exerted effects opposite to those of BDZ anxiolytics and came to be known as "inverse agonists" whose common feature was an antagonism to pharmacological BDZ effects. The first inverse agonists to be discovered were methyl-β-carboline-3-carboxylate (β-CCM) and methyl-6,7-dimethoxy-4-ethyl-β-carboline-3-carboxylate (DMCM) which produced convulsant and anxiogenic effects when given alone (229,230). The behavioral effects of β-CCM and DMCM are blocked by Ro 15-1788, a specific antagonist of $GABA_A$ receptor, indicating that these compounds indeed interact with BDZ receptors. Inverse agonists are thought to induce their effects by allosterically modulating the $GABA_A$ receptor complex by reducing the frequency of channel opening. According to this scheme, normal agonists positively modulate the $GABA_A$ receptor complex and thus potentiate GABA action, whereas inverse agonists negatively modulate the receptor and therefore lead to heightened anxiety and brain excitability.

1,2,3,4-Tetrahydroisoquinolines

The presence of TIQs in mammals including humans is well established. The first publications started in the early 1970s and were associated with alcohol intake. Biotransformation of ethanol by alcohol dehydrogenase to acetaldehyde may induce alterations in the metabolism of catecholamines (DA and noradrenaline), and produce aberrant aldehyde metabolites such as TIQ alkaloids (231). The endogenous catecholamines and their aldehyde oxidative metabolites may undergo a Pictet-Spengler type of condensation to yield TIQ alkaloids. TIQs are present in human brain at low levels (1–10 ng/g brain tissue), thus a highly sensitive and specific method is required to analyze TIQs in the brain (232,233). Interest in TIQs in PD was again aroused in the early 1980s owing to the discovery of MPTP as a parkinsonism-producing neurotoxin, especially in humans and monkeys (14,234) (Fig. 2). MPTP similarly to neurodegeneration observed in PD produced a severe loss of the pigmented DA neurons in the SN pars compacta (SNc), a marked reduction in the concentration of DA and its major metabolite HVA, and a significant

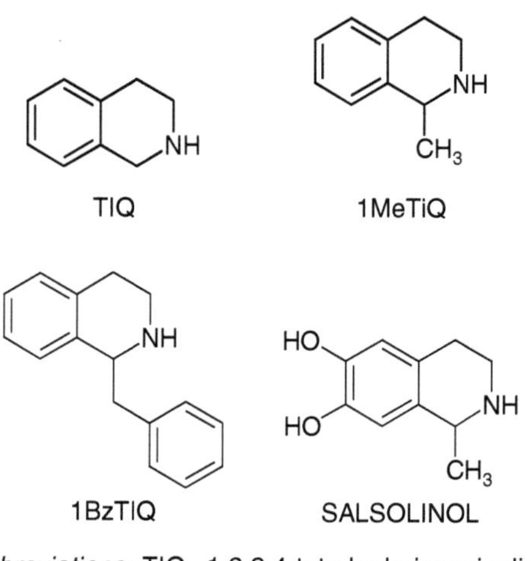

Figure 2 MPTP-like TIQs. *Abbreviations*: TIQ, 1,2,3,4-tetrahydroisoquinoline; 1MeTIQ, 1-methyl-1,2,3,4-tetrahydroisoquinoline, a neuroprotective compound; 1BzTIQ, 1-benzyl-1,2,3,4-tetrahydroisoquinoline, a neurotoxic compound; Salsolinol, 1-methyl-6,7-dihydroxy-1,2,3,4-tetrahydroisoquinoline.

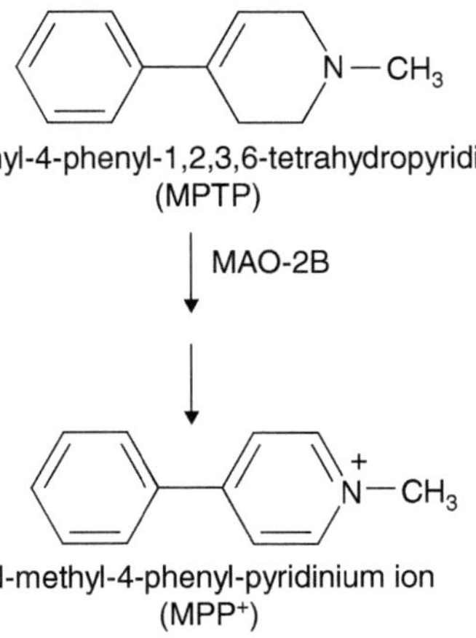

Figure 3 Bioactivation of MPTP to a neurotoxin ion MPP^+ by MAO-B.

decrease in DAT binding sites in the nucleus caudatus and putamen (235,236). Efforts have been made to find MPTP-like neurotoxins in the brain of parkinsonism patients. In this regard, it should be noted that MPTP per se is a pro-toxin oxidized by MAO-B to the overtly neurotoxic MPP^+ (237–239) (Fig. 3).

Neurotoxic Effects of TIQs and Their Mechanism of Action

It was assumed, that if any MPTP-like neurotoxins exist in the brain, they might be *N*-methylated and later oxidized by MAO to produce active MPP^+-like neurotoxins. TIQs are both endogenous and exogenous (i.e., in various foods such as banana, cheese, flour, egg, wine, beer, milk) (240). Substances related to the selective dopaminergic neurotoxin MPTP, have been suggested as potential etiologic factors that may contribute to the development of PD (241). The exogenously administered TIQs readily cross the blood-brain barrier. These alkaloids are presumably activated via *N*-methylation and are oxidized by MAO to form a neurotoxic *N*-methylisoquinolinium ion, the same as MPP^+ (242). Notably, however, while MPTP acts rapidly to produce irreversible neurotoxic changes mainly restricted to the nigrostriatal dopaminergic system after a single dose (243,245), TIQs do not produce immediate neurotoxicity (204,244).

Long-term exposure to a low concentration of neurotoxin is now regarded as a preferable means to model chronic neurological disorders such as PD. Initially it was found that only prolonged administration of TIQ and analogs (TIQ, salsolinol) at high dose (50–100 mg/kg) evoke behavioral and biochemical symptoms in monkeys and rodents, analogous to those in PD (201,204,244–246). Later, however, various members of this group were found to produce a range of effects—from potent neurotoxicity to neuroprotection, as exemplified by the highly neurotoxic 1BzTIQ and salsolinol (201,247–251), mildly neurotoxic TIQ (201,244,246), and the neuroprotective 1MeTIQ (202,204–206,252,253). Salsolinol and 1BzTIQ are two endogenous neurotoxins in which concentrations increase about three times in the CSF of patients with PD, and these TIQs have been proposed as etiological factors for PD (247,254). Recently it was shown by Shavali and coworkers (251,255) that 1BzTIQ activates apoptotic-signaling pathways, increases α-synuclein expression, significantly depletes glutathione levels, and induces cell death in human dopaminergic cells in culture. Interestingly, the proparkinsonian properties of 1BzTIQ were prevented by 1MeTIQ (256).

Apart from the propensity to form quaternary ions, the neurodegenerative effects of some TIQs may be caused by the facilitation of DA catabolism by *N*-oxidation. The oxidative MAO-dependent pathway of DA catabolism, leading to DOPAC, is known to produce potentially cytotoxic ROS (for instance, hydrogen peroxide, superoxide, and hydroxyl radicals), which may play an important role in the progressive and selective loss of the nigrostriatal dopaminergic neurons that occurs in aging and in PD (257–259). Conversely, the enhanced catabolism of DA through catechol-*O*-methyltransferase (COMT)-dependent *O*-methylation, leading to 3-methoxytyramine (3-MT), may be an important antioxidant defense mechanism (259).

In fact, recently it was shown that various TIQs differ in their effects on MAO-dependent and COMT-dependent DA catabolism. Neurotoxic 1BzTIQ shifts DA catabolism towards MAO-dependent *N*-oxidation, whereas neuroprotective 1MeTIQ strongly inhibits the MAO-dependent pathway and shifts DA catabolism towards COMT-dependent *O*-methylation (204). As MAO-dependent *N*-oxidation of DA generates free hydroxyl radicals, such differences in the biochemical effects of TIQs could explain the differences in their neurotoxic/neuroprotective profile (260). In this report in rats, simple noncatecholic TIQs (e.g., TIQ, 1MeTIQ, *N*-MeTIQ) acted as moderate or weak (1BzTIQ) MAO_B and MAO_A-inhibitors. 1MeTIQ inhibited more potently MAO_A than MAO_B, whereas catecholic TIQ, salsolinol, exerted a rather weak effect on MAO activity but markedly inhibited striatal tyrosine hydroxylase activity (260). MAO, present in the mitochondrial outer membrane, produces hydrogen peroxide by oxidation of monoamines, thereby inducing oxidative stress which is thought to predispose towards neuronal degeneration. MAO inhibition by some of the TIQs may protect neurons from cell death.

Neuroprotective Effect of TIQ and 1MeTIQ and Their Mechanism of Action

The pharmacological properties of TIQs suggest that these compounds may have much greater potential than being merely substances for studying PD. Particularly interesting are their properties as antidopaminergic agents with an atypical mechanism of action that suggests that they may specifically antagonize the active conformation of the DA receptor (201,203). This suggests that TIQs may possess a potential as either atypical antipsychotics or agents useful in preventing substance abuse disorders. Among them the most interesting and promising substance from a clinical vantage is 1MeTIQ. 1MeTIQ is a mixture of (*R*)- and (*S*)-enantiomers enzymatically synthesized from 2-phenylethylamine (PEA) and pyruvate by 1MeTIQ-synthesizing enzyme (a membrane-bound protein localized in the mitochondrial-synaptosomal fraction) (256,261). In a stereoselective study, it has been documented that the neuroprotective action of 1MeTIQ is closely correlated with the presence of (*R*)-1MeTIQ enantiomer as well as mixture of (*R*)- and (*S*)-enantiomers (262). In biochemical studies assessing activity of 1MeTIQ synthesizing enzyme (1MeTIQ-ase), it has been reported that parkinsonism-inducing substances (e.g., MPTP and BCs) considerably inhibited the activity of 1MeTIQ-ase (213,261). It is also well established by behavioral and biochemical study, that (*R,S*)-1MeTIQ demonstrates neuroprotective activity by antagonizing the behavioral and biochemical effects of dopaminergic neurodegeneration induced by numerous experimental neurotoxins such as MPTP, 1BzTIQ, and rotenone (205,206,247,252,263). Pharmacologically, 1MeTIQ and its analog lacking a methyl group, namely TIQ, inhibited MAO-dependent *N*-oxidation and the formation of DOPAC during DA degradation, and shifted catabolism of this neurotransmitter toward the COMT-dependent *O*-methylation—this resulted in the reduction of free radical production and production of neuroprotective activity (204,264). Recently, it was demonstrated that 1MeTIQ shares many activities with TIQ, including inhibition of glutamate-induced free-radical generation in an abiotic system, and reduction of indices of neurotoxicity (e.g., caspase-3 activity and lactate dehydrogenase release) in a mouse embryonic primary cell culture (202). However, in granular cell cultures obtained from seven-day-old rats, 1MeTIQ prevented glutamate-induced cell death and $^{45}Ca^{2+}$ influx, whereas TIQ did not. Such a profile of action for 1MeTIQ suggests a specific effect of this compound on excitatory amino acid (EAA) receptors. Additionally, it was shown in an in vivo microdialysis experiment that 1MeTIQ prevented kainate-induced release of EAAs from rat frontal cortex (202).

In comparing the chemical structure of 1MeTIQ with other compounds containing the TIQ skeleton and attendant molecular mechanisms of action, one can find similarities between 1MeTIQ and *N*-cetyl-1-(4-chlorophenyl)-6,7-dimethoxy-TIQ; and 1,1-pentamethylene-TIQ, derivatives which are non-competitive AMPA/kainate receptor antagonists and endowed with the ability to protect animals in maximal electroshock seizure, pentylenetetrazole seizure, and audiogenic DBA/2 mouse seizure models (265,266). In fact, 1MeTIQ exerts anticonvulsant effects by increasing the threshold for electroconvulsions and enhancing potentiation of the antiseizure action of carbamazepine and valproate against maximal electroshock (208).

In light of this series of experiments 1MeTIQ offers a unique and complex mechanism of neuroprotection in which inhibitory effects on MAO, resulting in free radical scavenging properties, and antagonism of the glutamatergic system may play important mechanistic roles. Other TIQ family members awaiting discovery may provide greater insight into their neurotoxic/neuroprotective properties, and provide additional opportunity towards utilizing TIQ analogs in pharmaceutical development.

SUMMARY

Dopaminergic nerves represent a system at risk for injury. The neurotransmitter DA per se is a putative neurotoxin in that its metabolism by MAO generates the ROS H_2O_2, which in the presence of iron (Fe^{2+}), generates the highly reactive hydroxyl radical, which is capable of rapid inactivation of enzymes, proteins, DNA, RNA, unsaturated lipids, and any—SH moiety with which it comes into contact. Conditions for this scenario are ripe in the substantia nigra, in which neuromelanin in DA nerves serves as a sink for Fe^{2+}. In addition, there are endogenously produced neurotoxic species for DA nerves, such as the DA metabolite aminochrome; or BCs, etc. Selective (exogenous) neurotoxins, of which there are many, replicate PD, the neurodegenerative disorder in which pars compacta substantia nigra DA neurons spontaneously degenerate with aging. MPTP has been used to model this disorder for more than two decades, 6-OHDA has been used even longer; and most recently, rotenone has been invoked to produce DA destruction to model PD. The life of a DA nerve is a precarious one, beset with internal challenges from cytoplasmic DA metabolism, as well as the attack by extraneuronal in situ DA-neurotoxins, and potential ultimate exposure to environmental (e.g., rotenone and such) or inadvertent intake of overt neurotoxins (e.g., MPTP).

REFERENCES

1. Ungerstedt U. Striatal dopamine release after amphetamine or nerve degeneration revealed by rotational behavior. Acta Physiol Scand 1971: 68.
2. Zigmond MJ, Stricker EM. Animal models of parkinsonism using selective neurotoxins: clinical and basic implications. Int Rev Neurobiol 1989; 31: 1–79.
3. Zigmond MJ, Keefe KA. 6-Hydroxydopamine as a tool for studying catecholamines in adult animals. In: Kostrzewa RM, ed. Highly Selective Neurotoxins: Basic and Clinical Applications. Totowa, NJ: Humana Press, 1998: 75–107.
4. Shaywitz BA, Klopper JH, Gordon JW. Paradoxical response to amphetamine in developing rats treated with 6-hydroxydopamine. Nature 1976; 261: 153–5.
5. Shaywitz BA, Yager RD, Klopper JH. Selective brain dopamine depletion in developing rats: an experimental model of minimal brain dysfunction. Science 1976; 191: 305–8.
6. Kostrzewa RM, Brus R, Kalbfleisch JH, Perry KW, Fuller RW. Proposed animal model of attention deficit hyperactivity disorder. Brain Res Bull 1994; 34: 161–7.
7. Kostrzewa RM, Kostrzewa JP, Kostrzewa RA, Nowak P, Brus R. Pharmacological models of ADHD. J Neural Transm 2008; 115: 287–98.
8. Brus R, Nowak P, Szkilnik R, et al. Serotoninergics attenuate hyperlocomotor activity in rats. Potential new therapeutic strategy for hyperactivity. Neurotox Res 2004; 6: 317–26.
9. Breese GR, Criswell HE, Duncan GE, et al. A dopamine deficiency model of Lesch-Nyhan disease—the neonatal-6-OHDA-lesioned rat. Brain Res Bull 1990; 25: 477–84. Review.
10. Breese GR, Knapp DJ, Criswell HE, et al. The neonate-6-hydroxydopamine-lesioned rat: a model for clinical neuroscience and neurobiological principles. Brain Res Brain Res Rev 2005; 48: 57–73. Review.
11. Breese CR, Breese GR. The use of neurotoxins to lesion catecholamine-containing neurons to model clinical disorders. In: Kostrzewa RM, ed. Highly Selective Neurotoxins: Basic and Clinical Applications. Totowa, NJ: Humana Press, 1998: 19–73.
12. Kostrzewa RM, Jacobowitz DM. Pharmacological actions of 6-hydroxydopamine. Pharmacol Rev 1974; 26: 199–288.
13. Lew R, Malberg JE, Ricaurte GA, Seiden LS. Evidence for and mechanism of action of neurotoxicity of amphetamine related compounds. In: Kostrzewa RM, ed. Highly Selective Neurotoxins: Basic and Clinical Applications. Totowa, NJ: Humana Press, 1998: 235–68.
14. Langston JW, Ballard PA, Tetrud JW, Irwin, I. Chronic parkinsonism in humans due to a product of meperidine analog synthesis. Science 1983; 219: 979–80.
15. Langston JW, Ballard PA. Parkinsonism induced by 1-methyl-4-phenyl-1,2,3,6-tetrahydropyridine (MPTP): implications for treatment and the pathogenesis of Parkinson's disease. Can J Neurol Sci 1984; 11: 160–65.
16. Manning-Bog AB, Langston JW. Model fusion, the next phase in developing animal models of Parkinson's disease. Neurotox Res 2007; 11: 219–40.

17. Betarbet R, Sherer TB, MacKenzie G, et al. Chronic systemic pesticide exposure reproduces features of Parkinson's disease. Nat Neurosci 2000; 3: 1301–6.
18. Fornai F, Schluter OM, Lenzi P, et al. Parkinson-like syndrome induced by continuous MPTP infusion: convergent roles of the ubiquitin proteasome system and α-synuclein. Proc Natl Acad Sci USA 2005; 102: 3413–8.
19. Senoh S, Witkop B. Formation and rearrangements of aminochromes from a new metabolite of dopamine and some of its derivatives. J Am Chem Soc 1959; 81: 6231–5.
20. Senoh S, Witkop B. Non-enzymatic conversions of dopamine to norepinephrine and trihydroxyphenethylamine. J Am Chem Soc 1959; 81: 6222–31.
21. Senoh S, Witkop B, Creveling CR, Udenfriend S. 2,4,5-tri-hydroxyphenethylamine, a new metabolite of 3,4-dihysroxyphenethylamine. J Am Chem Soc 1959; 81: 1768–9.
22. Tranzer JP, Thoenen H. Ultra-morphologische Veranderungen der sympathischen Nervendigunden der Katze nach Vorbehandlung mit 5- und 6-hydroxydopamin. Naunyn-Schmiedebergs Arch Pharmakol Exp Pathol 1967; 257: 343–4.
23. Thoenen H, Tranzer JP. Chemical sympathectomy by selective destruction of adrenergic nerve endings with 6-hydroxydopamine. Naunyn-Schmiedebergs Arch Pharmakol Exp Pathol 1968; 261: 271–88.
24. Ungerstedt U. 6-Hydroxydopamine-induced degeneration of central monoamine neurons. Eur J Pharmacol 1968; 5: 107–10.
25. Saner A, Thoenen H. Model experiments on the molecular mechanism of action of 6-hydroxydopamine. Mol Pharmacol 1971; 7: 147–51.
26. Heikkila RE, Mytilineou C, Cote L, Cohen G. Evidence for degeneration of sympathetic nerve terminals caused by the ortho- and para-quinones of 6-hydroxydopamine. J Neurochem 1973; 20: 1345–50.
27. Perumal AS, Gopal VB, Tordzro WK, Cooper TB, Cadet JL. Vitamin E attenuates the toxic effects of 6-hydroxydopamine on free radical scavenging systems in rat brain. Brain Res Bull 1992; 29: 699–701.
28. Kumar R, Agarwal AK, Seth PK. Free radical-generated neurotoxicity of 6-hydroxydopamine. J Neurochem 1995; 64: 1703–7.
29. Heikkila RE Cohen G. Inhibition of biogenic amine uptake by hydrogen peroxide: a mechanism for toxic effects of 6-hydroxydopamine. Science 1971; 122: 1257–8.
30. Heikkila RE, Cohen G. 6-Hydroxydopamine: evidence for superoxide radical as an oxidative intermediate. Science 1973; 181: 156–7.
31. Cohen G, Heikkila RE. The generation of hydrogen peroxide, superoxide radical, and hydroxyl radical by 6-hydroxydopamine, dialuric acid, and related cytotoxic agents. J Biol Chem 1974; 249: 2447–52.
32. Sachs C, Jonsson G. Mechanisms of action of 6-hydroxydopamine. Biochem Pharmacol 1975; 24: 1–8.
33. Hirsch EC, Brandel J-P, Galle P, Javoy-Agid F, Agid Y. Iron and aluminum increase in the substantia nigra of patients with Parkinson's disease: and X-ray microanalysis. J Neurochem 1991; 56: 446–51.
34. Mendez-Alvarez E, Soto-Otero R, Hermida-Ameijeiras A, Lopez-Martin ME, Labandeira-Garcia JL. Effect of iron and manganese on hydroxyl radical production by 6-hydroxydopamine: mediation of antioxidants. Free Radic Biol Med 2001; 15: 986–98.
35. Kienzl E, Jellinger K, Stachelberger H, Linert W. Iron as catalyst for oxidative stress in the pathogenesis of Parkinson's disease? Life Sci 1999; 65: 1973–76.
36. Ben-Shachar D, Eshel G, Finberg JPM, et al. The iron chelator desferrioxamine (desferal) retards 6-hydroxydopamine-induced degeneration of nigrostriatal dopamine neurons. J Neurochem 1991; 56: 1441–4.
37. Glinka YY, Youdim MBH. Inhibition of mitochondrial complexes I and IV by 6-hydroxydopamine. Eur J Pharmacol 1995; 292: 329–32.
38. Glinka YY, Tipton KF, Youdim MBH. Nature of inhibition of mitochondrial respiratory complex I by 6-hydroxydopamine. J Neurochem 1996; 66: 2004–10.
39. Linert W, Herlinger E, Jameson RF, Kienzl E, Jellinger K, Youdim MB. Dopamine, 6-hydroxydopamine, iron, and dioxygen-their mutual interactions and possible implication in the development of Parkinson's disease. Biochim Biophys Acta 1996; 1316: 160–8.
40. Wagner K, Trendelenburg U. Effect of 6-hydroxydopamine on oxidative phosphorylation and on momoamine oxidase activity. Naunyn-Schmiedebergs Arch Pharmakol Exp Pathol 1971; 269: 110–16.
41. Pletscher A, Bartholini G, Brydurer H, Burkard WP, Gey KF. Chlorinated arylalkylamines affecting the cerebral metabolism of 5-hydroxytriptamine. J Pharmacol Exp Ther 1964; 145: 344–50.
42. Harvey JA, Mc Master SE, Yunger, LM. p-Chloroamphetamine: selective neurotoxic action in the brain. Science 1975; 187: 841–3.

43. Koda LY, Gibb JW. Adrenal and striatal tyrosine hydroxylase activity after methamphetamine. J Pharmacol Exp Ther 1973; 185: 42–8.
44. Ricaurte GA, Guillery RW, Seiden LS, Schuster CR, Moore RY. Dopamine nerve terminal degeneration produced by high doses of metylamphetamine in the rat brain. Brain Res 1982; 235: 93–103.
45. Ellison GD, Eison MS, Huberman H. Stages of constant amphetamine intoxication: delayed appearance of abnormal social behaviours in rat colonies. Psychopharmacol 1978; 56: 293–9.
46. Morgan ME, Gibb JW. Short-term and long-term effects of methamphetamine on biogenic amine metabolism in extra-striatal dopaminergic nuclei. Neuropharmacology 1980; 19: 989–95.
47. Ricaurte GA, Schuster CR, Seiden LS. Long-term effects of repeated methylamphetamine administration on dopamine and serotonin neurons in the rat brain: a regional study. Brain Res 1980; 193: 153–63.
48. Seiden LS, Ricaurte GA. Neurotoxicity of methamphetamine and related drugs. In: Meltzer HY, ed. Psychopharmacology: The Third Generation of Progress. New York, NY: Raven Press, 1987: 359–66.
49. Logan BJ, Laverty R, Sandrson WD, Yee YB. Differences between rats and mice in MDMA (methylenedioxymethamphetamine) neurotoxicity. Eur J Pharmacol 1988; 152: 227–34.
50. O'Callaghan JP, Miller, DB. Neurotoxicity profiles of substituted amphetamines in the C57Bl/6J mouse. J Pharmacol Exp Ther 1994; 270: 741–51.
51. Cadet JL, Ladenheim B, Hirata H, et al. Superoxide radicals mediate the biochemical effects of methylendioxymethamphetamine (MDMA): evidence from using CuZn-superoxide dismutase transgenic mice. Synapse 1995; 21: 169–76.
52. Fornai F, Bassi L, Torracca MT, et al. Norepinephrine loss exacerbates methamphetamine-induced striatal dopamine depletion in mice. Eur J Pharmacol 1995; 283: 99–102.
53. Fornai F, Lenzi P, Gesi M, et al. Methamphetamine produces neuronal inclusions in the nigrostriatal system and in PC12 cells. J Neurochem 2004; 88: 114–23.
54. Fornai F, Lenzi P, Frenzilli G, et al. DNA damage and ubiquitinated neuronal inclusions in the substantia nigra and striatum of mice following MDMA (ecstasy). Psychopharmacology (Berl) 2004; 173: 353–63.
55. O'Shea E, Esteban B, Camarero J, Green AR, Colado MI. Effects of GBR 12909 and fluoxetine on the acute and long term changes induced by MDMA ("ecstasy") on the 5-HT and dopamine concentration in mouse brain. Neuropharmacology 2000; 40: 65–74.
56. Fornai F, Torracca MT, Bassi L, et al. Norepinephrine loss selectively enhances chronic nigrostriatal dopamine depletion in mice and rats. Brain Res 1996; 735: 349–53.
57. Fornai F, Torracca MT, Bassi L, et al. Norepinephrine loss selectively enhances chronic nigrostriatal dopamine depletion in mice and rats. Brain Res 1996; 735: 349–53.
58. Fornai F, Giorgi FS, Alessandrì MG, et al. Effects of pre-treatment with DSP-4 on methamphetamine-induced striatal dopamine loss and pharmacokinetics. J Neurochem 1999; 72: 777–84.
59. Hirata H, Ladenheim B, Rothman RB, Epstein C, Cadet JL. Methamphetamine-induced serotonin neurotoxicity is mediated by superoxide radicals. Brain Res 1995; 677: 345–7.
60. Suzuki O, Hattori H, Asano M, Oya M, Katsumata Y. Inhibition of monoamine oxidase by d-methamphetamine. Biochem Pharmacol 1980; 29: 2071–3.
61. Kuczenski R, Segal D. Concomitant characterization of behavioral and striatal neurotrasmitter response to amphetamine using in vivo microdyalisis. J Neurosci 1989; 9: 2051–65.
62. Fischer JF, Cho AK. Chemical release of dopamine from striatal homogenates: evidence for an exchange diffusion model. J Pharmacol Exp Ther 1979; 208: 203–9.
63. Zaczek R, Culp S, Goldberg H, McCann DJ, De Souza EB. Interactions of [^{3}H]amphetamine with rat brain synaptosomes. I. Saturable sequestration. J Pharmacol Exp Ther 1991; 257: 820–9.
64. Rudnick G, Wall SC. The molecular mechanism of MDMA: serotonin transporters are targets for MDMA-induced serotonin release. Proc Natl Acad Sci USA 1992; 89: 1817–21
65. Sulzer D, Chen T-K, Lau YY, et al. Amphetamine redistributes dopamine from synaptic vescicles to the cytosol and promotes reverse transport. J Neurosci 1995; 15: 4102–8.
66. O'Dell SJ, Weihmuller FB, Marshall, JF. Multiple methamphetamine injections induced marked increases in extracellular striatal dopamine which correlate with subsequent neurotoxicity. Brain Res 1991; 564: 256–60.
67. O'Dell SJ, Weihmuller FB, Marshall JF. Methamphetamine-induced dopamine overflow and injury to striatal dopamine terminals: attenuation by dopamine D_1 and D_2 antagonists. J Neurochem 1993; 60: 1792–99.
68. Wagner GC, Lucot J, Shuster C, Seiden L. α-Methyltyrosine attenuates and reserpine increases methamphetamine-induced neuronal changes. Brain Res 1983; 270: 285–8.
69. Graham DG. Oxidative pathways for catecholamines in the genesis of neuromelanin and cytotoxic quinones. Mol Pharmacol 1978; 14: 633–43.

70. Graham DG, Tiffany SM, Bell WR Jr, Gutknecht WF. Autoxidation versus covalent binding of quinones as the mechanism of toxicity of dopamine, 6-hydroxydopamine, and related compounds toward C1300 neuroblastoma cells in vitro. Mol Pharmacol 1978; 14: 644–53.
71. Seiden LS, Vosmer G. Formation of 6-hydroxydopamine in caudate nucleus of the rat brain after a single large dose of methylamphetamine. Pharmacol Biochem Behav 1984; 21: 29–31.
72. Nash JF, Yamamoto BK. Methamphetamine neurotoxicity and striatal glutamate release: comparison to 3,4-methylenedioxy-methamphetamine. Brain Res 1992; 581: 237–43.
73. Sonsalla PK, Manzino L, Sinton CM, et al. Inhibition of striatal energy metabolism produces cell loss in the ipsilateral substantia nigra. Brian Res 1997; 773: 223–6.
74. Schmidt CJ, Ritter JK, Sonsalla PK, Hanson GR, Gibb JW. Role of dopamine in the neurotoxic effects of methamphetamine. J Pharmacol Exp Ther 1985; 233: 539–44.
75. Fumagalli F, Gainetdinov RR, Valenzano KJ, Caron MG. Role of dopamine transporter in methamphetamine-induced neurotoxicity: evidence from mice lacking the transporter. J Neurosci 1998; 18: 4861–9.
76. Johnson M, Stone DM, Hanson GR, Gibb JW. Role of the dopaminergic nigrostriatal pathway in methamphetamine-induced depression of the neostriatal serotonergic system. Eur J Pharmacol 1987; 135: 231–4.
77. Davis GC, Williams AC, Markey SP, et al. Chronic parkinsonism secondary to intravenous injection of meperidine analogues. Psychiatry Res 1979; 1: 249–54.
78. Przedborski S, Jackson-Lewis V, Naini A, et al. 2001. The parkinsonian toxin 1-methyl-4-phenyl-1,2,3,6-tetrahydropyridine (MPTP): a technical review of its utility and safety. J Neurochem 2001; 76: 1265–74.
79. Javitch JA, D'Amato RJ, Strittmatter SM, Snyder SH. Parkinsonism-inducing neurotoxin, N-methyl-4-phenyl-1,2,3,6-tetrahydropyridine: uptake of the metabolite N-methyl-4-phenylpyridine by dopamine neurons explains selective toxicity. Proc Natl Acad Sci USA 1985; 82: 2173–7.
80. Ehringer H, Hornykiewicz O. Distribution of noradrenaline and dopamine (3-hydroxytyramine) in the human brain and their behavior in diseases of the extrapyramidal system. Klin Wochenschr 1960; 38: 1236–9.
81. Hornykiewicz O, Kish SJ. Biochemical pathophysiology of Parkinson's disease. Adv Neurol 1986; 45: 19–34.
82. Pollanen MS, Dickson DW, Bergeron CB. Pathology and biology of the Lewy body. J Neuropathol Exp Neurol 1993; 52: 183–91.
83. Forno LS. Neuropathology of Parkinson's disease. J Neuropathol Exp Neurol 1996; 55: 259–72.
84. Braak H, Del Tredici K, Rüba U, et al. Staging of brain pathology related to sporadic Parkinson's disease. Neurobiol Aging 2003; 24: 197–211.
85. Sachs C, Jonsson G. Degeneration of central noradrenaline neurons after 6-hydroxydopamine in newborn animals. Res Commun Chem Pathol Pharmacol 1972; 4: 203–20.
86. Kostrzewa RM. The blood-brain barrier for catecholamines—revisited. Neurotox Res 2007; 11: 261–71.
87. Breese GR, Baumeister A, Napier TC, et al. Evidence that D_1 dopamine receptors contribute to the supersensitive behavioral responses induced by L-dihydroxyphenylalanine in rats treated neonatally with 6-hydroxydopamine. J Pharmacol Exp Ther 1985; 235: 287–95.
88. Breese GR, Napier TC, Mueller RA. Dopamine agonist-induced locomotor activity in rats treated with 6-hydroxydopamine at differing ages: functional supersensitivity of D_1 dopamine receptors in neonatally lesioned rats. J Pharmacol Exp Ther 1985; 234: 447–55.
89. Breese GR, Duncan GE, Napier TC, et al. 6-Hydroxydopamine treatments enhance behavioral responses to intracerebral microinjection of D_1- and D_2-dopamine agonists into the nucleus accumbens and striatum without changing dopamine antagonist binding. J Pharmacol Exp Ther 1987; 240: 167–76.
90. Kostrzewa RM, Kostrzewa, JP, Brus R, Kostrzewa RA, Nowak P. Proposed animal model of severe Parkinson's disease: neonatal 6-hydroxydopamine lesion of dopaminergic innervation of striatum. J Neural Transm 2006; Suppl 70: 277–9.
91. Deumens R, Blokland A, Prickaerts J. Modeling Parkinson's disease in rats: an evaluation of 6-OHDA lesions of the nigrostriatal pathway. Exp Neurol 2002; 175: 303–17.
92. Taghzouti K, Le Moal M, Simon H. Suppression of noradrenergic innervation compensates for behavioural deficits induced by lesion dopaminergic terminals in the lateral septum. Brain Res 1991; 552: 124–8.
93. Fulceri F, Biagioni F, Lenzi P, et al. Nigrostriatal damage with 6-OHDA: validation of routinely applied procedures. Ann NY Acad Sci 2006; 1074: 344–8.

94. Hudson JL, van Horne CG, Stromberg I, et al. Correlation of apomorphine- and amphetamine-induced turning with nigrostriatal dopamine content in unilateral 6-hydroxydopamine lesioned rats. Brain Res 1993; 626: 167–74.
95. Andrew R, Watson DG, Best SA, et al. The determination of hydroxydopamines and other trace amines in the urine of parkinsonian patients and normal controls. Neurochem Res 1993; 18: 1175–7.
96. Jellinger K, Linert L, Kienzl E, Herlinger E, Youdim MBH. Chemical evidence for 6-hydroxydopamine to be an endogenous toxic factor in the pathogenesis of Parkinson's disease. J Neural Transm 1995; 46: 297–314.
97. Morelli M, Fenu S, Pinna A, et al. "Priming" to dopamine agonist-induced contralateral turning as a model of non-associative sensitization to the expression of the post-synaptic dopamine message. Behav Pharmacol 1993; 4: 389–97.
98. Araki T, Tanji H, Kato H, et al. Alterations of second messenger systems in the rat brain after 6-hydroxydopamine lesions of the medial forebrain bundle. Eur J Pharm Sci 1999; 8: 261–7
99. Doucet JP, Nakabeppu Y, Bedard PJ, et al. Chronic alterations in dopaminergic neurotransmission produce a persistent elevation of delta FosB-like protein(s) in both the rodent and primate striatum. Eur J Neurosci 1996; 8: 365–81.
100. Salin P, Hajji MD, Kerkerian-le Goff L. Bilateral 6-hydroxydopamine-induced lesion of the nigrostriatal dopamine pathway reproduces the effects of unilateral lesion on substance P but not on enkephalin expression in rat basal ganglia. Eur J Neurosci 1996; 8: 1746–57.
101. Agid Y, Javoy F, Glowinski J. Hyperactivity of remaining dopaminergic neurones after partial destruction of the nigro-striatal dopaminergic system in the rat. Nat New Biol 1973; 245: 150–1
102. Zigmond MJ, Abercrombie ED, Berger TW, Grace AA, Stricker EM. Compensations after lesions of central dopaminergic neurons: some clinical and basic implications. Trends Neurosci 1990; 13: 290–6.
103. Cenci MA, Lundblad M. Utility of 6-hydroxydopamnie lesioned rats in the preclinical screening of novel treatments for Parkinson's disease. In: LeDoux M, ed. Animal Models of Movement Disorders. Amsterdam: Elsevier/Academic Press, 2005: 193–209.
104. Seiden LS, Fischman MW, Schuster CR. Long-term methamphetamine induced changes in brain catecholamine in tolerant rhesus monkeys. Drug Alcohol Depend 1975; 1: 215–9.
105. Woolverton WL, Ricaurte GA, Forno LS, Seiden LS. Long-term effects of chronic methamphetamine administration in rhesus monkeys. Brain Res 1989; 486: 73–8.
106. Villemagne V, Yuan J, Wang DF, et al. Brain dopamine neurotoxicity in Baboons treated with doses of methamphetamine comparable to those recreationally abused by humans: evidence from [^{11}C] WIN-35,428 positron emission tomoography studies and direct in vitro determinations. J Neurosci 1998; 18: 419–27.
107. Melega WP, Lacan G, Harvey DC, Huang SC, Phelps ME. Dizocilpine and reduced body temperature do not prevent methamphetamine-induced neurotoxicity in the Vervet Monkey: [^{11}C]WIN-35,428 positron emission studies. Neurosci Lett 1998; 258: 17–20.
108. McCann UD, Wong DF, Yokoi F, et al. Reduced striatal dopamine transporter density in abstinent methamphetamine and methcathinone users: evidence from positron emission tomography studies with [^{11}C]WIN-35,428. J Neurosci 1998; 18: 8417–22.
109. Volkow ND, Chang L, Wang GJ, et al. Association of dopamine transporter reduction with psychomotor impairment in methamphetamine abusers. Am J Psychiatry 2001; 158: 377–82.
110. Volkow ND, Chang L, Wang GJ, et al. Loss of dopamine transporters in methamphetamine abusers recoves with protracted abstinence. J Neurosci 2001; 21: 9414–8.
111. McCann UD, Ricaurte GA. Amphetamine neurotoxicity: accomplishments and remaining challenges, Neurosci Biobehav Rev 2004; 27: 821–6.
112. Quan L, Ishikawa T, Michiue T, et al. Ubiquitin-immunoreactive structures in the midbrain of methamphetamine abusers. Leg Med 2005; 7: 144–50.
113. Heikkila RE, Nicklas WJ, Vyas I, Duvoisin RC. Dopaminergic toxicity of rotenone and the 1-methyl-4-phenylpyridinium ion after their stereotaxic administration to rats: implication for the mechanism of 1-methyl-4-phenyl-1,2,3,6-tetrahydropyridine toxicity. Neurosci Lett 1985; 62: 389–94.
114. Ernster L, Dallner G, Azzone GF. Differential effects of rotenone and amytal on mitochondrial electron and energy transfer. J Biol Chem 1963; 238: 1124–31.
115. Gutman M, Singer TP, Casida JE. Studies on the respiratory chain-linked reduced nicotinamide adenine dinucleotide dehydrogenase: XVII. Reaction sites of piericidin A and rotenone. J Biol Chem 1970; 245: 1992–7.
116. Kushnareva Y, Murphy AN, Andreyev A. Complex I-mediated reactive oxygen species generation: modulation by cytochrome c and $NAD(P)^+$ oxidation-reduction state. Biochem J 2002; 368(Pt. 2): 545–53.

117. Ferrante RJ, Schulz JB, Kowall NW, Beal MF. Systemic administration of rotenone produces selective damage in the striatum and globus pallidus, but not in the substantia nigra. Brain Res 1997; 753: 157–62.
118. Thiffault C, Langston JW, Di Monte DA. Increased striatal dopamine turnover following acute administration of rotenone to mice. Brain Res 2000; 885: 283–8.
119. Hoglinger GU, Feger J, Annick P, et al. Chronic systemic complex I inhibition induces a hypokinetic multisystem degeneration in rats. J Neurochem 2003; 84: 1–12.
120. Lapointe N, St-Hilaire M, Martinoli MG, et al. Rotenone induces non-specific central nervous system and systemic toxicity. FASEB J 2004; 18: 717–9.
121. Zhu C, Vourc'h P, Fernagut PO, et al. Variable effects of chronic subcutaneous administration of rotenone on striatal histology. J Comp Neurol 2004; 478: 418–26.
122. Sherer TB, Kim JH, Betarbet R, Greenamyre JT. Subcutaneous rotenone exposure causes highly selective dopaminergic degenerationand alpha-synuclein aggregation. Exp Neurol 2003; 179: 9–16.
123. Sherer TB, Betarbet R, Stout AK, et al. An in vitro model of Parkinson's disease: linking mitochondrial impairment to altered alpha-synuclein metabolism and oxidative damage. J Neurosci 2002; 22: 7016–26.
124. Akiyama U, Ito A, Koshimura K, et al. Effects of transient forebrain ischemia and reperfusion on function of dopaminergic neurons and dopamine reuptake in vivo in rat striatum. Brain Res 1991; 561: 120–7.
125. Cvejic V, Micic DV, Djuricic BM, et al. Monoamines and related enzymes in cerebral cortex and basal ganglia following transient ischemia in gerbils. Acta Neuropathol 1980; 51: 71–7.
126. Globus MY-T, Ginsberg MD, Harik SI, Busto R, Dietrich WD. Role of dopamine in ischemic striatal injury: metabolic evidence. Neurology 1987; 37: 1712–9.
127. Santiago M, Granero L, Machado A, Cano J. Complex I inhibitor effect in the presence and absence of nomifensine. Eur J Pharmacol 1995; 280: 251–6.
128. Burns RA, Chiueh CC, Markey SP, et al. A primate model of parkinsonism: selective destruction of dopaminergic neurons in the pars compacta of the substantia nigra by N-methyl-4-phenyl-l,2,3,6-tetrahydropyridine. Proc Natl Acad Sci USA 1983; 80: 4546–50.
129. Bankiewicz KS, Oldfield EH, Chiueh CC, et al. Hemiparkinsonism in monkeys after unilateral internal carotid artery infusion of 1-methyl-4-phenyl-1,2,3,6-tetrahydropyridine (MPTP). Life Sci 1986; 39: 7–16.
130. Zuddas A, Corsini GU, Schinelli S, et al. 1989. Treatment combined with ethanol or acetaldehyde selectively destroys dopaminergic neurons in mouse substantia nigra. Brain Res 1989; 501: 1–10.
131. Fornai F, Vaglini F, Maggio R, et al. Species differences in the role of excitatory amino acids in experimental Parkinsonism. Neurosci Biobehav Rev 1997; 21: 401–15.
132. Seniuk NA, Tatton WG, Greenwood CE. Dose-dependent destruction of the coeruleus-cortical and nigral-striatal projections by MPTP. Brain Res 1990; 27: 7–20.
133. Hornykiewicz O, Pifl C. The validity of the MPTP primate model for neurochemical pathology of idiopathic Parkinson's disease. In: Briley M, Marien M, eds. Noradrenergic Mechanisms in Parkinson's Disease. Boca Raton: CRC Press, 1994: 59–71.
134. Turski L, Bressler K, Rettig KJ, Losbmann PA, Wachtel H. Protection of substantia nigra from MPP^+ neurotoxicity by N-methyl-aspartate antagonists. Nature 1991; 349: 414–8.
135. Sonsalla PK, Zeevalk GD, Manzino L, Giovanni A, Nicklas WJ. MK-801 fails to protect against the dopaminergic neuropathology produced by systemic 1-methyl-4-phenyl-1,2,3,6-tetrahydropyridine in mice or intranigral 1-methyl-4-phenylpyridinium in rats. J Neurochem 1992: 58: 1979–82.
136. Zuddas A, Oberto G, Vaglini F, et al. MK-801 prevents 1-methyl-4-phenyl-l,2,3,6-tetrahydropyridine-induced Parkinsonism in primates. J Neurochem 1992; 59: 733–9.
137. Lange KW, Loschmann PA, Sofic E, et al. The competitive NMDA antagonist CPP protects substantia nigra neurons from MPTP-induced degeneration in primates. Naunyn Schmiedeberg's Arch Pharmacol 1993; 348: 586–92.
138. Marien M, Briley M, Colpaert F. Noradrenaline depletion exacerbates MPTP-induced striatal dopamine loss in mice. Eur J Pharmacol 1993; 236: 487–9.
139. Bing G, Zhang Y, Watanabe Y, et al. Locus coeruleus lesions potentiate neurotoxic effects of MPTP in dopaminergic neurons of the substantia nigra. Brain Res 1994; 668: 261–5.
140. Mavridis M, Degryse AD, Lategan AJ, Marien MR, Colpaert FC. Effects of locus coeruleus lesions on parkinsonian signs, striatal dopamine and substantia nigra cell loss after 1-methyl-4-phenyl-1,2,3,6-tetrahydropyridine in monkeys: a possible role for the locus coeruleus in the progression of Parkinson's disease. Neuroscience 1991; 41: 507–23.
141. Forno LS, Langston JW, DeLanney, et al. Locus coeruleus lesions and eosinophilic inclusions in MPTP-treated monkeys. Ann Neurol 1986; 20: 449–55.
142. Meredith GE, Totterdell S, Petroskec E. et al. Lysosomal malfunction accompanies alpha-synuclein aggregation in a progressive mouse model of Parkinson's disease. Brain Res 2002; 956: 156–65.

143. Yazdani U, German DC, Liang C-L, et al. Rat model of Parkinson's disease: chronic central delivery of 1-methyl-4-phenylpyridinium (MPP^+). Exp Neurol 2006; 200: 172–83.
144. Spillantini MG, Schmidt ML, Lee VM, et al. α-Synuclein in Lewy bodies. Nature 1997; 388: 839–40.
145. Bence NF, Sampat RM, Kopito RR. Impairment of the ubiquitin-proteasome system by protein aggregation. Science 2001; 292: 1552–5.
146. Chung KK, Dawson VL, Dawson TM. The role of the ubiquitin/proteasomal pathway in Parkinson's disease and other neurodegenerative disorders. Trends Neurosci 2001; 24: S7–S14.
147. Kitada T, Asakawa S, Hattori N, et al. Mutations in the parkin gene cause autosomal recessive juvenile Parkinsonism. Nature 1998; 392: 605–8.
148. Leroy E, Boyer R, Auburger G, et al. The ubiquitin pathway in Parkinson's disease. Nature 1998; 395: 451–2.
149. McNaught KS, Olanow CW, Hallywell B, Isacson O, Jenner P. Failure of the ubiquitin-proteasome system in Parkinson's disease. Nat Rev Neurosci 2001; 2: 589–94.
150. McNaught KS, Olanow CW. Proteolytic stress: a unifying concept for the etiopathogenesis of Parkinson's disease. Ann Neurol 2003; 53: S73–S86.
151. Fornai F, Lenzi P, Gesi M, et al. Fine structure and biochemical mechanisms underlying nigrostriatal inclusions and cell death after proteasome inhibition. J Neurosci 2003; 23: 8955–66.
152. McNaught KS, Perl DP, Brownell AL, Olanow CW. Systemic exposure to proteasome inhibitors causes a progressive model of Parkinson's disease. Ann Neurol 2004; 56: 149–62.
153. Fornai F, Schlüter, OM, Lenzi P, et al. Parkinson-like syndrome induced by continuous MPTP infusion: convergent roles of the ubiquitin-proteasome system and α-synuclein. Proc Natl Acad Sci USA 2005 (Feb 16); 102: 3413–8.
154. Cubells JF, Rayport S, Rajendran G, et al. Methamphetamine neurotoxicity involves vacuolation of endocytic organelles and dopamine-dependent intracellular oxidative stress. J Neurosci 1994; 14: 2260–71.
155. Larsen KE, Fon EA, Hasting TG, Edwards RH, Sulzer D. Methamphetamine-induced degeneration of dopaminergic neurons involves autophagy and up-regulation of dopamine synthesis. J Neurosci 2002; 22: 8951–60.
156. Conway KA, Rochet JC, Bienganski RM, et al. Kinetic stabilization of the alpha-synuclein protofibril by a dopamine α-synuclein adduct. Science 2001; 294: 1346–9
157. Sulzer D, Bogulavsky J, Larsen KE. Neuromelanin biosynthesis is driven by excess cytosolic catecholamines not accumulates by synthetic vesicles. Proc Natl Acad Sci USA 2002; 97: 11869–74.
158. Sulzer D. α-Synuclein and cytosolic dopamine: stabilizing a bad situation. Nat Med 2001; 7: 1280–2.
159. Sulzer D, Zecca L. Intraneuronal dopamine-quinone biosynthesis: a review Neurotoxicity Res 2000; 1: 181–95.
160. Moy LY, Zeevalk GD, Sonsalla PK. Role for dopamine in malonate-induced damage in vivo in striatum and in vitro in mesencephalic cultures. J Neurochem 2000; 74: 1656–65.
161. Gainetdinov RR, Fumagalli F, Wang YM, et al. Increased MPTP neurotoxicity in vesicular monoamine transporter 2 heterozygote knock-out mice. J Neurochem 1998; 70: 1973–8.
162. Miller GW, Gainetdinov RR, Levey AI, Caron MG. Dopamine transporter and neuronal injury. Trends Pharmacol Sci 1999; 20: 424–9.
163. Recchia A, Debetto P, Negro A, et al. α-Synuclein and Parkinson's disease. FASEB J 2004; 18: 617–26.
164. Kruger R, Kuhn W, Mueller T, et al. Ala30Pro mutation in the gene encoding α-synuclein in Parkinson's disease. Nat Genet 1998; 18: 106–8.
165. Polymeropoulos MH, Lavedan C, et al. Mutation in the α-synuclein gene identified in families with Parkinson's disease. Science 1997; 276: 2045–7.
166. Zarranz JJ, Alegre J, Gomez-Esteban JC, et al. The new mutation, E46K, of α-synuclein causes Parkinson and Lewy body dementia. Ann Neurol 2004; 55: 164–73.
167. Kirik D, Annett LE, Burger C, et al. Nigrostriatal α-synucleinopathy induced by viral vectormediated overexpression of human α-synuclein: a new primate model of Parkinson's disease. Proc Natl Acad Sci USA 2003; 100: 2884–29.
168. Stefanis L, Larsen KE, Rideout J, Sulzer D, Greene LA. Expression of A53T mutant but not wild-type α-synuclein in PC12 cells induces alterations of the ubiquitin-dependent degradation system, loss of dopamine release, and autophagic cell death. J Neurosci 2001; 21: 9549–60.
169. Tabrizi SJ, Orth M, Wilkinson JM, et al. Expression of mutant α-synuclein causes increased susceptibility to dopamine toxicity. Hum Mol Genet 2000; 9: 2683–9.
170. Rochet JC, Outeiro TF, Conway KA, et al. Interactions among alpha-synuclein, dopamine, and biomembranes: some clues for understanding neurodegeneration in Parkinson's disease. J Mol Neurosci 2004; 23: 23–34.
171. Lee HJ, Shin SY, Choi C, Lee YH, Lee SJ. Formation and removal of alpha-synuclein aggregates in cells exposed to mitochondrial inhibitors. J Biol Chem 2002; 277: 5411–7.

172. Xu J, Kao SY, Lee FJ, et al. Dopamine-dependent neurotoxicity of alpha-synuclein: a mechanism for selective neurodegeneration in Parkinson disease. Nat Med 2002; 8: 600–6.
173. Szechtman H, Dal H, Mustafa S, Einat H, Sullivan RM. Effects of dose and interdose interval on locomotor sensitization to the dopamine agonist quinpirole. Pharmacol Biochem Behav 1994; 48: 921–8.
174. Eilam D, Szechtman H. Dosing regimen differentiates sensitization of locomotion and mouthing to D_2 agonist quinpirole. Pharmacol Biochem Behav 1990; 36: 989–91.
175. Kostrzewa RM, Guo J, Kostrzewa FP. Ontogenetic quinpirole treatments induce vertical jumping activity in rats. Eur J Pharmacol 1993; 239: 183–7.
176. Kostrzewa RM, Hamdi A, Kostrzewa FP. Production of prolonged supersensitization of dopamine D_2 receptors. Eur J Pharmacol 1990; 183: 1411–2.
177. Kostrzewa RM, Brus R. Ontogenic homologous supersensitization of quinpirole-induced yawning in rats. Pharmacol Biochem Behav 1991; 39: 517–9.
178. Kostrzewa RM, Brus R, Rykaczewska M, Plech A. Low dose quinpirole ontogenically sensitizes to quinpirole-induced yawning in rats. Pharmacol Biochem Behav 1993; 44: 487–9.
179. Kostrzewa RM, Brus R, Kalbfleisch J. Ontogenetic homologous sensitization to the antinociceptive action of quinpirole in rats. Eur J Pharmacol 1991; 209: 157–61.
180. Brown RW, Gass JT, Kostrzewa RM. Ontogenetic quinpirole treatments produce spatial memory deficits and enhance skilled reaching in adult rats. Pharmacol Biochem Behav 2002; 72: 591–600.
181. Brus R, Szkilnik R, Nowak P, et al. Locomotor sensitization of dopamine receptors by their agonists quinpirole and SKF-38393, during maturation and aging in rats. Pharmacol Rev Commun 1998; 10: 25–30.
182. Brus R, Szkilnik R, Nowak P, et al. Sensitivity of central dopamine receptors in rats to quinpirole and SKF-38393, administered at their early stages of ontogenicity, evaluated by learning and memorizing a conditioned avoidance reflex. Pharmacol Rev Commun 1998; 10: 31–6.
183. Oswiecimska J, Brus R, Szkilnik R, Nowak P, Kostrzewa RM. 7-OH-DPAT, unlike quinpirole, does not prime a yawning response in rats. Pharmacol Biochem Behav 2000; 67: 11–15.
184. Nowak P, Brus R, Kostrzewa RM. Amphetamine-induced enhancement of neostriatal in vivo microdialysate dopamine content in rats, quinpirole-primed as neonates. Pol J Pharmacol 2001; 53: 319–29.
185. Brus R, Szkilnik R, Kostrzewa RM. Quinpirole-induced yawning behavior in rats neonatally pretreated with 6-hydroxydopamine (6-OHDA) and 5,7-dihydroxytryptamine (5,7-DHT). Med Sci Monitor 1997; 3: 324–7.
186. Brus R, Kostrzewa RM, Nowak P, et al. Ontogenetic quinpirole treatments fail to prime for D_2 agonist-enhancement of locomotor activity in 6-hydroxydopamine-lesioned rats. Neurotox Res 2003; 5: 329–38.
187. Brus R, Plech A, Kostrzewa RM. Enhanced quinpirole responses in rats lesioned neonatally with 5,7-dihydroxytryptamine. Pharmacol Biochem Behav 1995; 50: 649–53.
188. Kostrzewa RM, Kostrzewa JP, Brus R. Dopamine receptor supersensitivity: an outcome and index of neurotoxicity. Neurotox Res 2003; 5: 111–8.
189. Kostrzewa RM, Nowak P, Kostrzewa JP, Kostrzewa RA, Brus R. Dopamine D_2 agonist priming in intact and dopamine-lesioned rats. Neurotox Res 2004; 6: 457–62.
190. Kostrzewa RM, Kostrzewa JP, Brown R, Nowak P, Brus R. Dopamine receptor supersensitivity: development, mechanisms, presentation, and clinical applicability. Neurotox Res 2008; 14: 121–8.
191. Kostrzewa RM. Dopamine receptor supersensitivity. Neurosci Biobehav Reviews 1995; 19: 1–17.
192. Waddington JW, Gamble SJ. Neuroleptic treatment for a substantial proportion of adult life: behavioural sequelae of 9 months haloperidol administration. Eur J Pharmacol 1980; 67: 363–9.
193. Waddington JW, Cross AJ, Gamble SJ, Bourne RC. Spontaneous orofacial dyskinesia and dopaminergic function in rats after 6 months of neuroleptic treatment. Science 1983; 220: 530–2.
194. Huang NY, Kostrzewa RM, Li C, Perry KW, Fuller RW. Persistent spontaneous oral dyskinesias in haloperidol-withdrawn rats neonatally lesioned with 6-hydroxydopamine: absence of an association with the B_{max} for [^{3}H]raclopride binding to neostriatal homogenates. J Pharmacol Exp Ther 1997; 280: 268–76.
195. Matsubara K, Kobayashi S, Kobayashi Y, et al. β-Carbolinium cations, endogenous MPP^+ analogs, in the lumbar cerebrospinal fluid of patients with Parkinson's disease. Neurology 1995; 45: 2240–5.
196. Collins MA, Neafsey EJ. β-Carboline analogues of MPP^+ as environmental neurotoxins. In: Storch A, Collins MA, eds. Neurotoxic Factors in Parkinson's Disease and Related Disorders. New York: Kluwer-Plenum Publ., 2001: 115–30.
197. Nagatsu T. Amine-related neurotoxins in Parkinson's disease: past, present, and future. Neurotoxicol Teratol 2002; 24: 565–9.

198. Robinson ESJ, Anderson NJ, Crosby J, Nutt DJ, Hudson AL. Endogenous β-carbolines as clonidine-displacing substances. Ann NY Acad Sci 2003; 1009: 157–66.
199. Östergren A, Annas A, Skog K, Lindquist NG, Brittebo EB. Long term retention of neurotoxic β-carbolines in brain neuromelanin. J Neural Transm 2004; 111: 141–57.
200. Herraiz T, Chaparro C. Human monoamine oxidase enzyme inhibition by coffee and β-carbolines norharman and harman isolated from coffee. Life Sci 2006; 78: 795–802.
201. Antkiewicz-Michaluk L, Michaluk J, Romańska I, et al. Antidopaminergic effects of 1,2,3,4-tetrahydroisoquinoline and salsolinol. J Neural Transm 2000; 107: 1009–19.
202. Antkiewicz-Michaluk L, Lazarewicz JW, Patsenka A, et al. The mechanism of 1,2,3,4-tetrahydroisoquinolines neuroprotection: the importance of free radicals scavenging properties and inhibition of glutamate-induced excitotoxicity. J Neurochem 2006; 97: 846–56.
203. Vetulani J, Antkiewicz-Michaluk L, Nalepa I, Sansone M. A possible physiological role for cerebral tetrahydroisoquinolines. Neurotox Res 2003; 5: 147–55.
204. Antkiewicz-Michaluk L, Michaluk J, Mokrosz M, et al. Different action on dopamine catabolic pathways of two endogenous 1,2,3,4-tetrahydroisoquinolines with similar antidopaminergic properties. J Neurochem 2001; 78: 100–8.
205. Antkiewicz-Michaluk L, Karolewicz B, Romańska I, et al. 1-Methyl-1,2,3,4-tetrahydroisoquinoline protects against rotenone-induced mortality and biochemical changes in rat brain. Eur J Pharmacol 2003; 466: 263–9.
206. Antkiewicz-Michaluk L, Wardas J, Michaluk J, et al. Protective effect of 1-methyl-1,2,3,4-tetrahydroisoquinoline against dopaminergic neurodegeneration in the extrapyramidal structures produced by intracerebral injection of rotenone. Int J Neuropsychopharmacol 2004; 7: 155–63.
207. Antkiewicz-Michaluk L, Filip M, Michaluk J, et al. An endogenous neuroprotectant substance, 1-methyl-1,2,3,4-tetrahydroisoquinoline (1MeTIQ), prevents the behavioral and neurochemical effects of cocaine reinstatement in drug-dependent rats. J Neural Transm 2007; 114: 307–17.
208. Luszczki JJ, Antkiewicz-Michaluk L, Czuczwar SJ. 1-Methyl-1,2,3,4-tetrahydroisoquinoline enhances the anticonvulsant action of carbamazepine and valproate in the mouse maximal electroshock seizure model. Neuropharmacology 2006; 50: 133–42.
209. Wasik A, Romańska I, Antkiewicz-Michaluk L. The effect of an endogenous compound 1-methyl-1,2,3,4-tetrahydroisoquinoline on morphine-induced analgesia, dependence and neurochemical changes in dopamine metabolism in rat brain structures. J Physiol Pharmacol 2007; 58; 235–52.
210. Barker SA, Harrison REW, Monti JA, et al. Identification and quantitation of 1,2,3,4-tetrahydro-beta-carboline, 2-methyl-1,2,3,4-tetrahydro-beta-carboline, and 8-methoxy-1,2,3,4-tetrahydro-beta-carboline as in vivo constituents of rat brain and adrenal gland. Biochem Pharmacol 1981; 30: 9–17.
211. Bringmann G, God R, Feineis D, et al. The TaClo concept: 1-trichloromethyl-1,2,3,4-tetrahydro-β-carboline (TaClo), a new toxin for dopaminergic neurons. J Neural Transm 1995; Suppl 46: 235–44.
212. Kuhn W, Müller, TH, Grosse H, Rommelspacher H. Elevated levels of harman and norharman in cerebrospinal fluid of parkinsonian patients. J Neural Transm 1996; 103: 1435–40.
213. Matsubara K, Gonda T, Sawada H, et al. Endogenously occurring beta-carboline induces parkinsonism in nonprimate animals: a possible causative protoxin in idiopathic Parkinson's disease. J Neurochem 1998; 70: 727–35.
214. Gearhart DA, Neafsey EJ, Collins MA. Phenylethanolamine-N-methyltransferase has β-carboline 2N-methyltransferse activity: hypothetical relevance to Parkinson's disease. Neurochem Int 2002; 40: 611–20.
215. Matsubar, K, Aoyama K, Suno M, Awaya T. N-methylation underlying Parkinson's disease. Neurotoxicol Teratol 2002; 24: 593–8.
216. Neafsey EJ, Drucker G, Raikoff K, Collins MA. Striatal dopaminergic toxicity following intranigral injection in rats of 2-methyl-norharman, a β-carbolinium analog of N-methyl-4-phenylpyridinium ion (MPP^+). Neurosci Lett 1989; 105: 344–9.
217. Matsubara K, Collins MA, Akane A, et al. Potential bioactivated neurotoxicants, N-methylated β-carbolinium ions, are present in human brain. Brain Res 1993; 610: 90–6.
218. Pavlovic S, Schulze G, Wernicke C, et al. 2,9-dimethyl-β-carbolinium, a neurotoxin occurring in human brain, is a potent inducer of apoptosis as 1-methyl-4-phenylpyridinium. Neuroscience 2006; 139: 1525–37.
219. Sontag KH, Heim C, Sontag TA, et al. Long-term behavioural effects of TaClo (1-trichloro-methyl-1,2,3,4-tetrahydro-β-carboline) after subchronic treatment in rats. J Neural Transm 1995; Suppl 46: 283–9.

220. Grote C, Clement H-W, Wesemann W, et al. Biochemical lesions of the nigrostriatal system by TaClo (1-trichloromethyl-1,2,3,4-tetrahydro-β-carboline) and derivatives. J Neural Transm 1995; Suppl 46: 275–81.
221. Bonnet R, Pavlovic S, Lehmann J, et al. The strong inhibition of triosephosphate isomerase by the natural β-carbolines may explain their neurotoxic action. Neuroscience 2004; 127: 443–53
222. Lorenc-Koci E, Rommelspacher H, Schulze G, et al. Parkinson's disease-like syndrome in rats induced by 2,9-dimethyl-β-carbolinium ion, a β-carboline occurring in the human brain. Behav Pharmacol 2006; 17: 463–73.
223. Testa B, Naylor R, Costall B, Jenner P, Marsden CD. Does an endogenous methylpyridinium analogue cause Parkinson's disease? J Pharm Pharmacol 1985; 37: 679–80.
224. Drucker G, Raikoff K, Neafsey EJ, Collins MA. Dopamine uptake inhibitory capacities of β-carbolines and 3,4-dihydro-β-carboline analogs of N-methyl-4-phenyl-1,2,3,6-tetrahydropyridine (MPTP) oxidation products. Brain Res 1990; 509: 125–33.
225. Gearhart DA, Collins MA, Lee JM, Neafsey J. Increased β-carboline 9N-methyltransferase activity in the frontal cortex in Parkinson's disease. Neurobiol Dis 2000; 7: 201–11.
226. Janetzky B, God R, Bringmann G, Reichmann H. 1-Trichloromethyl-1,2,3,4-tetrahydro-β-carboline, a new inhibitor of complex I. J Neural Transm 1995; Suppl 46: 265–73.
227. Rausch WD, Abdel-Mohsen M, Koutsilieri E, Chan WW, Bringmann G. Studies of the potentially endogenous toxin TaClo (1-trichloromethyl-1,2,3,4-tetrahydro-β-carboline) in neuronal and glial cell cultures. J Neural Transm 1995; Suppl 46: 255–63.
228. Herraiz T, Guillen H, Galisteo, J. N-Methyltetrahydro-β-carboline analogs of 1-methyl-4-phenyl-1,2,3,6-tetrahydropyridine (MPTP) neurotoxin are oxidized to neurotoxic β-carbolinium cations by heme peroxidases. Biochem Biophys Res Commun 2007; 356: 118–23.
229. Braestrup C, Honore T, Nielsen M, et al. Ligands for benzodiazepine receptors with positive and negative efficacy. Biochem Pharmacol 1984; 33: 859–62.
230. Haefely WE. Pharmacology of the benzodiazepine receptor. Eur Arch Psychiatry Neurol Sci 1989; 238: 294–301.
231. Cohen G, Collins M. Alkaloids from catecholamines in adrenal tissue: possible role in alcoholism. Science 1970; 167: 1749–51.
232. Niwa T, Yoshizumi H, Tatematsu A, Matsuura S, Nagatsu T. Presence of tetrahydroisoquinoline, a parkinsonism-related compound in foods. J Chromatogr 1989; 493: 347–52.
233. Ohta S, Kohno M, Makino Y, Tachikawa O, Hirobe M. Tetrahydroisoquinoline and 1-methyl-tetrahydroisoquinoline are present in the human brain: relation to Parkinson's disease. Biomed Res 1987; 8: 453–56.
234. Heikkila RE, Hess A, Duvoisin RC. Dopaminergic neurotoxicity of 1-methyl-4-phenyl-1,2,3,6-tetrahydropyridine in mice. Science 1984; 224: 1451–53.
235. Niznik HB, Fogel EF, Fassos FF, Seeman P. The dopamine transporter is absent in parkinsonian putamen and reduced in the caudatenucleus. J Neurochem 1991; 56: 192–8.
236. Wilson JM, Levey AI, Rajput A, et al. Differential changes in neurochemical markers of striatal dopamine nerve terminals in idiopathic Parkinson's disease. Neurology 1996; 47: 718–26.
237. Chiba K, Trevor A, Castagnoli N Jr. Metabolism of the neurotoxic tertiary amine, MPTP, by brain monoamine oxidase. Biochem Biophys Res Commun 1984; 120: 574–8.
238. Glover V, Gibb C, Sandler M. Monoamine oxidase B (MAO-B) is the major catalyst for 1-methyl-4-phenyl-1,2,3,6-tetrahydropyridine (MPTP) oxidation in human brain and other tissue. Neurosci Lett 1986; 64: 216–20.
239. Fuller RW, Hemrick-Luecke SK, Perry KW. Deprenyl antagonizes acute lethality of 1-methyl-4-phenyl-1,2,3,6-tetrahydropyridine in mice. J Pharmacol Exp Ther 1988; 247: 531–5.
240. Makino Y, Ohta S, Tachikawa O, Hirobe M. Presence of tetrahydroisoquinoline and 1-methyl-tetrahydroisoquinoline in foods: compounds related to Parkinson's disease. Life Sci 1988; 43: 373–8.
241. Nagatsu T. Isoquinoline neurotoxins in the brain and Parkinson's disease. Neurosci Res 1997; 29: 99–111.
242. Naoi M, Matsuura S, Parvez H, et al. Oxidation of N-methyl-1,2,3,4-tetrahydroisoquinoline into the N-methyl-isoquinolinium ion by monoamine oxidase. J Neurochem 1989; 52: 653–5.
243. Burns RS, Lewitt PA, Ebert H, et al. The clinical syndrome of striatal dopamine deficiency: parkinsonism induced by 1-methyl-4-phenyl-1,2,3,6-tetrahydropyridine (MPTP). N Engl J Med 1985; 312: 1418–21.
244. Lorenc-Koci E, Antkiewicz-Michaluk L, Wardas J, Zapała M, Wierońska J. Effect of 1,2,3,4-tetrahydroisoquinoline administration under conditions of CYP2D inhibition on dopamine metabolism, level of tyrosine hydroxylase protein and the binding of [^{3}H]GBR 12,935 to dopamine transporter in the rat nigrostriatal, dopaminergic system. Brain Res 2004; 1009: 67–81.

245. Nagatsu T, Yoshida M. An endogenous substance of the brain, tetrahydroisoquinoline produces parkinsonism in primates with decreased dopamine, tyrosine hydroxylase and biopterin in the nigrostriatal regions. Neurosci Lett 1988; 87: 178–82.
246. Lorenc-Koci E, Śmiałowska M, Antkiewicz-Michaluk L, et al. Effect of acute and chronic administration of 1,2,3,4-tetrahydroisoquinoline on muscle tone, metabolism of dopamine in the striatum and tyrosine hydroxylase immunocytochemistry in the substantia nigra, in rats. Neuroscience 2000; 95: 1049–59.
247. Kotake Y, Tasaki Y, Makino Y, Ohta S, Hirobe M. 1-Benzyl-1,2,3,4-tetrahydroisoquinoline as a parkinsonism-inducing agent: a novel endogenous amine in mouse brain and parkinsonian CSF. J Neurochem 1995; 65: 2633–8.
248. Kotake Y, Yoshida M, Ogawa M, et al. Chronic administration of 1-benzyl-1,2,3,4-tetrahydroisoquinoline, an endogenous amine in the brain, induces parkinsonism in a primate. Neurosci Lett 1996; 217: 69–71.
249. Antkiewicz-Michaluk L, Romańska I, Papla I, et al. Neurochemical changes induced by acute and chronic administration of 1,2,3,4-tetrahydroisoquinoline and salsolinol in dopaminergic structures of rat brain. Neuroscience 2000; 96: 59–64.
250. Abe K, Taguchi K, Wasai T, et al. Biochemical and pathological study of endogenous 1-benzyl-1,2,3,-tetrahydroisoquinoline-induced parkinsonism in the mouse. Brain Res 2001; 907: 134–8.
251. Shavali S, Carlson EC, Swinscoe JC, Ebadi M. 1-Benzyl-1,2,3,4-tetrahydroisoquinoline, a parkinsonism-inducing endogenous toxin, increases α-synuclein expression and causes nuclear damage in human dopaminergic cells. J Neurosci Res 2004; 76: 563–72.
252. Tasaki Y, Makino Y, Ohta, S, Hirobe M. 1-Methyl-1,2,3,4-tetrahydroisoquinoline, decreasing in 1-methyl-4-phenyl-1,2,3,6-tetrahydropyridine-treated mouse, prevents parkinsonism-like behavior abnormalities. J Neurochem 1991; 57: 1940–3.
253. Abe K, Saitoh T, Horiguchi Y, et al. Synthesis and neurotoxicity of tetrahydroisoquinoline derivatives for studying Parkinson's disease. Biol Pharm Bull 2005; 28: 1355–62.
254. Moser A, Kompf D. Presence of methyl-6,7-dihydroxy-1,2,3,6-tetrahydroisoquinolines, derivatives of the neurotoxin isoquinoline, in parkinsonian lumbar CSF. Life Sci 1992; 50: 1885–91.
255. Shavali S, Ebadi M. 1-Benzyl-1,2,3,4-tetrahydroisoquinoline (1Bn-TIQ), an endogenous neurotoxin, induces dopaminergic cell death through apoptosis. Neurotoxicology 2003; 24: 417–24.
256. Yamakawa T, Kotake Y, Fujitani M, et al. Regional distribution of parkinsonism-preventing endogenous tetrahydroisoquinoline derivatives and an endogenous parkinsonism-preventing substance-synthesizing enzyme in monkey brain. Neurosci Lett 1999; 276: 68–70.
257. Adams JD Jr, Odunze IN. Oxygen free radicals and Parkinson's disease. Free Radic Biol Med 1991; 10: 161–9.
258. Olanow CW. An introduction to the free radical hypothesis in Parkinson's disease. Ann Neurol 1992; Suppl 32: S2–S9.
259. Miller J, Selhub J, Joseph J. Oxidative damage caused by free radicals produced during catecholamine autoxidation: protective effects of O-methylation and melatonin. Free Radic Biol Med 1996; 21: 241–9.
260. Patsenka A, Antkiewicz-Michaluk L. Inhibition of rodent brain monoamine oxidase and tyrosine hydroxylase by endogenous compounds—1,2,3,4-tetrahydroisoquinoline alkaloids. Pol J Pharmacol 2004; 56: 727–34.
261. Yamakawa T, Ohta, S. Biosynthesis of a parkinsonism-preventing substance, 1-methyl-1,2,3,4-tetrahydroisoquinoline, is inhibited by parkinsonism-inducing compounds in rat brain mitochondrial fraction. Neurosci Lett 1999; 259: 157–160.
262. Abe K, Taguchi K, Wasai T, et al. Stereoselective effect of (R)- and (S)-1-methyl-1,2,3,4-tetrahydroisoquinolines on a mouse model of Parkinson's disease. Brain Res Bull 2001; 56: 55–60.
263. Kotake Y, Taguchi R, Okuda K, et al. Neuroprotective effect of 1-methyl-1,2,3,4-tetrahydroisoquinoline on cultured rat mesencephalic neurons in the presence or absence of various neurotoxins. Brain Res 2005; 1033: 143–50.
264. Nappi AJ, Vass E. Hydroxyl radical formation via iron-mediated Fenton chemistry is inhibited by methylated catechols. Biochim Biophys Acta 1998; 1425: 159–67.
265. Gitto R, Barreca ML, De Luca L, et al. Discovery of a novel and highly potent noncompetitive AMPA receptor antagonist. J Med Chem 2003; 46: 197–200.
266. Ferreri G, Chimirri A, Russo E, et al. Comparative anticonvulsant activity of N-acetyl-1-aryl-6-7-dimethoxy-1,2,3,4-tetrahydroisoquinoline derivatives in rodents. Pharmacol Biochem Behav 2004; 77: 85–94.

6 Endocrine Disruption and Senescence: Consequences for Reproductive Endocrine and Neuroendocrine Systems, Behavioral Responses, and Immune Function

Mary Ann Ottinger, Emma T. Lavoie, Karen Dean, and Brandon D. Sitzmann

Department of Animal and Avian Sciences, University of Maryland, College Park, Maryland, U.S.A.

Michael J. Quinn, Jr.

U.S. Army Center for Health Promotion and Preventive Medicine, Directorate of Toxicology, Health Effects Research Program, Aberdeen Proving Ground, Maryland, U.S.A.

INTRODUCTION

The array of chemicals in the environment has increased exponentially in the last century with the widespread use of a range of chemicals for agriculture, industry, medicine, and residential use. Many of these chemicals, especially pesticides, herbicides, or phytochemicals have structural attributes which enable the compounds to interact with physiological systems. These endocrine disrupting chemicals (EDCs) often interact with endocrine and neuroendocrine systems by mimicking hormones. As such, EDCs can affect hormone sensitive neural systems, organs, and tissues including those from the immune and stress axes as recent evidence suggests. Many of these compounds are also toxicologically active and as such have additional deleterious actions on physiological systems. Investigations of EDCs have been conducted in a variety of animal models, including invertebrates, amphibians, reptiles, fish, birds, and rodents. In addition, there has been increased attention on short- and long-term effects of early exposure to EDCs, especially in terms of lifetime impacts. Specifically, if early EDC exposure affects developmental processes, alterations are likely to subsequently influence physiological function in adults and aging individuals. These long-term effects during aging may be expressed as early onset neural disease, premature reproductive senescence, impaired immunocompetence, and possibly reduced lifespan. Unfortunately, few data address potential EDC effects on the process of aging. This chapter will consider the effects of embryonic EDC exposure to select estrogenic or androgen active compounds on neuroendocrine systems, behavioral responses, and immune function, with focus on data collected in avian species. Additionally, aging processes in birds, rodent models of neural disease, and non-human primates will be discussed in the context of potential effects of EDCs on age-related declines in reproductive, metabolic, or immune axes and on cognitive senescence.

Aging is revealed via a complex choreography of changing functional responses that emerge over a wide range of physiological system. Understanding age-related changes in these interacting systems is further complicated by individual differences in rates of aging, which may be due to genetic as well as to environmental factors. The potential impact of environmental chemicals and aging has been broached in a variety of ways, including investigations of healthy aging, interventions to age-related changes in physiological systems, and in the study of disease states that may be triggered by early or lifetime exposure.

This chapter considers both documented and potential impacts of EDCs in a comparative framework with focus on the consequences of both early and/or lifetime exposure on characteristics of senescence, particularly reproductive and neural impacts. EDCs are biologically active compounds that often mimic endogenous hormones, thereby altering hormonally modulated responses. In the overview of EDC effects, most of the examples will be taken from avian species. Birds provide an excellent model for establishing the effects of EDCs because like the old "canary in the coal mine" adage, they are sensitive indicators of chemicals of concern in the environment, and like humans, they can be exposed to a wide array of contaminants in the environment. Migratory birds have the potential to be exposed to different types of environmental contamination across their lifetimes as they travel and forage across continents. The use of avian models in the laboratory allows for controlled injection studies to be conducted using eggs, which are independent of continued maternal influence. Further, the ability to visually monitor embryonic development as well as compare precocial and altricial species provides the opportunity of comparing species with varied developmental patterns. Finally, a great deal of data has been collected in avian species, both in laboratory and field studies, for a number of EDCs.

The comparative biology of aging will be considered relative to the cascade of events that are conserved across taxa followed by discussion of potential linkage of early EDC exposure and later effects. It must be pointed out that given the complexities of the EDCs, which span a vast array of chemical structures, modes of action, and toxicological effects as well as the often ill-defined mechanisms of action of suspected EDCs, it impossible to completely capture all the nuances of this issue. It is critical for scientifically sound research to provide the basis for understanding their potential impacts and for responsible regulation of the use of these compounds to maximize the positive applications and to minimize deleterious consequences to future generations of humans, domestic species, and wildlife.

A BRIEF OVERVIEW OF EDCS AND EFFECTS OF EXPOSURE

EDCs have been shown to disrupt aspects of embryonic development including organ development, sexual differentiation, reproduction and immune function, reproductive behavior, and metabolic processes. Biomedical and epidemiological data reveal clear linkages between EDC exposure, particularly early events and later disease or even epigenetic effects (1–5). In addition to biomedical data, field studies have correlated EDC exposure to immunosuppression in a number of taxa demonstrating that the effects of these compounds have conserved mechanisms and impact across a range of species, including invertebrates (6). Laboratory studies have been successful in establishing clear effects of various classes of EDCs on a number of physiological systems. Because many early EDCs of concern such as methoxychlor, dioxin, and some polychlorinated biphenyls (PCBs) have estrogenic activity, attention has focused on reproductive impact of these compounds in comparison to estradiol. However, wildlife species are actually exposed to many different types of EDCs at any one time with potentially different mechanisms of action (e.g., androgenic, thyroid active); it has been difficult to develop comprehensive risk assessment models. Furthermore, some EDCs, most notably DDE (dichlorophenyldichloroethylene) and vinclozolin, are biologically active in androgenic circuits by acting as androgen receptor (AR) blockers while other compounds such as trenbolone have frank androgenic activity. A brief overview will be provided below as a background for the later discussion of short- and long-terms consequences of exposure to EDCs at various stages of the life cycle.

A dizzying array of literature has amassed over the last two decades in the wake of public concern over the effects of DDT (1,1,1-trichloro-2,2-bis(*p*-chlorophenyl)ethane) on wildlife. The potential for deleterious results were reinforced with observations of effects on organisms in areas in which dioxin, PCB, or other chemical spills occurred. Consequently, EDCs have been defined functionally due to their influence on endocrine systems. Furthermore, the exposure of wildlife to environmentally relevant levels is often at sublethal concentrations, making the determination of effects of EDCs a challenging, especially when attempting to separate other interacting factors in the animal's environment (7). Additional variables may include environmental conditions, food availability, disease, or confounds such as simultaneous exposure to several chemicals. It is therefore important to develop reliable and sensitive measures that are appropriate for the endocrine or organ system that are the targets of the EDCs. A number of workshops and symposia have addressed these issues, generating publications and reports that raised awareness to the potential effects of EDCs on wildlife and humans (8–10). Other publications have documented impacts of EDCs and toxicants in laboratory studies and in wild populations in avian species (8,11–15). In 1998, the Endocrine Disruptor Screening and Testing Advisory Committee (EDSTAC) submitted a report that recommended a tiered approach to testing suspected EDCs (http://www.epa.gov/endo/pubs/edspoverview/finalrpt.htm) for the detection and evaluation of suspected EDCs. These recommendations and subsequent validation of potential testing protocols have provided a great deal of additional insight into the actions of EDCs. A basic issue is the development of the appropriate testing paradigm, including reliable end points. One source of confusion has been that avian toxicology tests currently in use basically include only reproductive end points as a measure of the fitness of the adults. Because these reproductive end points, such as fertility and hatching success are included in the testing paradigm, it has been assumed that these measures are sufficient. However, these reproductive end points may not necessarily be appropriate, reliable, or sensitive for detecting EDC effects. Therefore, it is important to distinguish the toxicology test from a test designed to detect endocrine disruption. In order to provide an understanding of the actions as well as issues of characterizing EDC effects, a few selected EDCs will be discussed including some pertinent literature collected in laboratory and field studies.

Examples of EDCs and Consequences of Exposure

A number of compounds and mixtures make up a lengthy list of likely suspects as candidate EDCs. A few selected compounds, including methoxychlor, PCB-126 (polychlorinated biphenyl 126), soy phytoestrogens, DDE [ethylene, 1,1-dichloro-2,2-bis(*p*-chlorophenyl)], and trenbolone represent only a few compounds specifically developed for a given purpose that have entered the environment in high enough concentrations to result in unexpected consequences. For example, when the insecticide DDT was found to be highly toxic to wild populations, a new pest control compound was required for protecting crops, so DDT was replaced with a compound believed to be less toxic. However, the problem only became exacerbated because of the toxicity of DDT's metabolites and methoxychlor as well as the original problem of DDT. Additionally, PCBs were never meant to be released directly into the environment and trenbolone's active metabolites were not expected to come into contact with non-domestic species for any appreciable time.

Each of these compounds will be discussed in more detail as there is information from laboratory and field studies relative to their physiological actions and potential long-term consequences from exposure. The reader will note that there is a rich literature for mammals and often for avian and aquatic species. However, little is known about effects of these compounds on non-human primates; biomedical information is just beginning to emerge via the epidemiological literature.

Most the literature that we will discuss pertains to effects on reproductive function, with ultimate implications for the fitness of the individual. Because many of the effects of EDCs target components of the reproductive axis due to the steroid-like action of these compounds, it is important to consider effects on the developing reproductive system. Therefore, to complement the emphasis of this review on EDC effects in birds, we provide below a brief overview of their reproductive axis development along with consideration of aging processes. This is intended to place into context life-long EDC effects, beginning with exposure during development and continuing throughout the lifespan. In addition, there is a heavy emphasis on the linkage between impacts on the reproductive axis and other deleterious effects of EDCs including immune system development and function. Again, there are implications for the organisms throughout the life cycle, with long term consequences of early exposure. Moreover, because most of the functional components of both systems are analogous to those of mammals, there are parallel implications for short and long-term consequences of EDC exposure for mammals, including primates.

Methoxychlor

Methoxychlor (MXC) is an organochlorine pesticide developed as a DDT substitute for use as a pesticide on fruit, vegetables, trees, home gardens, forage crops, and livestock. The highest intensity of use is in the northwest and eastern seaboard states (11,16). Endocrine disruption has been well documented for MXC in mammals including reproductive dysfunction (17,18). Studies in Japanese quail have demonstrated that females deposit MXC into eggs, almost exclusively into the yolk. This pattern of maternal deposition of MXC was similar to the increased estradiol deposition observed following either injected or implanted estradiol (19,20). Exposure of Japanese quail embryos to 1.5 or 3.0 mg/egg MXC resulted in a significant increase in gonadotropin releasing hormone-I (GnRH-I; 19) in female hatchlings and a significant decrease of GnRH-I in female adults (20). Although the impact of the altered GnRH-I content did not appear to ultimately affect total number of eggs produced by a hen over many weeks, there was a delay in the onset egg production in many of the females. Additionally, there were individual differences in the response to MXC, suggesting that individual sensitivity to EDCs is an important factor to consider in field populations. Adult males exposed to 1.5 or 3.0 mg/egg MXC had reduced sexual behavior; and exposure to 5 mg/egg MXC completely inhibited adult male sexual behavior (21,22). These data provide evidence that even at low dietary levels, EDCs that are considered relatively innocuous have the capacity to have long-term consequences for reproductive function and ultimately fitness.

p,p′-DDE

DDT and its metabolites are lipid soluble compounds, stored in organ lipids (23). In mammals, DDT is broken down into two main metabolites, DDD (1,1-(*p*-chlorophenyl)-2,2-di-chloro-ethane) and DDE, before being further broken down to the excretory metabolite, DDA (bis (*p*-chlorophenyl) acetic acid). The ranking of biological half-lives and affinity for storage in lipid tissue is: DDE > DDT > DDD and high in vivo persistence of DDE may result in a substantial body burden in adipose tissue until death (23). *o,p′*-DDT has estrogenic effects; however, *p,p′*-DDE acts as a potent AR antagonist as well as stimulates testosterone hydroxylase, resulting in increased production of androstenedione (24,25).

In Japanese quail, the two main metabolites are DDE (31%) and DDA (35%), with DDE more chemically stable and biologically persistent (26,27). Japanese quail embryos exposed to DDE (20 and 40 μg/egg) had significantly increased bursal weight consistent with the anti-androgenic effects of this compound (Fig. 1). However, sections of bursal tissue from DDE treated birds had lesions and abnormalities not observed in control birds. Moreover, females showed delayed sexual maturation, and the higher dose treated males showed impaired mating behavior (28) (Fig. 1). Other measures of reproductive and immune function also were altered in the DDE treated quail, supporting the deleterious effects of relatively low doses of DDE on both reproductive and immune function. These effects in field birds would impact individual fitness and potentially affect avian populations if exposure occurs in a number of individuals.

PCB-126

PCBs are industrial products, initially designed to provide a heat and degradation resistant lubricant. However, these characteristics have made these compounds extraordinarily persistent in the environment following spills and intentional dumping. PCBs have been the subject of intense investigation for many years as they have been identified as contaminants of

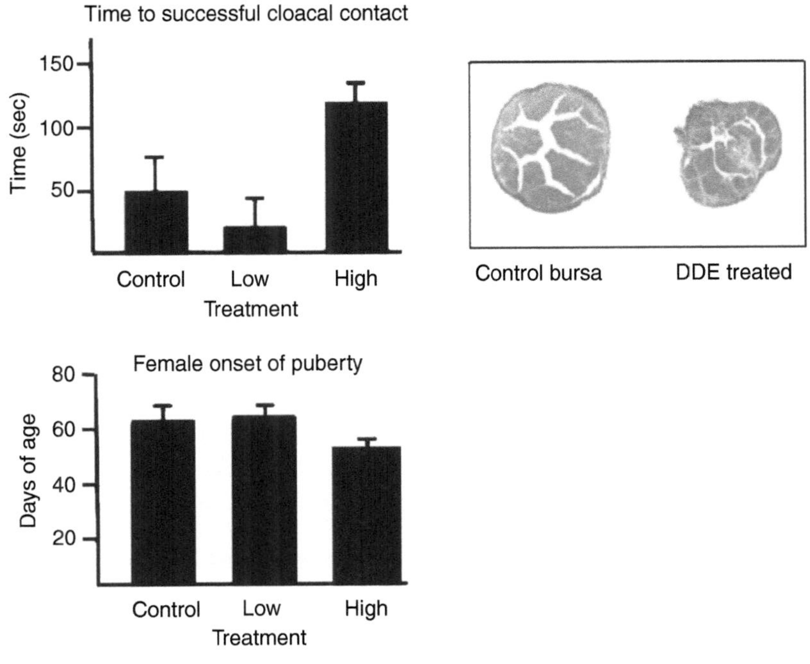

Figure 1 (*See color insert*) Representative bursal sections from control and DDE treated birds. Onset of puberty in females (first egg) and mating behavior in males were both impaired in high DDE (40 μg/kg) treatment groups. *Abbreviation*: DDE, dichlorophenyldichloroethylene. *Source*: From Ref. 28.

concern, especially in fish-eating birds (29,30). Characterizing PCBs is especially complex due to the number of possible congeners (209), the number in the environment and their persistence. The endocrine active effects of the PCB congener or mixture depend on the structure of the compound, resulting in effects on either ether reproductive or thyroid systems or both simultaneously (31). The way in which PCBs were manufactured has meant there is the potential for every batch to contain different ratios of congeners and thus change the impact on the ecosystem. Additionally, these impacts may change as these molecules degrade; incomplete dechlorination may result in the production of new congeners with different mechanisms or modes of action. PCBs have great potential for bioaccumulation and biomagnification, resulting in higher exposure in birds of prey.

Individual and mixtures of PCBs, well known for their estrogen and thyroid disrupting properties, also suppress androgens in both mammals and birds. Serum testosterone levels decreased significantly in adult chickens that received Aroclor 1254 (32), and significant decreases in testosterone concentrations were observed in vitro in porcine theca cells exposed to PCB 126. Aroclor 1248 suppressed testicular androgenesis in rats by an acute inhibition of 3β-hydroxysteroid dehydrogenase, 17α-hydroxylase/lyase, and 17β-hydroxysteroid dehydrogenase. This disruption of enzyme activities inhibited conversion of pregnenolone to progesterone, progesterone to testosterone, and δ(4)-androstenedione to testosterone (33). It appears that some immunotoxic effects associated with "multiple mechanisms of action" EDCs may be mediated by androgen disruption. PCBs are also biomagnified, resulting in higher exposure to birds of prey. For example, endogenous thyroxine and estradiol levels were reduced in kestrels associated with exposure to PCBs at relatively higher concentrations in the fatty tissue of their prey (34).

Some laboratory studies have shown specific immunotoxic effects of individual EDCs. Planar halogenated aromatic hydrocarbons, such as PCBs and 2,3,7,8-tetrachlorodibenzo-*p*-dioxin have been shown to have immunotoxic effects (35), including developmental effects on the bursa. PCB 126 caused decreases in bursa weight and numbers of developing B lymphocytes in the bursas of chicken embryos (13). PCBs and dioxins have also been linked to thymic atrophy and suppression of mitogen-induced proliferation, mixed lymphocyte responses, cell-mediated cytotoxicity, and humoral responses (reviewed in Ref. 35).

Trenbolone Acetate

Trenbolone acetate (TbA; 17β-acetoxyestra-4,9,11-triene-3-one) is a synthetic androgen used as a growth promoter for cattle. The bioactive metabolite of TbA is trenbolone-17β (trenbolone; 36). The affinity of trenbolone to recombinant human AR is similar to dihydrotestosterone (37). TbA as well as other synthetic and natural steroid hormones released from human and animal wastes have been measured in rivers and streams, sometimes at levels that exert endocrine disrupting effects on aquatic species (36,38).

In quail, embryonic exposure to trenbolone acetate had deleterious effects for hatchlings that persisted as birds matured. Effects occurred in both reproductive and immune measures, with delayed sexual maturation and impaired male mating behavior; hatchling bursa index was significantly reduced (39). One of the most interesting effects noted was observed in young chicks subjected to a "runway test" in which the chick would rejoin other chicks. In this test, the chick typically uses an alarm/separation call to locate other chicks, orients on the other chicks, and then runs to rejoin them. Trenbolone treated chicks exhibited a dose-related impairment in their calling behavior which was completely absent in birds exposed to 50 μg in ovo. Loss of calling behaviors could have serious consequences for birds in the field (Quinn and Ottinger, unpublished data).

Soy Phytoestrogens

Unlike the compounds mentioned above, soy phytoestrogens represent a different kind of EDC. They are naturally occurring and usually ingested by choice. Unsurprisingly plants produce chemicals that are recognized by steroid hormone receptors in vertebrates. In many instances these compounds are used as defense mechanisms to prevent ingestion, such as steroidal alkaloids found in plants such as potatoes, tomatoes, and deadly nighshade. In the short-term these types of compounds cause gastric upset and even some neurological effects; however, we know little about their long-term effects.

A case in point is the rich phytoestrogen source produced by the soybean plant. A plethora of dietary supplements are aimed at aging baby boomers that seek to avoid age-related symptoms, especially those associated with perimenopause. Many supplements for women are primarily soy phytoestrogen. Soy is already viewed as a non-harmful dietary component that provides a low fat, high protein source with desirable anti-aging effects assertions for which little scientific information is available (*Consumer Reports on Health*, May 2000). The prevalence of soy as a major dietary component reinforces the view of soy as healthy (40,41). There are contradictory data regarding potential benefits of exogenous sources of estrogen on both cognition and potential amelioration of stroke-induced neural damage (42–45). Further, as the public responds to the concerns about hormone replacement therapy (HRT), more women turn to soy-based supplements to obtain the promised benefits of HRT (46).

For women, dietary soy has been widely acclaimed as having sparing effects on the symptoms of menopause and aging due to estrogenic effects from the phytoestrogens. Two of the primary phytoestrogens, genistein and genistin, are contained in supplements currently on the market (Fig. 2) and can achieve relatively high levels of active phytoestrogens in the body. Based on the data, there are benefits from soy rich diets, which include potential benefits for circulatory system function and mild relief of perimenopausal symptoms (47–49). The primary biologically active isoflavone in soybeans is genistein, which is approximately 1000× less active than estradiol in vitro (50). Further, many of these supplements have relatively high levels of genistein or other phytoestrogens in their aglycone, glycosylated, and acylated forms (50,51–53). Therefore, supplements containing soy phytoestrogens provide these isoflavones in a highly absorbable form that establish significant levels.

Previous data collected in human and rodent studies on soy treatments have provided mixed and somewhat confusing results. More recently, Krikorian (54) reported improved response accuracy on a working memory task in postmenopausal women given 16 weeks of isoflavone supplementation, suggesting that isoflavones can improve cognitive function and memory. Experimental evidence in rodents showed that genistein, a potent soy isoflavone protected hippocampal neurons from damage (55,56). In addition to estrogenic actions of soy phytoestrogens, these isoflavones have antioxidant activities, potentially enabling them to

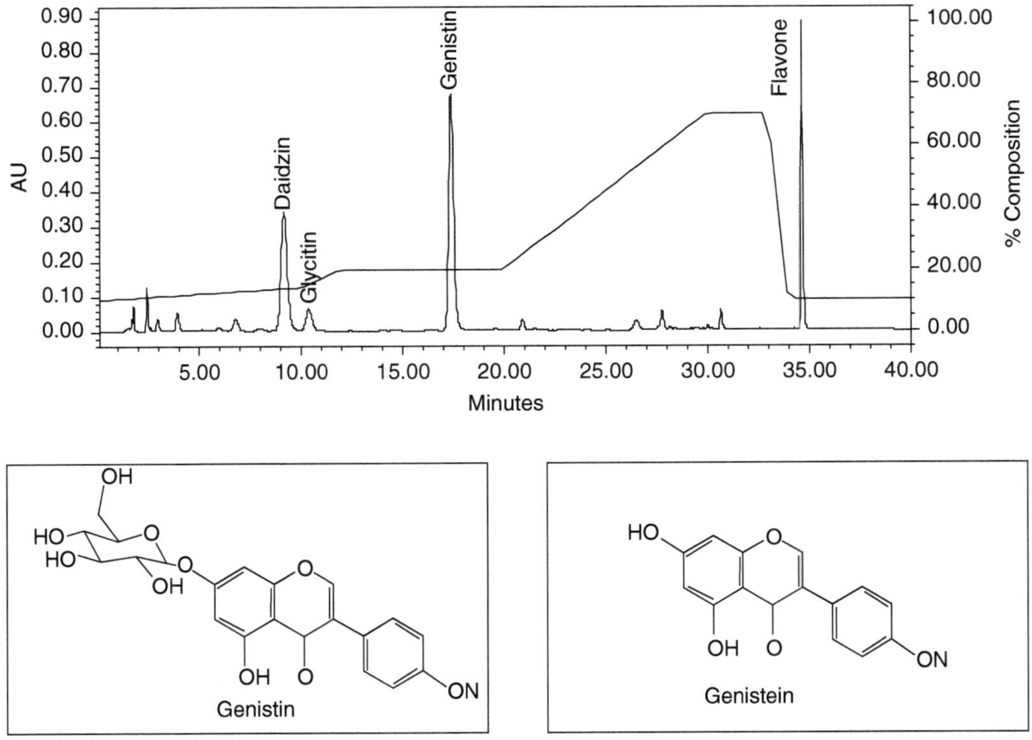

Figure 2 Profile of soy phytoestrogens in a commercially available isoflavone supplement (*top*). Structure of genistin and genistein, two soy phytoestrogens (*bottom*). *Source*: From Ref. 53.

counteract some of the age-related deleterious effects of oxidative damage (57). Isoflavones have been implicated in the prevention of colon and prostate cancers; whereas, effects on breast cancer are mixed (58–60). There are indications of additional health benefits of soy isoflavones in lowering the risk of cardiovascular disease and improving bone health (61). However the data are conflicting regarding benefits of isoflavones on insulin-like growth factor-I and its binding factors, and bone metabolism (61–63). The intriguing aspect of soy phytoestrogen action is that the primary target is the estradiol receptors beta (ERβ) and evidence points to ERβ as an important component in anxiety and aggressive behavior (64,65)

In birds, the primary route of exposure is via maternal deposition into the egg. Female quail given estradiol (E_2) implants deposited estradiol to offspring via the yolk (19). Once in the yolk, the EDCs constantly expose the developing embryo. Treated males also had impaired sexual behavior, similar to the effects of embryonic estradiol (21). Dietary sources of soy phytoestrogens are readily metabolized and genistein transfers into the yolk when ingested by the hen. The level of genistein in the egg yolk displays an exposure related response with an increase in levels at higher doses and with continued supplementation. With the removal of the dietary supplementation, the levels of genistein in the egg yolk show a rapid decline over a four- to five-day period (52,53). Genistein injected in the yolk at day 3 significantly affected some aspects of male copulatory behavior (neck grab, mount attempts, mount) and reduced arginine vasotocin expression (66). These data point to specific effects of genistein on sexual differentiation. Because genistein acts primarily via ERβ (67), this receptor becomes a primary focus for the mechanism of action. The distribution of the estradiol receptor (ER) subtypes in the adult quail brain showed that both receptor mRNAs were present in the medial preoptic region (POM), known to play a major role in activation of male sexual behavior, mRNA of both receptor types was expressed. The high density of ERβ transcripts in areas important for sexual behavior suggests a role of ERβ in male sexual behavior in quail. Some differences were observed in the localization of estradiol receptor alpha (ERα) and ERβ with only ERα mRNA transcripts in the latero-ventral POM. There was low expression of ERα in the nucleus of the stria terminalis (BSTm) and lateral septal region (SL) as compared to ERβ, which was markedly expressed in these areas. The BSTm and SL are sexually dimorphic in terms of vasotocin-immunoreactive (VT-ir) cells and steroid action within the BST, via afferents, could modulate the activity of neurons within the POM (68). Testosterone, following aromatization to estradiol, stimulates the expression of VT-ir material in neurons and fibers within the BSTm (69). High expression in the BSTm of ERβ mRNA compared with ERα mRNA suggests that this stimulation might be mediated via ERβ. Other studies in embryonic quail brain showed ERβ is expressed developmentally earlier than ERα (70). Further, ERβ mRNA was detected on embryonic day 9 (E9) and E17, whereas expression of ERα mRNA was only detected on E17. An intriguing possibility is that ERβ is the receptor responsible for mediating the effects of estrogen on sexual differentiation of the avian brain. The expression of ERβ in the preoptic area at E9, a time point where we found no expression of ERα, suggests that this receptor could play a key role in sexual differentiation of the brain.

INTERACTIONS OF THE REPRODUCTIVE AND IMMUNE SYSTEMS IN BIRDS: IS THERE AN EDC IMPACT?

It is clear that the role of steroid hormones in differentiation of sexually dimorphic endocrine, neuroendocrine, and behavioral responses differ in precocial and altricial birds. However, in both precocial and altricial birds, the pre- and peri-hatch steroid hormones impact the development of key components of the immune system. One of the primary components of the immune system is the bursa of Fabricius, a primary lymphoid organ unique to birds and for which B cells, which undergo maturation in the bursa, were originally named. The bursa of Fabricius is extremely sensitive to androgen exposure. Our studies and that of many other investigators have focused on development and function of the reproductive system in precocial birds. As discussed earlier, estrogen is critical for sexual differentiation of female behavior and both estradiol and androgens are involved in sexual differentiation of the hypothalamus-pituitary-gonadal axis (HPG axis). It is not surprising that androgens and other hormones such as glucocorticoids exert strong effects on development and function of the bursa. The neuroendocrine and immune systems interact possibly via gonadal steroid receptors on bursal epithelial cells and glucocorticoid receptors on bursal B cells (71,72). The bursa also produces a tripeptide growth factor, bursin, which acts to help balance the immune and adrenal systems. The effects of EDC exposure on bursin remain unknown.

Sex differences in immune response have been observed in many animals (73). In general, females tend to be more resistant to infection than males. Females also have a greater ability to reject skin grafts and are more efficient in antibody production in response to antigen challenges (74). These sex differences in immune function have been explained in several ways: (*i*) Activities related to sexual selection impose a resource drain on males that reduce the resources available for immune function, i.e., energy and nutrients that are spent on courtship and mating cannot be used to prevent or fight infection. (*ii*) Males may have higher levels of potential exposure to endo- and ectoparasites than females. This higher level of exposure may be associated with increased interactions between males in territory defense and matings with more than one female in polygamous species. (*iii*) Higher levels of androgens in males responsible for the development of secondary sexual characteristics and sexual displays might negatively impact immunocompetence (75,76). Conversely, just as higher levels of androgens in males appear to be immunosuppressive, estrogens enhance immune responses in females (74).

Many studies have shown that androgens have immunosuppressive action in birds. Serum lysozome activity, an index of macrophage activity, is reduced in birds treated with testosterone (77). Testosterone treatment also resulted in a reduction in the total number of leukocytes and lymphocytes in developing chicks (78). The first clue of an interaction between androgens and humoral immunity was the negative relationship between testicular maturation and bursal regression; the bursa begins to regress as the testes mature and steroidogenesis increases. Moreover, the bursa is exquisitely sensitive to testosterone (78,79) compared to the other primary lymphoid organs, such as the spleen and thymus, which appear to be unresponsive to exogenous androgens (77).

Administration of many different types of androgens, including androsterone, androstene-3, 17-dione, methylandrostene diol, 5α-dihydrotestosterone (5α-DHT), testosterone propionate (TP), and 19-nortestosterone, disrupts embryonic development of the bursa (78). DHT treatments resulted in reduced numbers of proliferating bursal cells in broiler chicks (74), and were found to stimulate testes growth (80). Bursas from chicken embryos treated with TP at E3 were considerably smaller than controls on E10 (81). Most of the mesenchymal cells did not differentiate into light and dark cells. A small number of dark cells that did develop were able to enter the epithelium, but lacked the characteristic cytoplasmic granules. These dark cells have been termed "bud inducers" because they are necessary for epithelial bud formation. The few granule-lacking bud inducing dark cells that formed under TP treatment and entered the epithelial tissue were not able to initiate bud formation. Although the epithelium did not appear to be altered by the TP treatment, it did not accept any hemopoietic stem cells for B cell maturation (81). Mase and Oishi (82) found that Japanese quail treated with testosterone developed wrinkled epithelium in the bursas. Also, the plicas, or buds, did not develop and the follicles were empty, rather than being filled with lymphocytes.

Some discrepancies have been found relative to the effects of testosterone on avian immunocompetence. Ros and coworkers (83) have shown that testosterone enhanced antibody titers in response to sheep red blood cells (SRBC) in black-headed gull chicks, but caused no effect on antibody titers in nine-month-old juveniles. One explanation for these effects is that testosterone may not always have as strong of an effect on the amount of antibody produced as it does on the quality of the antibodies produced. Testosterone has a negative effect on IgG production, and testosterone treated chicks have higher levels of IgM than controls (84). The inhibition of affinity maturation by testosterone may prevent normal isotype switch of IgM production to IgG production by lymphocytes. Although the overall amount of antibodies produced in response to foreign antigens may not significantly decrease under testosterone treatments, the antibodies produced may not be specific to the particular antigen.

Immunotoxic Effects of EDCs

A number of studies have shown an association between EDC exposure and immunosuppression (85). A key consideration is the mode of action of the EDC, specifically if the compound is estrogenic or androgen active. In birds, embryonic exposure to androgen active EDCs may impact humoral immunity by effects on bursal development. Moreover, exposed hatchlings are likely to have impaired neuroimmune responses as maturing and adult birds. Long-term effects of estrogenic EDCs on immunocompetence are less clear. However, embryonic exposure to exogenous estradiol resulted in a dose-related decrease in bursa index in hatchlings surprisingly similar to androgens (Quinn et al., unpublished data). This suggests that estrogenic EDCs may also have deleterious effects on components of the immune system. Because field birds are simultaneously exposed to multiple compounds with potentially differing actions, there is the possibility of impact on both reproductive and immune systems, depending on the target systems.

COMPARATIVE BIOLOGY OF REPRODUCTIVE AGING: CONSERVED PROCESSES AND POTENTIAL EDC EFFECTS

Avian Species: Aging Processes and Extremes in Lifespan

In order to understand the implications of EDC exposure on the process of reproductive aging, the functional components of the system and their development must be considered. As discussed earlier, most studies on the effects of sublethal EDC effects have concentrated on developmental consequences of exposure. In song birds, there has been extensive study of the differentiation of the sexually dimorphic song system (86). These studies have shown that there appears to be a complex steroid-dependent as well as steroid-independent differentiation process, ultimately organizing the anatomical components of the song system and the neural modulators of the song system (87). Studies on embryonic exposure to EDCs in songbirds have shown deleterious effects on the size and morphology of some the neural regions of the song system (88–90). Sexual differentiation of the reproductive axis appears to be more similar to that of other birds with interesting actions of both androgen and estradiol during development of the gonads, accessory sex structures, and brain regions regulating courtship and mating behavior (21,91).

There is a large literature on the comparative biology of aging across species that describes the characteristics of long and short-lived birds. Studies in the Japanese quail have informed us on lifespan traits and the timing and aging process experienced by a short-lived bird (16,20,92,93,141). Notable characteristics that distinguish the long-lived from many short-lived species include smaller clutch size with fewer offspring per year, slower rate of sexual maturation, apparent resistance to oxidative damage, and continued reproduction until extremely aged (94–96). Some large-scale studies are ongoing in field birds at this time to further characterize the immune and other physiological systems across a variety of species by Ricklefs and his colleagues. These studies will provide valuable data for understanding these amazing long-lived birds.

Most of the toxicological applications for regulatory monitoring have utilized species indigenous to the United States for testing the safety of the compounds, with the species of choice being the mallard duck and northern bobwhite quail (97). There are few studies that have considered aging as a factor in the impact of lifetime exposure to EDCs. Moreover, although the toxicological responses of these species have been tested with a variety of compounds, there is only now a growing background and baseline dataset for the reproductive endocrinology of these birds especially relative to EDC impact (7,21). Conversely, there are data available on the reproductive and metabolic endocrinology and on the neural regulation of reproductive behavior in Japanese quail, zebra finches, and doves (21,98,99). In addition to data on endogenous patterns in gonadal steroids during ontogeny and in adults, the effects of exogenous steroids have been examined in many of these species (21). Japanese quail have been under intense study as an excellent model for distinguishing the effects of EDCs (100). More recently with the recognition that multigenerational approaches are necessary to detect some of the more subtle effects of EDCs, the Japanese quail has received additional attention as a model test species. Moreover, since field exposures are often at low levels or in an unpredictable frequency, the impact may be more significant over several generations. As such, multigenerational testing also provides the necessary considerations over time and generation to detect effects of low dose exposure and to ascertain potential long-term impacts at the individual and population levels. Finally, in any toxicological testing paradigm, it is critical to consider the age of the breeders so that reproductive aging does not confound the test results.

Considering Rodent Models of Neural Disease for Determining the Effects of EDCs

A plethora of transgenic mice have been developed to study an array of neural diseases. These murine models provide a tremendous resource for asking very specific questions about the mechanism of effect of selected treatments on particular physiological systems. Some transgenic models have been developed to investigate Alzheimer's disease (AD). These transgenic mice provide the exciting potential for investigation of the early effects of EDC exposure on later development of neural disease and more detail is provided below about these models. Additionally, there is a rich and varied literature from studies conducted on non-transgenic rodents, which clearly informs us about many of the mechanisms of EDC effects in mammals (3,101,102). This literature provides critical data about the implications of EDC effects in basis for current and future investigation of EDC effects and examines the utility of emerging technologies for characterizing the "footprint of effects" for single or mixtures of these compounds (5,103–111).

Various mouse models for study of AD-like disease have been produced including a dtg and ttg mouse model for AD which appear to follow similar patterns in neuropathology, albeit on differing time lines. Moreover, the loss of locus coeruleus neurons in the dtg mouse is consistent with data in humans and complementary to data from Heneka and colleagues (112); these data point to the key role of loss of noradrenergic neurons in AD. The potential function of norepinephrine (NE) as an anti-inflammatory provides further support for a crucial component in the AD associated changes. Use of the various mouse models provides further power to our hypothesis. Heneka et al. (112) found clear increases in glial inflammation and other markers of AD-like disease in the APP23 transgenic mice subjected to NE depletion, thereby pointing to a supporting role of Aβ in the expression of the neuropathology due to inflammatory disease. Furthermore, DSP-4 lesioning exacerbated inflammatory processes that were attenuated by NE treatment (113). It should be noted that the authors found that NE depletion did not result in behavioral deficits in wild type mice. This suggests that a murine model of AD will have behavioral impairments analogous to that in humans with AD, with similar mechanisms operating in the mouse model. An impaired response was revealed in dtg AD mice (Pistell et al., unpublished data). Tyrosine hydroxylase containing (TH-ir) cells in the locus coeruleus significantly decreased (24%) in the dtg APP/PS1 mice by 15 to 22 months of age (114). Markers for AD neuropathology examined in the ttg model show striking similarities to human AD. The appearance of plaques and mutant tau revealed with immunohistochemical staining of Aβ and tau neurofibrillar protein that follow functional alterations in hippocampal physiology and thus reflect later stages of the disease process.

Primate Models: Deciphering Aging and Ascertaining if EDCs Have Potential Effects

The Perimenopausal Transition: A Time of Changing Hormones

As the perimenopausal transition progresses, there is declining function of the hypothalamic-pituitary-gonadal axis (HPG axis) ultimately resulting in cessation of ovarian function and menopause. The average age of menopause in women is approximately 51 years, resulting in a post-reproductive period that extends for nearly one-third of their lives (115). If menopause merely affected fertility, the study of ovarian aging would likely not be of such high priority. Unfortunately, a number of other physiological systems are also affected by the sudden withdrawal of hormonal support associated with menopause, including: bone density, cardiovascular health, cognition, and possibly some cancers (116,117). Moreover, these consequences of the changing hormonal milieu that accompanies the perimenopausal transition have not been well characterized in non-human primates.

The rhesus monkey *(Macaca mulatta)* has been a valuable biomedical model for reproductive studies. Females are pubertal by 2.5 to 3.5 years of age and exhibit menstrual cycles approximately 28 days in length, similar to humans. Furthermore, rhesus monkeys experience a reproductive decline much like that of human menopause at approximately 24 years (118). Urinary hormone profiles demonstrated that like women, menopause in rhesus monkeys is associated with amenorrhea, low urinary estrogen conjugates, and irregular patterns of urinary pregnanediol-3glucuronide (progesterone metabolites). Urinary FSH levels also increase in postmenopausal rhesus monkeys (119). Many other changes are associated with aging and are pertinent to the changes associated with the perimenopausal transition, including altered metabolism and disruption of circadian rhythms (120). In addition, motor activity and dexterity decline during aging in non-human primates, similar to humans (32). Post-menopausal women are also at risk for developing osteoporosis due to increased osteoclast activity (121). While calcium supplements, exercise, and weight lifting help, bone loss with estrogen loss appears inevitable. The benefits of HRT on bone health and potential amelioration of cardiovascular disease in women when administered during the perimenopausal transition or early postmenopause remains to be conclusively determined.

This intersection of age-related physiological changes with the falling ovarian steroids during perimenopause conspire to produce complex and rapidly changing physiological responses, which impact not only health and energy, but also cognitive function. As such, the interplay between changing ovarian steroids and the metabolic endocrine function becomes a critical element of the perimenopausal transition. Moreover, it is not clear if the lifetime body burden of endocrine disruptors has greater impact as endogenous steroid hormones decline. Therefore, it is critical to understand the impact of age-related change in endogenous steroids in terms of a background of body burden of environmental contaminants, which may mimic hormones.

Key Components of the Reproductive Decline

There has been extensive research on the hypothalamic changes that accompany the perimenopausal transition. Similarly, pituitary gonadotropins also show age-related changes with a transient increase in plasma luteinizing hormone (LH) and a dramatic increase in follicle stimulating hormone (FSH). Isolating age-related changes in hypothalamic response from ovarian aging is difficult; extensive studies in rodents have examined the response of the HPG axis at various stages in the life cycle of the female as well as the functional changes of the hypothalamus and pituitary gland with aging. Wise and colleagues (122) reported decreased amplitude and frequency in LH secretion in older rats and a decline in the number of activated GnRH neurons, despite normal cycles. These papers by Wise and colleagues and many others pointed to alterations in the systems that modulate GnRH as key elements of the reproductive decline. Overall, these studies indicated that hypothalamic response decreases, potentially in tandem with declining ovarian function (123,124). Data collected in female non-human primates and in humans suggest that there are similar mechanisms operating during the perimenopausal transition (118). An additional key component of aging in primates is inhibin produced by ovarian follicles and changes in circulating levels that accompany the perimenopausal transition. As levels of FSH increase in older cycling women throughout the menstrual cycle, estradiol-17β (E_2) and INHB levels decreased (124; Wu et al., unpublished data). These data confirm an inverse relationship between INHB and FSH as well as a general decline in inhibin B (INHB), inhibin A (INHA), and progesterone (P4) prior to detectable differences in circulating E_2 levels.

The Importance and Role of Steroid Hormones in Cognitive Function and Neuroprotection

In both males and females, there is a loss of gonadal steroids during the process of aging (124,125). In females, estradiol (E_2) has clear neuroprotective actions via action on microglia and astrocytes, as shown in stroke models (44). In addition, because neuropathology occurs more frequently in aging women, the loss of E_2 is viewed as a key element in the loss of cognitive function, especially if it is accompanied by neurobiological disease. Several epidemiological studies concluded that postmenopausal women on HRT have reduced risk for AD (126,127). Further, estradiol has demonstrated neuroprotective effects in rodent studies of the hippocampus, with modulation of microglial activation. Stereological analysis revealed a significantly higher number of microglia in the hippocampus of female compared to male C57BL/6 (B6) mice as well as an age-related increase in the females that was not observed in the males (128). These findings of significant age and gender effects on microglia in B6 mice led to the hypothesis that activation of glial cells in females may be triggered by the declining E_2 during aging in mice and, by analogy, at menopause in women (128–130).

THE AGING BRAIN ON EDCS; POSITIVE AND NEGATIVE ASPECTS

Any Evidence for Long-Term Effects Linked to EDCs?

The literature directly addressing this issue is limited. The most compelling evidence for long-term effects would be in reproductive health and issues, such as endometriosis. However, evidence for avian species has been limited due to the lack of long-term studies of birds in contaminated regions. Some notable studies of avian species in contaminated areas include those by the Custers (131). Moreover, avian species are even more complex due to migration and the stressors associated with that monumental biannual event for many birds. A relatively naïve representation of potential impacts is presented in Figure 3A and B. If the population is exposed to EDC(s) at varied concentrations and with varied frequencies, then the fitness of the population will be affected at graduated degrees (Fig. 3A and B). Conversely, if the return from migration is constant, then there may be an impact on reproductive capacity, which ultimately would have consequences for the population.

Are there Effective Interventions for the Age-Related Demise?

Effects of Calorie Restriction as an Intervention in Aging Processes

Calorie restriction (CR) has well established benefits on aging, including: extended life span, reduced incidence of age-related disease, sustained metabolic endocrine function, and delayed loss of reproductive function. A number of potential mechanisms exist and include enhanced physiological responses with low-level stress and/or diminished oxidative damage during aging. Further, there is evidence that the benefits of CR transcend phyla. Longitudinal studies in mammals have characterized a number of biological parameters that are favorably affected by 30–40% CR (120,132–135). In addition to extending lifespan, some of the most impressive effects of CR have been on delaying disease and age-related deterioration in endocrine function. Some specific effects of moderate CR have included reduced body fat, increased high density lipoproteins, lower fasting glucose and insulin levels, sparing of age-related apoptosis in liver and other tissues, decreased indicators of oxidative stress, delayed reproductive decline, and altered gene expression (120,136–139). Primates appear to benefit from moderate CR, in terms of lipid metabolism and delayed age-related diabetes (120,135). Therefore calorie restriction appears to have an overall beneficial effect on the physiology and health of the individual, which then translates into the potential for long lifespan and increased quality of life.

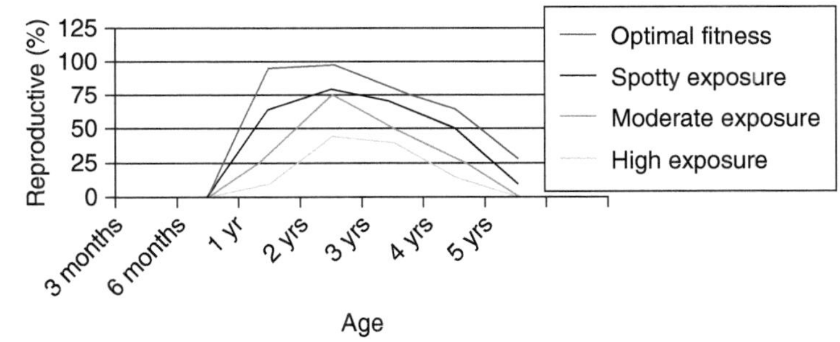

(A)

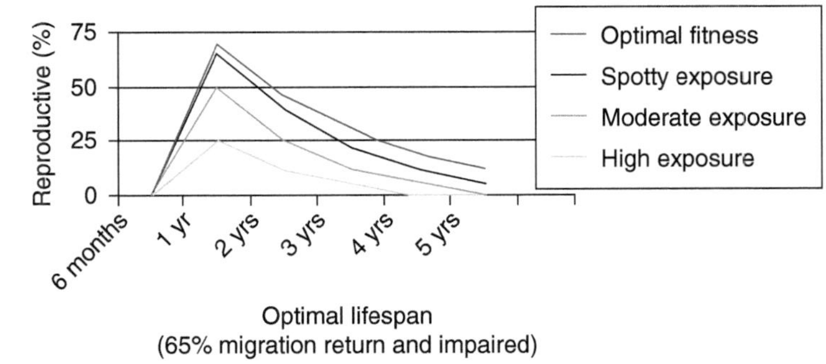

(B)

Figure 3 (*See color insert*) Diagrammatic reproduction addressing: (**A**) whether reproductive function is altered by EDC exposure at varied degrees due to variable timing of exposure or at different concentrations; or (**B**) considering a consistent rate of return from migration in combination with impaired reproduction due to EDC impacts.

Potential Interactions of CR and EDCs?

The question of early exposure to EDCs and later effects is one aspect of exposure to environmental chemicals that deserves attention. In addition, it is important to consider lifetime exposure leading to a body burden of resident EDCs, especially in lipids stored in fat cells or liver. If CR were initiated as adult or at middle age, it would seem likely that metabolism would involve significant mobilization of fat stores. Therefore, as many of us reach the middle-aged phases of life and embark on diets, it would be prudent to consider the effects of stored residues and clinical considerations may be warranted in the case of extreme weight loss. This issue becomes even more interesting when considering age-related sensitivity to pharmaceuticals. In fact aged individuals have greater sensitivity to medications and other chemicals due to slower metabolism. Therefore, sensitivity to the effects of environmental chemicals may be heightened in aging individuals and weight loss may exacerbate the response. Because we all carry a body burden of environmental chemicals, it is important to determine the potential impact of lifelong exposure and discern differences in response and sensitivity at varied stages of the life cycle.

Concluding Comments

It is clear that the complexities of development and aging are likely to be impacted by exposure to endocrine active compounds or EDCs throughout the lifespan (140). Points of particular vulnerability appear to be during development and potentially the aging process may prove to be altered by early as well as lifespan exposure to EDCs. At this time, only epidemiological evidence is available to address this question and there is some evidence of chemical exposure being a possible trigger for later neural and other disease progressions. The burden of proof is on us to determine if the aging animal becomes more vulnerable to either long- or short-terms EDC exposures. Moreover, it is imperative that we discern effects due to toxicological actions from that of the hormone mimic activity of EDCs. Overall, it is a fascinating and somewhat frightening prospect, given the ubiquitous nature and wide distribution of these compounds in our environment and in all aspects of our daily lives.

ACKNOWLEDGMENTS

This review relies on a rich literature that is steadily increasing. Our own research has been supported by the following: EPAR826134010 (Star Grant), NSF 9817024, and EPAR-2877801 (MAO). In addition, there have been a myriad of outstanding high school, undergraduate, graduate students, post doctoral fellows, and wonderful colleagues that have not only made our research possible, but also enabled us to do these demanding projects successful. We thank them and in particular recognize the contributions of Meredith Bohannan, Kasen Whitehouse, Marci Strauss, Kimya Davani, Ashley Barton, Moira McKernan, and Allegra Schafer.

REFERENCES

1. Anway MD, Skinner MK. Epigenetic programming of the germ line: effects of endocrine disruptors on the development of transgenerational disesase. Reprod Biomed Online 2008; 16: 23–5.
2. Buteau-Lozani H, Velasco G, Cristofari M, et al. Xenoestrogens modulate vascular endothelial growth factor secretion in breast cancer cells through an estrogen receptor-dependent mechanism. J Endocrinol 2008; 196: 399–412.
3. Phillips KP, Foster WG. Key developments in endocrine disruptor research and human health. J Toxicol Environ Health B Crit Rev 2008; 11: 322–44.
4. Prins G. Endocrine disruptors and prostate cancer risk. Endocr Relat Cancer 2008; Jun 4, PMID 18524946.
5. Soto AM, Vandenberg LN, Maffini MV, et al. Does breast cancer start in the womb. Basic Clin Pharmacol Toxicol 2008; 102: 125–33.
6. Yang G, Kille P, Ford AT. Infertility in a marine crustacean: have we been ignoring pollution impacts on male invertebrates. Aquat Toxicol 2008; 88: 81–7.
7. Ottinger MA, vom Saal FS. Impact of environmental endocrine disruptors on sexual differentiation in birds and mammals. In: Pfaff D, Arnold A, Etgen A, Rubin R, eds. Hormones and Behavior in Higher Vertebrates, Vol. 4. New York, NY: Academic Press, 2002: 325–83.
8. Fox GA. Epidemiological and pathobiological evidence of contaminant-induced alterations in sexual development in free-living wildlife. In: Colborn T, Clement C, eds. Chemically-Induced Alterations

in Sexual and Functional Development: The Wildlife/Human Connection. Princeton: Princeton Scientific Publishing, 1992: 147–58.
9. Kavlock RJ, Daston GP, DeRosa C, et al. Research needs for the risk assessment of health and environmental effects of endocrine disruptors: a report of the U.S. EPA-sponsored workshop. Environ Health Perspect 1996; 104(Suppl 4): 715–40.
10. EDSTAC, Endocrine Disruptor Screening and Testing Advisory Committee (EDSTAC) Final Report. Washington, DC: U.S. Environmental Protection Agency, 1998.
11. Fairbrother A, Landis WG, Dominquez S, et al. A novel nonmetric multivariate approach to the evaluation of biomarkers in terrestrial field studies. Ecotoxicology 1998; 7: 1–10.
12. Fry M. Reproductive effects in birds exposed to pesticides and industrial chemicals. Environ Health Perspect 1995; 103(Suppl 7): 165–71.
13. Fox LI, Grasman KA. Effects of PCB 126 on primary immune organ development in chicken embryos. J Toxicol Environ Health, Part A 1999; 58: 233–44.
14. Rattner BA, Pearson J L, Golden N H, et al. Contaminant exposure and effects in a terrestrial vertebrates database: trends and data gaps for Atlantic coast estuaries. Environ Monitoring Assessment 2000; 63: 131–42.
15. Golden NH, Rattner BA, McGowan PC, et al. Concentrations of metals in feathers and blood of nestling black-crowned night-herons (*Nycticorax nycticorax*) in Chesapeake and Delaware Bays. Bull Environ Contam Toxicol 2003; 70: 385–93.
16. Ottinger MA. Altered neuroendocrine mechanisms during reproductive aging. Poultry Science Reviews 1992; 4: 235–48.
17. Barnett KR, Tomic D, Gupta RK, et al. The aryl hydrocarbon receptor affects mouse ovarian follicle growth via mechanisms involving estradiol regulation and responsiveness. Biol Reprod 2007; 76: 1062–70.
18. Gupta RK, Aberdeen G, Babus JK, et al. Methoxychlor and its metabolites inhibit growth and induce atresia of baboon antral follicles. Toxicol Pathol 2007: 35: 649–56.
19. Adkins-Regan EK, Ottinger MA, Park J. Maternal transfer of estradiol to egg yolks alters sexual differentiation of avian offspring. J Exp Zool 1995; 271: 466–70.
20. Ottinger MA, Abdelnabi M, Li Q, et al. The Japanese quail: a model for studying reproductive aging of hypothalamic systems. Exp Gerontol 2004; 39: 1679–93.
21. Ottinger MA, Abdelnabi MA, Henry PFP, et al. Neuroendocrine and behavioral implications of endocrine disrupting chemicals in quail. Horm Behavior 2001; 40: 234–47.
22. Eroschenko VP, Amstislavsky SY, Schwabel H, et al. Altered behaviors in male mice, male quail, and salamander larvae following early exposures to the estrogenic pesticide methoxychlor. Neurotox Teratol 2002; 24: 29–36.
23. Clement International Corporation. Toxicological profile for DDT, DDE, and DDD, 1994, U.S. Dept. of Health and Human Services, Public Health Service, Agency for Toxic Substances and Disease Registry.
24. Kelce WR, Stone CR, Laws SC, et al. Persistent DDT metabolite p,p′-DDE is a potent androgen receptor antagonist. Nature 1995; 375: 581–85.
25. You L. p,p′-DDE: an endocrine-active compound with the potential of multiple mechanisms of action. CIIT Activities 2000; 20: 1–8.
26. Ahmed MM, Walker CH. The metabolism of DDT *in vivo* by the Japanese quail (*Coturnix coturnix japonica*). Pestic Biochem Physiol 1979; 10: 40–8.
27. Jefferies DJ, Walker CH. Uptake of p,p′-DDT and its postmortem breakdown in the avian liver. Nature (London) 1966; 212: 533.
28. Quinn MJ Jr, Summitt CL, Ottinger MA. The effects of androgen disruption by DDE on the development and function of the immune system in Japanese quail. Immunopharm Immunotox 2006; 28: 535–44.
29. Kubiac TJ, Harris HJ, Smith LM, et al. Microcontaminants and reproductive impairment of the Forster's tern on Green Bay, Lake Michigan. Arch Environ Comtam Toxicol 1983; 18: 706–27.
30. NRC. Hormonally Active Agents in the Environment. Washington, DC: National Academy Press, 1999.
31. Zoeller RT. Environmental neuroendocrine and thyroid disruption: relevance for reproductive medicine? Fertil Steril 2008; 89(Suppl 2): e99–100.
32. Zhang C, Fang C, Liu L, et al. Disrupting effects of polychlorinated biphenyls on gonadal development and reproductive functions in chickens. J Environ Sci Health, Part A 2002; 37: 509–19.
33. Andric SA, Kostic TS, Stojilkovic SS, et al. Inhibition of rat testicular androgenesis by a polychlorinated biphenyl mixture Aroclor 1248. Biol Reprod 2000; 62: 1882–8.
34. Quinn M, French JB, McNabb AF, et al. The effects of polychorinated biphenyls (Aroclor 1242) on thyroxine, estradiol, molt, and plumage characteristics in the American Kestral (*Falco sparverious*). Environ Toxicol Chem 2002; 21: 1417–22.

35. Grasman KA, Fox GA. Associations between altered immune function and organochlorine contamination in young Caspian terns (*Sterna caspia*) from Lake Huron, 1997–1999. Ecotoxicology 2001; 10: 101–14.
36. Schiffer B, Daxenberger A, Meyer K, et al. The fate of trenbolone acetate and melengestrol acetate after application as growth promoters in cattle: environmental studies. Environ Health Perspect 2001; 109: 1145–51.
37. Bauer ERS, Daxenberger A, Petri T, et al. Characterization of the affinity of different anabolics and synthetic hormones to the human androgen receptor, human sex hormone binding globulin and to the bovine progestin receptor. APMIS 2000; 108: 838–46.
38. Tilton F, Benson WH, Schlenk D. Evaluation of estrogenic activity from a municipal wastewater treatment plant with predominantly domestic input. Aquat Toxicol 2002; 61: 211–24.
39. Quinn MJ, Jr, McKernan M, Lavoie E, et al. Immunotoxicity of trenbolone acetate in Japanese quail. J Toxicol Environ Health, Part A 2007; 70: 88–93.
40. Aldercreutz H, Hamalainen E, Gorbach S, et al. Dietary phytoestrogens and the menopause in Japan. Lancet 1992; 339: 1233.
41. Penza M, Montani C, Romani A, et al. Genistein accumulates in body depots and is mobilized during fasting, reaching estrogenic levels in serum that counter the hormonal actions of estradiol and organochlorines. Toxicol Sci 2007; 97: 299–307.
42. Shear DA, Galani R, Hoffman SW, et al. Progesterone protects against necrotic damage and behavioral abnormalities caused by traumatic brain injury. Exp Neurol 2002; 178: 59–67.
43. Wise PM, Smith MJ, Dubal DB, et al. Neuroendocrine modulation and repercussions of female reproductive aging. Endocrinology 2002; 142: 969–73.
44. Wise PM. Estradiol exerts neuroprotective actions against ischemic brain injury. Endocrine 2003; 21: 11–15.
45. Rapp PR, Morrison JH, Roberts JA. Cyclic estrogen replacement improves cognitive function in aged ovariectomized rhesus monkeys. J Neurosci 2003; 23: 5708–14.
46. Washburn S, Burke GL, Morgan T, et al. Effect of soy protein supplementation on serum lipproteins, blood pressure, and menopausal symptoms in perimenopausal women. Menopause 1999; 6: 7–13.
47. Albertazzi P, Pansin F, Bonaccorsi G, et al. The effect of dietary soy supplementation on hot flushes. Obstet Gynecol 1998; 91: 6–11.
48. Duncan AM, Merz BE, Xu X, et al. Soy isoflavones exert modest hormonal effects in premenopausal women. J Clin Endocrinol Metab 1999; 84: 192–7.
49. Chiechi LM, Putignano G, Guerra V, et al. The effect of a soy rich diet on the vaginal epithelium in postmenopause: a randomized double blind trial. Maturitas 2003; 45: 241–6.
50. Setchell KDR. Phytoestrogens: the biochemistry, physiology, and implications for human health of soy isoflavones. Am J Clin Nutr 1998; 68: 1333S–46S.
51. Fukutake M, Takahashi M, Ishida K, et al. Quantification of genistein and genistin in soybeans and soybean products. Food Chem Toxicol 1996; 34; 457–61.
52. Lin F, Wu J, Abdelnabi MA, et al. Effects of dose and glycosylation on the transfer of genistein into the eggs of Japanese quail (*Coturnix japonica*). J Agric Food Chem 2004; 52: 2397–403.
53. Ottinger MA, Lin F, Wu JM, et al. Isoflavones-enriched eggs: Transfer of isoflavones into eggs and health benefits. In: Sim JS, ed. The Egg. Guelph, CA: University of Guelph Press, 2006: 155–70.
54. Krikorian R. Isoflavone effects on cognition in menopause, Abstract #23, 34th Annual Meeting of the American Aging Association, Oakland, CA, June 3–6, 2005, 20.
55. Zeng H, Chen Q, Zhao B. Genistein ameliorates β-amyloid peptide (25–35) induced hippocampal neuronal apoptosis. Free Radical Biol Med 2003; 36: 180–8.
56. Bang OY, Hong HS, Kim DH, et al. Neuroprotective effect of genistein against beta amyloid-induced neurotoxicity. Neurobiol of Dis 2004; 16: 21–8.
57. Gagne B, Gelinas S, Bureau G, et al. Effects of estradiol, phytoestrogens, and ginkgo biloba extracts against 1-methyl-4-phenyl-pyridine-induced oxidative stress. Endocrin 2003; 21: 89–95.
58. Hirano T, Oka K, Akiba M. Antiproliferative activity of synthetic and naturally occurring flavonoids on tumor cells of the human breast carcinoma cell line, ZR-75-1. Res Commun Chem Pathol Pharmacol 1989; 64: 69–78.
59. Helferich WG, Andrade JE, Hoagland MS. Phtoestrogens and breast cancer: a complex story. Inflammopharmacology 2008; September 26: e-print.
60. Lamartiniere CA, Moore J, Brown NM, et al. Genistein suppresses mammary cancer in rats. Carcinogenesis 1995; 16: 2833–40.
61. Adams KF, Newton KM, Chen C, et al. Soy isoflavones do not modulate circulating insulin-like growth factor concentrations in an older population in an intervention trial. J Nut 2003; 133: 1316–9.

62. Wangen KE, Duncan AM, Merz-Demlow BE, et al. Effects of soy isoflavones on markers of bone turnover in premenopausal and postmenopausal women. J. Clin Endocrin Metab 2000; 85: 3043–8.
63. Nagata C, Shimizu H, Takami R, et al. Soy product intake and serum isoflavonoid and estradiol concentrations in relation to bone mineral density in postmenopausal Japanese women. Osteoporos Int 2002; 13: 200–4.
64. Lund TD, Rovis T, Chung WC, et al. Novel actions of estrogen receptor-beta on anxiety-related behaviors. Endocrinol 2005; 146: 797–807.
65. Patisaul HB, Bateman HL. Neonatal exposure to endocrine active compounds or an ER beta agonist increases adult anxiety and aggression in gonadally intact male rats. Horm Behav 2008; 53: 580–8.
66. Panzica GC, Viglietti-Panzica C, Mura E, et al. Effects of xenoestrogens and xenoandrogens on the differentiation of behaviorally relevant neural circuits. Frontiers in Neuroendocrinology 2007; 28: 179–200.
67. Kostelac D, Rechkemmer G, Briviba K. Phytoestrogens modulate binding response of estrogen receptors alpha and beta to the estrogen response element. J Agric Food Chem 2003; 51: 7632–5.
68. Panzica GC, Aste N, Castagna C, et al. Steroid-induced plasticity in the sexually dimorphic vasotocinergic innervation of the avian brain: behavioral implications. Brain Res Rev 2001; 37: 178–200.
69. Viglietti-Panzica C, Balthazart J, Fratesi S, et al. Estradiol mediates effects of testosterone on vasotocin-immunoreactivity in the adult quail brain. Horm Behav 2001; 40: 445–61.
70. Axelsson J, Mattsson A, Brunstrom B, et al. Expression of estrogen receptor-alpha and -beta mRNA in the brain of Japanese quail embryos. Dev Neurobiol 2007; 67: 1742–50.
71. Sullivan DA, Wira CR. Sex hormone and glucocorticoid receptors in the bursa of Fabricius of immature chicks. J Immunol 1979; 122: 2617–23.
72. Marsh JA, Scanes CG. Neuroendocrine-immune interactions. Poul Sci 1994; 73: 1049–61.
73. Zuk M. Reproductive strategies and disease susceptibility, an evolutionary viewpoint. Parasitol Today 1990; 6: 231–33.
74. Novotny EA, Raveche ES, Sharrow S, et al. Analysis of thymocyte subpopulations following treatment with sex hormones. Clin Immunol Immunopathol 1983; 28: 205–17.
75. Moller AP, Sorci G, Erritoe J. Sexual dimorphism in immune defense. American Naturalist 1998; 152: 605–19.
76. Moller AP, de Lope F, Saino N. Reproduction and migration in relation to senescence in the barn swallow Hirundo rustica: a study of avian "centenarians." AGE 2006; 27: 301–18.
77. Al-Afaleq AI, Homeida AM. Effects of low doses of oestradiol, testosterone, and dihydrotestosterone on the immune response of broiler chicks. Immunopharmacology and Immunotoxicology 1998; 20: 315–27.
78. Glick B. Bursa of Fabricius. In: Farner DS, King JR, Parkes KC, eds. Avian Biology, Vol. 8. New York: Academic Press, Inc., 1983; 443–84.
79. Glick B. How it all began: the continuing story of the bursa of Fabricius. In: Toivanen A, Toivanen P, eds. Avian Immunology: Basis and Practice. Boca Raton: CRC Press, 1991; 31–42.
80. Leitner G, Landsman T, Blum O, et al. Effects of gonadal steroids and their antagonists on the humoral immune response of the immune-selected broiler chicks. Poult Sci 1996; 75: 1373–82.
81. Olah I, Glick B, Toro I. Bursal development in normal and testosterone-treated chick embryos. Poult Sci 1986; 65: 574–88.
82. Mase Y, Oishi T. Effects of castration and testosterone treatment on the development and involution of the bursa of Fabricius and the thymus in the Japanese quail. Gen Comp Endocrinol 1991; 84: 426–33.
83. Ros AFH, Groothuis TGG, Apanius V. The relation among gonadal steroids, immunocompetence, and behavior in young black-headed gulls (*Larus ridibundus*). Am Nat 1997; 150: 201–19.
84. Deyhim F, Moreng RE, Kienholz EW. The effect of testosterone propionate on growth of broiler chickens. Poult Sci 1992; 71: 1921–26.
85. Crisp TM, Clegg ED, Cooper RL, et al. Environmental endocrine disruption: an effects assessment and analysis. Environ Health Perspect 1998; 106(Suppl 1): 11–56.
86. Balthazart J, Adkins-Regan E. Sexual differentiation of brain and behavior in birds. In: Pfaff D, Arnold A, Etgen A, Rubin R, eds. Hormones and Behavior in Higher Vertebrates, Vol. 4. New York, NY: Academic Press, 2002; 223–302.
87. Wade J, Arnold AP. Sexual differentiation of the zebra finch song system. Ann NY Acad Sci 2004; 1016: 540–59.
88. Gee JM, Craig-Veit CB, Millam JR. Posthatch methoxychlor exposure adversely affects reproduction of adult zebra finches, Taeniopygia guttata. Bull Environ Contam Toxicol 2004; 73: 607–12.

89. Iwaniuk AN, Koperski DT, Cheng KM, et al. The effects of environmental exposure to DDT on the brain of a songbird: changes in structures associated with mating and song. Behav Brain Res 2006; 173: 1–10.
90. Markman S, Leitner S, Catchpole C, et al. Tollutants increase song complexity and the volume of the brain area HVC in a songbird. PloS ONE 2008; 3: e1674.
91. Balthazart J, Ottinger MA. 5 ß-reductase activity in the brain and cloacal gland of male and female embryos in the Japanese quail (*Coturnix coturnix japonica*). J Endocrinol 1984; 102: 77–81.
92. Ottinger MA, Reed E, Wu J, et al. Establishing appropriate measures for monitoring aging in birds: comparing short and long lived species. Exp Gerontol 2003; 38 : 747–50.
93. Ottinger MA. Neuroendocrine aging in birds: comparing lifespan differences and conserved mechanisms. Special issue of Aging Research Reviews entitled: Genetics of Aging in Vertebrates 2007; 6: 46–53.
94. Pamplona R, Portero-Otin M, Riba D, et al. Heart fatty acid unsaturation and lipid peroxidation, and aging rate, are lower in canary and the parakeet than in the mouse. Aging 1999; 11: 44–9.
95. Ottinger MA, Nisbet ICT, Finch CE. Aging and reproduction: comparative endocrinology of the common tern and Japanese quail. American Zoologist 1995; 35: 299–306.
96. Ogburn CE, Martin GM, Ottinger MA, et al. Exceptional cellular resistance to oxidative damage in long-lived birds requires active gene expression. J Gerontol Biol Sci 2001; 11: B468–74.
97. Bennett RS, Bentley R, Shiroyama T, et al. Effects of the duration and timing of dietary methyl parathion exposure on bobwhite reproduction. Environ Toxicol Chem 1990; 7: 1473–80.
98. Adkins-Regan E, Abdelnabi M, Mobarak M, et al. Sex steroid levels in developing and adult male and female zebra finches (*Poephila guttata*). Gen Comp Endocrinol 1990; 78: 93–109.
99. Abdelnabi MA, Ottinger MA. Hypothalamic indolamines during embryonic development and effects of steroid exposure. Gen Comp Endocrinol 2003; 130: 13–19.
100. Berg C, Halldin K, Fridolfson AK, et al. The avian egg as a test system for endocrine disruptors: effects of diethylstilbestrol and ethynylestradiol on sex organ development. Sci Total Environ 1999; 233: 57–66.
101. Gore AC. Neuroendocrine systems as targets for environmental endocrine disrupting chemicals. Fertil Steril 2008; 89(Suppl 2), e103–8.
102. Rider CV, Furr J, Wilson VS, et al. A mixture of seven antiandrogens induces reproductive malformations in rats. Int J Androl 2008; 31: 249–62.
103. Kortenkamp A. Ten years of mixing cocktails: a review of combination effects of endocrine disrupting chemicals. Environ Health Perspect 2007; 115(Suppl 1): 98–105.
104. Igushi T, Watanabe H, Katsu Y. Toxicogenomics and ecotoxicogenomics for studying endocrine disruption and basic biology. Gen Comp Endocrinol 2007; 153: 25–9.
105. Sheehan D. The potential of proteomics for providing new insights into environmental impacts on human health. Rev Enviorn Health 2007; 22: 175–94.
106. Weiss B. Can endocrine disruptors influence neuroplasticity in the aging brain? Neurotoxicology 2007; 28: 938–50.
107. Euling SY, Selevan SG, Pescovitz OH, et al. Role of environmental factors in the timing of puberty. Pediatrics 2008; 121(Suppl 3), S167–71.
108. Joensen UN, Jorgensen N, Rajpert-De Meyts E, et al. Testicular dysgenesis syndrome and Leydig cell function. Basic Clin Pharmacol Toxicol 2008; 102: 155–61.
109. Richter CA, Birnbaum LS, Farabollini F, et al. In vivo effects of bisphenol A in laboratory rodent studies. Reprod Toxicol 2007; 24: 199–224.
110. Rier SE. Environmental immune disruption: a comorbidity factor for reproduction? Fertil Steril 2008; 89(Suppl 2): e103–8.
111. Schoeters G, Den Hond E, Dhooge W, et al. Endocrine disruptors and abnormalities of pubertal development. Basic Clin Pharmacol Toxicol 2008; 102: 168–75.
112. Heneka MT, O'Banion MK. Inflammatory processes in Alzheimer's disease. J Neuroimmunol 2007; 184: 69–91.
113. Heneka MT, Gavrilyuk V, Landreth GE, et al. Noradrenergic depletion increases inflammatory responses in brain: effects on IkappaB and HSP 70 expression. J Neurochem 2002; 85: 387–98.
114. O'Neil JN, Mouton PR, Tizabi Y, et al. Catecholaminergic neuronal loss in locus coeruleus of aged female dtg APP/PS1 mice. J Chemical Neuroanatomy 2007; 34: 102–7.
115. Treloar AE. Menstrual cyclicity and the pre-menopause. Maturitas 1981; 3: 249–64.
116. Prior JC. Perimenopause: the complex endocrinology of the menopausal transition. Endocrine Rev 1998; 19: 397–428.
117. Sherwin BB. Estrogen and cognitive functioning in women. Endocr Rev 2003; 24: 133–51.
118. Bellino FL, Wise PM. Nonhuman primate models of menopause workshop. Reprod 2003; 68: 10–8.

119. Shideler SE, Gee NA, Chen J, et al. Estrogen and progesterone metabolites and follicle-stimulating hormone in the aged macaque female. Biol Reprod 2001; 65: 1718–25.
120. Roth GS, Mattison JA, Ottinger MA, et al. Rhesus monkeys: relevance to human health interventions. Science 2004; 305: 1423–26.
121. Riggs BL, Khosla S, Melton III LJ. Sex steroids and the construction and conservation of the adult skeleton. Endocrine Rev 2002; 23: 279–302.
122. Wise PM. Neuroendocrine influences on aging of the female reproductive system. Front Neuroendocrinol 1991; 12: 323–56.
123. Micevych P, Sinchak K, Mills RH, et al. The luteinizing hormone surge is preceded by an estrogen-induced increase of hypothalamic progesterone in ovariectomized and adrenalectomized rats. Neuroendocrinology 2003; 78: 29–35.
124. Wu JM, Zelinski-Wooten M, Ingram DK, et al. Ovarian aging and menopause: current theories and models. Exp Biol Med 1993; 230: 818–28.
125. Ottinger MA. Male reproduction: testosterone, gonadotropins, and aging. In: Mobbs CV, Hof PR, eds. Functional Endocrinology of Aging, Vol. 29. Switzerland, Basal: Karger Press, 1998: 105–26.
126. Athana S, Baker LD, Craft S, et al. High-dose transdermal estradiol improves cognition for women with AD: results of a randomized study. Neurology 2001; 57: 605–12.
127. Henderson VW, Paganini-Hill A, Miller BL, et al. Estrogen for Alzheimer's disease in women: randomized, double-blind, placebo-controlled trial. Neurology 2000; 54: 295–301.
128. Mouton PR, Long JM, Lei D-L, et al. Age and gender effects on microglia and astrocyte number in brains of mice. Brain Res 2002; 956: 30–5.
129. Garcia-Estrada J, Del Rio JA, Luquin S, et al. Gonadal hormones down-regulate reactive gliosis and astrocyte proliferation after a penetrating brain injury. Brain Res 1993; 628: 271–8.
130. Green PS, Simpkins JW. Neuroprotective effects of estrogens: potential mechanisms of action. Int J Devl Neurosci 2000; 18: 347–58.
131. Custer TW, Custer CM, Hines RK. Dixoins and congener-specific polychlorinated biphenyls in three avian species from the Wisconsin River, Wisconsin. EnvironPollut 2002; 119: 1–11.
132. Weindruch R, Walford RL. Dietary restriction in mice beginning at one year of age: effect on life span and spontaneous cancer incidence. Science 1982; 215: 1415–18.
133. Nelson JF, Gosden RG, Felicio LS. Effect of dietary restriction on estrous cyclicity and follicular reserves in aging C57BL/6J mice. Biol Reprod 1985; 32: 515–22.
134. Lane MA, Mattison J, Ingram DK, et al. Caloric restriction and aging in primates: relevance to humans and possible CR mimetics. Microscopy Res Tech 2002; 59: 335–38.
135. Lane MA, Ingram DK, Roth GS. Calorie restriction in nonhuman primates: effects on diabetes and cardiovascular risk. Toxicol Sci 1999; 52: 41–8.
136. McShane TM, Wise PM. Life-long moderate caloric restriction prolongs reproductive life span in rats without interrupting estrous cyclicity: effects on the gonadotropin-releasing hormone/luteinizing hormone axis. Biol Reprod 1996; 54: 70–5.
137. Lee CK, Klopp RG, Weindruch R, et al. Gene expression profile of aging and its retardation by caloric restriction. Science 1999; 285: 1390–92.
138. Ando KY, Higami T, Tsuchiya T, et al. Impact of aging and life-long calorie restriction on expression of apoptosis-related genes in male F344 rat liver. Micros Res Tech 2002; 59: 293–300.
139. Kim HJ, Jung KJ, Yu BP, et al. Modulation of redox-sensitive transcription factors by calorie restriction during aging. Mech Ageing Develop 2002; 123: 1589–95.
140. Ankley GE, Mihaich E, Stahl R, et al. Overview of a workshop on screening methods for detecting potential (anti-) estrogenic/androgenic chemicals in wildlife. EnvironToxicol Chem 1998; 17: 68–87.
141. Lavoie ET, Sorrell EM, Perez DR, et al. Immunosenescence and age-related susceptibility to influenza virus in Japanese quail. Dev Comp Immunol 2007; 31: 407–14.

7 White Fat Tissue, Obesity, and Possible Role in Neurodegeneration

Palaniyandi Ravanan* **and Rana Awada**

Laboratoire de Biochimie et Génétique Moléculaire, Groupe d'Etude sur l'Inflammation Chronique et l'Obésité (GEICO), Faculté des Sciences, Université de La Réunion, France

G. Jean Harry

Neurotoxicology Group, Laboratory of Molecular Toxicology, National Institute of Environmental Health Sciences, National Institutes of Health, Department of Health and Human Services, Research Triangle Park, North Carolina, U.S.A.

Christian Lefebvre d'Hellencourt

Laboratoire de Biochimie et Génétique Moléculaire, Groupe d'Etude sur l'Inflammation Chronique et l'Obésité (GEICO), Faculté des Sciences, Université de La Réunion, France

INTRODUCTION

Obesity is considered as one of the major health issue in the world. The prevalence of obesity worldwide has progressively increased over the past decades. In the past, it has only been considered as a problem in high-income countries. Recently, it is dramatically on the rise even in low- and middle-income countries. According to the World Health Organization (WHO) more than one billion adults are overweight and 300 million of them are obese. The WHO assumes that by the year of 2015, approximately 2.3 billion adults will be overweight and more than 700 million will be obese. Since 1980, obesity rates have risen more than three-fold in some areas of North America, the U.K., Eastern Europe, the Middle East, the Pacific Islands, Australia, and China. In 2002, obesity prevalence in the U.S.A. was 34.9%, which has grown to 39% in 2005 and the WHO estimates that by 2015, it will be 53%. Obesity prevalence is rapidly increasing among children as well as adolescents worldwide. At least 20 million children under the age of five were overweight worldwide in 2005 (http://www.who.int). Obesity is a direct result of an imbalance between energy intake and energy expenditure. The tendency toward obesity is most of the time multifactorial. The main causes are genetic and environmental factors such as decreased physical activity and consumption of foods that are high in fats and sugars but low in vitamins and minerals.

Different parameters can be used to estimate the fat content in human; the most cited measurement is the body mass index (BMI) corresponding to the ratio:

$$\text{BMI} = \frac{\text{weight (kg)}}{\text{height}^2\ (\text{m}^2)}$$

Applying this formula, a BMI under 18.5 is considered as underweight, between 18.5 and 24.9 corresponds to normal weight, between 25 and 29.9 is overweight, and above 30 is obese. BMI is an easy way to define overweight and obesity but has limitations, for example, BMI may be elevated for athletes. Other measurements are frequently used such as the waist circumference and waist to hip ratio (1).

ADIPOSE TISSUE

Adipose tissue is a specialized connective tissue that functions as a major storage of fat in the form of triglycerides. In mammals, there are two forms of adipose tissue. One is brown

---

*Present address: Burnham Institute for Medical Research, La Jolla, California, U.S.A.

adipose tissue which is present mainly in hibernating mammals or in human neonates and is involved in regulating the body temperature through non-shivering thermo genesis. Another form of adipose tissue is white adipose tissue (WAT), which is commonly spread, all over the body. It is a source of energy and is involved in heat insulation and mechanical cushion. WAT can be located in subcutaneous or in visceral area. Subcutaneous adipose tissue is largely defined as fat tissue between the skin and muscle, whereas visceral adipose is found within the main cavities of the body, primarily in the abdominal cavity. Abdominal visceral adipocytes are more metabolically active than abdominal subcutaneous adipocytes, as they have high lypolytic activity and release large amounts of free fatty acids. In "normal weight" people, WAT represents around 15–20% of body weight and in obese individuals it is up to 50%. WAT contains approximately 50–70% of mature adipocytes, 20–40% of stromal pre-adipocytes, 1–10% of endothelial cells, and 1–30% macrophages (Fig. 1) (2). All the above-mentioned cells are involved in adipose tissue homeostasis. Pre-adipocytes and endothelial cells make up the vasculature of tissues and enable processes such as adipose tissue growth and development. Secretion of proangiogenic factors by mature adipocytes such as vascular endothelial growth factor, contributes to ongoing angiogenesis within WAT at all times (3). Macrophages and monocytes are present in WAT and thought to aid in the clearance of necrotic adipocytes. Cells from the stromal vascular fraction of adipose tissue have shown an interesting plasticity, with the possibility to differentiate into several different lineages (Fig. 2). These cells produce a number of different inflammatory-related factors (Fig. 3) and bring hope for new therapeutics avenues (4–7).

As will be discussed later, excess of WAT increase the risk to develop pathologies, but WAT has an important role in the normal physiology. The role of adipocytes for health has been demonstrated with a model of genetically modified mouse (A-ZIP/F-1 mice) with virtually no WAT (8). This mouse model mimics human patients with severe lipoatrophic diabetes and shows insulin resistance, hyperglycemia, and fatty liver. Transplantation of a WAT from non-genetically modified mice to A-ZIP/F-1 mice reversed the phenotype to a normal situation (8,9).

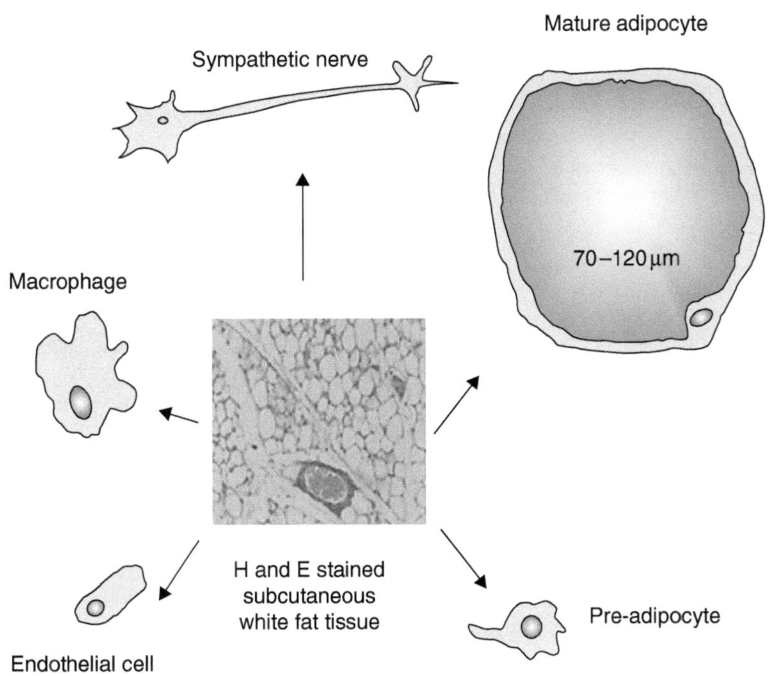

Figure 1 Cells present in the fat tissue. *Source*: Adapted from Ref. 3.

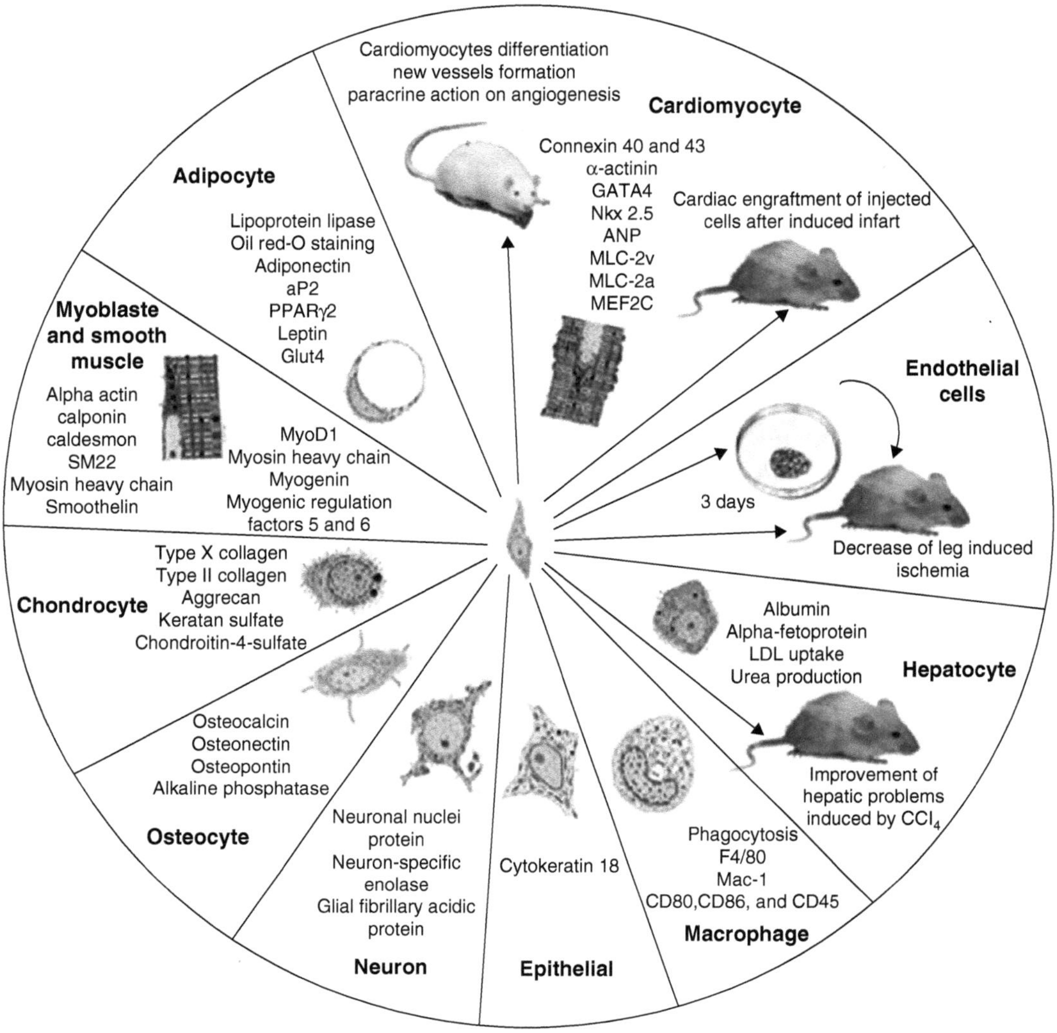

Figure 2 Differentiation potential in vitro and in vivo of adipose tissue derived SVF cells. All of the markers enabling the characterization of each cell type are presented. *Source*: From Ref. 4. *Abbreviation*: SVF, stromal vascular fraction.

Adipokines

WAT is the most familiar type of tissue in which triglycerides are stored and from there, lipids are mobilized for systemic utilization, when other tissues require energy. It was assumed in the past that, adipose tissue was just for storage of fat and it had no other function. However, in 1994, this concept was changed after the discovery of leptin, which is secreted from adipocytes to blood. Leptin acts on hypothalamus (10) and also on many other organs (11). A mutation in the gene coding for leptin (ob) induces obesity (10,12). Similar consequences are observed when the gene coding for the leptin receptor is not functioning (13). Leptin concentration is correlated to the fat tissue mass. This hormone can cross the blood–brain barrier (BBB) and interact with neurons involved in food intake (14). Leptin is produced by WAT, but also by placenta, bones, mammary glands, ovarian follicles, or stomach, and the leptin receptors are found in most of the tissues (15). Leptin activates anorexigenic neurons through a neuronal network in the hypothalamus nucleus and inhibits orexigenic neurons (16). In addition to its effects on food intake, leptin has immunomodulatory properties and is considered as a pro-inflammatory factor (17), for instance, it increases macrophages production of inflammatory cytokines (18). Besides leptin, adipocytes can produce several soluble factors, called adipokines (or adipocytokines), with

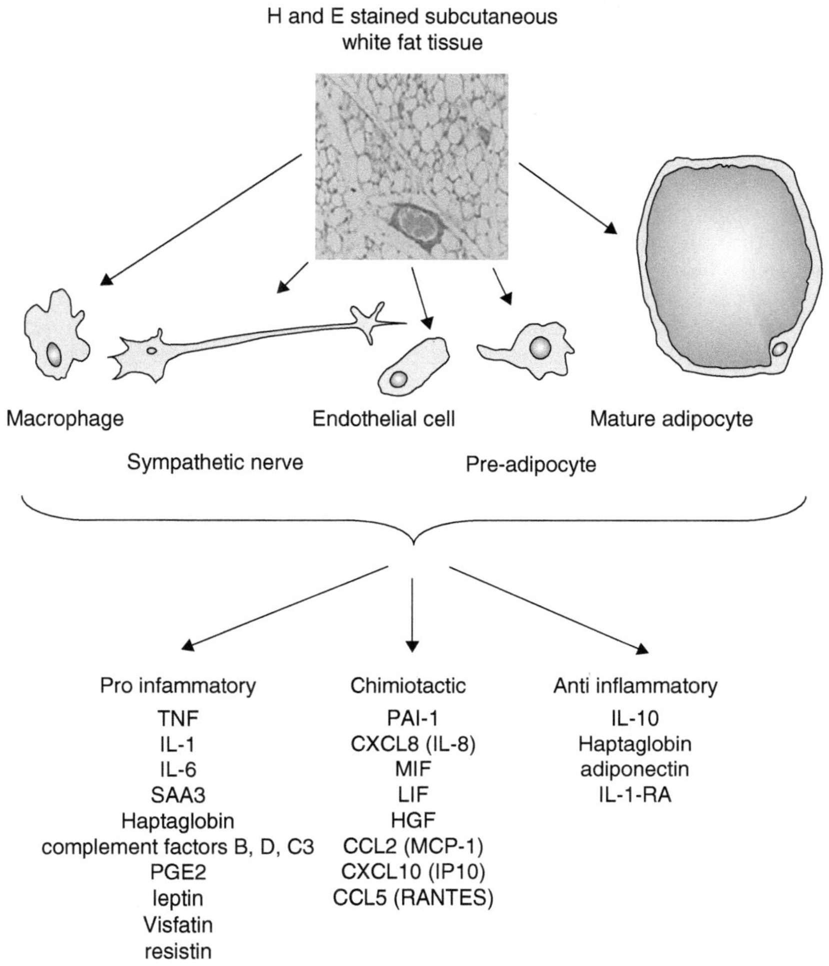

Figure 3 (*See color insert*) Inflammatory factors produced by fat tissue cells. These factors could be classified in pro-inflammatory or anti inflammatory based on literature (160–165). Chemokines produced by white adipose tissue may play a role in the monocyte/macrohage recruitment. *Source*: Adapted from Refs. 2, 7, and 20.

autocrine, paracrine, or endocrine actions. These include tumor necrosis factor α (TNF-α), plasminogen activator inhibitor-1 (PAI-1), interleukin (IL) -1β, IL-6, IL-8, IL-10, and IL-15, leukemia inhibitory factor, hepatocyte growth factor, serum amyloid A3 (SAA3), macrophage migration inhibitory factor, haptoglobin, complement factors B, D, C3, prostaglandin E2, adiponectin, and resistin (19–21). Some of these factors are involved in the regulation of the inflammatory response (Fig. 3), others such as omentin, apelin, or vaspin are produced by the fat tissue and could affect the inflammation (22–29).

In "normal" physiological conditions, about one-third of IL-6 circulating level is produced by WAT (30). As for leptin, IL-6 sera concentration is correlated to WAT mass. IL-6 induces a decrease of food intake and an increase of energy expenditure, in the CNS. IL-6 knock out mice develops obesity (31). TNF is another example of inflammatory cytokine produced by WAT. A correlation between sera TNF concentration, obesity level, and hyper insulinemia level has been described (32).

INFLAMMATION AND OBESITY

Acute or chronic inflammation can be defined as the protective response of tissues of the body to irritation or injury; characterized by redness, heat, swelling, or pain. More mechanistic definitions of inflammation have now been established, including invasion of circulating immune cells (lymphocytes and macrophages) and induction or activation of inflammatory mediators such as kinins, cycloxygenase products, and cytokines. Chronic inflammation is an inflammatory response of prolonged duration from weeks, months, or even indefinitely whereas acute inflammation is short, lasting only a few days. Acute inflammation may be due to physical damage, chemical substances such as environmental agents, microorganisms, or other agents. Inflammation leads to a generalized sequence of events as a host defence response to limit the proliferation of pathogens. That includes production of acute phase proteins by the liver, activation of the sympathetic nervous system, changes in cardiovascular function, activation of hypothalamic pituitary leading to the production of anti inflammatory steroids, behavioural changes which lead to energy conservation such as increased sleep, lethargy, reduced appetite, and fever which can limit the bacterial proliferation. Results of above-mentioned responses have clear benefits in infectious status. However, when these responses are not properly controlled or prolonged, it leads to excessive or inappropriate inflammation. This is the cause of many diseases including rheumatoid arthritis, psoriasis, inflammatory bowel disease, and the fundamental contributor to some types of cancer, diabetes, coronary and vascular disease, stroke, and cardiovascular diseases.

Several studies have shown that excess of fat leads to a chronic inflammation (9,33–36). Interestingly, in insect the main strategy of defence against infection rely on the fat body which drives an innate immune response. Bacteria and fungi are recognized by receptors, i.e., toll like receptor (TLR) present on the surface of fat body cells (37–40). The NF-kB signalling pathway is then activated and induces defence mechanisms. The fat body also has liver's functions and stores lipids (20,41). At some point during evolution, vertebrates split these metabolic duties between the liver and adipose tissue. However, it seems that some functions of the innate immune response have been kept in adipose tissue (42). In 1993, it was discovered that TNF-α expression was up-regulated in WAT of obese mice. This was the first study to show the link between obesity and inflammation (43). Recently we have shown the presence of TLRs on human adipocytes with expression levels of TLR type 2 (TLR2) and TLR type 4 (TLR 4) higher than those of the monocyte cell line THP-1 (44). The link between obesity and inflammation has been further illustrated in obese patients by the increased plasma levels of several pro-inflammatory markers including cytokines, fibrinogen or serum amyloids, and acute phase proteins like C-reactive protein (CRP) which was long known as a marker of inflammation, obesity, and cardiovascular diseases (45). Many of the inflammatory markers found in plasma of obese individuals appear to originate from adipose tissue (46). Adipose tissue was not usually thought of as an immune or inflammatory organ, however, evidence of a connection between obesity and inflammation has been found in the context of clinical weight loss studies. Whether the weight loss is attributable to decreased dietary intake, increased fuel use through exercise, or liposuction, loss of adipose tissue is associated with a decrease in markers of inflammation (47,48). Furthermore, the discovery of elevated secretion of these factors from obese adipose tissue provided the first evidence of a direct relationship between obesity and systemic inflammation. The observed increase in the concentrations of these adipocytokines is now suspected to play a determinant role in the development of most of the complications of obesity (36). From this perspective, obesity can be described as a pro-inflammatory condition.

The mechanisms responsible for the up regulation of inflammation in fat tissue are not totally clear. Macrophages as well as adipocytes are likely to be key players in the development of inflammation (49). One hypothesis proposed that accumulation of mature adipocytes without a sufficient vascularization lead some adipocytes to be in a hypoxic environment and stimulate an inflammatory response (69–72). Macrophages could then be recruited by the adipocytes-produced chemokine CCL2 and participate to the inflammation response (50,51). Other results suggest that interactions between free fatty acids and TLRs could activate the corresponding signalling pathways and produce inflammatory factors (52,53).

OBESITY ASSOCIATED DISEASES

Obesity is associated with number of disorders and increases the risk for a vast array of diseases, with significant impact on morbidity and mortality. The molecular basis of the link between obesity and diseases such as type 2 diabetes, fatty liver disease, atherosclerosis, hypertension, and stroke is well documented and studied. In comparison, less data are available concerning the link with diseases such as cancer, asthma, sleep apnoea, osteoarthritis, neurodegeneration or gall bladder disease (54).

Insulin Resistance and Type 2 Diabetes

Type 2 diabetes is characterized by insulin resistance which is strongly associated with obesity and physical inactivity. Insulin is released by the pancreatic beta cells in response to elevations in blood glucose. It promotes adipocyte triglycerides storage by a number of mechanisms, including fostering the differentiation of preadipocytes into mature adipocytes, stimulating glucose transport, and triglyceride synthesis as well as inhibiting lipolysis. Insulin also increases the uptake of fatty acids derived from circulating lipoproteins by stimulating lipoprotein lipase activity in adipose tissue. Insulin's metabolic effects are mediated by a broad array of tissue specific actions that involve rapid changes in protein phosphorylation and function, as well as changes in gene expression.

Insulin brings its metabolic responses by binding and activating a tyrosine kinase receptor that is present on the plasma membrane. Insulin receptor substrate (IRS) proteins, are efficiently phosphorylated on tyrosine on several sites leading to a cascade of intracellular events. Insulin activates a series of lipid and protein kinase enzymes which are linked to the glycogen synthesis or glucose transporters. The term "insulin resistance" corresponds to the resistance to insulin effect on glucose uptake, metabolism, or storage. Negative regulation of insulin resistance can be mediated via various pathways regulated by phosphorylation and dephosphorylation of IRS proteins (55,56). Increased production of inflammatory cytokines in WAT can affect the insulin signalling cascade. For example TNF-α and IL-6 are increased in obesity and affect the insulin signalling pathways (57–59).

On the one hand there are several evidences suggesting that a number of degenerative diseases including Alzheimer disease, Parkinson disease, Down syndrome/trisomy 21, Friedreich ataxia, Huntington disease, or Prader–Willi syndrome are the cause of an associated diabetes (60). On the other hand, diabetic neuropathy is a frequent consequence of diabetes type 2. High glucose concentration over time may damage peripheral nerves (61) and provokes numerous different symptoms such as dysesthesia, diarrhea, muscle weakness and paralysis, incontinence, or impotence among others. The control of glucose level in the blood and the duration of diabetes are important elements determining the severity of the diabetic neuropathy (60). The mechanisms of the diabetic neuropathy development are not totally elucidated, but increase of the advanced glycation end products (62), activation of the polyol pathway (63), activation of protein kinase C and mitogen activated protein kinase (64), and deficiency in growth factors (65) have been shown to be involved in the neuropathy. In addition to the diabetic neuropathy, an association between diabetes and neurodegenerative disorder such as Alzheimer disease, Parkinson, or Friedreich's Ataxia leading to dementia has been shown (66). A dysfunction of mitochondria could constitute a possible link between diabetes and these neurodegenerative disorders (67–69).

Cardiovascular and Metabolic Diseases

Cardiovascular disease is the major cause of death in industrialized countries. The excessive fat accumulation in adipose tissue strongly predisposes the whole body to the development of metabolic abnormalities that often accompany obesity, including hypertension, impaired glucose tolerance, insulin resistance leading to hyperinsulinemia, hyperglycaemia, and dyslipidemia. Collectively, these abnormalities have been clustered into the metabolic syndrome and individuals who are diagnosed with metabolic syndrome have significantly increased risk of developing cardiovascular disease (70). Fat distribution varies from individual to individual, and is also dependent of the gender, with an accumulation in the hip area for female and in abdomen area for men (71). Visceral obesity increases the relative risk of metabolic syndrome by more than 10 fold (70,72). Obesity increases the risk of vascular diseases by altering the vasculature through factors produced by the WAT (73). As already discussed, obesity is associated with a chronic inflammation which will affect the vasculature. Alteration of the vessel and circulation has an impact on different organs including the heart and the brain. There is an association between obesity and ischemic stroke independently of hypertension, diabetes and cholesterols level (74–76).

Environmental Factors and Fat Tissue

The high lipophilicity of many environmental factors or industrial chemicals, such as organochlorine pesticides and polychlorinated biphenyls (PCBs), favors their accumulation in the fat tissue (77–79). A recent study showed that PCBs promotes obesity and atherosclerosis, and induces adipocytes differentiation and pro inflammatory adipokines production (80). Similarly it has been suggested that exposure to dioxin could affect WAT and increase the development of type 2 diabetes (81). Furthermore, some environmental factors acting as endocrine disruptors can modulate the endocrine/paracrine production of WAT. This is the case for organotin, such as tributyltin (82–84), or diethylstilbestrol, which interact with retinoic receptor and affect adipocytes differentiation (85–87). Thus, it is possible that with the long-term bioaccumulation of toxic compounds, such as dioxin, PCB, or the polybrominated biphenyl flame retardant compounds, the WAT becomes, not only a sink for these compounds, but also a target site and source of pro-inflammatory cytokines with the continued exposure. The dramatic disproportion of WAT in obese subjects may represent an elevated risk factor for adverse effects of environmental chemical exposure based upon the production of pro-inflammatory cytokines. While concern is raised with regards to chronic inflammation, the impact of elevated cytokine production by WAT on acute toxicity is a significant consideration.

ANIMAL MODELS FOR OBESITY

Animal models of obesity have been very useful and important to understand the regulation of food intake and the imbalance in energy expenditure (Fig. 4). The first models described corresponded to spontaneous single gene mutation leading to the loss of the gene function (88). The oldest of these models described is the agouti mouse (89,90). The mutation of the agouti gene is associated with pleiotropic effects including obesity and increased susceptibility to diabetes (90). The agouti gene involved in these phenotypes has been cloned in 1992 (91). The cause of obesity in agouti mouse results of the agouti protein interaction with the melanocortin receptors (92). Other frequently used obesity models include invalidation of the leptin pathway either at the leptin level (ob/ob mouse) (10,93) or at the receptor level (db/db mouse), of the tubby or carboxypeptidase E genes (94,95). Around ten other genes have been characterized as single mutation (88). As several genes have been found to be involved in energy balance regulation and with the possibility of knocking in (KI) or out (KO) specific genes in mouse, the number of mouse models of obesity has dramatically increased (96). However, the limitations of these models are the same as in other research area, i.e., the selected gene may be needed for a correct development or compensated with others genes. These issues could be bypassed by the generation of conditional KO or KI. The gene of interest is then KO or KI after the development of the organism with system such as cre/lox. However, the generation of conditional mice remains difficult (97).

It should be stressed that in human single gene mutation remain rare, and most of the time obesity is polygenic. Because of the relative easiness of modifying mouse gene, mice are the most widely used model. In addition to rats, other species have been used to study obesity related issues. These include, pigs, chicken, and even bats (88,98,99). Experimental studies on lower evolved organisms such as worm, drosophila, or zebra fish have not been used often in obesity studies but their use could contribute to understanding the basic mechanisms of energy metabolism (100).

In addition to genetically modified or selected animals, several diet-induced obese models (DIO) have been described. However, as many different specific diets leading to obesity exist, comparisons between studies are difficult (101,102). In animal models, a basic indication of the fat quantity can be estimated by measuring the total weight or the weight of the WAT after dissection. However, dual energy X ray absorption with appropriate validations could be used to quantify the fat in vivo and allows to follow an animal over time (103). An example of such measurements with different fat contents in mice is shown in Figure 5. This demonstrates an imaging method for determining body fat as well as skeletal alterations.

NEURODEGENERATION AND OBESITY

Interactions between WAT and nervous system are well known and review articles on factors involved in communication between WAT and sympathetic nervous system or CNS have been

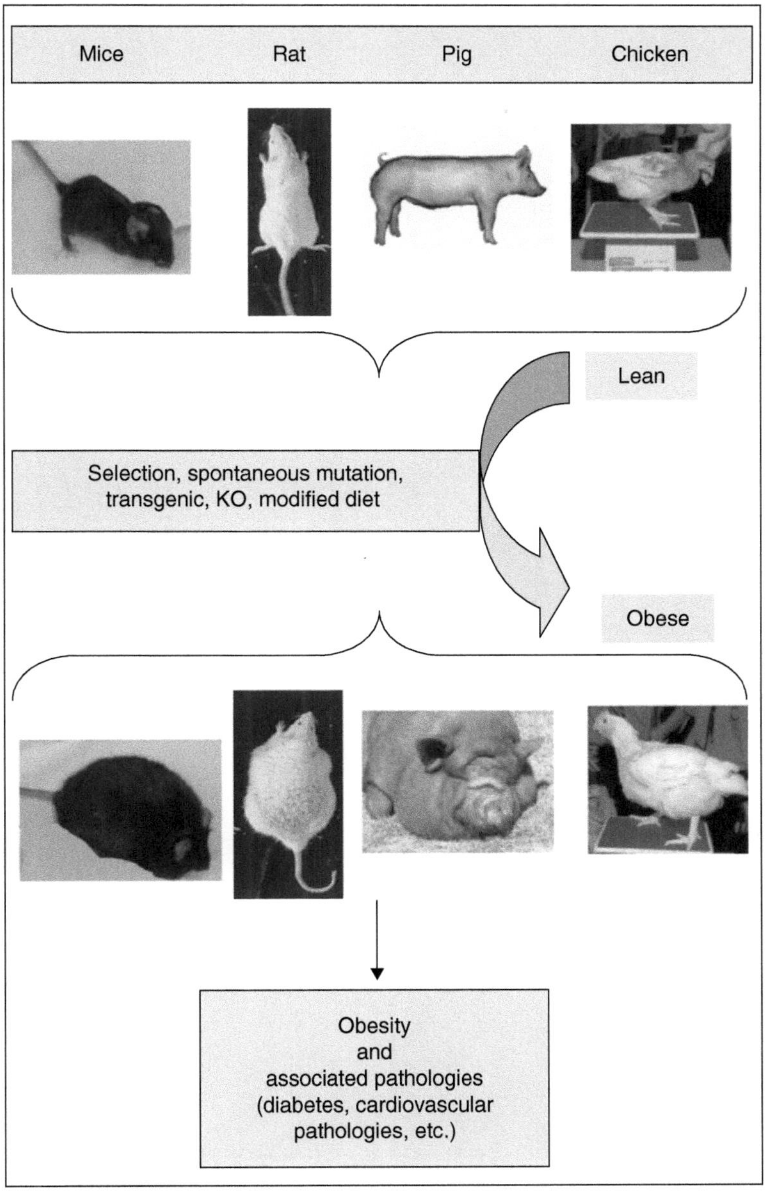

Figure 4 Animal models of obesity. Different animal species can be used to study obesity-related disorders including selection, spontaneous mutations, overexpression or impairing of genes, or modified diet. *Abbreviation*: KO, knock out.

published (104–106). As already seen, the cells present in the WAT can produce adipokines and cytokines playing a role in weight regulation and inflammation. Interactions between peripheral cytokines and brain have been reported, even in presence of an intact BBB (107). In case of brain or spinal cord injury, damage to the barriers would allow for an increased transport of blood borne substances and under these conditions it is likely that adipokines could then reach the brain parenchyma (108).

Neuronal damage can occur as a function of an acute injury or manifest over a prolonged period of time. Neurodegeneration can be characterized by either a structural or functional neuronal lost. Acute neurodegeneration occurs in timely exogenous events such

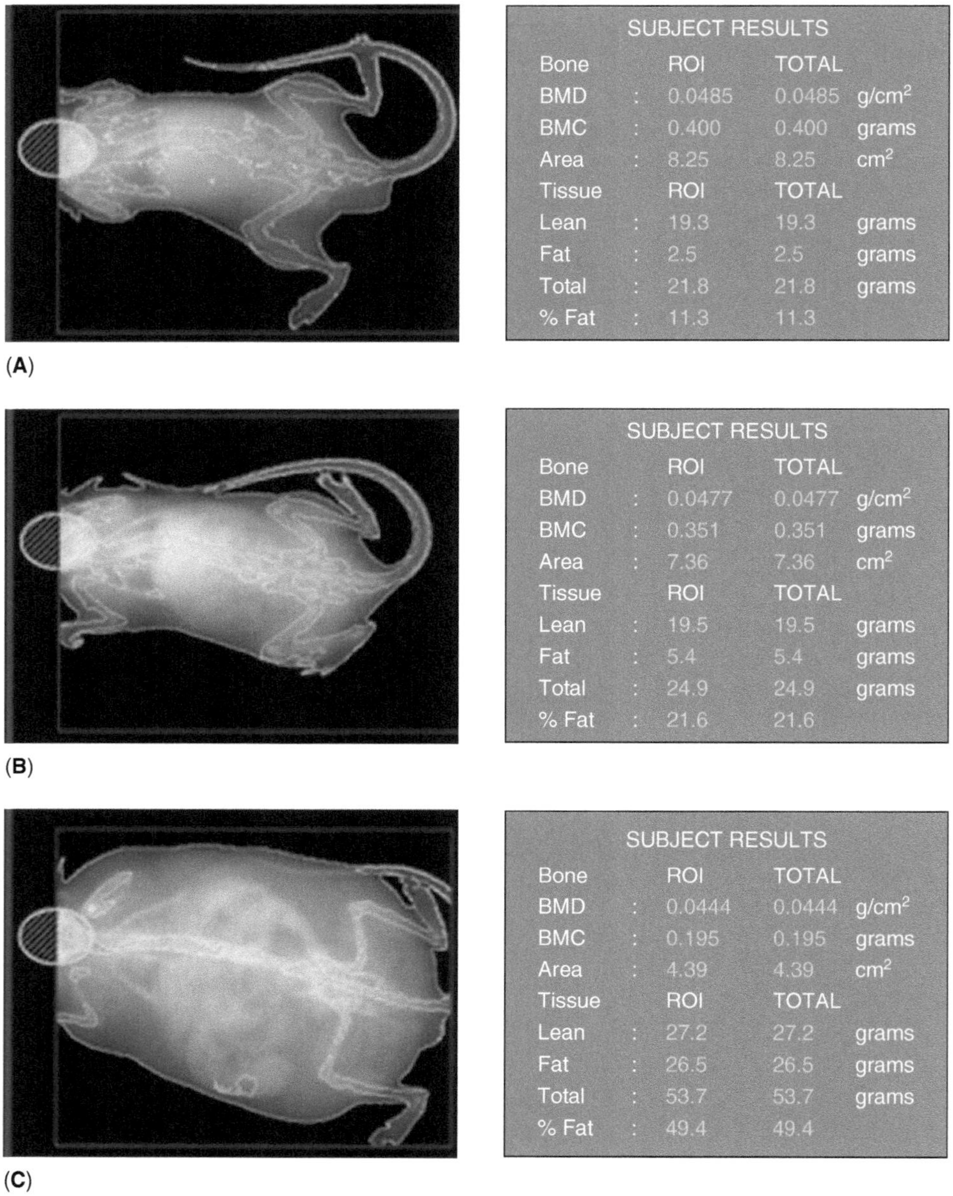

Figure 5 (*See color insert*) Fat measurement. Estimation of fat content by dual energy X-ray absorptiometry using a LUNAR PIXImus bone densitometer (GE Healthcare, Fairield, CT) as described previously (166) in 10 week-old male mouse (Jackson Laboratory, Bar Harbor, Maine, U.S.A.). (**A**) C57BL/6J mouse, (**B**) agouti (B6. Cg-A^y/J) mouse, (**C**) Ob/ob mouse (B6.V-*Lep*^{ob}/J). *Abbreviations*: BMD, bone mineral density; BMC, bone mineral content; ROI, region of interest.

as traumatic injury, chemical exposure, or endogenous events such as ischemia or cerebral vascular accidents. Chronic neurodegeneration is often associated with more progressive neurodegenerative diseases as seen in pathologies such as Alzheimer, Parkinson, or Huntington diseases.

Numerous studies have demonstrated a link between neuroinflammation and neurodegeneration. With brain injury and in neurodegenerative disease states there is an associated reaction of the brain non-neuronal cells. The glial cells', primarily the microglia and astrocytes, response can serve as both a source of growth factors and neurotoxic factors to the localized brain tissue. Based upon their localization and production of pro-inflammatory cytokines, microglia have been suggested to play a role in neurodegenerative diseases (109,110). Details of microglia and neuroinflammation are provided in a separate chapter. It is well accepted that activated microglia serve as an important source of inflammatory cytokines. While in vitro studies have demonstrated that the pro-inflammatory cytokine tumor necrosis factor alpha can induce death of oligodendroglia (the myelinating cells of the brain), other in vivo studies have suggested that this pro-inflammatory cytokine can initiate neuronal death in either acute organotin toxicity or in the prenumbra surrounding an ischemic infarct. As in periphery, inflammation can have both beneficial and deleterious effects depending upon the in vivo regulatory mechanisms (110). Inflammation is not always the origin of neuronal death, but even if it is a secondary event, it can influence the more long-term fate of the neuron (111). A recent study has demonstrated an induction of adipokines expression in the brain following traumatic brain injury (112). Given that "inflammatory factors" are maintained under strict regulatory control in the normal brain, any contributor however minor may be significant in the overall homeostatic, injury, and repair process. It is possible that the adipokines contribute in the overall induction and regulation of the inflammatory response in the brain.

Diet

Food and diet quantity are keys elements in overweight and obesity development, but food composition can also influence directly and indirectly the CNS functioning (Fig. 6). Epidemiologic results as well as experimental animal studies show that developed country diet type (i.e., rich in total fat, saturated fat, linoleic acid, cholesterol, and refined sugar) are associated with an increased risk for dementia and a decrease of neuronal plasticity and learning (113,114). Oxidative stress and inflammation in cerebral cortex and decrease of the neurotrophic factor brain derived neurotrophic factor (BDNF) could be a significant contributor to these negative effects (114,115). In comparison, a diet including foods rich in polyunsaturated fat (found in fish), could reduce the risk of neurodegenerative or vascular dementia (113,116). Presence of curcuma or feluric acid in the diet could also have neuroprotective effects (117). As an alternative to a diet rich in fat, intermittent fasting is more beneficial than caloric restriction in term of

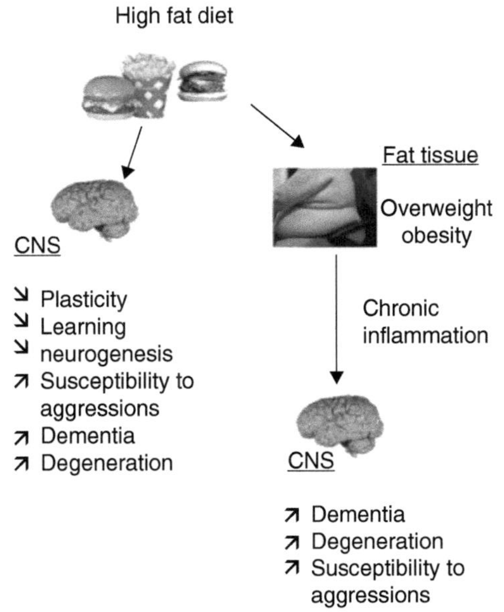

Figure 6 Diet and fat tissue influence on CNS. Diets rich in saturated fat, cholesterol, linoleic acid, or refined sugar directly affect the CNS and also contribute to the development of fat tissue and chronic inflammation. Obesity and overweight are associated with CNS disorders.

reducing serum glucose and insulin and increasing neuronal resistance to excitotoxic stress (118). As seen previously, diet influences CNS functioning and development of fat tissue due to disequilibrium in the energetic balance, will also affect the CNS. It is well accepted that obesity induces a chronic inflammation which influence the development of obesity associated pathologies (59,105). Several studies have shown the link between these pathologies and obesity, but very few data are available concerning the role of WAT excess on neurodegenerative disorders. It is possible that chronic inflammation would serve to change the normal constitutive level of circulating cytokines and modify the CNS response to injuries as suggested by a recent study showing that lipopolysaccharide induced systemic inflammation provokes a chronic neuroinflammation and a progressive neurodegeneration (119). As we will discuss in the last part, stem cell sources have been found in the CNS, however neurons regeneration capacities are often limited. The different cell type present in the CNS have variable sensibility to pathologies or traumatisms and it is possible that obesity could influence this susceptibility (Fig. 6) (120).

Human Epidemiologic Data

Epidemiologic studies suggest that obesity influence CNS functionality. Indeed obesity is associated with decrease cognitive functioning, hypertension, depression, and suicidal thoughts, in child and teens (121,122). Obesity is also associated with sleeping apnea, anxieties and bipolar disorders. Furthermore, numerous clinical studies have demonstrated that obese patients have a higher risk to develop neurodegenerative diseases, including Alzheimer disease (123). Pr D. Gustasfon from the Institute for Neuroscience and Physiology, (Göteborg University, Sweden) was one of the first to demonstrate, in a Swedish population sample, a relationship between overweight and risks to develop Alzheimer disease (124). The risk increase could be due to vascular disorders or to WAT secreted factors (125). It should be stressed that patients before entering in a dementia phase, have a tendency to lose weight, which changed their BMI and that a weight under normal is associated with Alzheimer disease (124). It is thus important in such epidemiologic studies to take this factor in account and to use an index estimating fat mass before the development of the pathology. Overweight or obesity effects are not limited to Alzheimer disease, as a relationship between BMI, cerebral atrophy and white matter lesions in elderly woman have also been described (126,127).

Animal Models of Neurodegenerative Diseases

Numerous animal models of neurodegeneration have been described, murine models being the most used. However, neurodegeneration models have also been developed with invertebrates, such as nematodes or drosophila and with low evolved vertebrate, such as zebra fish (128,129). Neurodegeneration can be induced by over expression or under expression of a gene, by a trauma or by chemical agents (Fig. 7).

In murine model of obesity, an increased susceptibility of CNS to trauma has been observed. High dietary fat is a significant risk for cerebral oxidative stress development, neuronal inflammation, vascular dementia, and Alzheimer disease (113,115). In genetic models, obesity is an aggravating factor in chemical-induced neurodegeneration. In mice deficient for the leptin gene (ob/ob), the effects of two neurotoxicants are exacerbated. Two different chemicals have been used in this study, methamphetamine, which affects dopaminergic neurons and kainic acid, affecting the hippocampus (130). Furthermore, an increase in the susceptibility of dopaminergic neurons in the substantia nigra in response of a treatment with the chemical 1-methyl-4-phenyl-1,2,3,6-tetrahydropyridine (MPTP) has been observed in high-fat diet-induced obesity. MPTP is often used in animal models of Parkinson disease. An increase in systemic inflammation, a decrease in an enzyme playing a role in free radical defence concentration in substantia nigra (super oxide dismutase), a modification in the lipid peroxidation and an increase of the striatal nitric oxide synthase (nNOAS) could play a role in this exacerbated response (131).

ADULT NEUROGENESIS AND OBESITY

It is now known that in the adult brain there exists the capacity for the generation of new neurons and glia. The idea of the presence of a population of stem cells capable of proliferation and differentiation in the adult CNS is not new, Altman and Bayer proposed this concept as

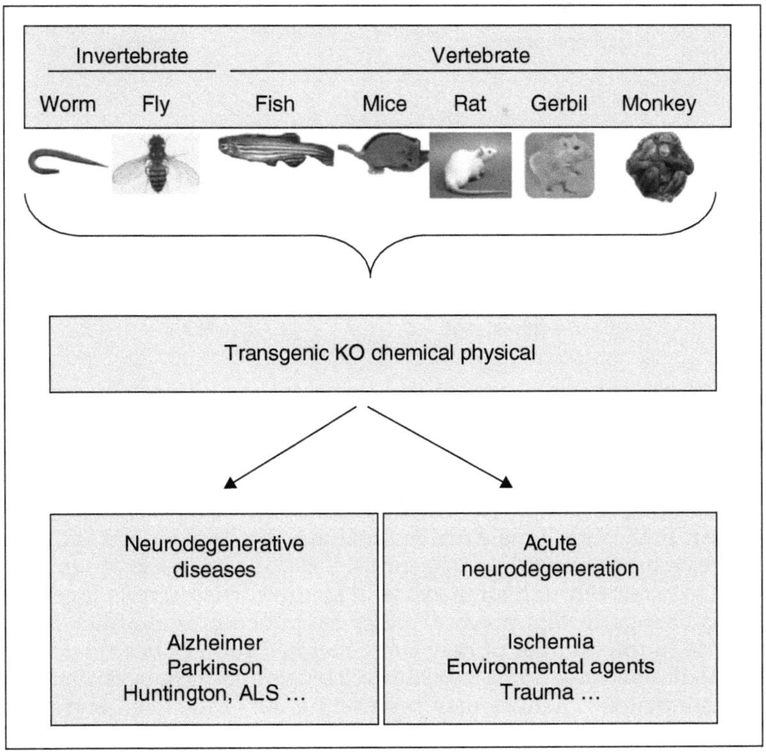

Figure 7 Animal models of neurodegeneration. Different animal species can be used to study neurodegenerative disorders induced by over expression or impairing of genes, by neurotoxic agents or physically. These models mimic acute or chronic neurodegeneration disorders. *Abbreviations*: ALS, amyotrophic lateral sclerosis; KO, knock out.

early as 1962 (132). However, it was not until the 1990s that several laboratories identified neurogenic regions within the adult rodent and avian brain. The presence of stem cells was demonstrated primarily in the subventricular zone (133). These stem cells migrate in the rostral migratory stream to reach the olfactory bulb where these cells differentiate into granular and periglomerular neurons. Further examination showed that the subgranular zone of the hippocampus represented another neurogenic region of the adult brain (134). These works open the way to a new and extensive field of research and to new therapeutic strategies based on brain neurogenesis (135). Neurogenesis in the adult CNS is not limited to the sub ventricular zone and the hippocampus regions, recent studies have shown that new neurons can be observed in the hypothalamus (136). It is now well accepted that a constitutive neurogenesis exists in adult, but several factors can modulate the production of new neurons (Fig. 8). For example, exercise or an enriched environment (cage with more space and enriched with stimulating elements such as tunnels or shelters) induces increases in the number of proliferating neurons in both the hippocampus and the subventricular zone (137,138). More importantly, with regards to inflammatory factors, injury to the brain such as ischemia (139), epilepsia (140), or chemically induced neurodegeneration (141) induce an increase in neurogenesis. This induction has been termed "injury-induced neurogenesis." This topic is covered in more detail in a separate chapter in this issue, however, it is briefly mentioned here as an alternative nervous system process that can be influenced by the underlying inflammatory state of the tissue. Addition of exogenous trophic factors including epidermal growth factor (EGF), fibroblast growth factor 2 (FGF2), BDNF, glial-derived neurotrophic factor (GDNF), or "insulin-like growth factors"-1 and -2 (IGF-1, IGF-2) also induce neurogenesis (142). Ciliary neurotrophic factor (CNTF) induces a hypothalamus neurogenesis and more specifically in the satiety centres,

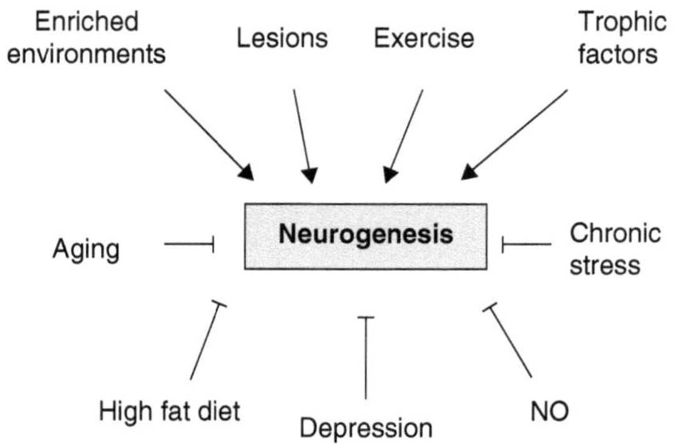

Figure 8 Factors modulating neurogenesis activity. ↓, neurogenesis augmentation; ⊥, neurogenesis inhibition.

inducing a weight loss persisting after the treatment (143). Diet has been shown to influence adult neurogenesis in that adult male rats maintained on a high fat diet show a decrease in the level of constitutive hippocampal neurogenesis. This occurs in the absence of overt obesity and is proposed to occur through an increase in serum corticosterone levels (144). Any long-term impact of a change in this process, either as an acute or chronic change leading to a decline or increase in the number of new cell generated has not been determined. However, the observation that diet can alter the constitutive levels and the susceptibility of these cells to metabolic or inflammatory changes may represent a new target cell and process within the brain that could be altered by the body fat status.

EXERCISE, OBESITY, AND NEURODEGENERATION

As previously mentioned, exercise has a positive effect on neurogenesis and brain health in animal model and human (145). The protective effect of exercise on neurodegenerative disorder is suggested in several studies on different pathologies (146) including spinal cord injury (147), stroke (148), ischemia (149), and spinal muscular atrophy (150,151). In model of Parkinson, exercise does not demonstrate a direct preventive type effect however it seems to offer an improvement of recovery of function (152,153). However the exercise intensity requirement remains to be determined (154), as intense or forced exercise can increase vulnerability in chemically induced neurodegeneration (155). The mechanisms of exercise inducing neurogenesis and neuroprotection may involve glia maturation factor (GMF) and BDNF (156,157). Interestingly, it has been recently demonstrated that exercise reduces resistin and inflammatory cytokines in diabetes type 2 patients (158,159).

SUMMARY AND CONCLUSIONS

In the context of life duration expenditure and pandemic development of obesity, it is important to study factors influencing the CNS response to chronic and acute aggressions. Epidemiologic as well as animal studies suggest a negative role of overweight and obesity on neurodegenerative diseases development as well as on acute neurodegeneration disorders.

A better understanding of the molecular mechanisms leading to this increased susceptibility of the CNS to aggressions should bring new perspectives on prevention and open the way to new therapeutic targets. Diet influences directly CNS functioning and specific diet could be beneficial in term of neurons resistance to chronic or acute aggressions. As exercise improves several obesity associated pathologies symptoms, as well as induces neurogenesis, it is also an interesting therapeutic option to further explore, but the parameters of this physical activity remain to be better determined. The effects of WAT excess on constitutive or induced neurogenesis are still to be explored and are probably an important factor to consider in order to develop reparation therapeutic strategies based on the use of endogenous or exogenous stem cells.

ACKNOWLEDGMENTS

R.P. has been funded by fellowships from "Conseil Régional de La Réunion." We also thank Dr. Emmanuel Bourdon and Florence Carayon-Lefebvre d'Hellencourt for reading and editing this chapter.

REFERENCES

1. Eckel RH, Grundy SM, Zimmet PZ. The metabolic syndrome. Lancet 2005; 365: 1415–28.
2. Hauner H. Secretory factors from human adipose tissue and their functional role. Proc Nutr Soc 2005; 64: 163–9.
3. Hauner H. The new concept of adipose tissue function. Physiol Beh 2004; 83: 653–8.
4. Roche R, Hoareau L, Mounet F, Festy F. Adult stem cells for cardiovascular diseases: the adipose tissue potential. Expert Opin Biol Ther 2007; 7: 791–8.
5. Noel D, Caton D, Roche S, et al. Cell specific differences between human adipose-derived and mesenchymal-stromal cells despite similar differentiation potentials. Exp Cell Res 2008; 314: 1575–84.
6. Schaffler A, Buchler C. Concise review: adipose tissue-derived stromal cells—basic and clinical implications for novel cell-based therapies. Stem Cells 2007; 25: 818–27.
7. Casteilla L, Dani C. Adipose tissue-derived cells: from physiology to regenerative medicine. Diabetes Metab 2006; 32: 393–401.
8. Gavrilova O, Marcus-Samuels B, Graham D et al. Surgical implantation of adipose tissue reverses diabetes in lipoatrophic mice. J Clin Invest 2000; 105: 271–8.
9. Greenberg AS, Obin MS. Obesity and the role of adipose tissue in inflammation and metabolism. Am J Clin Nutr 2006; 83: 461S–5S.
10. Zhang Y, Proenca R, Maffei M, et al. Positional cloning of the mouse obese gene and its human homologue. Nature 1994; 372: 425–32.
11. Friedman JM, Halaas JL. Leptin and the regulation of body weight in mammals. Nature 1998; 395: 763–70.
12. Ingalls AM, Dickie MM, Snell GD. Obese, a new mutation in the house mouse. J Hered 1950; 41: 317–18.
13. Chua SC Jr., Chung WK, Wu-Peng XS, et al. Phenotypes of mouse diabetes and rat fatty due to mutations in the OB (leptin) receptor. Science 1996; 271: 994–6.
14. Schwartz MW, Woods SC, Seeley RJ, et al. Is the energy homeostasis system inherently biased toward weight gain? Diabetes 2003; 52: 232–8.
15. Masuzaki H, Ogawa Y, Sagawa N, et al. Nonadipose tissue production of leptin, leptin as a novel placenta-derived hormone in humans. Nat Med 1997; 3: 1029–33.
16. Kalra SP, Dube MG, Pu S, et al. Interacting appetite-regulating pathways in the hypothalamic regulation of body weight. Endocr Rev 1999; 20: 68–100.
17. Lago R, Gomez R, Lago F, et al. Leptin beyond body weight regulation-current concepts concerning its role in immune function and inflammation. Cell Immunol 2008; 252: 139–45.
18. Loffreda S, Yang SQ, Lin HZ, et al. Leptin regulates proinflammatory immune responses. Faseb J 1998; 12: 57–65.
19. Trujillo ME, Scherer PE. Adipose tissue-derived factors: impact on health and disease. Endocr Rev 2006; 27: 762–78.
20. Berg AH, Scherer PE. Adipose tissue, inflammation, and cardiovascular disease. Circ Res 2005; 96: 939–49.
21. Fain JN, Madan AK, Hiler ML, et al. Comparison of the release of adipokines by adipose tissue, adipose tissue matrix, and adipocytes from visceral and subcutaneous abdominal adipose tissues of obese humans. Endocrinology 2004; 145: 2273–82.
22. Schaffler A, Neumeier M, Herfarth H, et al. Genomic structure of human omentin, a new adipocytokine expressed in omental adipose tissue. Biochim Biophys Acta 2005; 1732: 96–102.
23. de Souza Batista CM, Yang RZ, Lee MJ. Omentin plasma levels and gene expression are decreased in obesity. Diabetes 2007; 56: 1655–61.
24. Yang RZ, Lee MJ, Hu H, et al. Identification of omentin as a novel depot-specific adipokine in human adipose tissue: possible role in modulating insulin action. Am J Physiol Endocrinol Metab 2006; 290: E1253–61.
25. Tatemoto K, Hosoya M, Habata Y, et al. Isolation and characterization of a novel endogenous peptide ligand for the human APJ receptor. Biochem Biophys Res Commun 1998; 251: 471–6.
26. Boucher J, Masri B, Daviaud D, et al. Apelin, a newly identified adipokine up-regulated by insulin and obesity. Endocrinology 2005; 146: 1764–71.
27. Castan-Laurell I, Boucher J, Dray C, et al. Apelin, a novel adipokine over-produced in obesity: friend or foe? Mol Cell Endocrinol 2005; 245: 7–9.

28. Daviaud D, Boucher J, Gesta S, et al. TNFalpha up-regulates apelin expression in human and mouse adipose tissue. FASEB J 2006; 20: 1528–30.
29. Hida K, Wada J, Eguchi J, et al. Visceral adipose tissue-derived serine protease inhibitor: a unique insulin-sensitizing adipocytokine in obesity. Proc Natl Acad Sci USA 2005; 102: 10610–15.
30. Fried SK, Bunkin DA, Greenberg AS. Omental and subcutaneous adipose tissues of obese subjects release interleukin-6: depot difference and regulation by glucocorticoid. J Clin Endocrinol Metab 1998; 83: 847–50.
31. Wallenius V, Wallenius K, Ahren B, et al. Interleukin-6-deficient mice develop mature-onset obesity. Nat Med 2002; 8: 75–9.
32. Hotamisligil GS, Arner P, Caro JF, et al. Increased adipose tissue expression of tumor necrosis factor-alpha in human obesity and insulin resistance. J Clin Invest 1995; 95: 2409–15.
33. Cancello R, Clement K. Is obesity an inflammatory illness? role of low-grade inflammation and macrophage infiltration in human white adipose tissue. BJOG 2006; 113: 1141–7.
34. Das UN. Is obesity an inflammatory condition? Nutrition 2001; 17: 953–66.
35. Fantuzzi G. Adipose tissue, adipokines, and inflammation. J Allergy Clin Immunol 2005; 115: 911–19; quiz 920.
36. Fontana L, Eagon JC, Trujillo ME, et al. Visceral fat adipokine secretion is associated with systemic inflammation in obese humans. Diabetes 2007; 56: 1010–13.
37. Zhang D, Zhang G, Hayden MS, et al. A toll-like receptor that prevents infection by uropathogenic bacteria. Science 2004; 303: 1522–6.
38. Hoffmann JA. The immune response of Drosophila. Nature 2003; 426: 33–8.
39. Gobert V, Gottar M, Matskevich AA, et al. Dual activation of the Drosophila toll pathway by two pattern recognition receptors. Science 2003; 302: 2126–30.
40. Ulevitch RJ. Therapeutics targeting the innate immune system. Nat Rev Immunol 2004; 4: 512–20.
41. Rolff J, Siva-Jothy MT. Invertebrate ecological immunology. Science 2003; 301: 472–5.
42. Mortensen RF. C-reactive protein, inflammation, and innate immunity. Immunol Res 2001; 4: 163–76.
43. Hotamisligil GS, Shargill NS, Spiegelman BM. Adipose expression of tumor necrosis factor-alpha: direct role in obesity-linked insulin resistance. Science 1993; 259: 87–91.
44. Bes-Houtmann S, Roche R, Hoareau L, et al. Presence of functional TLR2 and TLR4 on human adipocytes. Histochem Cell Biol 2007; 127: 131–7.
45. Vidal H. [Obesity and inflammation: the adipocytokines]. Ann Endocrinol (Paris) 2003; 64: S40–4.
46. Trayhurn P, Wood IS. Signalling role of adipose tissue: adipokines and inflammation in obesity. Biochem Soc Trans 2005; 33: 1078–81.
47. Nicoletti G, Giugliano G, Pontillo A, et al. Effect of a multidisciplinary program of weight reduction on endothelial functions in obese women. J Endocrinol Invest 2003; 26: RC5-8.
48. Heilbronn LK, Noakes M Clifton PM. Energy restriction and weight loss on very-low-fat diets reduce C-reactive protein concentrations in obese, healthy women. Arterioscler Thromb Vasc Biol 2001; 21: 968–70.
49. Weisberg SP, McCann D, Desai M, et al. Obesity is associated with macrophage accumulation in adipose tissue. J Clin Invest 2003; 112: 1796–1808.
50. Neels JG, Olefsky JM. Inflamed fat: what starts the fire? J Clin Invest 2006; 116: 33–5.
51. Weisberg SP, Hunter D, Huber R, et al. CCR2 modulates inflammatory and metabolic effects of high-fat feeding. J Clin Invest 2006; 116: 115–24.
52. Kim F, Pham M, Luttrell I, et al. Toll-like receptor-4 mediates vascular inflammation and insulin resistance in diet-induced obesity. Circ Res 2007; 100: 1589–96.
53. Nguyen MT, Favelyukis S, Nguyen AK. A subpopulation of macrophages infiltrates hypertrophic adipose tissue and is activated by free fatty acids via Toll-like receptors 2 and 4 and JNK-dependent pathways. J Biol Chem 2007; 282: 35279–92.
54. Hotamisligil GS. Inflammation and metabolic disorders. Nature 2006; 444: 860–7.
55. Kido Y, Nakae J, Accili D. Clinical review 125: The insulin receptor and its cellular targets. J Clin Endocrinol Metab 2001; 86: 972–9.
56. White MF. IRS proteins and the common path to diabetes. Am J Physiol Endocrinol Metab 2002; 283: E413–22.
57. Rajala MW, Scherer PE. Minireview: the adipocyte—at the crossroads of energy homeostasis, inflammation, and atherosclerosis. Endocrinology 2003; 144: 3765–73.
58. Ravussin E, Smith SR. Increased fat intake, impaired fat oxidation, and failure of fat cell proliferation result in ectopic fat storage, insulin resistance, and type 2 diabetes mellitus. Ann N Y Acad Sci 2002; 967: 363–78.
59. Xu H, Barnes GT, Yang Q, et al. Chronic inflammation in fat plays a crucial role in the development of obesity-related insulin resistance. J Clin Invest 2003; 112: 1821–30.

60. Ristow M. Neurodegenerative disorders associated with diabetes mellitus. J Mol Med 2004; 82: 510–29.
61. Tomlinson DR, Gardiner NJ. Glucose neurotoxicity. Nat Rev Neurosci 2008; 9: 36–45.
62. Sugimoto K, Yasujima M, Yagihashi S. Role of advanced glycation end products in diabetic neuropathy. Curr Pharm Des 2008; 14: 953–61.
63. Oates PJ. Aldose reductase, still a compelling target for diabetic neuropathy. Curr Drug Targets 2008; 9: 14–36.
64. Chan L, Terashima T, Fujimiya M, Kojima H. Chronic diabetic complications: the body's adaptive response to hyperglycemia gone awry? Trans Am Clin Climatol Assoc 2006; 117: 341–52.
65. Leinninger GM, Vincent AM, Feldman EL. The role of growth factors in diabetic peripheral neuropathy. J Peripher Nerv Syst 2004; 9: 26–53.
66. Pasquier F, Boulogne A, Leys D, Fontaine P. Diabetes mellitus and dementia. Diabetes Metab 2006; 32: 403–14.
67. Moreira PI, Santos MS, Seica R, Oliveira CR. Brain mitochondrial dysfunction as a link between Alzheimer's disease and diabetes. J Neurol Sci 2007; 257: 206–14.
68. Sas K, Robotka H, Toldi J, Vecsei L. Mitochondria, metabolic disturbances, oxidative stress and the kynurenine system, with focus on neurodegenerative disorders. J Neurol Sci 2007; 257: 221–39.
69. Calabrese V, Lodi R, Tonon C, et al. Oxidative stress, mitochondrial dysfunction and cellular stress response in Friedreich's ataxia. J Neurol Sci 2005; 233: 145–62.
70. Mathieu P, Pibarot P, Larose E, et al. Visceral obesity and the heart. Int J Biochem Cell Biol 2008; 40: 821–36.
71. Vague J. [Obesity and thinness; obesity, diabetes and arteriosclerosis.]. Vie Med 1954; 35: 497–502.
72. Park YW, Zhu S, Palaniappan L, et al. The metabolic syndrome: prevalence and associated risk factor findings in the US population from the Third National Health and Nutrition Examination Survey, 1988–1994. Arch Intern Med 2003; 163: 427–36.
73. Fantuzzi G, Mazzone T. Adipose tissue and atherosclerosis: exploring the connection. Arterioscler Thromb Vasc Biol 2007; 27: 996–1003.
74. Hankey GJ. Potential new risk factors for ischemic stroke: what is their potential? Stroke 2006; 37: 2181–8.
75. Kurth T, Gaziano JM, Berger K, et al. Body mass index and the risk of stroke in men. Arch Intern Med 2002; 162: 2557–62.
76. Kurth T, Gaziano JM, Rexrode KM, et al. Prospective study of body mass index and risk of stroke in apparently healthy women. Circulation 2005; 111: 1992–8.
77. Mullerova D, Kopecky J. White adipose tissue: storage and effector site for environmental pollutants. Physiol Res 2007; 56: 375–81.
78. Kutz FW, Wood PH, Bottimore DP. Organochlorine pesticides and polychlorinated biphenyls in human adipose tissue. Rev Environ Contam Toxicol 1991; 120: 1–82.
79. Quintana PJ, Delfino RJ, Korrick S, et al. Adipose tissue levels of organochlorine pesticides and polychlorinated biphenyls and risk of non-Hodgkin's lymphoma. Environ Health Perspect 2004; 112: 854–61.
80. Arsenescu V, Arsenescu RI, King V, et al. Polychlorinated biphenyl-77 induces adipocyte differentiation and proinflammatory adipokines and promotes obesity and atherosclerosis. Environ Health Perspect 2008; 116: 761–8.
81. Fujiyoshi PT, Michalek JE, Matsumura F. Molecular epidemiologic evidence for diabetogenic effects of dioxin exposure in U.S. Air force veterans of the Vietnam war. Environ Health Perspect 2006; 114: 1677–83.
82. Grun F, Watanabe H, Zamanian Z, et al. Endocrine-disrupting organotin compounds are potent inducers of adipogenesis in vertebrates. Mol Endocrinol 2006; 20: 2141–55.
83. Grun F, Blumberg B. Environmental obesogens: organotins and endocrine disruption via nuclear receptor signaling. Endocrinology 2006; 147: S50–S55.
84. Kanayama T, Kobayashi N, Mamiya S, et al. Organotin compounds promote adipocyte differentiation as agonists of the peroxisome proliferator-activated receptor gamma/retinoid X receptor pathway. Mol Pharmacol 2005; 67: 766–74.
85. Newbold RR, Padilla-Banks E, Jefferson WN, Heindel JJ. Effects of endocrine disruptors on obesity. Int J Androl 2008; 31: 201–8.
86. Newbold RR, Padilla-Banks E, Snyder RJ, Jefferson WN. Perinatal exposure to environmental estrogens and the development of obesity. Mol Nutr Food Res 2007; 51: 912–17.
87. Newbold RR, Padilla-Banks E, Snyder RJ, Developmental exposure to endocrine disruptors and the obesity epidemic. Reprod Toxicol 2007; 23: 290–6.

88. Speakman J, Hambly C, Mitchell S, Krol E. Animal models of obesity. Obes Rev 2007; 8 (Suppl 1): 55–61.
89. Dickies MM. A new viable yellow mutation in the house mouse. J Hered 1962; 53: 84–6.
90. Yen TT, Gill AM, Frigeri LG, et al. Obesity, diabetes, and neoplasia in yellow A(vy)/- mice: ectopic expression of the agouti gene. FASEB J 1994; 8: 479–88.
91. Bultman SJ, Michaud EJ, Woychik RP. Molecular characterization of the mouse agouti locus. Cell 1992; 71: 1195–204.
92. Ellacott KL, Cone RD. The role of the central melanocortin system in the regulation of food intake and energy homeostasis: lessons from mouse models. Philos Trans R Soc Lond B Biol Sci 2006; 361: 1265–74.
93. Carroll L, Voisey J, van Daal A. Mouse models of obesity. Clin Dermatol 2004; 22: 345–9.
94. Kleyn PW, Fan W, Kovats SG, et al. Identification and characterization of the mouse obesity gene tubby: a member of a novel gene family. Cell 1996; 85: 281–90.
95. Naggert JK, Fricker LD, Varlamov O, et al. Hyperproinsulinaemia in obese fat/fat mice associated with a carboxypeptidase E mutation which reduces enzyme activity. Nat Genet 1995; 10: 135–42.
96. Rankinen T, Zuberi A, Chagnon YC, et al. The human obesity gene map: the 2005 update. Obesity 2006; 14: 529–644.
97. Rossant J, McMahon A. "Cre"-ating mouse mutants-a meeting review on conditional mouse genetics. Genes Dev 1999; 13: 142–5.
98. Hen G, Yosefi S, Simchaev V, et al. The melanocortin circuit in obese and lean strains of chicks. J Endocrinol 2006; 190: 527–35.
99. Dietrich HM. Housing, breeding and selecting chickens of the Obese strain (OS) with spontaneous autoimmune thyroiditis. Lab Anim 1989; 23: 345–52.
100. Schlegel A, Stainier DY. Lessons from "lower" organisms: what worms, flies, and zebrafish can teach us about human energy metabolism. PLoS Genet 2007; 3: e199.
101. Buettner R, Parhofer KG, Woenckhaus M. Defining high-fat-diet rat models: metabolic and molecular effects of different fat types. J Mol Endocrinol 2006; 36: 485–501.
102. Buettner R, Scholmerich J, Bollheimer LC. High-fat diets: modeling the metabolic disorders of human obesity in rodents. Obesity 2007; 15: 798–808.
103. Nagy TR, Clair AL. Precision and accuracy of dual-energy X-ray absorptiometry for determining in vivo body composition of mice. Obes Res 2000; 8: 392–8.
104. Bartness TJ, Kay Song C, Shi H, et al. Brain-adipose tissue cross talk. Proc Nutr Soc 2005; 64: 53–64.
105. Cancello R, Tounian A, Poitou C, Clement K. Adiposity signals, genetic and body weight regulation in humans. Diabetes Metab 2004; 30: 215–27.
106. Turtzo LC, Lane MD. Completing the loop: neuron-adipocyte interactions and the control of energy homeostasis. Horm Metab Res 2002; 34: 607–15.
107. Pan W, Kastin AJ. Interactions of cytokines with the blood-brain barrier: implications for feeding. Curr Pharm Des 2003; 9: 827–31.
108. Pan W, Kastin AJ, Rigai T, et al. Increased hippocampal uptake of tumor necrosis factor alpha and behavioral changes in mice. Exp Brain Res 2003; 149: 195–9.
109. Rogers J, Mastroeni D, Leonard B, et al. Neuroinflammation in Alzheimer's disease and Parkinson's disease: are microglia pathogenic in either disorder? Int Rev Neurobiol 2007; 82: 235–46.
110. Wyss-Coray T, Mucke L. Inflammation in neurodegenerative disease—a double-edged sword. Neuron 2002; 35: 419–32.
111. Zipp F, Aktas O. The brain as a target of inflammation: common pathways link inflammatory and neurodegenerative diseases. Trends Neurosci 2006; 29: 518–27.
112. Brown R, Thompson HJ, Imran SA, et al. Traumatic brain injury induces adipokine gene expression in rat brain. Neurosci Lett 2008; 432: 73–8.
113. Kalmijn S. Fatty acid intake and the risk of dementia and cognitive decline: a review of clinical and epidemiological studies. J Nutr Health Aging 2000; 4: 202–7.
114. Molteni R, Barnard RJ, Ying Z, et al. A high-fat, refined sugar diet reduces hippocampal brain-derived neurotrophic factor, neuronal plasticity, and learning. Neuroscience 2002; 112: 803–14.
115. Zhang X, Dong F, Ren J, et al. High dietary fat induces NADPH oxidase-associated oxidative stress and inflammation in rat cerebral cortex. Exp Neurol 2005; 191: 318–25.
116. Solfrizzi V, D'Introno A, Colacicco AM, et al. Dietary fatty acids intake: possible role in cognitive decline and dementia. Exp Gerontol 2005; 40: 257–70.
117. Calabrese V, Guagliano E, Sapienza M, et al. Redox regulation of cellular stress response in aging and neurodegenerative disorders: role of vitagenes. Neurochem Res 2007; 32; 757–73.

118. Anson RM, Guo Z, de Cabo R, et al. Intermittent fasting dissociates beneficial effects of dietary restriction on glucose metabolism and neuronal resistance to injury from calorie intake. Proc Natl Acad Sci USA 2003; 100: 6216–20.
119. Qin L, Wu X, Block ML, et al. Systemic LPS causes chronic neuroinflammation and progressive neurodegeneration. Glia 2007; 55: 453–62.
120. Palaniyandi R. Modulation of inflammation in fat tissue: Potential involvement in chemically induced neurodegeneration [Thèse de doctorat]. Saint Denis: Université de La Réunion. 2007: 171.
121. Zametkin AJ, Zoon CK, Klein HW, Munson S. Psychiatric aspects of child and adolescent obesity: a review of the past 10 years. J Am Acad Child Adolesc Psychiatry 2004; 43: 134–50.
122. Li Y, Dai Q, Jackson JC, Zhang J. Overweight Is Associated With Decreased Cognitive Functioning Among School-age Children and Adolescents. Obesity 2008, (Silver Spring).
123. Whitmer RA. The epidemiology of adiposity and dementia. Curr Alzheimer Res 2007; 4: 117–22.
124. Gustafson D, Rothenberg E, Blennow K, et al. An 18-year follow-up of overweight and risk of Alzheimer disease. Arch Intern Med 2003; 163: 1524–8.
125. Gustafson D. Adiposity indices and dementia. Lancet Neurol 2006; 5: 713–20.
126. Gustafson D, Lissner L, Bengtsson C, et al. A 24-year follow-up of body mass index and cerebral atrophy. Neurology 2004; 63: 1876–81.
127. Gustafson DR, Steen B, Skoog I. Body mass index and white matter lesions in elderly women. An 18-year longitudinal study. Int Psychogeriatr 2004; 16: 327–36.
128. Birman S. La drosophile, un modèle génétique pour l'étude des maladies neurodégénératives. Med Sci (Paris) 2000; 16: 164–70.
129. Schmidt E, Seifert M, Baumeister R. Caenorhabditis elegans as a model system for Parkinson's disease. Neurodegener Dis 2007; 4: 199–217.
130. Sriram K, Benkovic SA, Miller DB, O'Callaghan JP. Obesity exacerbates chemically induced neurodegeneration. Neuroscience 2002; 115: 1335–46.
131. Choi JY, Jang EH, Park CS, Kang JH. Enhanced susceptibility to 1-methyl-4-phenyl-1,2,3,6-tetrahydropyridine neurotoxicity in high-fat diet-induced obesity. Free Radic Biol Med 2005; 38: 806–16.
132. Altman J. Are new neurons formed in the brains of adult mammals? Science 1962; 135: 1127–8.
133. Richards LJ, Kilpatrick TJ, Bartlett PF. De novo generation of neuronal cells from the adult mouse brain. Proc Natl Acad Sci USA 1992; 89: 8591–5.
134. Gage FH, Coates PW, Palmer TD, et al. Survival and differentiation of adult neuronal progenitor cells transplanted to the adult brain. Proc Natl Acad Sci USA 1993; 92: 11879–83.
135. Okano H, Sakaguchi M, Ohki K, et al. Regeneration of the central nervous system using endogenous repair mechanisms. J Neurochem 2007; 102: 1459–65.
136. Kokoeva MV, Yin H, Flier JS. Evidence for constitutive neural cell proliferation in the adult murine hypothalamus. J Comp Neurol 2007; 505: 209–20.
137. Brown J, Cooper-Kuhn CM, Kempermann G, et al. Enriched environment and physical activity stimulate hippocampal but not olfactory bulb neurogenesis. Eur J Neurosci 2003; 17: 2042–6.
138. van Praag H, Christie BR, Sejnowski TJ, Gage FH. Running enhances neurogenesis, learning, and long-term potentiation in mice. Proc Natl Acad Sci USA 1999; 96: 13427–31.
139. Liu J, Solway K, Messing RO, Sharp FR. Increased neurogenesis in the dentate gyrus after transient global ischemia in gerbils. J Neurosci 1998; 18: 7768–78.
140. Bengzon J, Kokaia Z, Elmer E, et al. Apoptosis and proliferation of dentate gyrus neurons after single and intermittent limbic seizures. Proc Natl Acad Sci USA 1997; 94: 10432–7.
141. Harry GJ, McPherson CA, Wine RN, et al. Trimethyltin-induced neurogenesis in the murine hippocampus. Neurotox Res 2004; 5: 623–7.
142. Chen H, Tung YC, Li B, et al. Trophic factors counteract elevated FGF-2-induced inhibition of adult neurogenesis. Neurobiol Aging 2007; 28: 1148–62.
143. Kokoeva MV, Yin H, Flier JS. Neurogenesis in the hypothalamus of adult mice: potential role in energy balance. Science 2005; 310: 679–83.
144. Lindqvist A, Mohapel P, Bouter B, et al. High-fat diet impairs hippocampal neurogenesis in male rats. Eur J Neurol 2006; 13: 1385–8.
145. Cotman CW, Berchtold NC, Christie LA Exercise builds brain health: key roles of growth factor cascades and inflammation. Trends Neurosci 2007; 30: 464–72.
146. Carro E, Trejo JL, Busiguina S, Torres-Aleman I. Circulating insulin-like growth factor I mediates the protective effects of physical exercise against brain insults of different etiology and anatomy. J Neurosci 2001; 21: 5678–84.

147. Rojas Vega S, Abel T, Lindschulten R, et al. Impact of exercise on neuroplasticity-related proteins in spinal cord injured humans. Neuroscience 2008; 153: 1064–70.
148. Guo M, Cox B, Mahale S, et al. Pre-ischemic exercise reduces matrix metalloproteinase-9 expression and ameliorates blood-brain barrier dysfunction in stroke. Neuroscience 2008; 151: 340–51.
149. Ding Y, Li J, Luan X, et al. Exercise pre-conditioning reduces brain damage in ischemic rats that may be associated with regional angiogenesis and cellular overexpression of neurotrophin. Neuroscience 2004; 124: 583–91.
150. Grondard C, Biondi O, Armand AS, et al. Regular exercise prolongs survival in a type 2 spinal muscular atrophy model mouse. J Neurosci 2005; 25: 7615–22.
151. Biondi O, Grondard C, Lecolle S, et al. Exercise-induced activation of NMDA receptor promotes motor unit development and survival in a type 2 spinal muscular atrophy model mouse. J Neurosci 2008; 28: 953–62.
152. O'Dell SJ, Gross NB, Fricks AN, et al. Running wheel exercise enhances recovery from nigrostriatal dopamine injury without inducing neuroprotection. Neuroscience2007; 144: 1141–51.
153. Tillerson JL, Caudle WM, Reveron ME, Miller GW. Exercise induces behavioral recovery and attenuates neurochemical deficits in rodent models of Parkinson's disease. Neuroscience 2003; 119: 899–911.
154. McCrate ME, Kaspar BK. Physical activity and neuroprotection in amyotrophic lateral sclerosis. Neuromolecular Med 2008; 10: 108–17.
155. Ramsden M, Berchtold NC, Kesslak JP, et al. Exercise increases the vulnerability of rat hippocampal neurons to kainate lesion. Brain Res 2003; 971: 239–44.
156. Oliff HS, Berchtold NC, Isackson P, Cotman CW. Exercise-induced regulation of brain-derived neurotrophic factor (BDNF) transcripts in the rat hippocampus. Brain Res Mol Brain Res 1998; 61: 147–53.
157. Zaheer A, Haas JT, Reyes C, et al. GMF-knockout mice are unable to induce brain-derived neurotrophic factor after exercise. Neurochem Res 2006; 31: 579–84.
158. Kadoglou NP, Iliadis F, Angelopoulou N. The anti-inflammatory effects of exercise training in patients with type 2 diabetes mellitus. Eur J Cardiovasc Prev Rehabil 2007; 14: 837–43.
159. Kadoglou NP, Perrea D, Iliadis F, et al. Exercise reduces resistin and inflammatory cytokines in patients with type 2 diabetes. Diabetes Care 2007; 30: 719–21.
160. Moschen AR, Kaser A, Enrich B, et al. Visfatin, an adipocytokine with proinflammatory and immunomodulating properties. J Immunol 2007; 178: 1748–58.
161. Beltowski J. Apelin and visfatin: unique "beneficial" adipokines upregulated in obesity? Med Sci Monit 2006; 12: RA112–19.
162. Bokarewa M, Nagaev I, Dahlberg L, et al. Resistin, an adipokine with potent proinflammatory properties. J Immunol 2005; 174: 5789–95.
163. Nagaev I, Bokarewa M, Tarkowski A, Smith U. Human resistin is a systemic immune-derived proinflammatory cytokine targeting both leukocytes and adipocytes. PLoS ONE 2006; 1: e31.
164. Quaye IK. Haptoglobin, inflammation and disease. Trans R Soc Trop Med Hyg 2008.
165. Anderson PD, Mehta NN, Wolfe ML, et al. Innate immunity modulates adipokines in humans. J Clin Endocrinol Metab 2007; 92: 2272–9.
166. Myers LK, Bhattacharya SD, Herring PA, et al. The isozyme-specific effects of cyclooxygenase-deficiency on bone in mice. Bone 2006; 39: 1048–52.
167. Leonard BE. Inflammation, depression and dementia: are they connected? Neurochem Res 2007; 32: 1749–56.
168. Lee YH, Tharp WG, Maple RL, et al. Amyloid Precursor Protein Expression Is Upregulated in Adipocytes in Obesity. Obesity (Silver Spring), 2008.

8 Microglia: Neuroprotective and Neurodestructive Properties

G. Jean Harry

Neurotoxicology Group, Laboratory of Molecular Toxicology, National Institute of Environmental Health Sciences, National Institutes of Health, Department of Health and Human Services, Research Triangle Park, North Carolina, U.S.A.

INTRODUCTION

In the mid-nineteenth century, Virchow provided an early description of neuroglial. Following this, others observed non-neural cells in the brain that were thought to be due to the infiltration of the developing nervous system by mesodermally derived cells. In 1899, Nissl identified cells in the brain with rod-shaped nuclei and named them Staebchenzellen (1). Ramon y Cajal (2) identified this group of cells as the third element of the central nervous system as they were morphologically distinct from neurons and astrocytes. These cells were further distinguished as microglia and oligodendrocytes by del Rio Hortega. Microglia were identified as a distinct cell type (3) originating from mononuclear cells of the circulating blood (4). The current view is that microglia are derived from circulating monocytes or precursor cells in the monocytes-macrophage lineages, which originate in the bone marrow (5–8). The early work of del Rio Hortega suggested that microglia have the ability to transform from a resting ramified phenotype into an amoeboid phagocytic phenotype. In this early work on microglia, morphological classifications fell into three types: ramified, intermediate forms, and amoeboid (4,9). More recent studies have used a similar classification but these terms have lead to the attachment of an additional functional connotation with resting, activated, and amoeboid phagocytic microglia (10,11). The accumulation of more recent data suggests that it is likely that the earlier classification relying on morphological features will end up being the more correct form of a classification system given the heterogeneity of functional activities of the cells (Fig. 1)

IDENTIFICATION OF MICROGLIA

In his initial work, del Rio Hortega (12) used a weak silver carbonate method to selectively stain microglia that required very specific fixation procedures. While this method can be used for electron microscopy examination, it is variable with regards to the number of microglia stained and across species. Enzyme histochemical methods have been successfully employed for staining microglia. Thiamine pyrophosphatase activity is associated with the plasma membrane of microglial cells and with blood vessels in the CNS (13). Nonspecific esterase is an enzyme used to identify microglia in cultures (14); however, it has limited use in tissue sections. For example, it does not label reactive microglia in the hypoglossal nucleus after peripheral nerve transection (15), yet it can be found in brain macrophages within a stab-wound. This distinct staining pattern in vivo raised concerns that microglia isolated and cultured represent cells that have transformed into brain macrophages as a consequence of the culture process (see following section). Activated phagocytic microglia (macrophages) both in vivo and in culture can be detected based upon increased enzyme activity. Such enzymes include acid phosphtase, 5′-nucleotidase, and oxidoreductase (16). Activated microglia can express other markers such as nitric oxide synthase, lysosomal proteinases, plasminogen activator, lysozyme, urine nucleoside phosphorylase, and elastase (17–19); however, these enzymes are not specific for microglia and can be expressed by other glial cell types.

Immunohistochemical methods to detect microglia are often based on the fact that the microglial plasma membrane contains a large number of receptors and adhesion molecules. Immunological detection of surface antigens on microglia is primarily based upon expression on cells of the immune system, macrophages, thymocytes, and lymphocytes. They share phenotypic characteristics and lineage-related properties with bone-marrow-derived monocytes and macrophages. Such properties include the ability to secrete cytokines common to immune accessory cells and, while significantly less than that seen in peripheral organs, microglia serve

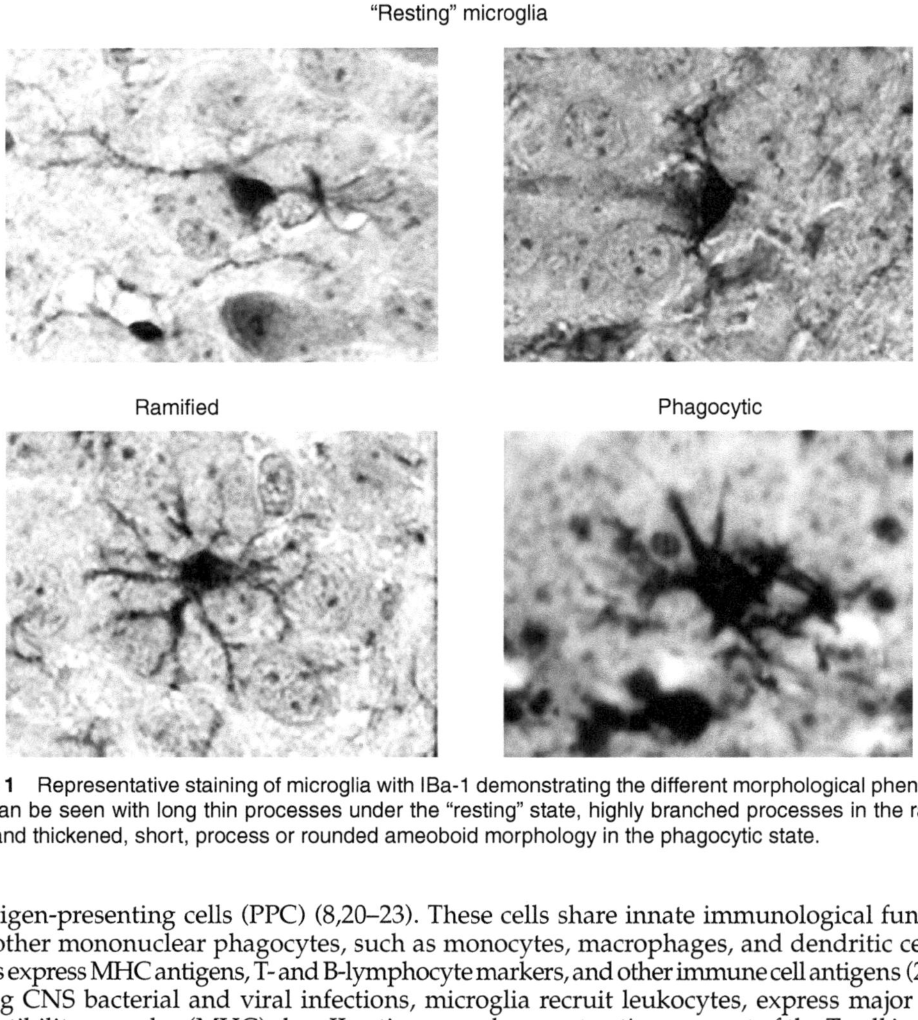

Figure 1 Representative staining of microglia with IBa-1 demonstrating the different morphological phenotypes. Cells can be seen with long thin processes under the "resting" state, highly branched processes in the ramified state, and thickened, short, process or rounded ameoboid morphology in the phagocytic state.

as antigen-presenting cells (PPC) (8,20–23). These cells share innate immunological functions with other mononuclear phagocytes, such as monocytes, macrophages, and dendritic cells, as well as express MHC antigens, T- and B-lymphocyte markers, and other immune cell antigens (24,25). During CNS bacterial and viral infections, microglia recruit leukocytes, express major histocompatibility complex (MHC) class II antigens, and present antigen as part of the T-cell immune response. However, while microglia serve as the most likely candidate for the antigen-presenting cell (APC) in the CNS, this capability is significantly less as compared with that seen in APCs from peripheral organs (21–23). The other macrophage of the CNS, the perivascular cell, lies at the interface between the CNS parenchyma and the blood (26) and, by contrast, is a strong APC (8). Microglia constitutively express the beta2-integrins (CD11a—the ligand for ICAM, and the complement proteins—CD11b, CD11c, and CD18). The monoclonal antibody F4/80 can detect a mouse macrophage specific antigen on resting microglial (27,28). Immunoglobulin Fc gamma receptors are highly expressed on phagocytic cells, including phagocytic microglia, as has been identified by antibodies to 2.4G2, and complement receptors by antibody to Mac-1. Ramified microglia in the rat brain can be demonstrated with OX-42 antibody against the CR3 complement receptor (29); however, this method is not successful in the mouse brain.

Tyrosine kinases are known to be associated with surface receptors, thus the detection of microglia with antibodies to phosphotyrosine (30) is consistent with the multiple cell surface receptors on microglia. They express the leukocyte common antigen (LCA; CD45) representative of a family of tyrosine protein phosphatase receptors. Antigens of the MHC have been localized to microglia (31–33). Reactive microglia express MHC class I glycoproteins, e.g., HLA-ABC, Beta-2-microglobulin. MCH class II glycoprotein (e.g., HLA-DR, HLA-DP, and HLA-DQ) expression is increased with pathology and with a systemic injection of gamma interferon (34–36); they are also expressed on endothelial cells and cells within the walls of the cerebral blood vessels. The multi-antigenicity of the microglia membrane has lead to the use of a number of antibodies for detection (37). However, specificity for resting or resident microglia is defeated by the cross reactivity to other macrophages. Thus, after much effort it can be stated that an antibody that binds to a resting microglia will also stain an activated microglia and a brain macrophage. The reverse, however, is not true in that an antibody for a brain macrophage will not automatically detect a reactive or activated microglia or a resting microglia. An example is the observation that resting microglia are vimentin negative; however, they quickly and transiently upregulate this intermediate filament protein as part of their response to injury (38). Other markers have included the multiarched ribonucleoprotein particles, vaults (which are enriched in microglia during development but absent in the mature cell), as well as the iron storage protein, ferritin, suggesting that microglia are involved in iron metabolism. In fact, recent work suggests that ferritin-staining pattern in microglia changes with aging and is associated with functional changes in the cell (39). Lectin histochemistry has been very successful in staining microglia by detecting the lectin-binding glycoproteins on the cell surface (40).

Amoeboid microglia in the corpus callosum have been labeled with an intraperitoneal injection of the fluorescent dye, rhodamine isothiocyanate (41). Other labeling efforts have used tracer substances within neurons to label the phagocytic microglia such as carbocyanine dye, DiI (42), and Fluoro-Gold (43).

DEVELOPMENT AND DISTRIBUTION OF MICROGLIA

Microglia can change their morphology during development, in response to acute injury, and with pathological states. Microglia are highly heterogeneous within the healthy CNS. They comprise only 10% of the total cell population of the brain but have multiple morphological and possible functional profiles as influenced by their environment. During brain development, the resident population of microglia changes as a function of maturation. Prior to the formation of the blood-brain-barrier, blood-borne monocytes are able to migrate into the brain, maturing into resident microglia. In the human, microglia are present as early as 13 weeks of gestation (44), and between 13 and 18 weeks of gestation there is a functional and morphological variability in the cells (45). In the cortex, they display a ramified morphology while, in the germinal matrix, they display a more amoeboid morphology. Developmentally, this is a period of active brain remodeling and the microglia are believed to participate in the clearance of apoptotic cells (46). A more aggressive role of microglia is seen in inducing neuronal death during normal development in the cerebellum (47). With birth and during the first few postpartum weeks, the number of nascent round and amoeboid microglia decrease and an increase is seen in highly ramified cells bearing long, thin, branched processes (48,49). In the rodent, microglia develop primarily after birth and mainly between postnatal days 5 to 20, the time when microglia staining is increasing throughout the entire CNS (7,50).

The classical bone-marrow-derived microglial cells have a small soma with arborized or ramified processes and reside in the brain parenchyma. Under normal conditions, these cells do not display macrophage-like functions and serve as resident immune cells with a very slow turnover rate. The developmental maturation of microglia has also been considered related to the developmental regulation of cytokine gene expression that occurs in a region specific manner in the brain and influences normal cell maturation (51–53). However, a pro-inflammatory cytokine response to injury has also been demonstrated in the perinatal brain (54,55) suggesting that, even in the developmental stages, at least a subpopulation of microglia serve in a more mature role.

A very extensive examination by Lawson (50) showed a relatively uniform distribution of microglia in the adult rodent brain, (with the exception of a higher relative number of microglia with similar phenotype in dentate gyrus of the hippocampus, the substantia nigra, and portions of the basal ganglia), with the highest number of microglia in the olfactory telencephalon. In addition, the cells modify their morphology and expression of cell surface antigens depending upon their environment (50). Ramified microglia in gray matter show processes extending in multiple directions while, in the white matter, the cells align their cytoplasmic extensions in parallel, or at right angles to nerve processes. In the normal brain, microglia immunopositive for MHC class II or CD4 antigens are localized preferentially in the white matter (10,56). In addition, microglia cells show a different phenotype in brain regions lacking a blood-brain-barrier, such as the circumventricular organs (57), suggesting that serum proteins can have an influence on the resident microglia population.

In the normal brain, "resting" microglia are ramified (highly branched) with a small amount of perinuclear cytoplasm and a small dense and heterochromatic nucleus. These cells constitutively express CR3 complement receptor (which can be recognized by the OX-42 antibody in the rat) and bind lectins with specificity for glactose residues. These cells are located outside of the vascular basement membrane; however, the cytoplasmic processes can be found intermingled with astrocytic foot processes in the perivascular glia limitans (58). Microglia found in the vicinity of blood vessels have been often termed perivascular microglia. These microglia are not perivascular cells, which are not ramified but show an elongated shape and are not part of the parenchyma but rather are components of the vascular wall (59). However, both cell types are phagocytic and may express MHC antigens (10). This can often present difficulty in distinguishing between the two cell types in pathological conditions especially if there is damage to the vascular cell wall.

Thus, the brain has two indigenous sources of brain macrophages—a general term encompassing all phagocytic cells, including blood-derived monocytes entering the brain upon vascular injury. Macrophage cells exist in various states of activation from pro-inflammatory, alternatively activated, and even anti-inflammatory. In addition, within any specific stage of the inflammatory response, macrophages modify their phenotype (60). The functional changes of activated microglia are often accompanied by a morphological transformation leading from cells with thin, ramified processes to cells with larger somata and shorter and coarser cytoplasmic processes. This can eventually progress to amoeboid cells with morphology similar to macrophages. There is additional evidence to suggest that individual cells are not required to follow this progression but rather can immediately differentiate into amoeboid cells. In vivo, the transformation of resident microglia into those with a phagocytic phenotype is strictly regulated and occurs in response to cell death or accumulated debris. While the morphology can be important in determining a response and role of the microglia, these cells can also rapidly shift their functional phenotype in the absence of any change in morphology. In addition, there are a multitude of changes that can occur including secretion of soluble products, as well as, electrophysiological changes. Unfortunately, with the exception of a fully activated microglia showing an amoeboid morphology consistent with phagocytic activity, to date there is little functional information to accompany these morphological differences. However, this is beginning to change with the increased awareness of microglia heterogeneity and knowledge that the ability to directly translate effects seen in culture systems to the in vivo state is limited. Characterizing the heterogeneity and temporal pattern of such changes becomes important in interpreting data on the nature of the microglia response.

MICROGLIA AGING AND SENESCENCE

The progressive, age-related change in the expression of MHC II antigens by microglia led to the assumption that microglial activation occurs with aging. The pattern of expression was reported to be similar to what occurs in experimental animal models of injury. However, in the aged brain, neuronal injury and death, which stimulate microglia activation, are not apparent (61). Further investigation showed morphological changes of microglia cytoplasmic structure in the aged human brain reflective of dystrophy and senescence (62). Thus, while the cells change as a function of age, the morphological phenotype is suggestive of a decreased ability to mount a normal host-resistance response rather than a transformation to an activated cell. With an increasing focus on aged animals, such a shift is beginning to be reported. It has been suggested that age-related neurodegeneration might not only be due to a loss of neuroprotective properties, but also the actual loss of microglia (63). A recent hypothesis put forth to explain the phenotypic changes in aging microglia suggest that microglial senescence is triggered by an intracellular oxidative stress response due to enhanced intracellular accumulation of iron (63). Under this hypothesis, a prolonged oxidative stress response would lead to senescence rather than activation. Additional support for microglial replicative senescence in normal and pathological aging comes from the report that microglial cells exhibit telomere shortening and decreased telomerase activity with aging (64). Whether this occurs with age related diseases remains in question however, preliminary work suggests that microglia from AD brains have shorter telomerases as compared to control brains (65). This would be in line with the proposed mechanism that β-amyloid may promote microglial deterioration and accelerate senescence (39). If microglia are functionally impaired by senescence they would also show decreased ability to produce neurotrophic factors, impaired phagocytosis, impaired protein clearance, and, thus, increased neurotoxicity (63,66). As an example, hippocampal damage induced by the organometal, trimethyltin (TMT), in mice expressing the apolipoprotein E4 as a risk factor for AD, showed no difference in the microglia and associated cytokine response to injury in adolescent mice. However, by eight months of age a genotype related shift was observed in that both the *APOE* knockout mice and the *APOE4* transgenic mice showed a muted TNFα and MIP-1α response with neuronal injury (67). This was suggestive of a decreased ability to mount an appropriate response to injury.

INFILTRATING CELLS THAT INFLUENCE MICROGLIA

The CNS is unique in two regards to the immune system in that there is the lack of a lymphatic system to capture potential antigens and the presence of the blood-brain-barrier, a vascular endothelial system interposed between blood borne cells and the brain parenchyma sealed with tight junctions. Under normal condition, this blood-brain barrier prevents molecules in the brain extracellular space from gaining access to the vascular lumen. With a physical injury such as trauma or ischemia, or during an autoimmune disease, the blood brain barrier is disrupted and blood-borne immune cells gain access to the brain parenchyma. The extravasating peripheral APCs then present brain antigens to lymphocytes. Upon activation, T cell can enter the CNS (68) regardless of their antigen specificity but only T cells that recognize CNS antigen persist (69). It has been proposed that upon T-cells entering the brain, a series of events is initiated. This process includes the activation of endothelial cells, induction of pro-inflammatory cytokines, such as TNFα, IL-1, and IFNγ, and recruitment of additional lymphocytes and monocytes in an antigen-independent manner. With ischemia, the predominant infiltrating cells are polymorphonuclear neutrophils (PMNs) that contribute to tissue damage by the release of oxygen radicals, proteases, and proinflammatory cytokines, like tumor necrosis factor alpha (TNFα). While some studies had demonstrated that the prevention of PMN infiltration into the injured parenchyma is neuroprotective (70) other studies have failed to demonstrate a clear cause-effect relationship (71). More recently, the interaction between microglia and PMNs in ischemia has been characterized with demonstration of the direct engulfment of invading neutrophil granulocytes by resident microglia, thus providing a significant neuroprotective action (72).

While, the entry into the brain is limited and delayed as compared to peripheral tissue (73), blood-derived macrophages may enter brain tissue, react upon it, and then return to the circulation in the presence or absence of injury (74,75) or they may differentiate to the microglial morphology and remain in the brain tissue for an extended period of time. In this state they can persist until destroyed by senescence or prompted to move back into the circulation.

The inability to discriminate between the two cell types of brain macrophages limits our understanding of the role and impact of the resident microglia. In an attempt to discriminate the response of resident microglia/macrophages from those of blood-borne origin, Sedgwick et al. (76) developed a flow cytometry method to show that both cell types expressed CD11b but that infiltrating cells express high levels of leukocyte common antigen (CD45) while the resident microglia express CD45 at a low level. This discrimination between the two populations was confirmed in later studies (77,78) and has lead to further understanding of the nature of the two cell populations. An additional approach to identify the contribution of infiltrating cells is the generation of bone-marrow chimeria mice using fluorescently tagged bone marrow cells for reconstitution after body radiation. Within four to six months, a full reconstitution will then allow one to determine the contribution of infiltrating bone-marrow derived cells to an injury site. While these types of studies are very powerful with regards to the possibility of genetic manipulation they are complicated and require a significant level of experimental control for accurate data interpretation (79). However, they have demonstrated that the majority of the models on which the field has relied to examine a microglia response and neuroinflammation are compromised with an alteration in the blood-brain-barrier and the infiltration of blood-borne cells.

The recruitment of monocytes from the blood through the blood brain barrier and their differentiation into a microglia phenotype could have direct implications for the etiology of many neurotoxic events and neurodegenerative diseases. These cells of the peripheral immune system provide an enriched source of cytokine and inflammatory factors as compared to resident microglia. An impact can be driven not only from the direct influence of infiltrating cells but also by any modifications of the peripheral immune cells as a result of exposure or disease. As in other organ systems, there are multiple contributions to, as well as outcomes from, an immune-mediated response in the CNS. For example, we now know that molecules of the systemic innate immune system are able to stimulate immune cells of the brain as well as directly impact neuronal populations. While the exact level of influence this has on the intact brain has not been clearly determined, the fact that the CNS is not impervious to cellular or soluble pathogens and immunogens from the circulation presents a different framework than originally held. Within this framework, one could envision that systemic modulating factors such as stress, immune-status, and genetic background may influence the normal neuroimmune interactions and shift the normal response to injury.

FUNCTIONS OF MICROGLIA

Monitoring Sentinal and Scavenger

Rather than maintaining a "resting" state, normal microglia dynamically survey the CNS (80–83). They maintain a critical relationship with other cells in the brain and are in intimate contact with neurons for which they serve important maintenance functions. In the brain, microglia play a critical role as resident immunocompetent and phagocytic cells (84,85). Parenchymal microglia express scavenger receptors and apoptosis-recognition components. For example, surveillant microglia express P2Y12, a receptor for ATP and ADP, that mediates the cell's process extension in reaction to tissue damage (86). Microglia serve a critical function in scavenging and clearing unwanted proteins and material from the brain including apolipoprotein E, amyloid beta (Aβ), Huntintin, parkin, and prion. With regards to the clearance of normal protein products in the brain, phagocytosis of amyloid beta by cultured microglia cells is dependent upon the recruitment of the CD35 receptor to membrane rafts (87). This process is normally studied by examining the active phagocytosis of latex beads by microglia in culture following stimulation with the bacterial LPS, fungal zymozan, or alternative compounds of interest, such as isolated myelin proteins. Further confirmation of this process in vivo has been provided by various studies that have examined the phagocytosis of fluorescent tagged neuronal processed by microglia in addition to those that have identified morphologically distinct characteristics of phagocytizing macrophages within the brain. This general approach has been employed to study the cellular dynamics with regards to the phagocytic process.

Microglia respond rapidly and play an important role during injury and infection in the CNS (88–96). In the event of infection, inflammation, trauma, ischemia, and neuronal death, these cells serve in a scavenger role (25,97,98). After ischemic or traumatic injury and in demyelinating diseases, microglia remove cellular debris by phagocytosis (99,100). In each of these models, it is now known that the infiltrating T-cells participate in the local response to an acute or chronic CNS injury, in addition to resident microglia and infiltrating macrophages. In their role as phagocytic cells of the brain, microglia can facilitate the apoptosis and phagocytosis of infiltrating T-cells. They do this through various signaling pathways. By eliminating the presence and signals from T-cells this then contributes to a subsequent down regulation of the immune activation of microglia (101). However, T-cells primed to react to specific myelin antigens promote neuronal survival in animal models of nerve crush and other physical injuries (102) and have been shown to protect hippocampal neurons against chemical exposure to the organometal, TMT (103). Overall, these studies demonstrate that CNS responses previously considered to be the result of CNS tissue specific cells such as microglia may involve contributions from multiple cell types whose actions are significantly influenced by the systemic state of the organism. The reader is directed to the following references for further reading on the influence of systemic inflammation of neuroinflammation (104–107).

Role in Synapse Remodeling

An additional role proposed for microglia is their contribution in synapse remodeling with an active participation in synapse stripping. In an interesting series of studies, microglial motility was examined in a model of peripheral axotomy of the facial nerve (Schiefer 1999). In brain slices containing the axotomized facial nucleus, microglia cells accumulated at the surface of the slice. These microglia remained stationary but the cell soma developed undulating pseudopods of different shape and sizes that were used for phagocytosis of cell debris. Within the core of the slice, single microglial cells were seen to move along the proximal parts of neuronal dendrites maintaining cell-to-cell contact. These studies suggested a microglia component to the loss of synapses of these motorneurons. Additionally, Culhein and Thams (109) suggested that microglia contribute to plasticity after lesions by similar mechanisms to those used for pruning of excess synapses during development. In mice lacking specific MHC class I determinants, synaptic pruning is aberrant (110). Recent evidence suggests that the complement components C1q and C3 have a role in the selective tagging of supernumerary synapses in the developing visual system (111). Similar types of defects in visual system synaptic refinement in mice lacking complement components C1q or C3 (112) as those seen in mice deficient in MHC class I determinants (110).

Role in Promoting Repair

From the expanding number of studies on microglia functions we are beginning to appreciate the diverse nature of these cells within the various brain regions and their multiple responses to maintain their role of surveillance and rapid responder to regulate the neuronal environment (113). Considerable evidence demonstrates that microglia triggered by injured/dying neurons mediate a reduction of neuronal damage and induce tissue repair (114–118). For example, we now know that activated microglia serve as a source of trophic and growth factors (119,120). Brain derived neurotrophic factor, BDNF, can be released by activated microglia (121,122). This can support the survival and sprouting of dopaminergic neurons in the striatum (121). Microglia offer protection during excitotoxic injury (122,123) and microglial supernatants protect dissociated neurons from excitotoxic injury (124). The contact of neurons with microglia in culture changed the phenotype of microglia from toxic to protective (125). Microglia produce insulin like growth factor (IGF-1), which has been demonstrated to have neuroprotective properties. In addition, TNFα release induced by P2X7 receptor is neuroprotective (126). The grafting of microglia into injured spinal cord promotes neurite outgrowth and neuronal regeneration (127). The restricted expression of 5′-nucleotidase to perineural microglia, as identified by the 5N4-2 antibody suggested a function of the adhesion molecule on microglia participating in regeneration processes after axotomy (128).

In vivo, any neuroprotective effect of microglia is dependent upon a rapid response. It has been proposed that such a rapid response can occur, if the cells are activated by a controlled level of pro-inflammatory T-cells and their production of interferon gamma (IFN-g) (129). In addition, the pro-inflammatory cells would serve to recruit additional immune cells to serve in defense of the tissue. For example, recruitment of neuroprotective T-cells and microglial activation is associated with increased neuronal survival in an optic nerve crush injury (130,131). These examples serve as representative of the diverse conditions in which microglia, or a microglia factor, are found to provide significant beneficial effects.

MICROGLIA IN TISSUE CULTURE

Activation State of Microglia

A large amount of the data available on functional properties of microglia has been obtained from cells in culture. Yet, given that these are monocyte derived cells that have phagocytic properties, the simple act of culturing the cells results in cells that are different from those within the in vivo condition. In addition to the traumatic act of culturing the cells, there are a number of factors that contribute to this change. For example, the preparation of primary mixed glial cells from which microglia are derived results in a significant amount of tissue debris that can induce a phagocytic phenotype. The majority of the current isolation protocols rely on the previous methods established by McCarthy and DeVellis (132) as a model for the isolation of brain glia from post-natal day 1 to 2 pups and those of Giulian for additional purity of the individual cell types (133). These procedures employ multiple series of filtration steps in the initial tissue preparation to remove excess debris. In addition, the in vitro characterization from the McCarthy and DeVellis (132) procedure focuses on a distinct sub-dissection of the cortex in an attempt to minimize white matter contamination. An expansion of this has been to limit cell isolation from only the cortex. While this would combine both white and gray matter glia it prevents the contamination of the deeper brain structures. The collection of microglia from the mixed glia preparation is conducted based upon differential adhesion of the microglia such that they can be removed by vigorously shaking the culture preparation. The culture age for microglia collection can be critical as the cells change over time in the mixed culture. In addition, attempts to obtain additional cells by allowing the remaining mixed culture prep to incubate additional days will lead to a low yield of microglia in a "second shake" and cells that can respond differently to stimuli as compared to the initial isolation.

Serum Influence

Another consideration is the serum content of the growth medium for cultured cells. In contrast to other macrophages, microglia are shielded from serum proteins. There is recent evidence that serum constituents can selectively activate microglia (134,135). In addition, serum contains protein factors necessary for binding of LPS to its receptor CD14 (136). While serum is critical to the establishment of the glial cultures, between the presence of debris and serum, the cultured microglia transform into a brain macrophage phenotype. As in most culture systems of immune-cells, the endotoxin level within the serum is critical not only to the response of the cells but also survival of the cells. Work by Janke et al. (137) established an optimal range of endotoxin for the establishment of a microglia culture, yet maintaining a process bearing cell with very low TNFα mRNA and protein levels. Gebicke-Haerter et al. (138) reported that prolonged culturing of microglia in lipopolysaccharide-free media can result in process bearing cells. However, the processes are significantly less that that seen in vivo. Concern have been raised regarding the impact of serum and the possibility of endotoxin contamination in commercially available substances. For these reasons, the use of a macrophage defined media and confirming that test compounds are endotoxin free are recommended to ensure specificity in the microglia response.

Proliferation and Migration

Much of the information available on microglia has been obtained from culture and, while some of it has been confirmed in vivo, the exact level and relevance or generality of the responses are not necessarily translatable from in vitro to in vivo. The observation of increased microglial staining at an injury site led to the assumption that the cells maintained a capacity for proliferation and migration. For example, in order to be effective at the site of injury, it has been assumed that cell migration allows for a significant increase in microglia. While there are limited in vivo studies using ^{3}H-thymidine or bromodeoxyuridine (BrdU) to label proliferating microglia, these features have primarily been demonstrated in culture. With injuries that damage the blood-brain-barrier, the changes that have been attributed to microglia such as migration and proliferative capacity may be reflective of the infiltrating monocytes rather than the resident microglia. In any case, studies have identified specific microglia mitogens including: macrophage colony-stimulating factor (M-CSF), granulocyte/monocyte colony-stimulating factor (GM-CSF), brain-derived neurotrophic factor (BDNF), neurotrophin (NT)-3, interleukin (IL)-1, chemokines such as CX3CL1/fractalkine, and the purinergic receptor P2X7. Proliferative activity can be inhibited by other signaling molecules of the immune system and growth factors, including transforming growth factor (TGF)β1. Thus, the outcome depends on the cumulation of multiple stimulatory and regulatory actions.

Migration of microglia has been demonstrated in slice cultures and in acute brain slices (chapter 9). In these models, the cells retract their initial ramifications and extend new pseudopodia for migration (139,140). Two chamber assay systems and genetically modified mice have demonstrated that complement factor C5a, monocyte chemoattractant protein (MCP)-1, the chemokine CXCL10, and epidermal growth factor (EGF) can direct the chemotaxis process (141–143). Some of these factors, like MCP-1 and CXCL10, are expressed in neurons (144,145), with the corresponding CXCR3 receptor expressed on the microglia (146). To what degree and to what distance microglia migration occurs in vivo within the parenchyma has not been fully determined. While the existing data would suggest that this occurs, additional data demonstrating a rather dense population of microglia throughout the brain suggests that, while migration may occur, there are sufficient microglia within any specific brain structure for an initial response.

Cell Surface and Secreted Factors

In addition to these physical features, culture systems have provided a significant amount of information regarding cell surface expression and secreted factors. In vitro, microglia have been shown to be capable of releasing several potentially cytotoxic substances, e.g., free oxygen intermediates, nitric oxide, proteases, arachidonic-acid derivates, excitatory amino acids, quinolinic acid, prostaglandins, cytokines, and other factors that, if occurring in vivo, may contribute to the exacerbation of neurons already compromised by disease or injury. In fact, reactive oxygen signaling has been considered as a unifying mechanism underlying multiple forms of neuroinflammation and microglia responses. The generation of ROS, leading to the induction of the pro-inflammatory product, inducible nitric oxide synthase (iNOS) (147), which enhances nitric oxide (NO) production from glial and endothelial cells. In excess concentrations, NO forms peroxynitrite, and, through nitrosylation of cell signaling messengers, serves in immune regulation (148). However, many of these studies have been conducted in vitro and raise the question of how to translate the data to the in vivo condition. In fact, several reports suggest that astrocytic and neuronal factors are key in determining the toxic potential of microglia or macrophages (149,150).

Culture systems have also provided information of the ionic properties of microglia. In vivo, microglia are capable of responding to subtle changes in the microenvironment such as ion homeostasis. This responsiveness is possibly due to the membrane channels on microglia such as an inward-rectifying K+ channel (151,152), as well as the presence of receptors for ATP (153,154), calcitonin gene-related peptide (CGRP) (155), acetylcholine, and noradrenaline (156). Other work in culture suggests that microglia produced a significantly high amount of glutamate (157). However, further studies showed that when microglia were grown in a serum-free, glutamine-free defined medium the release of glutamate was minimal. Rather the addition of glutamine in the absence of glutamate uptake resulted in an increase in glutamate accumulation with time and cell density (158). Under serum free conditions, a number of microglia activating agents were tested and found to not induce glutamate production. With the addition of only 1% serum, 100 ng/ml LPS caused a 2–3-fold accumulation. In addition, the extensive work from Giulian et al. to identify the major neurotoxic component of microglia from brains subjected to trauma or ischemia did not identify glutamate as a neurotoxic factor (159).

Microglia Cytokine Expression and Regulation

One significant contribution from the in vitro studies was the observation that microglia secrete a number of cytokines such as interleukin 1 (IL-1) (160), interleukin 6 (IL-6) (161), and tumor necrosis factor (TNF) (162). They have also shown that microglia are responsive to growth factors such as granulocyte-macrophage colony stimulating factor (GM-CSF) and colony stimulating factor (CSF-1) (163,164). Microglia are capable of exhibiting a M1 proinflammatory phenotype as occurs following activation with lipopolysacharide (LPS), characterized by the secretion of proinflammatory cytokines, nitric oxide, and superoxide. They can also exhibit a M2 anti-inflammatory phenotype as occurs with interleukin (IL)-4, characterized by the secretion of neurotrophic factors such as insulin like growth factor-1 (IGF-1) and IL-10 (165). Balance between activation and suppression of immune cells is critical in the host response against insult and prolonged injury. Excessive or prolonged activation may result in hypersensitivity resulting in a more severe response to subsequent injury. In many cases, the level of production induced from primed cells is at a much greater level than that which would have been induced directly.

The role of the microglial cell as the immune cell of the brain requires a slight discussion of some of the related characteristics and regulatory factors that require consideration in determining the nature of the cytokine response. While both in vitro and in vivo studies have demonstrated microglia as a primary cell type expressing the pro-inflammatory cytokines, the translation of findings with regards to induction or with regards to impact upon the brain requires that the entire system be considered and if possible examined. For example, cytokines can induce or modulate a broad spectrum of cellular responses including cell adhesion, migration, survival, differentiation, replication, secretory function, and cell death. In the same cell, a cytokine can either induce death or promote survival or proliferation dependent upon the function of the context in which it acts. Cell death can be actively triggered by a cytokine (166–168) as a physiologically necessary role during development, differentiation, and in many host defense responses to injury and subsequent repair. However, the cytokine can be providing a critical survival factor thus, its absence can result in cell death at a specific point in the cell cycle.

To understand the impact of a cytokine response, the basal level and regulatory mechanisms need to be considered. Under normal homeostatic conditions, the majority of cytokines are either not expressed or, if they are, are expressed at only low levels. In the early phase, host defense responses to injury are triggered by microbial antigens, extracellular matrix antigens, or by activation products or preformed components of humoral defense systems. This process consists of cell recruitment and the induction of those cytokines that contribute to activation of proinflammatory, immunostimulatory, and catabolic responses. The onset of inflammation is often characterized by the release of pro-inflammatory cytokines, e.g., TNF, IL-1, as well as adhesion molecules. The rapid increase in cytokine expression following injury requires an active process involving transcription, posttranscription, translation, and the conversion of latent precursors to biologically actives forms. Cytokine mRNA transcripts have a short half life of 10 to 30 minutes, thus, cytokines are rapidly translated and secreted into the extracellular space producing a burst of cytokine release. Some cytokines are secreted upon complete intracellular processing while others, at one stage, are stored intracellularly and additional stimuli are required to trigger secretion. For example, transcription without translation can be observed for IL-1 following the adherence of blood monocytes to surfaces or exposure to calcium ionophore. In these cases, the level of steady state mRNA levels for IL-1β are comparable with those following induction by endotoxin however, there is no translation into protein. The half-life of the mRNA is unchanged suggesting that the failure of translation is not due to increased destruction of the message (169). Alternatively, the system may effectively regulate the downstream effects of IL-1 by the induction of IL-1 receptor antagonist (IL-1RA) to bind the protein and prevent receptor activation. Cells that contain IL-1 mRNA are considered "primed" and small amounts of other stimuli rapidly trigger translation. It is possible that, in certain models, initial brain injury and a response of microglia could increase the transcription of pro-inflammatory cytokines resulting in a "primed" system. The basal level of the system would then be altered and primed cells would over respond upon subsequent stimulation.

Cytokines regulate the expression of cytokine receptors that, in soluble form, are involved in the modulation of cytokine activity. For induction, the extracellular stimuli binds to a specific cell surface receptor to activate intracellular signals, protein kinases, or phosphatases that regulate the activity of DNA binding proteins. These then recognize specific binding motifs in cytokine genes, such as binding sites for the transcription factors, AP-1, NF-kB, CREB, SP-1, and IRF-1. In general, protein binding to several of these sites is required for high level of cytokine gene expression. It is also possible that, while activated forms of transcription factors may be required for maximal expression of cytokine gene transcription, they may not be a critical regulatory step. Once transcription is successful it does not necessarily equate to protein synthesis. For several of the cytokines, translational regulation is an important regulator of expression. Cytokines can be either membrane-associated or secreted. While some cytokines are secreted upon complete intracellular processing, others are stored intracellularly and additional stimuli are required to trigger secretion. When secreted from the cell, most cytokines are biologically active. However, some (e.g., TNF and TGFβ) are secreted in a biologically inactive or latent form and require further extracellular processing by proteases to be biologically active.

Many of these secreted factors influence the system in both an autocrine and paracrine fashion. For example, IL-1 and TNF are pleiotrophic and trigger general responses that are not specific for the initiating stimulus or antigen. They stimulate a set of chemotactic factors whose primary function is to promote recruitment of inflammatory cells. In addition to the activation of local glia and often the recruitment of blood-borne leukocytes, local synthesis of inflammatory-related cytokines, (e.g., IL-1 and TNFα) elicits a wide range of effects including, cell adhesion, migration, survival, differentiation, replication, secretory function, and cell death. For IL-1 and TNFα, these effects are mediated primarily by IL-1 type I receptor (IL-1R1) and TNFR1 (p55) or 2 (p75), respectively. IL-6 and other members of this family—oncostatin M (OSM), leukemia inhibitory factor (LIF), IL-11, and ciliary neurotrophic factor (CNTF)—are induced by IL-1 and TNF and protect against the pro-inflammatory and catabolic effects.

While a pathophysiological role for IL-1 as part of the inflammatory process in the brain has been established, exactly how the signaling process may contribute to neuronal death and protection is still unclear. However, data support its role as a key mediator of inflammation and neuronal death in acute brain injuries, such as stroke and trauma (170). Microglia functions related to an innate immune response are associated with TNF signaling and its regulation of both inflammation and apoptosis. TNF is a multipotent, inflammatory cytokine that can induce apoptosis via activation of receptors containing a homologous cytoplasmic sequence identifying an intracellular death domain. This includes TNFR1 and CD95 (APO-1/Fas) with their corresponding death ligands, TNF and the structurally related type II transmembrane protein, FasL. The release of TNFα and FasL shedding by microglia are implicated in neurotoxicity (171). Signaling by TNF is dependent upon constitutive and induced target cell expression of TNF receptors (TNFRs), a 55 kDa type-1 receptor (TNFp55R), and a 75 kDa type-2 receptor (TNFp75R) (172). Membrane receptor mechanisms of apoptosis are implicated in neuronal death involving downstream intracellular death-signaling complexes, e.g., AP-1, NF-kB, and caspases. However, the soluble form of TNFR1 can serve in a protective fashion by binding of TNF and preventing further receptor signaling. Signal transduction through the death receptors can induce apoptosis via activation of caspases. TNFR1 is thought to bind to the TNFR1-associated death domain protein, TRADD (173). Following membrane activation, internalized TNFR1 (TNF receptosomes) rapidly recruits TRADD (174) that can interact directly with FADD (173–175). Binding to the adaptor proteins forms a death inducing signaling complex allowing for caspase-8 recruitment and activation to initiate apoptosis. This in turn activates downstream effector caspases (173,174). Upon activation of caspases, a signaling process is initiated within the cell. While this process can be modified, the activation of caspase 3 is considered a final step in the commitment to cell death and a key mediator of neuronal death. It has been proposed that TNFR1 activation provides a molecular mechanism for the rapid apoptosis of injured or sick neurons through a caspase 3-mediated pathway (176). The tissue distribution of the receptor and the differentiation state of the target cell influence the cellular response to cytokines. Although neither TNFα nor IL-1ß has been demonstrated to cause neuronal death in healthy brain tissue or normal neurons (177) and normal cellular architecture is maintained in mice deficient in pro-inflammatory cytokines or receptors, this mechanism supports the hypothesis that localized pro-inflammatory cytokine activation could initiate neuronal death. Increases in TNFα and IL-1ß have been observed prior to neuronal death and, as mentioned, recent studies suggest that, indeed, the activation of immune factors such as TNF can participate in causation of neuronal death. Recent studies examining cell death and survival following an ischemic insult (178) or induced by a systemic injection of TMT (179) suggest that the level of TNFα produced by microglia at a specific site and the neuronal expression pattern of TNF receptors can be determining factors for neuronal death or survival in an inflammatory environment.

Cytokine Inhibition

Inhibition of cytokine expression represents an important regulatory mechanism by which differentiation of cellular responses and potentially detrimental cytokine activities are controlled. In order to counter pro-inflammatory signaling pathways, cells utilize multiple mechanisms including concurrent production of anti-inflammatory factors. There are a number of endogenous molecules and mechanisms that inhibit cytokine systems or action. TGFβ, IL-4, IL-10 (180–184), and IL-13 (185) are the best-characterized inhibitors of cytokine production. Other anti-inflammatory cellular processes include the activation of protein kinase A to hinder the expression of iNOS. Furthermore, upon exposure to pro-inflammatory cytokines, cells rapidly upregulate their normally low levels of a family of proteins called suppressors of cytokine signaling (SOCS) (186). These signaling molecules can negatively regulate the response of immune cells, either by inhibiting the activity of JAK or by competing with signaling molecules for binding to the phosphorylated receptor. Data suggests that the up-regulation of SOCS represents a step towards suppressing glial inflammation via negative regulation of the JAK-STAT pathway (187). Inhibition also occurs with cytokine-induced iNOS expression. Both rapid transcription of SOCS 1 and 3, as well as the phosphorylation of SHP2, down-regulate the JAK-STAT pathway. This leads to the suppression of inflammatory responses (187). Specific anti-inflammatory cytokines in the CNS such as IL-10, IL-13, and IL-4 down-regulate the inflammatory process by stimulating biosynthesis of pro-inflammatory cytokine inhibitors, such as soluble receptors (188). They also intercept signals that are generated by pro-inflammatory cytokine, receptor-ligand complexes.

TGFβ can inhibit IL-1 or LPS-induced TNF or IL-6 synthesis in monocytes, though stimulatory or bifunctional effects of TGFβ on cytokine synthesis can occur in other cell types. IL-4 is a potent inhibitor of monocyte activation and interferes with IL-1, TNF, IL-6, and IL-8 production. IL-10 inhibits IFNγ production by T lymphocytes and cytokine production by monocytes leading to a down-regulation of the inflammatory response. The induction of IL-6 by TNFα and IL-1α (189) can produce a number of responses including anti-inflammatory responses to modulate TNFα and IL-1α (190,191). The receptor antagonist, IL-1ra is present in both peripheral and brain tissue and upon induction it inhibits many of the central actions of IL-1. Cytokines themselves can have opposing actions in the brain. For example IL-4, IL-10, and TGFβ can downregulate induction of IL-1 and stimulate IL-1ra synthesis (192,193).

The natural course of an acute injury response or chronic inflammatory diseases is the result of the balance between cytokines that activate versus those that deactivate immune and inflammatory responses. Both are simultaneously activated and their balance is a manifestation of the multiple factors that are responsible for etiology and pathogenesis.

MICROGLIA RESPONSES IN INJURY AND DISEASE MODELS

Microglia are tuned to detect disturbances in their microenvironment and thus, are often referred to as sensors of pathology (194). It has been suggested that microglial responses are tailored in regional and insult-specific manners (195) and that they are dictated by the nature of the stimulus, the receptor repertoire engaged, and the prior state of the cell at the time of stimulus (196). While microglia are not solely responsible for the inflammatory or immune mediated responses in the brain, they are poised to be able to rapidly respond to environmental changes. The capability of microglia to rapidly sense homeostatic disturbances and recognize danger signals triggering a transformation from a "naïve" state (resting) into an activated state represents a significant defense mechanism for the nervous system. The presence of activated microglia localized to regions of brain injury was initially considered as a sign of pathology and, as such, considered for use as a sensitive marker to identify injury sites predestined for imminent tissue destruction.

The detection of activated microglia within injured brain regions and in post-mortem tissue from patients with various neurodegenerative diseases, has led to the initial assumption that all reactive microglia would contribute to an adverse and degenerative process. This makes the assumption that a brain resident microglial cell transforms into an aggressive effector cell and that, in this mode, it can attack healthy neurons either physically, as by phagocytosis, or via secreted factors. However, rather than this form of transformation, microglia seem to alter their morphology dependent upon the specific task at hand. Such resident immune responders may be beneficial in the healing phases of CNS injury by actively monitoring and controlling the extracellular environment, walling off areas of the CNS from non-CNS tissue, and removing dead, damaged, or dysfunctional cells. A strong microglia reaction occurs at the primary site of injured often temporally and spatially progressing to recruit responses in areas that are projection sites to and from the area of damage. While anterograde and retrograde degeneration of neuronal projections most likely causes a microglial reaction in the first-order projection areas, other mechanisms likely drive the microglia response in more distant and remote areas. For example, neuronal excitability, either direct or following removal of an inhibitory signal, can stimulate microglia due to changes in ionic balance.

The hematopoietic lineage of microglia ensures that they respond to antigens and acquire a phagocytic phenotype, especially in culture. The earlier studies developing the hypothesis that microglia secreted neutoxic factors were based upon culture studies where media from stimulated isolated microglia was placed on isolated neuronal cultures resulting in cell death. These observations served to foster the idea of a "bystander lysis" effect of microglia in progressive neurodegeneration. However, similar studies have also shown that microglia can secrete factors that serve to protect neurons from degeneration. Thus, to take this to the in vivo environment one must consider than any immune/inflammatory process is not driven by or the result of any one cell type but rather a complex interaction between cells. In addition, this interaction is often not only between resident cells of the nervous system but, depending upon the disease or injury, can involve multiple cell types infiltrating from the periphery. Thus, identifying reactive or activated microglia in proximity to injured tissue does not automatically imply a causal relationship with the pathology. The cellular pattern can just as easily represent the clearance and healing process.

Acute Injury

The complex interdependency of neuronal injury and glial activation makes it difficult to distinguish between an initiating, prerequisite activation versus a compensatory, functional response of the glia. In an acute injury, there are multiple stages of degeneration. There is the cell death that occurs in the primary injury site, the secondary degeneration of neurons in close proximity, and degeneration/response of neurons in the projection sites. Such models are of interest for microglia biologists in that the major class of cells that populate the site of injury are microglia. With acute or short-term injury models, the possibility of determining the individual cellular contributions and relevant processes is significantly increased.

Experimentally, a stab wound injury to the cortex of the brain has been used as a model for examining a glial response to injury. The cellular response is primarily limited to the needle tract and the surrounding parenchyma. With a slightly deeper penetration to include the corpus callosum, responses along the myelinated tracts can be observed. The nature of the injury involves recruitment of blood-borne cells to the reaction including leukocytes, granulocytes, and monocytes (89). Within the lesion itself, round hematogenous macrophages are primarily found. At the edge of the lesion, process-bearing, hypertrophic intrinsic microglia are observed. This initial response occurs within the first 10 hrs or so, peaks between days 2 and 3, then declines until approximately 10 days post-injury (89). At the peak time of response (day 2), IL-1 is elevated and may stimulate the astrocyte response to wall-off the injury site. Both IL-6 and TGFβ1 are elevated in the lesion site (197,198). At about five-days post-injury, the hypertrophied microglia express MHC class II and the CD4 antigen (199). Activated cells can show an upregulation of complement type 3 receptor (CR3) and lectin binding proteins (200). Several studies demonstrate that, under acute injury conditions, microglia can be destructive or protective (200–202). Activation of the adaptive immune response evoked by injury, and the CD4+ T-cells influence the microglial phenotype and effects on neurons. A controlled T-cell response can change the profile of molecules released by microglia (203). With activation following acute injury, microglia express pro-inflammatory mediators and display a phagocytic phenotype. Interactions with T-cells will result in fewer neurotoxic factors produced and an increase in anti-inflammatory and growth mediators. In addition, microglia that were activated by T-cells reverse the glutamate transporters resulting in a buffering rather than a release of glutamine (202,204).

Over the years concern has been raised with regards to the impact that serum components may have on microglia as well as the differential impact of resident versus infiltrating macrophages. The majority of animal models used to address the response and contribution of microglia are compromised by the fact that, in the nature of the insult, changes occur in the vasculature and the blood brain barrier loses some of its protective function. For example, models of ischemia, head trauma, and physical injury all physically compromise the blood brain barrier (BBB). In addition, models using kainic acid or MPTP also result in damage to the BBB. Thus, an approach of using nerve transection models of deafferentation lesions has been employed. Two different models have been predominantly employed. One is the transection of the enthorinal cortex from the hippocampus (205). The other model is the transection of the facial motor nerve (206–208). This allows for the actual transection of the cranial nerve outside of the CNS, yet, with dying back retrograde degeneration, the motor neurons die. These models have started to provide significant information regarding both the response of the resident microglia, the contribution from infiltrating macrophages, and the influence of infiltrating T-cells.

Microglia in Chronic Degenerative Disease

Alzheimer's Disease

As early as 1927, Bolsi reported argentophilic reactive cells surrounding senile plaques in human brain tissue (209). Identification of the cellular type occurred when reactive microglia immunopositive for the histocompatibility marker (HLA-DR) were described within the corona of senile plaques (210). While inflammation is not a hallmark of AD, recent data suggests that both Aβ deposition and neurofibrillary tangles activate a potentially pathological innate immune response (211,212). Microglia are considered key players in AD (213–215). Inflammation has been associated in the brains of AD patients with abnormal levels of cytokines such as IL-1β, IL-6, and TGFβ. In post-mortem brain tissue of AD patients, an atypical inflammation occurs with both activation of the resident microglia (216) and infiltrating monocytes (217). These cells surround dense-core amyloid-β1-40/42 plaques (218) and show increased expression of cell-surface antigens. One of the earliest examinations of the possible contribution of microglia in AD was in the report of Griffin et al. (219) showing that microglia surrounding amyloid plaques produce interleukin-1. In animal models of AD, microglia are a major component of the amyloid plaques (220–222). However, while they are components of the dense cored plaques they are not associated with the diffuse plaques. One hypothesis has been put forth based upon the proximity of the two structures is that microglia potentiate the toxicity of Aß peptide (223). In the AD brain, microglia are in contact with plaques yet, they are not stimulated to phagocytize and effect clearance. While the actual mechanism is not known, the alterations in the normal ability of microglia to efficiently clear ß-amyloid $(Aß)_{42}$ (214) may also be a major contributor to Alzheimer's disease (AD). This may be due to a senescence process in microglia, an overload of the cell, or to an altered cell signaling. Clearance of debris is facilitated by the expression of highly conserved pattern recognition receptors (PRRs). On microglia, these include the complement receptors (224,225), Toll-like receptors, and scavenger receptors (226,227). TLR2 in microglia can act like a natural innate immune response to clear amyloid beta (228). Aberrant proteins and peptides can trigger the activation of the complement cascade leading to a local inflammatory response leading to the phagocytosis of excess material. For example, complement component 1q (C1q) co-localizes with amyloid deposits and triggers the complement cascade (229). Complement proteins are induced and associated with Aβ plaques in brains of AD patients and in AD animal models (230–232). Complement activation can lead to both detrimental, such as cell lysis (233,234), and beneficial effects, such as C1qb and C3b promoting clearance of cellular debris and enhance cell survival (232,235). It is the balance of the two that determines the final outcome (236). Immune cells are the primary source for complement proteins however in the brain, microglia, astrocytes, as well as, neurons can produce complement proteins upon stimulation (237,238). Data obtained from cultured microglia suggest that activation of the complement system can lead to increased phagocytosis of Aβ (239). While the phagocytic activation serves in a beneficial manner to eliminate aggregated and toxic proteins from the brain, it has been proposed that an extended activation or dysregulation of this system can lead to an adverse environment and subsequent toxicity to surrounding cells. A recent study using the Tg-SwDI mice as a model for AD, demonstrated that activated microglia localized around microvascular amyloid deposits increased synthesis of native complement proteins C1q, C3, and C4 (240). In transgenic mouse models of AD, less neuropathology is demonstrated in the absence of C1q (233); however, inhibition of complement activation diminishes microglial activity, increases amyloid load, and enhances neuronal degeneration (232).

Parkinson Disease

In Parkinson disease (PD), the nigrostriatal pathway, dopaminergic, and non-dopaminergic neurons are compromised contributing to the motor related symptoms. A number of mechanisms have been proposed for the death of dopamine-containing neurons in the substantia nigra (SN) pars compacta. Multiple mechanisms of neuronal death have been proposed including mitochondrial dysfunction, oxidative stress, and the impairment of protein degradation (241). The detection of elevated levels of pro-inflammatory cytokines and evidence of oxidative stress-mediated damage in postmortem PD brains gave rise to the hypothesis of a microglia involvement. However, the loss of neurons is accompanied by inflammatory changes in microglia and the infiltration of T-lymphocytes. Early work demonstrated the presence of cytotoxic T-cells in the SN of a patient with PD (242) and an increased density in IFNγ+ cells in the brains of PD patients (243). These observations suggested a role for the cellular arm of the adaptive immune system.

These cells have also been seen in the 1-methyl-4-phenyl-1,2,3,6-tetrahydropyridine (MPTP)-induced model of dopaminergic neuronal death (244). A direct injection of MPTP produced an increase in cells expressing the proliferating cell nuclear antigen (PCNA) and the myeloid marker, CD11b+ suggesting that infiltrated macrophages and/or resident microglial cells multiply at the injury site. This response preceded astrogliosis and the appearance of CD4+ T-cells suggesting a recruitment role for the glia response. Of additional interest, is that MPTP induced a widespread and transitory serum protein leakage in the brain prior to the cell type and region-specific leukocyte extravasation process. Using two different strains of immunodeficient mice lacking mature T lymphocytes (*Tcrb–/–* and *Rag1–/–* mice), dopaminergic cell injury was reduced following administration of MPTP. Further work demonstrated that CD4+ T-cells were the population of T-cells mediating the cytotoxicity of MPTP. In addition, these cells require the expression of a functional proapoptotic FasL but not the inflammatory cytokine IFN-γ (245). The infiltration of CD4+/FasL+ T-cells could activate microglia to secrete pro-inflammatory cytokines (246). While, these studies suggest that CD4+T-cells mediate cytotoxicity possibly via the activation of microglia cells to release free radicals and proinflammatory cytokines, the recent description of an IL-17 secreting Th17 lymphocyte suggested an alternative cellular source for proinflammatory cytokines. These cells mediate inflammation by stimulating the production of inflammatory cytokines such as TNFα, IL-1B, and IL-6 (247) and influence CD4+ T-cell-induced cytotoxicity. One could envision that with progressive changes in the neuron, a microglia response to α-synuclein can occur, that may recruit CD4+ T-cells, leading to an induction of pro-inflammatory cytokines and nitric oxide secretion by microglia alternatively, or in conjunction with the infiltration of TH17 lymphocytes.

The SN, along with the dentate gyrus, is rich in microglia cells under normal conditions (50). In addition, dopaminergic neurons in the SN have reduced levels of the intracellular thiols rendering them vulnerable to oxidative stress insults (248–250). This combination has lead to the hypothesis that microglia activation can have a direct effect on dopaminergic neurons leading to cell death. When a progressive model of Parkinson's disease was examined with reference to the adverse impact of microglial activation and inflammation on damage to dopaminergic neurons, Wright and colleagues (251) clearly demonstrated in vivo that, while the immunosuppressant drug, tacrolimus (FK506), reduced the initial stage of neuronal death in the SN, it did not alter the morphological response of microglia.

MICROGLIA IN CHEMICAL-INDUCED NEUROTOXICITY

Research efforts in neuroinflammation and neuroimmunology have significantly expanded over the last decade. Quite a bit of this work has been directed toward understanding the mechanisms involved in immune signaling and the contribution of inflammation to neuronal and white matter degradation with autoimmune diseases, e.g., multiple sclerosis, human immunodeficiency virus, or brain infections. More recently, consideration has been directed toward the possible role of such responses in the neurotoxicity of pharmacological and environmental agents. While there is a significant amount of interest in the microglial response and contribution to neuronal dysfunction, there is surprisingly little work actually conducted within the area of environmental neurotoxicology. This may be due to the evolving nature of the field, the rapidly moving pace of the basic science of microglia cells, and the questions raised with regards to translation of findings obtained in isolated microglia cells to the responses observed in vivo.

In regards to neurotoxicology of environmental agents, this is a newly evolving field and quite often the studies are based on a limited number of endpoints, neuropathology inclusive of a microglia response, or the generation of oxidative stress by-products and pro-inflammatory cytokines, and quite often generated in an isolated in vitro system. Many of the initial studies initiated in chemical-induced brain injury over a decade ago were focused on establishing that a host response and pro-inflammatory cytokine response would occur in the brain. Consistent with the heterogeneity of responses of microglia, the continuation and expansion of these studies also identified unique differences in the microglia responses suggesting a combination of neurotoxic and neuroprotective actions. Some of the initial work has focused on heavy metals such as mercury and inorganic lead that are well known to have neurotoxic properties and induce an alteration in the systemic immune system. Some of the initial work on mercury demonstrated a transient increase in microglia in the neuropil following acute exposure of adult rats (252). Mercury deposits were demonstrated in microglia and, with additional studies in non-human primates (253), the accumulation in astrocytes and microglia was suggested to identify a primary location for the demethylation of mercury. Further work to determine if the mercury-induced activation of microglia was involved in neuronal apoptosis showed no relationship. With exposure to low levels of mercury, a neuroprotective effect was observed as a result of the microglial reactivity and the induction and release of IL-6 from astrocytes (254,255). More recent work (256) utilizing a continuous (10–50 days) application of the heavy metals, lead and mercury, to aggregating brain-cell cultures demonstrated an increase in amyloid precursor protein with lead and the formation of insoluble Aß by mercury. In each case, a microglia response was observed that would be consistent with normal phagocytic activity of these two classic stimuli. A significant amount of research has examined the role of microglia in neurotoxicity produced by the organometal, TMT (257–270). These studies have varied from identifying microglia as a sensitive marker for the neurotoxicity to efforts to identify the mechanism of cells death, and the use of the model system to test compounds for neuroprotective actions. Interestingly, this injury model has provided some of the first evidence that the induction of TNFα by microglia may directly contribute to the neuropathology (179,265–267). Work examining the osteopetrotic mouse deficient in (CSF-1) and accordingly severely deficient in macrophages and microglia, demonstrated that the physical presence of microglia was not the deciding factor in TMT-induced toxicity but that rather the significant overproduction of TNFα was the major contributing factor (265). Additional in vivo work continues to demonstrate heterogeneity of the microglia response, even within targeted brain structures and individual classes of chemical or drug-induced injuries, continuing to demonstrate heterogeneity of microglia responses (179). Thus, there is a body of literature for assessment of neurotoxicity of environmental agents that, to date supports the diversity and heterogeneity of the microglia and neuroimmune or neuroinflammatory response.

Recent interest in environmental exposure and inflammation in brain tissue has arisen from in vivo and in vitro studies showing that nearly all of the proposed etiologies for Parkinson's disease, including bacteria, viruses, pesticides, drug contaminants, and head trauma, are known to produce an inflammatory response. However, a mechanism by which microglia would specifically target healthy dopamine neurons remain un-identified (271). Speculation of a role for TNF and microglia activation has been proposed in the acute MPTP model of dopaminergic neuronal loss (272). These studies have gained interest due to the linkage between oxidative stress, microglia, and susceptibility of dopaminergic neurons. In addition, the structural similarity of rotenone to MPTP and the loss of dopaminergic neurons with exposure suggest a possible link between early chemical exposure and loss of dopamine neurons. Susceptibility to rotenone was increased in dopamine neurons from parkin deficient mice and this could be reduced by minocycline (273); however, minocycline enhances MPTP toxicity to dopaminergic neurons (274). Further work examining the effects of microglia showed that iptakalim alleviates rotenone-induced degeneration of dopaminergic neurons by inhibition of the microglia response (275). Paraquat/maneb can activate microglia (276) and Purisai et al. suggested that the activation of microglia represented a priming event contributing to dopaminergic neuronal degeneration following paraquat (277). Some of the initial work in neuroimmunotoxicology was conducted by McCann et al. (278) looking at the production of reactive oxygen species by microglia following exposure to the organochlorinated pesticide, dieldrin. More recent work demonstrated that exposure to polychlorinated biphenyls induces a systemic inflammatory response in mice with an associated reduction in striatal dopamine related proteins, potentially related to IL-6 expression (279). It has more recently been shown that microglia express functional ryanodine receptors that may contribute to the diverse effects of these cells with regards to chemical neurotoxicity (280).

An induction of microglia in the deep brainstem was noted following acute inhalation of carbon monoxide. This increase in process bearing microglia occurred in the absence of neuronal death (281). In vitro studies have suggested that microglia can directly cause the death of dopaminergic neurons following exposure to nanometer size diesel exhaust particles (282). Given the phagocytic nature of microglia one would speculate that any exposure to a physical particle could initiate this response for clearance and that the resulting products could be toxic to dopamine neurons. The limitation for many of these studies is the fact that microglia are activated with the induction of neuronal death. Thus, any conclusion based upon the presence of microglia with neuronal death, or the absence of microglia with no neuronal death are limited with regards to determining causative factors derived from the microglia

MICROGLIA AND INJURY INDUCED NEUROGENESIS

The complex process of inflammation can augment or suppress the generation of new neurons after injury. One difficulty in interpreting experimental findings is the fact that the level of neurogenesis induced is dependent upon the severity of the injury. Thus, any modulation of the initial injury will significantly influence the associated microglia response and the induction of neurogenesis. Inflammation mediated varying effects on neurogenesis (283–285). Initial work suggested that inflammation is detrimental for neurogenesis (286,287). However, more recent work suggests that immune cells contribute to the maintenance of neurogenesis (288,289). The delivery of non-steroid anti-inflammatory drugs prevented the impairment of neurogenesis following either irradiation or injection of bacterial LPS (287). This was taken to imply that inflammation suppressed the neurogenic response in the adult brain. As an alternative approach, Ekdahl et al. (285,286) used minocycline to block microglia activation and also showed an attenuation of the impairment of neurogenesis following LPS. Minocycline is a second-generation, semi-synthetic tetracycline analog that can easily penetrate the blood-brain barrier. While it does block microglia activation, it also has a direct effect to block caspase mediated apoptosis and thus, given the relationship between neuronal death, microglia activation, and induction of adult neurogenesis, a direct causative role for microglia cannot be concluded. In addition, it has been demonstrated that neurogenesis occurs within an inflammatory environment created by the death of dentate granule neurons by trimethyltin (290–292) and that microglia proliferation precedes neurogenesis following alcohol abstinence (293).

Within a culture environment, mouse microglia release soluble factors that are capable of supporting prolonged neurogenesis and direct migration of neural stem/precursor cells (294,295). These factors also promote the differentiation of neural stem cells to a neuronal lineage (294). Recent studies examining microglia within the subgranular zone of the hippocampus suggest that insulin-like growth factor-1 (IGF-1) produced by microglia may serve as such a neurotrophic factor (296). These data suggest that microglia promote CNS lesion repair by promoting the neural stem/progenitor cells. When this process was examined in transgenic mice expression mutant forms of human presenilin-1 (PS1) the normal induction of hippocampal neurogenesis by environmental enrichment was not observed. When the cells were isolated from the hippocampus of these mice, proliferation and differentiation were equivalent to that seen in the wildtype mice. However, microglia from mice expressing mutant PS1 were co-cultured with wild-type neural stem/progenitor cells the proliferation and neuronal differentiation process was impaired. Further studies demonstrated that these effects were due to microglial-derived secreted factors (297). It has been suggested that the outcome is dependent upon the means by which the microglia, macrophages, and/or astrocytes are activated and the duration of the inflammatory response (298).

SUMMARY

This plethora of responses is representative of the dichotomy of microglia reactivity in promoting neuronal survival or degeneration. However, to understand the impact and contribution of inflammation in the brain one needs to understanding the diverse nature of the responses, the potential contribution of infiltrating cells, the development and aging of resident immune cells, and the multitude of endogenous regulatory processes. How these cells interpret the cellular changes in their environment will determine whether a local immune response to a CNS insult will be harmful or beneficial. These distinct features of the inflammatory response and microglia will impact the design and interpretation of experimental models to assess effects within a neurotoxicology framework. Thus, an entirely negative role for microglia is probably incorrect. However, this normal contribution of microglia cells in maintaining the homeostatic balance in the brain may serve as a target process for either environmental agents, genetic alterations, disease processes, or in aging.

REFERENCES

1. Nissl F. Ueber einige Beziehungen zwishcen Nerven zellerkrankungen und glioesen Erscheinungen bei verschiedenen Psychosen. Arch Psychiat 1899; 32: 1–21.
2. Cajal RY. Contribucion al conocimiento de la meuroglia del cerebro humano. Trab Lab Invest Biol 1913; 11: 254.
3. del Rio Hortega P, Penfield W. Cerebral cicatrix. The reaction of neuroglia and microglia to brain wounds. Bull Johns Hopkins Hosp 1927; 41: 278–82.
4. del Rio Hortega P. Microglia. In: Penfield W, ed. Cytology and Cellular Pathology of the Nervous System. New York: Hoeber, 1932: 481–584.
5. Ling EA, Penney D, Leblond DP. Use of carbon labeling to demonstrate the role of blood monocytes as precursors of the "amoeboid cells" present in the corpus callosum of postnatal rats. J Comp Neurol 1980; 193: 631–57.

6. Barron KD. The microglial cell. A historical review. J Neurol Sci 1995; 134(Suppl): 57–68.
7. Perry VH, Hume DA, Gordon S. Immunohistochemical localization of macrophages and microglia in the adult and developing mouse brain. Neurosci 1985; 15: 313–26.
8. Hickey WF, Kimura H. Perivascular microglial cells of the CNS are bone marrow-derived and present antigen in vivo. Science 1988; 239: 290–2.
9. Kershman J. Genesis of microglia in the human brain. Arch Neurol Psych 1939; 41: 24–50.
10. Streit WJ, Graeber MB, Kreutzberg GW. Expression of Ia antigens on perivascular and microglial cells after sublethal and lethal neuronal injury. Exp Neurol 1989; 105: 115–26.
11. Sasaki A, Nakanishi Y, Nakazato Y, Yamaguchi H. Application of lectin and B-lymphocyte-specific monoclonal antibodies for the demonstration of human microglia in formalin-fixed paraffin-embedded brain tissue. Virchows Arch A Pathol Anat 1991; 419: 291–9.
12. del Rio Hortega P. El tercer elemento de los centros nerviosos. Poder fagocitario y movilidad de la microglia. Biol Soc Esp Biol Ano 1919; ix: 154–66.
13. Murabe Y, Sano Y. Thiaminepyrophoshatase activity in the plasma membrane of microglia. Histochem 1981; 71: 45–52.
14. Sawada M, Suzumura A, Yamamoto H, Marunouchi T. Activation and proliferation of the isolated microglia by colony stimulating factor-1 and possible involvement of protein kinase C. Brain Res 1990; 509: 119–24.
15. Schelper RL, Adrian EK. Non-specific esterase activity in reactive cells in injured nervous tissue labeled with ^{3}H-thymidine or ^{125}iododeoxyuridine injected before injury. J Comp Neurol 1980; 194: 829–44.
16. Oehmichen M. Enzyme-histochemical differentiation of neuroglia and microglia: a contribution to the cytogenesis of microglia and globoid cells. Pathol Res Pract 1980; 168: 344–73.
17. Castellano B, Gonzales B, Finsen BR, Zimmer J. Histochemical demonstration of purine nucleoside phosphorylase (PNPase) in microglial and astroglial cells of adult rat brain. J Histochem Cytochem 1990; 38: 1535–9.
18. Nakajima K, Shimojo M, Hamanoue M, et al. Identification of elastase as a secretory protease from cultured rat microglia. J Neurochem 1992; 58: 1401–8.
19. Banati RB, Gehrmann J, Schubert P, Kreutzberg GW. Cytotoxicity of microglia. Glia 1993; 7: 111–18.
20. Frei K, Fontana A. Antigen presentation in the CNS. Mol Psychiatry 1997; 2: 96–8.
21. Carson MJ, Reilly CR, Sutcliff JG, et al. Mature microglia resemble immature antigen-presenting cells. Glia 1998; 22: 72–85.
22. Carson MJ, Sutcliffe JG, Campbell IL. Balancing function vs. self defense, the CNS as an active regulator of immune responses. J Neurosci Res 1999; 55: 1–8.
23. Flugel A, Labeur MS, Grasbon-Frodl EM, et al. Microglia only weakly present glioma antigen to cytotoxic T cells. Int J Dev Neurosci 1999; 17: 547–56.
24. Williams K, Ulvestad E, Antel J. Immune regulatory and effector propereties of human adult microglia studies in vitro and in situ. Adv Neuroimmunol 1994; 4: 273–81.
25. McGeer PL, McGeer EG. The inflammatory response system of brain: implications for therapy of Alzheimer and other neurodegenerative diseases. Brain Res Brain Res Rev 1995; 21: 195–218.
26. Graeber MB, Streit WJ. Perivascular microglia defined. Trends Neurosci 1990; 13: 366.
27. Hume DA, Perry VH, Gordon S. Immunohistochemical localization of a macrophage-specific antigen in developing mouse retina: phagocytosis of dying neurons and differentiation of microglial cells to form a regular array in the plexiform layers. J Cell Biol 1983; 97: 243–7.
28. Perry VH, Hume DA, Gordon S. Immunohistochemical localization of macrophages and microglia in the adult and developing mouse brain. Neurosci 1985; 15: 313–26.
29. Graeber MB, Streit WJ, Kreutzberg GW. Axotomy of the rat facial nerve leads to increased CR3 complement receptor expression by activated microglial cells. J Neurosci Res 1988; 21: 18–24.
30. Tillotson ML, Wood JG. Phosphotyrosine antivodies specifically label ameboid microglia in vitro and ramified microglia in vivo. Glia 1989; 2: 412–19.
31. Craggs RI, Webster HdeF. Ia antigens in the normal rat nervous system and in lesions of experimental allergic encephalomyelitis. Acta Neuropathol 1985; 68: 263–72.
32. Hayes GM, Woodroofe MN, Cuzner ML. Microglia are the major cell type expressing MHC class II in human white matter. J Neurol Sci 1987; 80: 25–37.
33. Streit WJ, Graeber MB, Kreutzberg GW. Expression of Ia antigens on perivascular and microglial cells after sublethal and lethal neuronal injury. Exp Neurol 1989; 105: 115–26.
34. McGeer PL, Itagaki S, McGeer EG. Expression of the histocompatibility glycoproein HLA-DR in neurological disease. Acta Neuropathol 1988; 76: 550–7.
35. Steiniger B, van der Meide PH. Rat expendyma and microglia cells express class II MHC antigens after intravenous infusion of recombinant gamma interferon. J Neuroimmunol 1988; 19: 111–18.
36. Streit WJ, Graeber MB, Kreutzberg GW. Functional plasticity of microglia: a review. Glia 1988; 1: 301–7.

37. Flaris NA, Densmore TL, Molleston MC, Hickey WF. Characterization of differential expression of molecules using standard and novel monoclonal antibodies in normal CNS and in for models of parenchymal reaction. Glia 1993; 7: 34–40.
38. Graeber MB, Streit WJ, Kreutzberg GW. The microglial cytoskeletan: vimentin is localized within activated cells in situ. J Neurocytol 1988; 17: 573–80.
39. Streit WJ, Sammons NW, Kunhs AJ, et al. Dystrophic microglia in the aging human brain. Glia 45: 2004; 208–12.
40. Mannoji H, Yeger H, Becker LE. A specific histochemical marker (lectin Ricinus Communis agglutinin-I) for normal human microglia, and application to routine histopathology. Acta Neuropathol (Berl) 1986; 71: 341–430.
41. Leong SK, Ling EA. Amoeboid and ramified microglia: their interrelationship and response to brain injury. Glia 1992; 6: 39–47.
42. Thanos S. Specific transcellular carbocyanin-labelling of rat retinal microglia during injury-induced neuronal degeneration. Neurosci Lett 1991; 127: 108–12.
43. Rinamen L, Milligan CE, Levitt P. Persistence of Fluoro-Gold following degeneration of labeled motoneurons is due to phagocytosis by microglia and macrophages. Neurosci 1991; 44: 765–76.
44. Billiards SS, Haynes RL, Folkerth RD, et al. Development of microglia in the cerebral white matter of the human fetus and infant. J Comp Neurol 2006; 497: 199–208.
45. Chan WY, Kohsaka S, Razaie P. The origin and cell lineage of microglia—new concepts. Brain Res Rev 2007; 53: 344–54.
46. Polazzi E, Contestabile A. Reciprocal interactions between microglia and neurons. From survival to neuropathology. Rev Neurosci 2002; 13: 221–42.
47. Marín-Teva JL, Dusart I, Colin C, et al. Microglia promote the death of developing Purkinje cells. Neuron 2004; 41: 535–47.
48. Monier A, Evrard P, Gressens P, et al. Distribution and differentiation of microglia in the human encephalon during the first two trimesters of gestation. J Comp Neurol 2006; 499: 565–82.
49. Wu CH, Wen CY, Shieh JY, et al. Down-regulation of membrane glycoprotein in amoeboid microglia transforming into ramified microglia in postnatal rat brain. J Neurocytol 1994; 23: 258–69.
50. Lawson LJ, Perry VH, Dri P, Gordon S. Heterogeneity in the distribution and morphology of microglia in the normal adult mouse brain. Neurosci 1990; 39: 151–70.
51. Giulian D, Young DG, Woodward J, et al. Interleukin-1 is an astroglial growth factor in the developing brain. J Neurosci 1988; 8: 709–14.
52. Brenneman DE, Schultzberg M, Bartfai T, et al. Cytokine regulation of neuronal survival. J Neurochem 1992; 58: 454–60.
53. Kahn MA, Ellison JA, Speight GJ, et al. CNTF regulation of astrogliosis and the activation of microglia in the developing rat central nervous system. Brain Res 1995; 685: 55–67.
54. Hagan P, Poole S, Bristow AF, et al. Intracerebral NMDA injection stimulates production of interleukin-1 beta in perinatal rat brain. J Neurochem 1996; 67: 2215–18.
55. Silverstein FS, Barks JDE, Hagan P, et al. Cytokines and perinatal brain injury. Neurochem Int 1997; 30: 375–83.
56. Mattiace LA, Davies P, Dickson DW. Detection of HLA-DR on microglia in the human brain is a function of both clinical and technical factors. Am J Pathol 1990: 136.
57. Perry VH, Gordon S. Modulation of CD4 antigen on macrophages and microglia in rat brain. J Exp Med 1987; 166: 1138–43.
58. Lassmann H, Zimprich F, Vass K, Hickey WF. Microglial cells are a component of the perivascular glia limitans. J Neurosci Res 1991; 28: 236–43.
59. Graeber MB, Streit WJ. Perivascular micorglia defined. Trends Neurosci 1990; 13: 366.
60. Stout RD, Jiang C, Matta B, et al. Macrophages sequentially change their functional phenotype in response to changes in microenvironmental influences. J Immunol 2005; 175: 342–9.
61. Finch CE, Morgan TE, Rozovsky I, et al. Microglia and aging in the brain. In: Streit WJ, ed. Microglia in the Regenerating and Degenerating CNS. New York: Springer Verlag, 2002: 275–305.
62. Ma L, Morton AJ, Nicholson LF. Microglia density decreases with age in a mouse model of Huntington's disease. Glia 2003; 43: 274–80.
63. Streit WJ, Miller KR, Lopes KO, et al. Microglial degeneration in the aging brain—bad news for neurons? Front Biosci 2008; 13: 3423–38.
64. Flanary BE, Streit WJ. Progressive telomere shortening occurs in cultured rat microglia, but not astrocytes. Glia 2004; 45: 75–88.
65. Flanary BE, Sammons NW, Nguyen C, Walker D, Streit WJ. Evidence that aging and amyloid promote microglial cell senescence. Rejuv Res 2007; 10: 61–74.
66. Richartz E, Stransky E, Batra A, et al. Decline of immune responsiveness: a pathogenetic factor in Alzheimer's disease? J Psychiatr Res 2005; 39: 535–43.

67. Harry GJ, Lefebvre d'Hellencourt C, Bruccoleri A, Schmechel D. 2000 Age-dependent cytokine responses: trimethyltin hippocampal injury in wild-tyep, APOE knockout, and APOE4 mice. Brain Behav Immun 14: 288–304.
68. Fabry Z, Raine CS, Hart MN. Nervous tissue as an immune compartment, the dialect of the immune response in the CNS. Immunol Today 1994; 15: 218–24.
69. Hickey WF, Hsu BL, Kimura H. T-lymphocyte entry into the central nervous system. J Neurosci Res 1991; 28: 254–60.
70. Beray-Berthat V, Croci N, Plotkine M, et al. Polymorphonuclear neutrophils contribute to infarction and oxidative stress in the cortex but not in the striatum after ischemia-reperfusion in rats. Brain Res 2003; 987: 32–8.
71. Beray-Berthat V, Palmier B, Plotkine M, et al. Neutrophils do not contribute to infarction, oxidative stress, and NO synthase activity in severe brain ischemia. Exp Neurol 2003; 182: 446–54.
72. Harris AK, Ergul A, Kozak A, et al. Effect of neutrophil depletion on gelatinase expression, edema formation and hemorrhagic transformation after focal ischemic stroke. BMC Neurosci 2005; 6: 49.
73. Andersson PB, Perry VH, Gordon S. Intracebral injection of proinflammatory cytokines or leucocyte chemotaxins induces minimal myelomonocytic cell recruitment to the parenchyma of the CNS. J Exper Med 1992; 176: 255–9.
74. Matsumoto Y, Fujiwara M. Absence of donor-type major histocompatibility complex class I antigen bearing microglia in the rat central nervous system of radiation bone marrow chimeras. J Neuroimmunol 1987; 17: 71–82.
75. Lassmann H, Schmied M, Vass K, Hickey WF. Bone marrow derived elements and resident microglia in brain inflammation. Glia 1993; 7: 19–24.
76. Sedgwick JD, Schwender S, Imrich H, et al. Isolation and direct characterization of resident microglial cells from the normal and inflamed central nervous system. Proc Natl Acad Sci U S A 1991; 88: 7438–42.
77. Renno T, Zeine R, Girard JM, et al. Selective enrichment of Th1 CD45RBlow CD4+ T cells in autoimmune infiltratesin experimental allergic encephalomyelitis. Int Immunol 1994; 6: 347–54.
78. Carson MJ, Reilly CR, Sutcliffe JG, Lo D. Mature microglia resembleimmature antigen-presenting cells. Glia 1998; 22: 72–85.
79. Ransohoff RM. Microgliosis: the questions shape the answers. Nature Neurosci 2007; 10: 1507–9.
80. Kreutzberg GW. Microglia, a sensor for pathological events in the CNS. Trends Neurosci 1996; 19: 312–18.
81. Nimmerjahn A, Kirchhoff F, Helmchen F. Resting microglial cells are highly dynamic surveillants of brain parenchyma in vivo. Science 2005; 308: 1314–18.
82. Davalos D, Gruntzendler J, Yang G, et al. ATP mediates rapid microglial response to local brain injury in vivo. Nat Neurosci 2005; 8: 752–8.
83. Raivich G. Like cops on the beat, the active role of resting microglia. Trends Neurosci 2005; 28: 571–3.
84. Perry VH, Gordon S. Modulation of CD4 antigen of macrophages and microglia in rat brain. J Exp Med 1987; 166: 1138–43.
85. Kreutzberg GW. Autoradiographische Untersuchung uber die Beteiligung von Gliazellen and der axonalen Reaktion im facialiskern der ratte. Acta Neuropathol 1996; 7: 149–61.
86. Haynes SE, Hollopeter G, Yang G, et al. The P2Y12 receptor regulates microglial activation by extracellular nucleotides. Nat Neurosci 2006; 9: 1512–19.
87. Persaud-Sawin D, Banach L, Harry GJ. Raft aggregation with specific receptor recruitment is required for microglial phagocytosis of Abeta42. Glia 2008; 57: 320–35.
88. Streit WJ, Graeber MB, Kreutzberg GW. Functional plasticity of microglia, a review. Glia 1988; 1: 301–7.
89. Giulian D, Chen J, Ingeman JE, et al. The role of mononuclear phagocytes in wound healing after traumatic injury to adult mammalian brain. J Neurosci 1989; 9: 4416–29.
90. Dickson DW, Mattiace LA, Kure K, et al. Microglia in human disease, with an emphasis on the acquired immune deficiency syndrome. Lab Invest 1991; 64: 135–56.
91. Perry VH, Andersson P-B, Gordon S. Macrophages and inflammation in the central nervous system. Trends Neurosci 1993; 16: 268–73.
92. Gehrmann J, Bonnekoh P, Miyazawa T, et al. Immunocytochemical study of an early microglial activation in ischemia. J Cer Blood Flow Metab 1992; 12: 257–69.
93. Gehrmann J, Mies G, Bonnekoh P, et al. Microglial reaction in the rat cerebral cortex induced by cortical spreading depression. Brain Pathol 1993; 3: 11–18.
94. Gehrmann J, Matsumoto Y, Kreutzberg GW. Microglia, intrinsic immuneffector cell of the brain. Brain Res Rev 1995; 20: 269–87.
95. Gebicke-Haerter PJ, VanCalker D, Norenberg W, et al. Molecular mechanisms of microglial activation. A. Implications for regeneration and neurodegenerative diseases. Neurochem Int 1996; 29: 1–12.

96. Kreutzberg GW. Microglia, a sensor for pathological events in the CNS. Trends Neurosci 1996; 19: 312–28.
97. Thomas WE. 1992. Brain microphages: evaluation of microglia and their functions. Brain Res Brain Res Rev 17: 61–74.
98. Gonzalez-Scarano F, Baltuch G. Microglia as mediators of inflammatory and degenerative diseases. Annu Rev Neurosci 1999; 22: 219–40.
99. Carson MJ. Microglia as liaisons between the immune and central nervous systems, functional implications for multiple sclerosis. Glia 2002; 40: 218–31.
100. Danton GH, Dietrich WD. Inflammatory mechanisms after ischemia and stroke. J Neuropathol Exp Neurol 2003; 62: 127–36.
101. Streit WJ. Microglia as neuroprotective, immunocompetent cells of the CNS. Glia 2002; 40: 133–9.
102. Schwartz M, Moalem G, Leibowitz-Amit R, et al. Innate and adaptive immune responses can be beneficial for CNS repair. Trends Neurosci 1999; 22: 295–9.
103. Kurkowska-Jastrzebska I, Joniec I, Zaremba M, et al. Anti-myelin basic protein T cells protect hippocampal neurons against trimethyltin-induced damage. Neuroreport 2007; 18: 425–9.
104. Pavlov VA, Wang H, Czura CJ, et al. The cholinergic anti-inflammatory pathway, a missing link in neuroimmunomodulation. Mol Med 2003; 9: 125–34.
105. Perry VH. The influence of systemic inflammation on inflammation in the brain, implications for chronic neurodegenerative disease. Brain Behav Immmun 2004; 18: 407–13.
106. Carson MJ, Lo DD. Perspective is everything. An irreverent discussion of CNS-immune system interactions as viewed from different scientific traditions. Brain Behav Immun 2007; 21: 367–73.
107. Perry VH, Cunningham C, Holmes C. Systemic infections and inflammation affect chronic neurodegeneration. Nature Rev 2007; 7: 161–7.
108. Schiefer J, Kampe K, Dodt HU, Zieglgänsberger W, Kreutzberg GW. Microglial motility in the rat facial nucleus following peripheral axotomy. J Neurocytol 1999; 28: 439–531.
109. Cullheim S, Thams S. The microglial networks of the brain and their role in neuronal network plasticity after lesion. Brain Res Rev 2007; 55: 89–96.
110. Fourgeaud L, Boulanger LM. Synapse remodeling, compliments of the complement system. Cell 2007; 131: 1034–6.
111. Perry VH, O'Connor V. C1q: the perfect complement for a synaptic feast? Nat Rev Neurosci 2008; 9: 807–11.
112. Stevens B, Allen NJ, Vazquez LE, et al. The classical complement cascade mediates CNS synapse elimination. Cell 2007; 131: 1164–78.
113. Streit WJ. Microglia as neuroprotective, immunocompetent cells of the CNS. Glia 2002; 40: 133–9.
114. Kohl A, Dehghani F, Korf HW, et al. The bisphosphonate clodronate depletes microglial cells in excitotoxically injured organotypic hippocampal slice cultures. Exp Neurol 2003; 181: 1–11.
115. Kitamura Y, Takata K, Inden M, et al. Intracerebroventricular injection of microglia protects against focal brain ischemia. J Pharmacol Sci 2004; 94: 203–6.
116. Hayashi Y, Tomimatsu Y, Suzuki H, et al. The intra-arterial injection of microglia protects hippocampal CA1 neurons against global ischemia-induced functional deficits in rats. Neurosci 2006; 142: 87–96.
117. Lalancette-Hebert M, Gowing G, Simard A, et al. Selective ablation of proliferating microglial cells exacerbates ischemic injury in the brain. J Neurosci 2007; 27: 2596–605.
118. Oshiro S, Kawamura K, Zhang C, et al. Microglia and astroglia prevent oxidative stress-induced neuronal cell death, implications for aceruloplasminemia. Biochim Biophys Acta 2008; 1782: 109–17.
119. Elkabes S, DiCicco-Bloom EM, Black I.B. Brain microglia/macrophages express neurotrophins that selectively regulate microglial proliferation and function. J Neurosci 1996; 16: 2508–21.
120. Magnus T, Chan A, Savill J, et al. Phagocytotic removal of apoptotic, inflammatory lymphocytes in the central nervous system by microglia and its functional implications. J Neuroimmunol 2002; 130: 1–9.
121. Batchelor PE, Liberatore GT, Wong JY, et al. Activated macrophages and microglia induce dopaminergic sprouting in the injured striatum and express brain-derived neurotrophic factor and glial cell line-derived neurotrophic factor. J Neurosci 1999; 19: 1708–16.
122. Stadelmann C, Kerschensteiner M, Misgeld T, et al. BDNF and gp145trkB in multiple sclerosis brain lesions: neuroprotective interactions between immune and neuronal cells? Brain 2002; 125: 75–85.
123. López-Redondo F, Nakajima K, Honda S, et al. Glutamate transporter GLT-1 is highly expressed in activated microglia following facial nerve axotomy. Brain Res Mol Brain Res 2000; 76: 429–35.
124. Watanabe H, Abe H, Takeuchi S, et al. Protective effect of microglial conditioning medium on neuronal damage induced by glutamate. Neurosci Lett 2000; 289: 53–6.

125. Zietlow R, Dunnett SB, Fawcett JW. The effect of microglia on embryonic dopaminergic neuronal survival in vitro, diffusible signals from neurons and glia change microglia from neurotoxic to neuroprotective. Eur J Neurosci 1999; 11: 1657–67.
126. Suzuki T, Hide I, Ido K, et al. Production and release of neuroprotective tumor necrosis factor by P2X7 receptor-activated microglia. J Neurosci 2004; 24: 1–7.
127. Rabchevsky AG, Streit WJ. Grafting of cultured microglial cells into the lesioned spinal cord of adult rats enhances neurite outgrowth. J Neurosci Res 1997; 47: 34–48.
128. Schoen SW, Graeber MB, Kreutzberg GW. 5 -Nucleotidase immunoreactivity of perineural microglia responding to rat facial nerve axotomy. Glia 1992; 6: 314–17.
129. Kipnis J, Mizrahi T, Yoles E, et al. Myelin specific Th1 cells are necessary for post-traumatic protective autoimmunity. J Neuroimmunol 2002; 130: 78–85.
130. Schwartz M, Kipnis J. Self and non-self discrimination is needed for the existence rather than deletion of autoimmunity, the role of regulatory T cells in protective autoimmunity. Cell Mol Life Sci 2004; 61: 2285–9.
131. Shaked I, Tchoresh D, Gersner R, et al. Protective autoimmunity, interferon-gamma enables microglia to remove glutamate without evoking inflammatory mediators. J Neurochem 2005; 92: 997–1009.
132. McCarthy KD, DeVellis J. Preparation of separate astroglial and oligodendroglial cell cultures from rat cerebral tissue. J Cell Biol 1980; 85: 890–902.
133. Giulian D, Vaca K, Corpuz M. Brain glia release factors with opposing actions on neuronal survival. J Neurosci 1993; 13: 29–37.
134. Hooper C, Pinteaux-Jones F, Fry VA, et al. Differential effects of albumin on microglia and macrophages; implications for neurodegeneration following blood-brain barrier damage. J Neurochem 2009; DOI 10.1111/j.1471–4159.2009.05953.x.
135. Adams RA, Bauer J, Flick MJ, et al. The fibrin-derived γ377–395 peptide inhibits microglia activation and suppresses relapsing paralysis in central nervous system autoimmune disease. J Exp Med 2007; 204: 571–82.
136. Wright SD, Ramos RA, Tobias PS, Ulevitch RJ, Mathison JC. CD14, a receptor for complexes of lipopolysaccharide (LPS) and LPS binding protein. Science 1990; 249: 1431–3.
137. Jahnke GD, Brunssen S, Maier WE, Harry GJ, Neurotoxicant-induced elevation of adrenomedullin expression in hippocampus and glia cultures. J Neurosci Res 2001; 66: 464–74.
138. Gebicke-Haerter PJ, Bauer J, Schobert A, Morthoff H. Lipopolysaccharide-free conditions in primary astrocyte cultures allow growth and isolation of microglial cells. J Neurosci 1989; 9: 183–94.
139. Hailer NP, Jarhult JD, Nitsch R. Resting microglial cells in vitro, analysis of morphology and adhesion molecule expression in organotypic hippocampal slice cultures. Glia 1996; 18: 319–31.
140. Stence N, Waite M, Dailey ME. Dynamics of microglial activation, a confocal time-lapse analysis in hippocampal slices. Glia 2001; 33: 256–66.
141. Nolte C, Möller T, Walter T, et al. Complement 5a controls motility of murine microglial cells in vitro via activation of an inhibitory G-protein and the rearrangement of the actin cytoskeleton. Neurosci 1996; 73: 1091–107.
142. Flugel G, Hager A, Horvat C, et al. Neuronal MCP-1 expression in response to remote nerve injury. J Cereb Blood Flow Metab 2001; 21: 69–76.
143. Rappert A, Biber K, Nolte C, et al. Secondary lymphoid tissue chemokine (CCL21) activates CXCR3 to trigger a Cl- current and chemotaxis in murine microglia. J Immunol 2002; 168: 3221–6.
144. Zhang J, De Koninck Y. Spatial and temporal relationship between monocyte chemoattractant protein-1 expression and spinal glial activation following peripheral nerve injury. J Neurochem 2006; 97: 772–83.
145. Babcock AA, Kuziel WA, Rivest S, et al. Chemokine expression by glial cells directs leukocytes to sites of axonal injury in the CNS. J Neurosci 2003; 23: 7922–30.
146. Rappert A, Bechmann I, Pivneva T, et al. CXCR3-dependent microglial recruitment is essential for dendrite loss after brain lesion. J Neurosci 2004; 24: 8500–9.
147. Pawate S, Shen Q, Fan F, et al. Redox regulation of glial inflammatory response to lipopolysaccharide and interferon-gamma. J Neurosci Res 2004; 77: 540–51.
148. Guix FX, Uribesalgo I, Coma M, et al. The physiology and pathophysiology of nitric oxide in the brain. Prog Neurobiol 2005; 76: 126–52.
149. Zietlow R, Dunnett SB, Fawcett JW. The effect of microglia on embryonic dopaminergic neuronal survival in vitro, diffusible signals from neurons and glia change microglia from neurotoxic to neuroprotective. Eur J Neurosci 1999; 11: 1657–67.
150. Smits HA, van Beelen AJ, de Vos NM, et al. Activation of human macrophages by amyloid-beta is attenuated by astrocytes. J Immunol 2001; 166: 6869–76.

151. Kettenmann H, Hoppe D, Gottmann K, et al. Cultured mciroglial cells have a distinct pattern of membrane channels different from peritoneal macrophages. J Neurosci Res 1990; 26: 278–87.
152. Kettenmann H, Banati R, Walz W. Electrophysiological behavior of microglia. Glia 1993; 7: 93–101.
153. Walz W, Ilschner S, Ohlemeyer C, Banati R, Kettenmann H. Extracellular ATP activates a cation conductance and a K+ conductance in cultured microglial cells from mouse brain. J Neurosci 1993; 13: 4403–11.
154. Langosch JM, Gebicke-Haerter PJ, Nörenberg W, Illes P. Characterization and transduction mechanisms of purinoceptors in activated rat mciroglia. Brit J Pharmacol 1994; 113: 29–34.
155. Priller J, Haas CA, Reddington M, et al. Calcitonin gene-related peptide and ATP induce immediate early gene expression in cultured rat microglia cells. Glia 1995; 15: 447–57.
156. Whittemore ER, Korotzer AR, Etebari A, et al. Carbachol increases intracellular free calcium in cultured rat microglia. Brain Res 1993; 621: 59–64.
157. Piani D, Frei K, Do KQ, et al. Murine brain macrophages induce NMDA receptor mediated neurotoxicity in vitro by secreting glutamate. Neurosci Lett 1991; 133: 159–62.
158. Patrizio M, Levi G. Glutamate production by cultured microglial: differences between rat and mouse, enhancement by lipo-polysaccharide and lack of HiV coat protein gp120 and depolarizing agents. Neurosci Lett 1994; 178: 184–8.
159. Giulian D, Corpuz M, Chapman S, Manzouri M, Robertson C. Reactive mononuclear phagocytes release neurotoxins after ischemic and traumatic injury to the central nervous system. J Neurosci Res 1993; 36: 681–93.
160. Giulian D, Baker TJ, Shih LN, Lachman LB. Interleukin 1 of the central nervous system is produced by ameboid microglia. J Exp Med 1986; 164: 594–604.
161. Frei K, Malipiero UV, Leist TP, et al. On the cellular source and function of interleukin 6 produced in the central nervous system in viral diseases. Eur J Immunol 1989; 19: 689–94.
162. Sawada M, Kondo N, Suzumura A, Marunouchi T. Production of tumor necrosis factor-alpha by microglia and astrocytes in culture. Brain Res 1989; 509: 119–24.
163. Giulian D, Ingeman JE. Colony-stimulating factors as promoters of ameboid microglia. J Neurosci 1988; 8: 4707–17.
164. Suzumaur A, Sawada M, Yamamoto H, Marunouchi T. Effects of colony stimulating factors on isolated microglia in vitro. J Neuroimmunol 1990; 30: 111–20.
165. Perrier P, Martinez FO, Locati M, et al. Distinct transcriptional programs activated by interleukin-10 with or without lipopolysaccharide in dendritic cells: induction of the B cell-activating chemokine, CXC chemokine ligand 13. J Immunol 2004; 172: 7031–42.
166. Alderson MR, Armitage RJ, Maraskovsky E, et al. Fas transduces activation signals in normal human T lymphocytes. J Exper Med 1993; 178: 2231–5.
167. Hernandez-CasellesT, Stutman O. Immune functions of tumor necrosis factor. I. Tumor necrosis factor induces apoptosis of mouse thymocytes and can also stimulate or inhibit IL-6-induced proliferation depending on the concentration of mitogenic costimulation. J Immunol 1993; 151: 3999–4012.
168. Kessler JA, Ludlam WH, Freidin MM, et al. Cytokine-induced programmed death of cultured sympathetic neurons. Neuron 1993; 11: 1123–32.
169. Schindler R, Clark BD, Dinarello CA. Dissociation between interleukin-1b mRNA and protein synthesis in human peripheral blood mononuclear cells. J Biol Chem 1990; 265: 10232–7.
170. Allan SM, Tyrrell PJ, Rothwell NJ. Interleukin-1 and neuronal injury. Nat Rev Immunol 2005; 5: 629–40.
171. Taylor DL, Jones F, Kubota ES, et al. Stimulation of microglia metabotropic glutamate receptor mGlu2 triggers tumor necrosis factor α-induced neurotoxicity in concert with microglial-derived fas ligand . J Neurosci 2005; 25: 2945–64.
172. Tartaglia LA, Goeddel DV. Two TNF receptors. Immunol Today 1992; 13: 151–3.
173. Hsu H, Shu HB, Pan MG, et al. TRADD-TRAF2 and TRADD-FADD interactions define two distinct TNF receptor 1 signal transduction pathways. Cell 1996; 84: 299–308.
174. Schneider-Brachert W, Tchikov V, Neumeyer J, et al. A novel protein that interacts with the death domain of Fas/APO1 contains a sequence motif related to the death domain. J Biol Chem 1995; 270: 7795–8.
175. Chinnaiyan AM, O'Rourke K, Tewari M, et al. FADD, a novel death domain-containing protein, interacts with the death domain of Fas and initiates apoptosis. Cell 1995; 81: 505–12.
176. Yang L, Lindholm K, Konishi Y, et al. Target depletion of distinct tumor necrosis factor receptor subtypes reveals hippocampal neuron death and survival through different signal transduction pathways. J Neurosci 2002; 22: 3025–32.
177. Gendelman HE, Folks DG. Innate and acquired immunity in neurodegenerative disorders. J Leukoc Biol 1999; 65: 407–8.

178. Lambertsen KL, Clausen BH, Babcock AA, et al. Microglia protect neurons against ischemia by synthesis of tumor necrosis factor. J Neurosci 2009; 29: 1319–30.
179. Harry GJ, Lefebvre d'Hellencourt C, McPherson CA, et al. Tumor necrosis factor p55 and p75 receptors are involved in chemical-induced apoptosis of dentate granule neurons. J Neurochem 2008; 106: 281–98.
180. Ding A, Nathan CF, Graycar J, et al. Macrophage deactivating factor and transforming growth factors-b1, -b2, and -b3 inhibit induction of machrophage nitrogen oxide synthesis by IFN-g. J Immunol 1990; 145: 940–4.
181. Schluesener HJ. Transforming growth factors type beta 1 and beta 2 suppress rat astrocytes autoantigen presentation and antagonize hyperinduction of class II major histocompatibility complex antigen expression by interferon-gamma and tumor necrosis factor-alpha. J Neuroimmunol 1990; 27: 41–7.
182. Merrill JE, Zimmerman RP. Natural and induced cytotoxicity of oligodendrocytes by microglia is inhibitable by TGF beta. Glia 1991; 4: 327–31.
183. Lodge PA, Sriram S. Regulation of microglial activation by TGF-beta, IL-10 and CSF-1. J Leukocyte Biol 1996; 60: 502–8.
184. Bogdan C, Nathan C. Modulation of macrophage function by transforming growth factor beta, interleukin-4 and interleukin-10. Ann N Y Acad Sci 1993; 685: 713–39.
185. Zurawski G, deVries E. Interleukin 13, an interleukin 4-like cytokine that acts on monocytes and B cells, but not on T cells. Immunol Today 1994; 15: 19–26.
186. Leroith D, Nissley P. Knock your SOCS off! J Clin Invest 2005; 115: 233–6.
187. Park EJ, Park SY, Joe EH, et al. 15d-PGJ2 and rosiglitazone suppress Janus kinase-Stat inflammatory signaling through induction of suppressor of cytokine signaling 1 (SOCS1) and SOCS3 in glia. J Biol Chem 2003; 278: 14747–52.
188. Burger D, Dayer JM. Inhibitory cytokines and cytokine inhibitors. Neurology 1995; 45: S39–43.
189. Norris JG, Tang LP, Sparacio SM, et al. Signal transduction pathways mediating astrocite IL-6 induction by IL-1 beta and tumor necrosis factor-alpha. J Immunol 1994; 152: 841–50.
190. Aderka D, Le JM, Vilcek J. IL-6 inhibits lipopolysaccharide-induced tumor necrosis factor production in cultured human monocytes, U937 cells, and in mice. J Immunol 1989; 143: 3517–23.
191. Tilg H, Trehu E, Atkins MB, et al. Interleukin-6 as an anti-inflammatory cytokine, induction of circulating IL-1 receptor antagonist and soluble tumor necrosis factor receptor p55. Blood 1994; 83: 113–18.
192. Turner M, Chantry D, Katsikis P, et al. Induction of the interleukin 1 receptor antagonist protein by transforming growth factor-beta. Eur J Immunol 1991; 21: 1635–9.
193. Vannier E, Miller LC, Dinarello CA. Coordinated antiinflammatory effects of interleukin 4: interleukin 4 suppresses interleukin 1 production but up-regulates gene expression and synthesis of interleukin 1 receptor antagonist. Proc Natl Acad Sci U S A 1992; 89: 4076–80.
194. Kreutzberg GW. Microglia: a sensor for pathological events in the CNS. Trends Neurosci 1996; 19: 312–18.
195. Carson MJ, Bilousova TV, Puntambekar SS, et al. A rose by any other name? The potential consequences of microglial heterogeneity during CNS health and disease. Neurotherapeutics 2007; 4: 571–9.
196. Gordon S. Alternative activation of macrophages. Nat Rev 2003; 3: 23–35.
197. Woodroofe M, Sarna G, Wadha M, et al., Detection of IL-1 and IL-6 in adult rat brain, following mechanical injury, by in vivo microdialysis: evidence of a role for microglia in cytokine production. J Neuroimmunol 1991; 33: 227–36.
198. Lindholm D, Castrén E, Kiefer R, Zafra F, Thoenen H. Transforming growth factor-beta 1 in the rat brain: increase after injury and inhibition of astrocyte proliferation. J Cell Biol 1992; 117: 395–400.
199. Flaris NA, Densmore TL, Molleston MC, Hickey WF. Characterization of microglia and macrophages in the central nervous system of rats: definition of the differential expression of molecules using standard and novel monoclonal antibodies in normal CNS and in four models of parenchymal reaction. Glia 1993; 7: 34–40.
200. Nelson PT, Soma LA, Lavi E. Microglia in diseases of the central nervous system. Ann Med 2002; 34: 491–500.
201. Butovsky O, Hauben E, Schwartz M. Morphological aspects of spinal cord autoimmune neuroprotection: colocalization of T cells with B7–2 (CD86) and prevention of cyst formation. FASEB J 2001; 15: 1065–7.
202. Butovsky O, Talpalar AE, Ben-Yaakov K, Schwartz M. Activation of microglia by aggregated beta-amyloid or lipopolysaccharide impairs MHC-II expression and renders them cytotoxic whereas IFN-gamma and IL-4 render them protective. Mol Cell Neurosci 2005; 29: 381–93.

203. Shaked I, Porat Z, Gersner R, Kipnis J, Schwartz M. Early activation of microglia as antigen-presenting cells correlates with T cell-mediated protection and repair of the injured central nervous system. J Neuroimmunol 2004; 146: 84–93.
204. Shaked I, Tchoresh D, Gersner R, et al. Protective autoimmunity: interferon-gamma enables microglia to remove glutamate without evoking inflammatory mediators. J Neurochem 2005; 92: 997–1009.
205. Jensen MB, Hegelund IV, Poulsen FR, et al. Microglial reactivity correlates to the density and the myelination of the anterogradely degenerating axons and terminals following perforant path denervation of the mouse fascia dentata. Neuroscience 1999; 93: 507–18.
206. Byram SC, Carson MJ, DeBoy CA, et al. CD4-positive T cell-mediated neuroprotection requires dual compartment antigen presentation. J Neurosci 2004; 24: 4333–9.
207. Sanders VM, Jones KJ. Role of immunity in recovery from a peripheral nerve injury. J Neuroimmune Pharmacol 2006; 1: 11–19.
208. Huang Z, Ha GK, Petitto JM. IL-15 and IL-15R alpha gene deletion: effects on T lymphocyte trafficking and the microglial and neuronal responses to facial nerve axotomy. Neurosci Lett 2007; 417: 160–4.
209. Bolsi D. Placche senile e microgla. Riv Pat Nerv Ment 1927; 32: 65.
210. McGeer PL, Itagaki S, McGeer EG. Expression of the histocompatibility glycoprotein HlA-DR in neurological disease Acta Neuropathol 1988; 76: 550–7.
211. Gaskin F, Finley J, Fang Q, et al. Human antibodies reactive with beta-amyloid protein in Alzheimer's disease. J Exp Med 1993; 177: 1181–6.
212. Hyman BT, Smith C, Buldyrev I, et al. Autoantibodies to amyloid-beta and Alzheimer's disease. Ann Neurol 2001; 49: 808–10.
213. Eikelenboom P, Bate C, Van Gool WA, et al. Neuroinflammation in Alzheimer's disease and prion disease. Glia 2002; 40: 232–9.
214. Rogers J, Strohmeyer R, Kovelowski CJ, Li R. Microglia and inflammatory mechanisms in the clearance of amyloid beta peptide. Glia 2002; 40: 260–9.
215. Kim SU, de Vellis J. Microglia in health and disease. J Neurosci Res 2005; 81: 302–13.
216. McGeer EG, McGeer PL. Inflammatory processes in Alzheimer's disease. Prog Neuropsychopharmacol Biol Psych 2003; 27: 741–9.
217. Simard AR, Soulet D, Gowing G, et al. Bone marrow-derived microglia play a critical role in restricting senile plaque formation in Alzheimer's disease. Neuron 2006; 49: 489–502.
218. D'Andrea MR, Cole GM, Ard MD. The microglial phagocytic role with specific plaque types in the Alzheimer disease brain. Neurobiol Aging 2004; 25: 675–83.
219. Griffin WS, Stanley LC, Ling C. Brain IL-1 and S-100 immunoreactivity are elevated in Down syndrome and Alzheimers disease. Proc Natl Acad Sci U S A 1989; 86: 7611–15.
220. Aisen PS. Inflammation and Alzheimer disease. Mol Chem Neuropathol 1996; 28: 83–8.
221. Bornemann KD, Staufenbiel M. Transgenic mouse models of Alzheimer's disease. Ann N Y Acad Sci 2000; 908: 260–6.
222. Emmerling MR, Watson MD, Raby CA, Spiegel K. The role of complement in Alzheimer's disease pathology. Biochim Biophys Acta 2000; 1502: 158–71.
223. Streit WJ, Conde JR, Fendrick SE, Flanary BE, Mariani CL. Role of microglia in the central nervous system's immune response. Neurol Res 2005; 27: 685–91.
224. Gasque P, Dean YD, McGreal EP, et al. Complement components of the innate immune system in health and disease in the CNS. Immunopharmacology 2000; 49: 171–86.
225. McGeer PL, McGeer EG. Inflammation, autotoxicity and Alzheimer disease. Neurobiol Aging 2001; 22: 799–809.
226. Husemann J, Loike JD, Anankov R, et al. Scavenger receptors in neurobiology and neuropathology, their role on microglia and other cells of the nervous system. Glia 2002; 40: 195–205.
227. Kielian T, Mayes P, Kielian M. Characterization of microglial responses to Staphylococcus aureus, effects on cytokine, costimulatory molecules, and Toll-like receptor expression. J Neuroimmunol 2002; 130: 86–99.
228. Richard KL, Filali M, Préfontaine P, et al. Toll-like receptor 2 acts as a natural innate immune receptor to clear amyloid beta 1-42 and delay the cognitive decline in a mouse model of Alzheimer's disease. J Neurosci 2008; 28: 5784–93.
229. Shen Y, Lue L, Yang L, et al. Complement activation by beta-amyloid in Alzheimer disease. Proc Natl Acad Sci U S A 1992; 89: 10016–20.
230. Strohmeyer R, Shen Y, Rogers J. Detection of complement alternative pathway mRNA and proteins in the Alzheimer's disease brain. Brain Res 2000; 81: 7–18.
231. Yasojima K, Schwab C, McGeer EG, McGeer PL. Up-regulated production and activation of the complement system in Alzheimer's disease brain. Amer J Pathol, 1999; 154: 927–36.

232. Wyss-Coray T, Yan F, Lin AH, et al. Prominent neurodegeneration and increased plaque formation in complement-inhibited Alzheimer's mice. Proc Natl Acad Sci U S A 2002; 99: 10837–42.
233. Fonseca MI, Zhou J, Botto M, Tenner AJ. Absence of C1q leads to less neuropathology in transgenic mouse models of Alzheimer's disease. J Neurosci 2004; 24: 6457–65.
234. Shen Y, Halperin JA, Benzaquen L, Lee CM. Characterization of neuronal cell death induced by complement activation. Brain Res 1997; 1: 186–94.
235. Rus H, Cudrici C, Niculescu F, Shin ML. Complement activation in autoimmune demyelination: dual role in neuroinflammation and neuroprotection. J Neuroimmunol 2006; 180: 9–16.
236. Shen Y, Meri S. Yin and Yang: complement activation and regulation in Alzheimer's disease. Prog Neurobiol 2003; 70: 463–72.
237. Shen Y, Li R, McGeer EG, McGeer PL. Neuronal expression of mRNAs for complement proteins of the classical pathway in Alzheimer brain. Brain Res 1997; 769: 391–5.
238. Stoltzner SE, Grenfell TJ, Mori C, et al. Temporal accrual of complement proteins in amyloid plaques in Down's syndrome with Alzheimer's disease. Amer J Pathol 2000; 156: 489–99.
239. Webster SD, Galvan MD, Ferran E, et al. Antibody-mediated phagocytosis of the amyloid beta-peptide in microglia is differentially modulated by C1q. J Immunol 2001; 166: 7496–503.
240. Fan Ran R, DeFilippis K, Van Nostrand WE. Induction of complement proteins in a mouse model for cerebral microvascular Ab deposition. J Neuroinflammation 2007; 4: 22 doi:10.1186/1742-2094-4-22.
241. Dauer W, Przedborski S. Parkinson's disease: mechanisms and models. Neuron 2003; 39: 889–909.
242. McGeer PL, Itagaki S, Akiyama H, McGeer EG. Rate of cell death in Parkinsonism indicates active neuropathological process. Ann Neurol 1988; 24: 574–6.
243. Hunot S, Dugas N, Faucheux B, et al. FcεRII/CD23 is expressed in Parkinson's disease and induces, in vitro, production of nitric oxide and tumor necrosis factor-α in glial cells. J. Neurosci 1999; 19: 3440–7.
244. Kortekaas R, Leenders KL, van Oostrom JCH, et al. Blood-brain barrier dysfunction in parkinsonian midbrain in vivo. Ann Neurol 2005; 57: 176–9.
245. Brochard V, Combadiere B, Prigent A, et al. Infiltration of CD4+ lymphocytes into the brain contributes to neurodegeneration in a mouse model of Parkinson disease. J Clin Invest 2009; 119: 182–92.
246. Park DR, Thomsen AR, Frevert CW, et al. Fas (CD95) induces proinflammatory cytokine responses by human monocytes and monocyte-derived macrophages. J Immunol 2003; 170: 6209–16.
247. Kebir H, Kreymborg K, Ifergan I, et al. Human TH17 lymphocytes promote blood-brain barrier disruption and central nervous system inflammation. Nat Med 2007; 13: 1173–5.
248. Mirza B, Hadberg H, Thomsen P, Moos T. The absence of reactive astrocytosis is indicative of a unique inflammatory process in Parkinson's disease. Neurosci 2000; 95: 425–32.
249. Koutsilieri E, Scheller C, Grunblatt E, et al. Free radicals in Parkinson's disease. J Neurol 2002; 249(Suppl 2): II1–5.
250. Hald A, Lotharius J. Oxidative stress and inflammation in Parkinson's disease: is there a causal link? Exp Neurol 2005; 193: 279–90.
251. Wright AK, Miller C, Williams M, et al. Microglial activation is not prevented by tacrolimus but dopamine neuron damage is reduced in a rat model of Parkinson's disease progression. Brain Res 2008; 1216: 78–86.
252. Gajkowska B, Szumanska G, Gadamski R. Ultrastructural alterations of brain cortex in rat following intraperitoneal administration of mercuric chloride. J Hirnforsch 1992; 33: 471–6.
253. Charleston JS, Body RL, Mottet NK, et al. Autometallographic determination of inorganic mercury distribution in the cortex of the calcarine sulcus of the monkey Macaca fascicularis following long-term subclinical exposure to methylmercury and mercuric chloride. Toxicol Appl Pharmacol 1995; 132: 325–33.
254. Monnet-Tschudi F. Induction of apoptosis by mercury compounds depends on maturation and is not associated with microglial activation. J Neurosci Res 1998; 53: 361–7.
255. Eskes C, Honegger P, Juillerat-Jeanneret L, et al. Microglial reaction induced by noncytotoxic methylmercury treatment leads to neuroprotection via interactions with astrocytes and IL-6 release. Glia 2002; 37: 43–52.
256. Monnet-Tschudi F, Zurich MG, Boschat C, et al. Involvement of environmental mercury and lead in the etiology of neurodegenerative diseases. Rev Environ Health 2006; 21: 105–17.
257. Bruccoleri A, Brown HW, Harry GJ. Cellular localization and temporal elevation of tumor necrosis factor-a, interleukin-1a, and transforming growth factor-β1 mRNA in hippocampal injury response induced by trimethyltin. J Neurochem 1998; 71: 1577–87.
258. Viviani B, Corsini E, Galli CL, et al. Glia increase degeneration of hippocampal neurons through release of tumor necrosis factor-alpha. Toxicol Appl Pharmacol 1998; 150: 271–6.

259. Little AR, Benkovic SA, Miller DB, et al. Chemically induced neuronal damage and gliosis: enhanced expression of the proinflammatory chemokine, monocyte chemoattractant protein (MCP)-1, without a corresponding increase in proinflammatory cytokines. Neuroscience 2002; 115: 307–20.
260. Eskes C, Juillerat-Jeanneret L, Leuba G, et al. Involvement of microglia-neuron interactions in the tumor necrosis factor-alpha release, microglial activation, and neurodegeneration induced by trimethyltin. J Neurosci Res 2003; 71: 583–90.
261. Monnet-Tschudi F, Zurich MG, Pithon E, van Melle G, Honegger P. Microglial responsiveness as a sensitive marker for trimethyltin (TMT) neurotoxicity. Brain Res 1995; 690: 8–14.
262. Maier WE, Brown HW, Tilson HA, Luster MI, Harry GJ. Trimethyltin increases Il-1α, IL-6, and TNFα mRNA levels in rat hippocampus. J Neuroimmunol 1995; 59: 65–75.
263. Bruccoleri A, Brown HW, Harry GJ. Cellular localization and temporal elevation of TNF, IL-1, and TGFβ1 mRNA in hippocampal injury response induced by trimethyltin. J Neurochem 1998; 71: 1577–87.
264. Maier WE, Bartenbach MJ, Brown HA, Tilson HA, Harry GJ. Induction of tumor necrosis factor alpha in cultured glial cells by trimethyltin. Neurochem Int 1997; 30: 385–92.
265. Bruccoleri A, Harry GJ. Chemical-induced hippocampal neurodegeneration and elevations in TNFα, TNFβ, IL-1β, IP-10, and MCP-1 mRNA in osteopetrotic (op/op) mice. J Neurosci Res 2000; 62: 146–55.
266. Harry GJ, Tyler K, Lefebvre d'Hellencourt C, et al. Morphological alterations and elevations in tumor necrosis factor-alpha, interleukin (IL)-1alpha, and IL-6 in mixed glia cultures following exposure to trimethyltin, modulation by proinflammatory cytokine recombinant proteins and neutralizing antibodies. Toxicol Appl Pharmacol 2002; 180: 205–18.
267. Figiel I, Dzwonek K. TNF-alpha and TNF receptor 1 expression in the mixed neuronal-glial cultures of hippocampal dentate gyrus exposed to glutamate or trimethyltin. Brain Res 2007; 1131: 17–28.
268. Mao H, Fang X, Floyd KM, et al. Differential activation of microglia and astrocytes following trimethyl tin-induced neurodegeneration. Brain Res 2007; 1186: 267–74.
269. Haga S, Haga C, Aizawa T, Ikeda K. Neuronal degeneration and glial cell-responses following trimethyltin intoxication in the rat. Acta Neuropathol 2002; 103: 575–82.
270. Koda T, Kuroda Y, Imai H. Rutin supplementation in the diet has protective effects against toxicant-induced hippocampal injury by suppression of microglial activation and pro-inflammatory cytokines: protective effect of rutin against toxicant-induced hippocampal injury. Cell Mol Neurobiol 2009; 29: 523–31.
271. Rogers J, Mastroeni D, Leonard B, et al. Neuroinflammation in Alzheimer's disease and Parkinson's disease, are microglia pathogenic in either disorder? Int Rev Neurobiol 2007; 82: 235–46.
272. Sriram K, Matheson JM, Benkovic SA, et al. Deficiency of TNF receptors suppresses microglial activation and alters the susceptibility of brain regions to MPTP-induced neurotoxicity, role of TNF-alpha. FASEB J 2006; 20: 670–82.
273. Casarejos MJ, Menéndez J, Solano RM, et al. Susceptibility to rotenone is increased in neurons from parkin null mice and is reduced by minocycline. J Neurochem 2006; 97: 934–46.
274. Yang L, Sugama S, Chirichigno JW, et al. Minocycline enhances MPTP toxicity to dopaminergic neurons. J Neurosci Res 2003; 74: 278–85.
275. Zhou F, Wu JY, Sun XL, et al. Iptakalim alleviates rotenone-induced degeneration of dopaminergic neurons through inhibiting microglia-mediated neuroinflammation. Neuropsychopharmacol 2007; 32: 2570–80.
276. Saint-Pierre M, Tremblay ME, Sik A, Gross RE, Cicchetti F. Temporal effects of paraquat/maneb on microglial activation and dopamine neuronal loss in older rats. J Neurochem 2006; 98: 760–72.
277. Purisai MG, McCormack AL, Cumine S, et al. Microglial activation as a priming event leading to paraquat-induced dopaminergic cell degeneration. Neurobiol Dis 2007; 25: 392–400.
278. McCann MJ, O'Callaghan JP, Martin PM, Bertram T, Streit WJ. Induction of microglial reactive oxygen species production by the organochlorinated pesticide dieldrin. Neuroscience 1996; 72: 273–81.
279. Goodwill MH, Lawrence DA, Seegal RF. Polychlorinated biphenyls induce proinflammatory cytokine release and dopaminergic dysfunction, protection in interleukin-6 knockout mice. J Neuroimmunol 2007; 183: 125–32.
280. Klegeris A, Choi HB, McLarnon JG, McGeer PL. Functional ryanodine receptors are expressed by human microglia and THP-1 cells: their possible involvement in modulation of neurotoxicity. J Neurosci Res 2007; 85: 2207–15.
281. Brunssen SH, Harry GJ, Parham FM, Morgan DL. Carbon monoxide neurotoxicity: transient inhibition of avoidance esponse and delayed microglia reaction in the absence of neuronal death. Toxicology 2003; 194: 51–63.

282. Block ML, Wu X, Pei Z, et al. Nanometer size diesel exhaust particles are selectively toxic to dopaminergic neurons: the role of microglia, phagocytosis, and NADPH oxidase. FASEB J 2004; 18: 1618–20.
283. Whitney NP, Eidem TM, Peng H, Huang Y, Zheng JC. Inflammation mediates varying effects in neurogenesis: relevance to the pathogenesis of brain injury and neurodegenerative disorders. J Neurochem 2009; 108: 1343–59.
284. Jakubs K, Bonde S, Iosif RE, et al. Inflammation regulates functional integration of neurons born in adult brain. J Neurosci 2008; 28: 12477–88.
285. Ekdahl CT, Kokaia Z, Lindvall O. Brain inflammation and adult neurogenesis: the dual role of microglia. Neuroscience 2009; 158: 1021–9.
286. Ekdahl CT, Classen JH, Bonde S, Kokaia Z, Lindvall O. Inflammation is detrimental for neurogenesis in adult brain. Proc Natl Acad Sci U S A 2003; 100: 13632–7.
287. Monje ML, Toda H, Palmer TD. Inflammatory blockade restores adult hippocampal neurogenesis. Science 2003; 302: 1760–5.
288. Ziv Y, Ron N, Butovsky O, et al. Immune cells contribute to the maintenance of neurogenesis and spatial learning abilities in adulthood. Nat Neurosci 2006; 9: 268–75.
289. Kim BJ, Kim MJ, Park JM, et al. Reduced neurogenesis after suppressed inflammation by minocycline in transient cerebral ischemia in rat. J Neurol Sci 2009; 279: 70–5.
290. Harry GJ, McPherson CA, Wine RN, Atkinson K, Lefebvre d'Hellencourt C. Trimethyltin-induced neurogenesis in the murine hippocampus. Neurotox Res 2004; 5: 623–7.
291. Ogita K, Nishiyama N, Sugiyama C, et al. Regeneration of granule neurons after lesioning of hippocampal dentate gyrus: evaluation using adult mice treated with trimethyltin chloride as a model. J Neurosci Res 2005; 82: 609–21.
292. Corvino V, Geloso MC, Cavallo V, et al. Enhanced neurogenesis during trimethyltin-induced neurodegeneration in the hippocampus of the adult rat. Brain Res Bull 2005; 65: 471–7.
293. Nixon K, Kim DH, Potts EN, He J, Crews FT. Distinct cell proliferation events during abstinence after alcohol dependence: microglia proliferation precedes neurogenesis. Neurobiol Dis 2008; 312: 218–29.
294. Aarun J, Sandberg K, Haeberlein S, Persson MA. Migration and differentiation of neural precursor cells can be directed by microglia. Proc Natl Acad Sci U S A 2003; 100: 15983–8.
295. Butovsky O, Ziv Y, Schwartz A, et al. Microglia activated by IL-4 or IFN-gamma differentially induce neurogenesis and oligodendrogenesis from adult stem/progenitor cells. Mol Cell Neurosci 2006; 31: 149–60.
296. Choi YS, Cho HY, Hoyt KR, Naegele JR, Obrietan K. IGF-1 receptor-mediated ERK/MAPK signaling couples status epilepticus to progenitor cell proliferation in the subgranular layer of the dentate gyrus. Glia 2008; 56: 791–800.
297. Choi SH, Veeraraghavalu K, Lazarov O, et al. Non-cell-autonomous effects of presenilin 1 variants on enrichment-mediated hippocampal progenitor cell proliferation and differentiation. Neuron 2008; 59: 568–80.
298. Cacci E, Ajmone-Cat MA, Anelli T, Biagioni S, Minghetti L. In vitro neuronal and glial differentiation from embryonic or adult neural precursor cells are differently affected by chronic or acute activation of microglia. Glia 2008; 56: 412–25.

9 The Slice Culture Model: Considerations for Neurotoxicity Assessment

Andrew D. Kraft

Laboratory of Molecular Toxicology, National Institute of Environmental Health Sciences, National Institutes of Health, Research Triangle Park, North Carolina, U.S.A.

INTRODUCTION

Brain responses involve coinciding contributions from multiple cell types sharing a highly organized microenvironment. In vitro cell culture techniques have the capacity to address mechanistic questions concerning cell-specific responses within the nervous system. The majority of these studies have relied heavily on dissociated primary cultures enriched in one or more distinct, neural cell types in isolation. While these types of preparations allow for an examination of specific mechanistic questions, they are significantly hindered with regard to translation to the in vivo environment. Accurate modeling of neuronal toxicity, in particular, is highly dependent upon the cellular interactions available to the target cells in question. As such, the dynamic cell-cell interactions of the nervous system need to be considered to appropriately interpret and translate changes in neural activity and toxicity. Thus, many in vitro models have been designed with a goal towards maintaining endogenous functional interactions, including normal synaptic connections, between both neuronal and non-neuronal cell types. One approach to incorporate the three-dimensionality of in vivo systems into a culture model has been the use of organotypic slice cultures (OSCs). Such slice systems can maintain integrated components of in vivo neural architecture for extended periods of time in vitro.

Slice neurons retain much of their functionality and endogenous cellular contacts. As a result, studies to date have used slice cultures to ask questions focused on the neuronal populations. Recent work, however, has expanded to include an understanding of the actions of non-neuronal cell populations. Organotypic cultures contain physiological levels of myelinating oligodendrocytes, neuro-supportive astrocytes, and the brain-resident immune cells, microglia. These glia respond to changes in neurons, as well as to subtle alterations in the surrounding tissue, and can serve a varied array of functions capable of modifying neuronal behavior. Data from various disease and brain injury models bears out the observation that the contribution of non-neuronal cells is of much greater impact than originally considered. Glia can act as direct and distinct target sites, often serving diverse roles as inhibitors and effectors of subsequent neuronal loss.

Organotypic preparations are considered a more integrated in vitro approach to examining the nervous system, as compared to isolated cell cultures. The ability to access and manipulate this platform for electrophysiological recording and real-time microscopic imaging of viable cells within a natural matrix has considerable potential for examining processes which are difficult to observe on an in vivo scale. However, while there are many positive attributes of slice cultures, there are also several limiting features. This type of culture system continues to have a significant impact upon the neurotoxicological research field. Thus, it is important to understand both the positive and the negative attributes of this system, as well as how these compare to the biology of the in vivo situation. The following chapter will attempt to provide a review of selected cellular functions observable in slices, ways in which researchers have attempted to use these cultures as in vitro models of neuronal function and injury, perspectives on model development, and potential confounders inherent to these preparations.

THE SLICE CULTURE METHOD

The technique of using rapidly dissected, non-dissociated tissue to study the three-dimensional anatomical structure of the brain is not a recent one. Henry McIlwain first refined OSCs in the 1950s (1–3). OSCs are thick sections (250–400 μm) of early postnatal tissue. They can be generated from many regions of the nervous system, such as the spinal cord, cortex, cerebellum, and the hippocampus. While much of this work has been conducted with rodent tissue, similar technical procedures have been used to prepare slice cultures from other species. Depending on the orientation of the slice, it is possible to obtain cultures containing multiple regions of the brain inclusive of neurons as well as their projection sites. More complex slice models have been generated through the use of co-culture systems that consist of multiple, distinct brain regions or cell types. These models are constantly being fine-tuned in an effort to better maintain or re-establish endogenous cellular networks.

Methodology

Brain regions that are carefully and gently dissected from the brain are sliced into flat-sided sections with a fine blade. Methods for slicing include the use of an automated tissue sectioner, such as a McIlwain tissue chopper, which works via rapid, vertical strokes on wet tissue, or a vibratome, which uses a slowly moving, rapidly oscillating blade to cut tissues immersed in solution. These rapidly isolated tissue slice preparations can be used immediately or maintained using either the roller tube (4,5) or interface (6) method. Using the roller tube technique, the slice is attached to a coverslip and embedded in a plasma clot, which breaks down over subsequent days in vitro (DIV). The tube rotates continuously within the incubator, alternately bathing the explant in medium and removing it to the gas phase. Within a few weeks, the tissue thins down to a monolayer of cells with intact circuitry. This method allows for direct observations with standard light microscopy and direct experimental manipulations of individual neurons. The more commonly used (and more easily prepared) "interface" culture procedure relies upon a semi-permeable membrane to facilitate the maintenance of a static gas-medium interface for the appropriate diffusion of gases and medium nutrients throughout the tissue. By 14DIV, interface cultures decrease to approximately 100 μm (7), resulting in a more physiological, three-dimensional morphology than roller tube cultures (typically, 4–8 cell layers deep), and thick slices more amenable to biochemical analyses.

The extended period of time spent in culture allows for the resolution of many changes that occur simply as a result of explantation. Thus, the term "organotypic slice culture" is often reserved for long-term culture studies, where experimental treatments do not begin until after the organotypic tissue has matured for upwards of 14DIV. However, slice cultures are also routinely employed in acute studies of the developing brain over the first 1–10DIV. In this chapter, the term OSC will encompass acute and mature slice models. Both the slice architecture and the functional activity of neurons and glia change dramatically over the time in vitro. These variations may portend long-term alterations to the responsiveness of OSC cells, compromising essential components of their in vivo actions. The culture preparation-induced changes are often self-resolving and can be incorporated into many neurobiological endpoints, but, if not considered, may pose a difficulty during data interpretation.

FEATURES OF OSCS: TRANSLATION TO IN VIVO NEUROBIOLOGY

Organotypic cultures maintain many of the endogenous synaptic contacts present in vivo and preserve the spatial localization of neuronal fields and white matter tracts. Glia remain intimately associated and intricately placed, at least for a time, in close proximity to neurons in a slice. The cellular makeup and location of the previously existing vasculature and ventricular systems persist transiently in vitro. In addition to these important similarities, however, there are notable differences. Most of these differences are initiated in response to the loss of signaling afferents from outside of the cultured brain region. Any culture-induced alterations have the potential to cause in vitro results to deviate from observations in vivo. Thus, wherever possible, these attributes need to be identified so that their impact can be minimized. The following section highlights important cellular phenotypes of OSCs that are of interest in the context of in vivo cellular behavior. The timecourse of some notable alterations to cellular behavior that result from OSC preparation are summarized in Figure 1. The majority of the features present in vivo are conserved in in vitro OSC systems, some differ in a transient, self-resolving manner, while a handful of attributes necessary for normal brain function in vivo are known to remain altered throughout the culture life of a slice.

Acute Responses in Slice Cultures

The trauma involved in generating the slice and the loss of afferent and efferent signals results in the loss of neurons. In essence, the preparation itself is a model of axon lesioning,

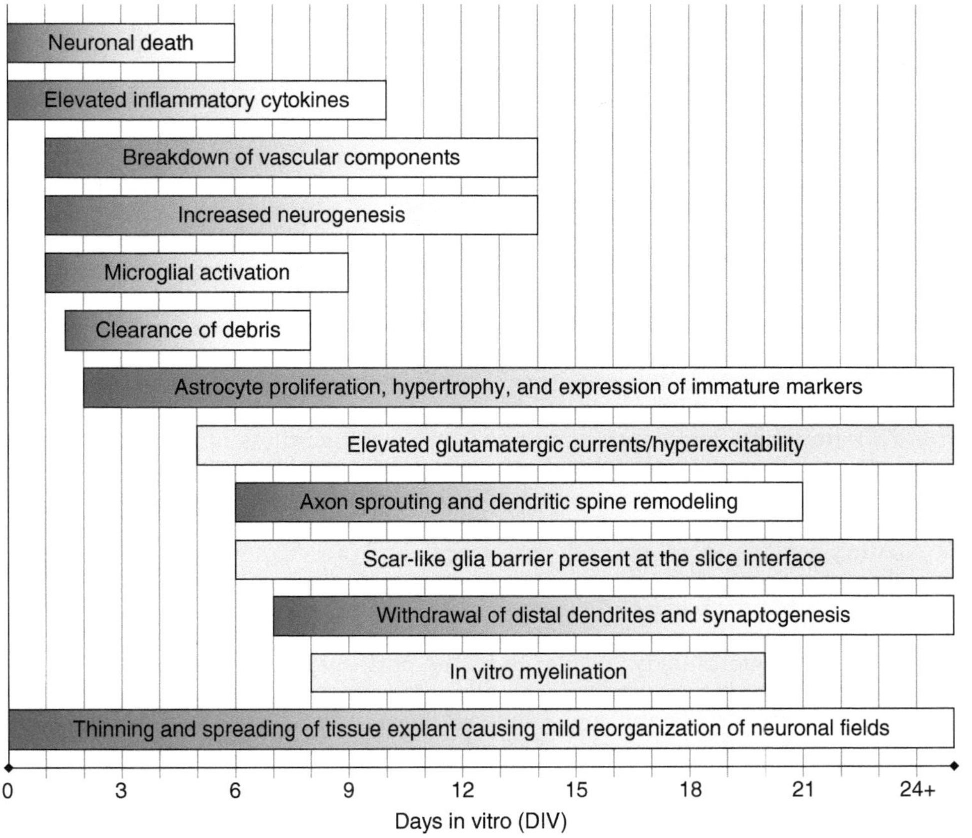

Figure 1 Timeline of events in slice preparations following explantation (7,8,11,16,19,20,22,23,25,33,54,127). Horizontal bars indicate the approximate duration of select events following explantation at 0DIV. Bars with shading show events that have a gradient of intensity (i.e., possess a more severe phenotype at earlier DIV). These changes are more applicable to organotypic preparations cultured according to the interface method than the roller tube method. *Abbreviation*: DIV, days in vitro.

complete with synaptic reorganization, glial activation, and cell loss due to stimulus withdrawal. The isolated nature of the slice system also prevents the survival of brain systems dependent on extra-parenchymal influence, such as the vascular and ventricular systems. The adaptive responses of a slice after explant, including direct reactions to physical displacement and damage as well as reorganization after being placed in a culture vessel, allow it to function in vitro. A consideration of these alterations is essential to understanding the strengths and limitations of applications using either long or short-term cultures of organotypic tissue.

Slice Neurons

In part owing to the loss of synaptic connections, the slice culture preparation has a profound effect on the survivability of neurons in vitro, at least over the short term. Certain subsets of the neurons are lost, while others persist and thrive over their time in vitro. As early as four hours after culturing, degenerating neurons and swollen fibers are observable (8). Extensive neuronal cell death continues up to at least 6DIV (9), followed by a low-level, sporadic and spontaneous neuronal degeneration that may persist beyond 28DIV (10). The majority of the axonal and neuronal debris created is effectively cleared from the neuronal microenvironment by approximately 8DIV, encouraging the successful growth, restoration, and maturation of functional synapses.

Synaptic Connections

Electrophysiological recordings are particularly variable during the active period of synaptic reorganization that immediately follows culture preparation. It is important to acknowledge and consider this phenomenon when conducting physiological studies in OSCs. The loss of signaling afferents causes a mild reorganization of the synaptic circuitry. In cultures derived from the hippocampus, the most prominent structural reorganization that occurs is the establishment of supragranular collaterals by mossy fibers, similar to the normal reaction of the dentate gyrus to deafferentation seen during in vivo lesion studies (11,12) This aberrant axon sprouting in the molecular layer of the dentate gyrus begins at 6-8DIV and persists, increasing in intensity over the first 21DIV, though these effects can be reduced via co-culture with entorhinal cortex (13). In addition, pyramidal CA3 and CA1 regions form new, functional synapses to the dentate granule cell layer (14,15). These connections are unique to organotypic cultures and therefore physiologically anomalous. The formation of these excess excitatory contacts is likely responsible for an elevation in glutamatergic currents in slices witnessed from 7–21DIV (16). This elevation parallels a hyperexcitability of granule cells that increases as the slices mature (17). Importantly, synaptic connections are lost regardless of the region of the brain cultured, necessitating the deafferented culture neurons to sprout and recreate vital synaptic components. This represents a physiological effort to re-establish anterograde and retrograde signaling gradients.

Neurotransmitter receptors, both postsynaptic and presynaptic, inhibitory and excitatory, are maintained at comparable levels in slices as compared to in vivo, with the possible exception of gamma- aminobutyric acid (GABA) receptors (18). In OSCs, the expression of synaptic marker proteins follows a similar trend to that observed during in vivo development, as described by Buckby et al. (19). Interestingly, in this study, the postsynaptic marker, PSD95, is noticeably increased compared to in vivo preparations. Altogether, these changes may represent a compensatory increase in the activity and connectivity of the synaptic components that persist in slices in order to maintain an approximation of the in vivo developmental timecourse.

The Glia Population within Slice Cultures

Glial cells outnumber neurons ~10:1 and are critical for homeostatic maintenance, appropriate monitoring of the environment, cell-cell signaling events, neurotrophic factor production, and repair-oriented brain processes. Thus, for a slice culture to be evaluated as a potential surrogate for in vivo studies, the glial cells, not only the neurons, need to be considered with regards to distribution and functionality. For example, researchers are able to record and evaluate the firing properties of an individual neuron in a slice. However, this process is mediated, in part, by the actions of proximal glia. In addition to influencing the survival and connectivity of neurons, the trauma of preparing organotypic cultures causes long-term changes in the morphology and function of glia. Del Rio noted early on that there appears to be a profound "reorganization of glia" in organotypic cultures (20). Use of the slice preparation as an in vitro model intended to comprehensively portray in vivo neuronal networks is partially dependent on an understanding of how neuroglia within the in vitro environment may be responding differently than their in vivo counterparts.

Oligodendroglia

The least studied of the glia persisting in brain OSCs are the oligodendrocytes. These glia are the myelinating cells of the brain, each one capable of helping to insulate over 50 axons, facilitating signal propagation. Oligodendrocytes differentiate later than all other neural cells types. As a result, the impact of soluble signals within slices on the maturation profile of oligodendrocytes is presumed to be the greatest of all glial subtypes. Multiple studies indicate that slice cultures maintain both the appropriate physiological ratio of oligodendrocytes, as well as normal precursor cell proliferation over several weeks in vitro. Slice oligodendrocytes have been shown to effectively transition from immature precursors to non-proliferative, morphologically distinct (small soma and elaborately ramified processes) and functionally mature (expressing myelin basic protein, MBP) cells in slice cultures (21). Myelination does continue in vitro, albeit at reduced levels and typically only after the OSCs pass some sort of critical maturation window (~7DIV). This feature can be exploited in organotypic cultures as a neurotoxicological endpoint (22). Unfortunately, aside from this, minimal information is currently available concerning whether or not the survival, functional capacity, timecourse of maturation, and ratio of mature to immature cells appropriately relates to the in vivo situation.

Astroglia

In contrast to the knowledge base on oligodendrocytes, a great deal of information on the morphological and functional changes of astrocytes in organotypic preparations has accrued over recent years. Astrocytes directly influence the health of neurons. They serve as mediators between the vascular system and neurons, providing glucose, lactate, and other nutrients essential for neuronal function. They are adjacent to neuronal membranes and help to regulate synaptic transmission by controlling the levels of extracellular glutamate. Additionally, they provide structural support and physical protection, contributing to the establishment and maintenance of the blood brain barrier (BBB) and limiting diffusion of toxicants or other soluble factors within the neuronal microenvironment. In the brain, astrocytes have the highest levels of enzymes associated with drug detoxification and conjugation. Thus, the activity of these cells in slice preparations is an important consideration for any studies targeting pharmacological effects.

In slice cultures, these cells, in particular, seem to behave in a manner dissimilar to those of the uninjured brain. Over the culture timecourse, they display a phenotype more routinely seen following traumatic injury in vivo. There is a proliferation and potent hypertrophy of GFAP-positive astrocytes beginning around 2DIV that completely covers the top and bottom of the slice by 3DIV, spreads to encompass the entire slice by 7DIV, peaks around 10DIV, and persists until at least 20DIV (23–25). This astrocyte reaction is responsible for the formation of a scar-like layer, particularly over damaged regions such as the dentate gyrus in hippocampal preparations (23). In addition, a large number of slice astrocytes migrate to the outer rim of the organotypic tissue, an event which may be mediated, in part, by release of IL-1β from microglia (20,26). Perhaps more interestingly, slice astrocytes maintain an immature phenotype, according to morphology and continued expression of markers absent from mature astrocytes, such as vimentin and nestin (27–29).

Microglia

Of particular interest with regard to neuroinflammatory studies in slices are the actions of the macrophages of the brain, microglia. Microglia are now recognized to be the most rapid responders to insult in the mature CNS, existing in a maintained state of vigilance even when morphologically "resting" (30). Microglia produce a wide range of chemokines and cytokines capable of inducing both pro- and anti-inflammatory effects; they facilitate the phagocytic removal of debris; and they may help to mediate neuroimmune interactions via antigen presentation. All of these actions are necessary to sustain neuronal function, both basally and following insult; however, the full range of properties possessed by microglia is still expanding and constantly being clarified. Microglia have a prolific presence across recent studies of neurodegenerative conditions and stand poised to fill gaps in the current understanding of early mechanistic changes in these conditions. This linkage makes them an appealing target for pharmaceutical design. Their actions within isolated organotypic preparations, in the absence of any influence from the systemic immune system, can provide important insights into their endogenous behavior.

During OSC preparation, microglia react immediately to the trauma of explantation by initiating an orchestrated series of morphogenic changes that are known correlates of functional alterations. During the morphological sequelae to injury of the slice tissue, microglia progress from their quiescent, ramified, in vivo appearance to an activated and ameboid, likely phagocytic, phenotype, and back again. The conversion of microglia from resting to ameboid occurs as early as two hours in vitro (31), and when further insults to the organotypic tissue (such as ischemia) are induced, can occur in as little as seven minutes (32). After responding to the cut edge within hours of culture, the majority of microglia become ameboid by 1-2DIV and remain that way until at least 4–5DIV (33). Between 7-14DIV, many microglia in the center of the slice regain their ramified phenotype (24,34). However, there remain sporadic, ameboid microglia evenly distributed throughout the culture after 10DIV (23).

In addition to morphological changes, slice microglia routinely display increased proliferation, which is witnessed in vivo only following pathological insult (25,35). The increased numbers of microglia identified in slices over DIV may reflect an increased detection threshold of microglia markers that are upregulated due to activation. However, this observation more likely represents denervation-induced proliferation, as is seen with in vivo lesions (36). This population of microglia would be expected to be able to rapidly increase in number and subsequently downregulate to effect rapid, localized damage resolution.

It is widely held that microglia multiply, become activated, and then become motile in an effort to facilitate, and subsequently resolve, inflammatory reactions in damaged tissue. In slices, microglia phagocytose a considerable amount of debris and damaged nuclei over the first 6DIV (37,38) before their phagocytic response is downgraded. However, at 12DIV, ~20% of microglia can still be found actively phagocytosing axons in unmanipulated cultures (39). It is important to recognize that microglial phagocytosis of apoptotic debris is accompanied by a downregulation of proinflammatory cytokines in an effort to limit tissue damage (40). The pre-explant location of these cells may influence their in vitro phenotype. In an interesting study, Grossman et al. (41) identified and tracked "juxtavascular" microglia versus those not directly in contact with blood vessels. In his slice culture system, these microglia make up 10–30% of the total microglial population and display a phenotype suggestive of a more robust activation response than non-juxtavascular microglia. These cells migrate in contact with the remaining vessels, display more dynamic movements of their protrusions, and are twice as likely to be locomotory than parenchymal microglia.

Overall, the observed functions of organotypic microglia appear indispensable to quelling the potential neurodegenerative processes initiated during OSC preparation. In these cultures, microglial processes can be seen separating neuronal processes, and cell bodies, presumably aiding synapse stripping in order to facilitate the removal of afferent projections, as would be necessary for any eventual synaptic reorganization and repair (42,43). These activation characteristics of slice microglia are important to recognize. Long-lived or excessive microglial responses may prove to be a means by which subtle alterations in slice health can be recognized and avoided. Identifying and, possibly, reducing or normalizing some of these changes prior to experimental manipulations may improve the quality of the resultant data.

Vascular and Ventricular Systems

The flow of fluid within the brain is known to influence both neuronal function and glial responsiveness. The brain vasculature and ventricular systems serve as sources for cytokines and nutrients via active transport, and as filters and depositories for waste substances such as debris or long-lived soluble factors that can prove harmful to the brain parenchyma over the long term. They provide pressure to aid in tissue perfusion and the distribution of trophic factors, and serve as sites of entry for surveilling, peripheral immune cells that sample the parenchyma on a regular basis in the intact brain (44). Although it appears that organotypic preparations adequately compensate for the loss of these systems, the altered cellular functionality that must take place for this to occur remains unclear.

OSCs lose rat endothelial cell antigen-1 positive and endothelial nitric oxide synthase (eNOS) positive vascular structures (capillaries) after two weeks in vitro, although some inducible components may persist over the short term, presumably via the actions of surviving endothelial cells (7,45). In organotypic cultures, in addition to the loss of much of the vasculature, tight junctions are lost, although, again, some inducible components may persist, since stimulation with fibroblast growth factor-2 (FGF2) can increase their presence (46). Junction-forming glia at sites previously containing vascular elements maintain contacts with vessels that disappear in vitro, eliminating a direct source of trophic factors for these glia. In part, this lack of endogenous trophic support necessitates the need to replace vascular-derived glucose with high glucose media in vitro, an alteration not without its own concerns (some of which are discussed in subsequent sections). This disruption of the normal BBB has the potential to affect the function of glia normally implicit in this barrier. It remains unclear how the lack of a true BBB alters their normal structure, cellular interactions, and metabolism. One phenotype widely thought to result from the loss of vascular influence is a 95% decrease in N-acetylaspartate by 20DIV in OSCs. This free amino acid is synthesized by neurons and is normally increased 200% during postnatal brain maturation (47).

A second, related consideration is the lack of cerebrospinal fluid flow within the OSCs. Morphologically, the ventricular system remains relatively intact and choroid plexus and ependymal cells do appear to persist in vitro. However, it is clear that these cells do not maintain their in vivo function and organization. The choroid plexus cannot make CSF in the absence of blood flow and the tight junctions of the blood-CSF barrier are lost. It remains unclear how these non-functional cells contribute to the OSC environment. Taken together with the loss of the vascular system, these losses represent significant changes to the structural dynamics of the brain system in slices.

Cytokines

Glial proliferation, astrocyte hypertrophy, microglial migration, antigen recognition, and engulfment of material are all regulated through the release of soluble mediators and the coordinating expression of specific, cell surface proteins. It is known that CNS insult causes glia to upregulate and secrete proteins normally produced by leukocytes, including major histocompatability complex (MHC) II, toll-like receptors (TLRs), chemokines and cytokines (48). These changes typically result in a transient, self-resolving sequence of events necessary for recovery following CNS injury. As the basic functions of neurons and glia remain relatively conserved from in vivo to the slice environment, it is unsurprising that most of these soluble signals have similar effects on neurons in organotypic culture and those in the intact brain. However, there are some noteworthy exceptions.

One family of small proteins that is deserving of consideration in this regard is the cytokine family of small molecules, critical to innate and adaptive immune-type responses. The presence or overabundance of cytokines can exacerbate cellular injury; however, evidence does exist suggesting that pre-exposure to these same molecules is neuroprotective against toxicity paradigms such as excitotoxicity and ischemia. Cytokines that are normally almost undetectable in vivo, including interleukin (IL)-1β, IL-6, and tumor necrosis factor (TNF) α, are expressed at increased levels in slices. The preparation of organotypic cultures causes temporary increases in secreted IL-6 and TNFα, which return to baseline, in vivo-like levels after 4DIV (33). Interestingly, these factors are not induced any further with mechanical injury to slices maintained in vitro, perhaps owing to a saturation of some type of stimulus-specific activation already initiated in response to the cut edge of the slice (43). Conversely, in vitro stimulation of the pro-inflammatory cascade of microglia within in vitro-maintained slices via chronic application of the bacterial endotoxin, lipopolysaccharide (LPS), transiently increases IL-1β, and IL-6 (49). Prior to 10DIV, microglia express IL-1β, regardless of whether they possess a ramified or ameboid morphology (23). This mediator is thought to contribute to many of the transneuronal changes witnessed following lesion-induced reorganization, and may directly interfere with synaptic function (39). The long-term consequence of the temporary elevations in cytokine signaling molecules during the developmental maturation of OSCs has not been examined in detail.

Unique Characteristics of Established OSCs

After several weeks in vitro, the viable neurons in organotypic cultures continue to retain a morphological phenotype representative of the region from which the cultures were isolated. These cells are similar in size to their in vivo counterparts, they produce the same neurotransmitters, and they express the majority of the same proteins (50,51). However, their extended culture period does result in subtle deviations from comparable neurons in vivo. Once the slice environment has progressed beyond an initial sensitive period, it seems that the majority of activation phenotypes, including glial reactivity and pro-inflammatory cytokine expression, have returned to endogenous levels. One notable exception to this is the observation of a prolonged astrocyte response. On the whole, mature organotypic slices have the advantage over acute preparations as regards the resolution of trauma-induced changes. However, the compensatory processes initiated to resolve this trauma, as well as the unique artifactual variations present in mature OSCs as a result of these processes, have not been well characterized.

Neuronal Elements

One of the most desirable and unique features of OSCs is that the functional activity of neurons within the slice is maintained over many weeks in vitro. Even after such a long time in culture, the activity of slice neurons has been shown to be nearly indistinguishable from recordings of acutely-isolated brain slices. Elaborate synaptic connections are present throughout mature OSCs and closely resemble those contacts existing in the intact animal. Many of these connections are not only maintained in vitro, but have also been created there. Aside from their electrical activity, neurons maintained in culture manifest minor, but distinct, morphological changes and long-term variations in networking.

Over the timecourse of slice culture, there are noticeable changes to the size and morphology of dendritic arbors, which display a greater intricacy than those found in vivo. Specifically, an increased complexity of higher order dendrite branching has been noted in CA1 neurons (16,52) After 20DIV, there are notable decreases in overall dendrite length and primary dendrite branching (53,54). The ability to more closely image neurons in organotypic cultures has allowed for the identification of changes in the density and morphology of the spines on these dendrites. In vitro synaptic reorganization appears to involve a compensatory increase in the total number of synapses, despite a noticeable decrease in spine density (16). In fact, the density of synaptic spines on the distal portion of the dendrite layer normally receiving connections from outside of the cultured region declines and remains lower throughout the lifespan of the OSC (54,55), whereas more proximal dendrites undergo a rapid and robust period of activity-dependent development. Thus, as would be expected, this overall increase in synapses on proximal dendrites appears to reflect a response of intrinsic neurons to newly formed fibers associated with lesion-induced sprouting. Alongside this is a concomitant withdrawal of peripheral dendrites due to a loss of synaptic partners from neuronal nuclei outside of the cultured region of the slice. These changes continue over multiple weeks in vitro, as the remaining synaptic contacts continue to strengthen.

Many important attributes of OSCs have to do with the synaptic properties that persist after an initial period of acclimation to the traumatic damage that accompanies explantation. Mature organotypic cultures are known to exhibit the spectrum of known recordable potentials (51). Despite all of these adaptive changes in mature organotypic cultures, however, there remains evidence of a persisting immaturity in select neurotransmitter effects, particularly GABAergic signaling (56). This may reflect slice preparations retaining some aspects of GABAergic transmission in its putative role as an excitatory transmitter during early postnatal life (57). In addition, the aforementioned hyperexcitability caused by forced synaptic reorganization can lead to spontaneous seizures in OSCs beyond 35DIV (58,59). Aside from the formation of a few aberrant connections, though, most neurons in OSCs continue to synapse appropriately. For example, mossy fibers synapse only on CA3 cells, not CA1 cells, and CA1 cells do not establish contacts with other pyramidal cells, both imitating in vivo properties of the hippocampal network. Further, co-culture studies reveal that slice neurons preserve a great deal of selectivity in projection site targeting in vitro (60). These observations intimate the preservation of integral patterning properties and/ or neuronal guidance cues. In OSCs, it is clear that guidance cues persist, either within the specific neuronal populations themselves, or within the regional architecture of the organotypic slice. As all of these responses persist over an extended time in vitro, most reflecting a developmental increase in the degree of connectivity, in combination these features allow for detailed studies of synapse function and plasticity.

Mature Organotypic Glia

The "scar"-like barrier of astrocytes and microglia persists over the long term in organotypic preparations. These cells at the interface areas of the slice retain a highly activated and motile phenotype regardless of time in vitro. Barrier glia are thought to be responsible for the resistance seen when attempting to pharmacologically manipulate mature cultures. Throughout the life of the OSC, GFAP-positive cells remain in a reactive form, with thickened processes that are fewer in number and have minimal branching. Nearly all of these cells also express nestin in vitro (28), a marker observed in reactive astrocytes following nervous system injury initiated by seizures, trauma, or ischemia, but absent from mature glia in uninjured animals. Perhaps, though, not all astrocytes in slices are immature and reactive. Recently, using particle mediated transfection, Benediktsson et al. showed that a population of astrocytes within the deep layers of the slice eventually assume a complex 3D morphology with fine processes, more reminiscent of in vivo astrocytes (61). When considering chemical manipulations of OSCs, long-term activation of astrocytes, the direct effectors of environmental detoxification and barrier formation, can be dire.

INVESTIGATIVE STUDIES USING OSCS: NEUROBIOLOGY

In light of their limitations, slice preparations appear well suited for studies emphasizing a direct effect on neuronal signaling or circuitry, such as synapse maturation, but less so for analyses highlighting elements such as neuron-glial interactions or propagation of soluble signals. Astrocytes and microglia, at early DIV in particular, display modified profiles when compared to their counterparts in the uninjured brain, while the longevity and diffusion of soluble cues in a slice environment devoid of vascular and ventricular systems remains largely unexamined. Organotypic cultures exhibit phenotypes relevant to developmental studies of neuronal plasticity, examinations into the properties of intact neuronal circuits and neuronal connectivity, and observations of neuronal morphology and differentiation. For example, researchers have been able to observe dendritic spine development and pruning in OSCs, including the novel finding that spine retraction may not irrevocably lead to loss of synapses (62). OSCs have also been useful for tracking developmental events, such as newly-generated neuronal precursors during in vitro maturation (25) and the expression of neurotransmitter receptors in the hippocampus (63). Similar to observations of slice astrocytes, the latter study did identify increased kainate binding sites, particularly in the CA3, in OSCs compared to age-matched in situ hippocamppi, indicating a more immature, or "specialized," organotypic tissue. As with any in vitro system, it appears that only simplified aspects of the human condition can be accurately and completely modeled with OSCs. Some studies of this type will be examined in the following sections.

In Vitro Synapse Maturation and Brain Development

One of the most widely studied and successful topics to which organotypic tissue has been applied is the in vitro evaluation of the development and maintenance of synaptic connections. Perhaps more directly than any other observations made in cultured brain tissue, synaptic potentials and changes in dendritic spine morphology have translated directly to in vivo phenotypes. With improved technologies, the testing of these endpoints in routine toxicology studies using OSCs may become feasible. As it stands, the tools require a high degree of specialization not commonly available for routine toxicity testing, although these tools are becoming more accessible [(64) for a protocol on imaging dendritic spines in situ].

Synaptogenesis and Synapse Strengthening

Electrophysiological recordings of maturing organotypic tissue have borne out the initial observation that these tissues develop in vitro at a rate that approximates the in vivo maturation of postnatal tissue. As mentioned previously, OSCs exhibit a dynamic range of known potentials, including the reproducible induction of long-term changes to synaptic strength such as long-term potentiation and depression [LTP and LTD, reviewed in (51)]. The use of organotypic preparations allows for mechanistic studies into these important memory-related processes, as illustrated by the observed alteration to glutamate receptor expression in response to p38MAPK activation by neuroplastin-65, a protein integral in establishing LTP (65). A concern can be raised about fictitious potentials being generated in response to the isolation of OSCs from external afferents and the compensatory generation of intrinsic connections not typically made in vivo. However, the vast majority of slice neurons remain synaptically similar to their in vivo counterparts (51). Notably, though, synaptic markers PSD-95 and synaptophysin increase dramatically between 7 and 21DIV in these cultures (19,66), reflecting an active period of synaptogenesis in slices. Overexpression of PSD-95 in cortical OSCs increases the number of AMPAR-expressing synapses, leading to LTD and conclusively establishing a role for this protein in the regulation of synaptic plasticity (67).

Neurotransmitter receptors and postsynaptic proteins necessary for transmission are clustered in dendritic spines. The arborization and dynamics of these protrusions during synapse development and drug exposure or insult are of key concern, with particular relevance to learning and memory. Nägerl et al. detailed a time-lapse, two-photon study of the early timepoints following activity-dependent, or LTP-induced, spinogenesis and the synapse maturation which occurs slowly over the subsequent 19 hrs (68). Further supporting this linkage between neuronal activity, plasticity, and morphological changes in dendritic spines, mature OSCs (>21DIV) exhibit spines which are morphologically comparable to those found in the adult brain (69). This activity-dependent type of modulation to synapse structure has been a recent focus in slice culture research (68,70,71).

Presynaptic inputs are retracted from axotomized neurons in brain OSCs shortly after explantation. The detailed, cell-specific contributions to this remodeling and active synapse stripping remains only a minor focus of studies to date. This synapse removal appears to reflect a protective mechanism initiated in response to deafferentation. A fundamental understanding of this process in real-time may provide potential targets for regenerative drugs. Further, alterations to this process, such as the absence of MHC class I-type receptors on microglia (72), could have significant effects on the maturation and plasticity of slice synapses, thus providing a unique model for mechanistic studies of such processes.

Network Development

An important functional characteristic of neurons derived from specific brain regions is their appropriate proliferation, migration, and integration into circuits in agreement with that seen in the developing animal. Cellular functions during normal development, and alterations to these processes following an insult, can be quite difficult to visualize in vivo. OSC neurons have proven quite successful at adhering to their appropriate temporal organization into networks. In fact, the maturation of these cultures has been so closely likened to the development of neuronal networks in vivo that many researchers model in vivo development using the in vitro development of isolated OSCs. In some cases, reasonable in vivo postnatal age equivalents have been defined for OSCs grown in vitro (16).

Neurogenesis, such as occurs during development, is affected by hypoxia and trauma, which are inherent to slice cultures, as well as by increases in neuronal activity, microglial activation, and availability of target neurons with which to integrate. Of specific interest, microglial activity has been shown to correlate with enhanced adult neurogenesis in vivo, dependent on the status of the surrounding neural tissue (73,74). This observation suggests the possible value of OSCs in dissecting what appears to be a tight balance between the pro- and anti-proliferative effects of microglia and microglial cytokines (75). A difficulty can be foreseen in the fact that OSCs exhibit pronounced in vitro proliferation of microglia in the early DIV, as well as a potent astrogliogenesis that persists throughout the entire culture timecourse (25). It is unclear how this glial proliferation affects in vitro neurogenesis. In addition, with fewer connections being maintained in the immature slice, albeit transiently, new neurons would have a reduced number of sites at which to integrate. These neurons, which fail to become established within active neuronal networks, are likely to be eliminated. Thus, although some of the newly-generated neurons in maturing OSCs may positively contribute towards the plasticity of developing neural networks in vitro, the impact of new neurons which fail to integrate, the compensatory neuronal responses to axonal reorganization, as well as the multiplicity and increased proliferative capacity of slice glia, all combine to make studies of developmental neurogenesis difficult to interpret.

Synaptic networks can be induced to develop and mature in vitro, even in the absence of cues present in vivo. One example is given in an elegant series of studies relating to visual experience-dependent synapse formation in OSCs derived from the visual cortex (76). The authors observed synapse formation in the absence of thalamic input. This synapse maturation seemed to mirror in vivo development. However, the formation of these new synapses proved to be highly dependent on increasing the neuronal activity of the slice. Conversely, blocking neuronal activity in the slice, through the use of tetrodoxin to inhibit Na^{2+}-dependent action potentials, for example, results in gradual cell death (77).

Migrational studies are often carried out using OSCs to provide a matrix environment similar to that encountered during developmental migration in vivo. Typically, these studies involve transgenic animals with selectively labeled, migratory cell populations, or in vitro-labeled cells, and subsequent implantation of these into the OSC tissue matrix. One example of the former, using short-term explant cultures, highlights migratory interneurons in embryonic neocortex OSCs from animals at varying ages. Using this migrational model, the authors described a threshold age necessary for permissivity of the transplanted cell population (78). Similarly, examination of subventricular zone (SVZ) progenitors in organotypic cortex, using GFP-labeled progenitors transplanted in vivo two days prior to the preparation of organotypic cultures, revealed that these transplanted cells migrate in vitro at rapid speeds along the rostral migratory stream towards the cortical surface, where they decelerate and become stationary before differentiating into GABAergic interneurons (79). This study emphasizes the continued role of the specific neuronal microenvironment, or niche, in the induction of proliferation, migration, or differentiation in vitro. Importantly, the migration and integration of progenitors was much greater in OSCs than that witnessed in vivo, arguing for augmentation due to the slice procedure. Thus, it is likely that these slice paradigms more closely represent in vitro models of in vivo brain injury-induced neuroprogenitor recruitment and differentiation, rather than descriptors of basal brain processes.

Potential Impact of Explantation

While studies such as these demonstrate important properties of neuronal network development, the aggravating effects of generating the slice culture have the potential to confound interpretations. These tissue regions have already experienced an insult. Thus, subtle changes in neuronal responsiveness within the slice, or the altered in vitro development of glia and their coinciding efforts to re-establish basal conditions, can be difficult to distinguish from an experimental effect. In order to conduct accurate studies, it is important to establish the appropriate behavior of newly-formed neurons and in vitro-generated synaptic connections. The severe cellular loss witnessed in response to the destruction of extrinsic signaling afferents immediately following explantation, as evidenced by elevated levels of secreted lactate dehydrogenase (LDH) until at least 4DIV (33), assuredly influences the subsequent development of the tissue. This neuron loss is assumed to be extremely sensitive to changes in the septo-temporal cut of the tissue and variable across slices due to differences in the location and intensity of released, neurotoxic mediators. How this degeneration affects the long-term maturation of these slices is impossible to interpret.

In OSCs, the absence of sensory input at a developmental period critical for experience-dependent learning also engenders considerable concern. Although these cultures increase their neuronal activity over time in vitro to affect migration, differentiation, and maturation on a timecourse paralleling development in vivo (16), there remains the likelihood that these networks do not evolve in a manner identical to the in vivo situation, particularly as regards soluble cues. OSC glia are highly active at this time of neuronal network maturation, and are well-known to participate in the formation, maintenance, and functional development of synapses. As such, it is unclear how their enhanced activity alters functional components of neuronal signaling, aside from electrophysiologically measurable phenotypes. Moreover, there may be additional, unforeseen consequences stemming from the inability of slices to sustain neuronal input and output beyond their closed environment. Initial observations in OSCs are encouraging, but the full complexity of in vivo neuronal systems, including learning, attention, and memory-like phenotypes, which often involve intricate afferent and efferent connections that are lost in OSCs, has yet to be conclusively shown in culture. Allusions to complex in vivo functions reliant on data drawn from organotypic preparations should be moderated.

Glia and Inflammation

Almost all neuronal injury paradigms involve a glial component, either in the toxicity profile or during damage resolution. Astrocytes regulate the neuronal environment, providing trophic factors and removing excess toxins, such as extracellular glutamate. In addition, these support cells of the brain play an integral part in forming and sustaining the tripartite synapse, providing assistance in signaling processes between pre- and post-synaptic regions. Microglia facilitate many of the essential, immune-type responses in the brain parenchyma. In a quiescent state, these cells remain vigilant, primed to rapidly resolve any disruptions to the neural environment. Together, these two cell types mediate the vast majority of inflammatory responses in the brain. As the dysregulation of these processes appears to play a causative role in many brain degenerative conditions, a fundamental understanding of the mechanisms underlying proper glial function is vital to preventing and/or resolving these pathologies. Thus, the potential to model these actions in OSCs is of particular interest to neurotoxicologists.

Microglial Function

Microglial cells are the most rapid responders to pathological insults. In vivo, microglia help in synapse remodeling and play a visible role in both innate and adaptive immune responses, most notably antigen presentation (80). Microglia participate directly in signaling with neurons, and appear to either regulate or assist astrocytes in walling off damaged and sensitive neural regions from injurious foci. Perhaps most recognizably, these cells have a well-established role in the removal of cellular debris. The various clearance processes of microglia are well described in a review by Napoli and Neumann (81). These cells prune axonal connections and engulf apoptotic neurons produced in excess during CNS development, and selectively target inappropriate synaptic connections throughout adulthood. Microglial activity is tightly regulated in vivo. This is likely to hold in vitro as well, as similar neurochemical cues persist in the OSC parenchyma. The activation and proliferation of these cells, particularly over the early time-course of OSC, appears to be a protective function necessary for the survival of slice neurons. Typically, microglial activation is associated with negative effects on brain neuron function and survival; however, this stigma has been widely challenged in recent research. Currently, there is an emphasis on the chronic dysfunction of these cells, or the actions of their morphological mimics, infiltrating macrophages, as the true culprits behind many of the deleterious actions previously attributed to all "reactive" or "activated" microglia.

Microglial morphogenics and phagocytosis are important functions that are incompletely understood within the context of an environment relevant to the in vivo brain parenchyma. Studies in primary culture are limited in regards to the extracellular cues present. Multiple studies on isolated microglia highlight their morphological responsivity to co-culture with various other cell types, including a ramification when contacting primary neurons and astrocytes. The activation of microglia in OSCs has been shown to involve the replacement of ramified branches with new, more dynamic, protrusions, signifying a functional change to one more closely related to that of microglia in dissociated primary cell preparations before 5DIV, and one more closely akin to the in vivo situation after 10DIV (31,82). These morphological changes initiated in response to the explantation injury may parallel electrophysiological changes, such as those observed between ramified slice microglia and reactive cells found after facial nerve axotomy (83). These early changes likely indicate trauma-induced release of soluble factors, including ATP, from damaged neurons that stimulate resting microglia on cell surface proteins, such as the purinergic family of receptors. Treatment of acute slices with purinergic ligands reduces microglial activation (84) and microglia in slices from $P2Y_{12}$ knockout mice fail to exhibit chemotaxis following bath application of ADP or ATP and display a delayed response to local tissue damage caused by laser ablation in vivo (85). This chemotaxis to sites of injury is similar to that observed in the large numbers of slice microglia that migrate to neuronal cell bodies damaged during explantation and to the cut edge of organotypic preparations. Future studies of the electrophysiological properties of slice microglia during tissue damage resolution could prove quite revealing.

At early DIV, microglia in OSCs undergo a variety of changes at their cell surface that identify an immune-like reaction. The costimulatory molecule, B7-2, normally displayed on antigen presenting cells (APCs) for T cell stimulation, is upregulated by slice microglia over the first 7DIV, possibly representing a non-destructive neuroprotective mechanism and means by which microglia initiate phagocytosis of myelin debris (86,87). Adding support to this, microglia in organotypic tissue that are in proximity to degenerating neurons, but not those in areas with intact neuronal architecture, can be induced to express MHC II in response to interferon gamma (88). Further, activated microglia in organotypic tissue express different integrin adhesion molecules, specifically lymphocyte function associated molecule-1 (LFA-1, the receptor for intercellular adhesion molecule-1, ICAM-1), and $\alpha_4\beta_1$ integrin—molecules typically associated with T cell/ APC interactions (24,34). The expression of integrin-like molecules would allow microglia to recognize neurons, initiating processes including synapse stripping and phagocytosis (89). As described in previous sections, once the slice environment progresses beyond an initial sensitive period, it seems capable of quelling microglial activation phenotypes and then maintaining most of these cells in a relatively quiescent state. After microglia establish this "mature" in vitro state, they still retain their ability to respond to in vitro tissue injury (90), though these cells have been shown to have a reduced phagocytic response to latex beads (82). Increasingly, attention is being paid to the functional progression of these "inactive" cells to become migratory or phagocytic in situ. The implantation of microglia with long-term fluorescent labels into OSCs should offer future insights into these processes.

An approach deserving of more attention given the growing association of brain microglia with neurodegeneration and neuroprotection, is implantation of OSCs with primary microglia. Implanted BV-2 microglia have been shown to penetrate deeply into OSCs, where they subsequently protect against oxygen glucose deprivation, or OGD (91). Interestingly, BV-2 microglia which overexpress macrophage colony stimulating factor receptor (M-CSFR, a phenomena observed in microglia surrounding plaques in Alzheimer's disease, AD, in vivo) proliferate, release cytokines, and exhibit enhanced phagocytosis, protecting OSCs against N-methyl-D-aspartate (NMDA)-induced excitotoxicity. Despite the fact that the in vivo relevance of these cell lines is of much greater concern than in primary preparations, surprisingly few studies examine the outcome of adding exogenous, primary microglia to brain OSCs. Addition of primary microglia or microglia-conditioned medium to OSCs has been shown to potentiate NMDA-mediated, but not AMPA or KA-mediated, synaptic responses (92). Implantation of fluorescently-labeled polymorphonuclear neutrophils (PMNs) into 10DIV slices enhances OGD, according to propidium iodide (PI) incorporation and the loss of Thy1-driven YFP expression as indicators of cellular loss. When exogenous microglia were implanted in combination with the PMNs, the exacerbation of the OGD neurotoxicity was abrogated via direct engulfment of the invading PMNs by both endogenous and exogenously-added microglia (93). It should be noted that, due to the difficulty inherent in clearly visualizing the actual engulfment interaction in intact OSCs, the authors translated the observation of engulfed PMNs inside of microglia to observations of the actual "chasing" and engulfment using co-cultures of dissociated primary cells. This technical consideration should be kept in mind for the replacement of difficult visualization processes carried out in OSCs during real-time imaging techniques that require a high degree of resolution.

Astrocyte Actions

Astrocytes have the capability to directly modulate excitatory and inhibitory synaptic transmission through glutamate and ATP uptake and release. They also facilitate neuronal synchronization and long-range network signaling, likely through intracellular Ca^{2+} levels. Astrocytes play a vital role in the detoxification of the immediate neuronal and synaptic microenvironment. In addition, they possess a well-described, putative role as mediators of brain inflammatory reactions [for a review of astrocyte function, see (94)]. Complicating this, however, is the fact that a considerable portion of the recent, fundamental understanding of the aforementioned functions of astrocytes has been discovered using slice preparations (94).

Developmentally, astrocytes in vivo have a well-established role in providing guidance cues for migration. The migration of astrocyte precursors injected into slices appears to be regulated directionally by cues specific to the organotypic tissue region, whereas migrational distance is determined by the age of the host tissue (95). This, again, appears to reflect developmental permissivity, as migration was abrogated in tissues derived from >P7 animals. In roller tube, midbrain slices, astrocyte processes were shown to precede not only neuronal migration, but also oligodendrocyte and microglial migration (96). Of note is the observation that bromodeoxyuridine (BrdU)-positive astrocytes can be seen within a glia scaffold beyond 20DIV in roller tube slices, but only over the first seven days in vivo (20). Astrocytes in brain OSCs continue to provide guidance cues for dopaminergic fiber outgrowth (97). Studies such as these are typically carried out shortly after explantation, during periods of robust glial proliferation and activity.

Preservation of neuronal health and function are, possibly, the most important features performed by astrocytes in the intact brain. Astrocytes in OSCs, as well as those in vivo, regulate extracellular glutamate concentrations (98). Another means by which astrocytes may influence synaptic efficacy is via process extension and retraction to motile, postsynaptic dendritic spines, as shown in hippocampal OSCs (99). In hippocampal slices, gap junction communication by astrocytes, a signaling mechanism easily manipulated in these cultures, helps to reduce damage from oxidative toxicity induced following the application of kainate or $FeSO_4$ (100). Both microglia and astrocytes have an active role contributing to synapse elimination during CNS development, plasticity of the adult brain, and, likely, the re-establishment of functional circuitry in OSCs following explantation. Glial ensheathment and synapse stripping by astrocytes has been shown in "transplanted" cerebellar OSCs (101), where cells endogenous to the slice were inactivated and depleted, then astrocytes were repopulated with transplanted replacements at 9DIV.

Unique to slices, the scar-like barrier that forms at the interface may be a neuroprotective mechanism (20). Failure to regenerate injured axons in vivo is known to involve the formation of a glial scar by, predominantly, reactive astrocytes. This physical barrier is beneficial in many ways (102). The scar protects neurons spared in the initial insult from the spread of released toxic factors, demarcates the lesion site to focus immune-type responses, and may help to regulate inury-induced cellular proliferation. The graded migration and activation of glia towards the slice interface most logically indicates an attempt to shield the sensitive neuronal fields in the deep, inner layers of the slice tissue from elevated concentrations of harmful byproducts of the explant injury, such as excitatory amino acids. Importantly, these activities are almost assuredly dependent on neuronal activity (103). A phenomenon that may be uniquely observable using slices, astrocytes appear to downregulate the activity of cells entering the slice environment. Infiltrating macrophages, when transplanted atop organotypic preparations, lose their ameboid morphology as they pass astrocytes at the slice interface; these observations were confirmed and further dissected in dissociated cell cultures (104).

Considerations of OSC Glial Activity

The amplified responsiveness of slice glia is of great concern as a potential confounder unique to the organotypic preparation. The elevated numbers of active slice glia can have disruptive effects on multiple aspects of experiments in OSCs. They can decrease the visibility of the neuronal fields, degrade extracellular matrix materials, and migrate out from the three-dimensionally conserved explant, where they are still capable of altering soluble signals, despite being segregated from slice neurons and in vivo-like tissue. These cells can be further transformed by experimental approaches that modify slice neurons, thus creating aberrantly over-activated or damaged astrocytes with little physiological relevance. Metabolically, these reactive cells are typically very active and represent a potential impediment to drug or toxicant treatments. Further, their activation presages the need for drug or DNA-based manipulations to be carried out at very early days in vitro in order to achieve effective labeling without causing ancillary trauma to fragile tissue at later timepoints. Any physical change to the OSC tissue, such as a needle track, partial separation from the culture substrate due to agitation, or prolonged exposure to harmful wavelengths of light can all cause surface glia in the vicinity of the disruption to become motile and activated, potentially invalidating experimental endpoints. Even cellular implants and vital dyes, despite requiring the addition of only a small volume of liquid atop the fluid barrier at the slice interface, can sometimes induce this response (observations above from author's unpublished data). Slice cells are sensitive to hypoxic conditions and OSC: membrane matrix interactions are easily disrupted, particularly at early ages in vitro, so the addition of even a small amount of medium to the surface of the slice may have significant effects on the function and viability of slice neurons. In short, it is clearly evident that the increased presence of microglia and astrocytes has the capacity to disrupt experimental manipulations and endpoints in slice cultures.

One of the main concerns with this model is the inability of astrocytes to progress beyond a relatively immature phenotype. Astrocytes in organotypic cultures proliferate, become hypertrophied, extend processes into the slice matrix, and migrate a considerable distance from the borders of the slice (23,66). The expression of nestin observed within slice astrocytes is thought to reflect a prolonged activation state resulting from the trauma of explantation and represents a differential between glial maturation in vivo and in vitro (28). The authors hypothesized that this altered maturation is caused, in part, due to the lack of a functional vascular system and a corresponding absence of trophic interactions between astrocytes and endothelial cells. The long-term, functional consequences of this differential in vitro maturation could be far-reaching. In OSCs, it has yet to be shown what the consequences of the maintenance of glia in a pseudo-activated state may be, or what signals are lacking in OSCs that inhibit their maturation. At the least, the possibility of alterations to processes and downstream endpoints under astrocyte regulation should be considered when translating slice data to the in vivo situation.

The predisposition of microglia towards an activated phenotype is worrisome. Microglia have been shown to be directly responsible for neuronal damage that results from axon damage during in vitro entorhinal cortex lesion (39). The interfacing edges of organotypic preparations are often closely apposed to the neuronal nuclei being interrogated, increasing the likelihood for crosstalk between these highly active cells at the culture edge and nearby, viable neurons. In vivo, microglia are activated at projection sites from the region lesioned, as well as at the injury site itself (105), presenting the problem of damage along in-plane projections from neurons affected by the explantation lesion. Further, activation limits the ability of microglia to physically survey the tissue parenchyma and impedes communication with other cell types via short-lived soluble mediators, as is typically seen with the close association of branched microglial processes and neurons in vivo (30). Thus, this temporary in vitro activation can significantly blunt their responsiveness to subtle alterations in the neuronal microenvironment. As morphologically "activated" microglia downregulate purinergic receptor expression (85), changes such as these effectively reduce their long-range migratory ability and any corresponding participation in the resolution of subsequent injuries to the organotypic tissue.

Reducing the number or inhibiting the activity of microglia in slices has met with variable results. Treatment of hippocampal slice cultures with astroglial factors that deactivate microglia improves neuronal survival following excitotoxic injury (106). Conversely, if IL-1β is applied to slices after NMDA injury, the number of microglia is increased and there is an exacerbation of the damage (107). The authors went on to show that treatment with IL-1 receptor antagonist after injury had the opposite effect, improving neuronal survival and decreasing the number of microglia. It is important to note, however, that treatment with IL-1β in uninjured slices was shown to cause an increase in the number and activation state of slice microglia without any neuronal death. This is in disagreement with the assumption that activated microglia are intrinsically harmful to neurons. In a study of microglial depletion with clodronate, a bisphosphonate that depletes cells of monocyte lineage and also reduces astrocyte proliferation (108), Kohl and colleagues showed that clodronate-induced elimination of microglia increased neuronal damage following NMDA treatment (109). In vitro entorhinal cortex lesions (ECLs) provide another model for these types of studies. Inhibition of inflammatory processes with transforming growth factor β (TGFβ), which reduces IL-1β release from microglia, or direct elimination of microglia with the lysomotrophic agent L-leucine methyl ester (LLME) both help to maintain dendritic density after ECL and enhance sprouting of the remaining afferents (39). It is interesting to note that TGFβ conversely enhances neuronal damage caused by NMDA (106). With ECL of slices derived from CXCR3 knockout mice, which lack a receptor thought to be required for microglial activation and migration to neuronal injury sites following brain injury, denervated dendrites are not removed, as microglial recruitment to the zone of axonal degeneration is impeded (110). With the injury caused by explantation as a model for microglial mobilization, activation, and phagocytosis, Kurpius et al. showed that application of apyrase, an enzyme that degrades extracellular ADP/ATP, reduced branch extension and migration towards the edges of the slice following bath ADP/ATP application (111). One notable finding from this study was that the behavior of microglia within the slice varied depending on their distance from the neuronal cell bodies injured by explantation. These studies stress the sensitivity and specificity of microglia in their response to stimuli, and urge caution in assumptions about downstream effects of inhibiting their activity.

As many glia in these cultures have migrated towards the slice interface or to sites of lesion-induced cell loss, in addition to having an increased activity and a reduced maturity level, these cells are not localized to their proper in vivo location. As our knowledge of glial cells expands, it is becoming increasingly likely that, in vivo, microglia and astrocytes have regionally defined functions, dependent on the local microenvironment. It remains to be seen how this cellular relocation in slices would alter the glia response to, and resolution of, subsequent tissue injury. However, it is important to recognize that IL-1β has been brought forward as one of the key cytokines necessary for the recovery from a lesion-induced injury (112). This inflammatory molecule is expressed at elevated levels for the majority of the OSC duration. Interestingly, IL-1β induces the proliferation and migration of astrocytes in vitro and in vivo (97,113,114). It may be that the priming-type activation of glia by the organotypic preparation allows these cells to respond more quickly to any subsequent local injury, effecting a more rapid clearance of debris by microglia and a walling off of damaged or sensitive tissue by astrocytes. In this way, glial activation at early DIV may simply be an impetus for recovery, allowing for a resolution of tissue damage and the subsequent production or rejuvenation of functional synapses. Even so, the functional maturity and activation history of slice glia makes them distinct from in vivo cells.

Because of the myriad known functions of glia cells, it is difficult to dissect whether the phenotype of these cells in vitro is relevant to the actions of their in vivo counterparts. It is entirely possible that the prolonged "activation," "reactivity," or, at the least, "heightened vigilance" of these cells influences their subsequent response to inflammatory or pathological stimuli. Similar to the preconditioning-type responses witnessed in brain neurons, the injury stimulus of organotypic culture preparation may serve to create adaptive changes to microglia and astrocyte function or responsiveness. The probability of regional or subtype differences in microglia and astrocytes in OSCs, some of which are likely to translate to in vivo phenotypes, adds increasing importance to elaborating structure-function linkages between the observed glial morphological changes and definitive functional changes.

Studying Immune Cell Influence

In vivo, the vascular and ventricular systems provide access for immune cell infiltration of the brain, which serves a policing-type function during normal brain activity and in an effector-type capacity during times of severe injury. A limited number of circulating immune cells, as well as soluble signals from the periphery, penetrate the brain parenchyma in vivo. Systemic immune factors can modify neuronal function and glial activity during insults such as stroke or traumatic injury, and, additionally, can modulate the in vivo profile of several drugs. This infiltration can have both beneficial and deleterious effects, depending on the initiating stimulus, the timing, and the magnitude of the cellular or soluble infiltrate. In the uninjured brain, these responses remain under tight regulatory control.

OSCs represent a unique CNS tissue system with which to examine the actions of peripherally-derived immune cells and soluble factors through exogenous implantation or application, respectively. Immune cells that penetrate the BBB or blood-CSF barrier have been found to survey the parenchyma in both active and inactive states. Unlike the naïve brain, OSCs are protected from the influence, either cellular or soluble, of factors typically found in the circulation. As previously mentioned, slice cultures exhibit a profoundly altered expression of MHCII following application of IFN-gamma that is dependent on neuronal degeneration and activity (88). IFN-gamma is a pro-inflammatory cytokine that increases MHCII molecules on astrocytes and microglia during disease and in lesioned areas of the CNS, facilitating immune cell extravasation into the brain parenchyma (115). MHC molecules are present at very low or undetectable levels in the normal CNS, so an increase in MHCII suggests an increased immune reactivity (116). In response to CpG-DNA, a mimic of unmethylated DNA found circulating in autoimmune conditions such as systemic lupus erythematosus, OSC neurons are damaged by microglia stimulated through TLR9. Interestingly, these neurotoxically-stimulated microglia act first on neurites in their immediate vicinity, initiating axonal loss, and subsequently interact with the neuronal somata (117). Thus, it appears that organotypic cells retain their responsiveness to immune-related signals in vitro.

Using slices, it has been clearly shown that naïve T cells can enter the in vitro parenchyma and be activated by local APCs, most likely CD11c+ dendritic cells, not microglia (118). Conversely, as discussed earlier, splenic macrophages morphologically "deactivate," becoming less ameboid, when transferred to OSCs (104). Co-culture studies with T cells have revealed a soluble, subtype-dependent neuroprotective function of MBP-specific Th2 cells when cultured beneath membrane inserts growing entorhinal–hippocampal OSCs (119). This protective function does not appear to persist if T cells are allowed to establish direct cell-to-cell contacts within the organotypic tissue, as T cell interactions with slice microglia can result in axonal loss (120), microglial activation, and a downregulation of T cell-induced inflammation, depending on Th1 versus Th2 expression (121). These culture preparations offer a unique avenue through which immune cells can be selectively included when dissecting a toxicant effect on intact neural tissue.

Concerns Related to a Lack of Systemic Influence

In slice cultures, the absence of endogenous, non-resident brain cells, including T cells and infiltrating monocytes from the periphery, has not been described in detail. With the loss of blood flow and, at least partially, a compromised brain barrier system, it is not clear whether or not perivascular macrophages lining the blood vessels of the brain are completely prevented entry into the slice during culture preparation. Further, due to the age of animals at culture, most OSCs are prepared without perfusion. Thus, although the presence of a robust number of these cells would almost assuredly have been noticed in previous studies, a relatively small population of non-resident immune cells persisting within OSCs cannot be completely ruled out. Even in small numbers, immune cells are known to have vast and varied effects on cytokine release and cellular damage. Complicating matters, peripherally-derived macrophages share nearly all of the same cell surface antigens as brain-resident microglia and are almost impossible to distinguish morphologically.

A similar concern is raised when considering the so-called "clean-up" hypothesis (122). In this theory, angiogenic factors are upregulated after injuries such as stroke in an effort to open capillaries and make new microvessels. This process is initiated so that macrophages can infiltrate, destroy, and phagocytose necrotic brain tissue to limit injury and facilitate rapid recovery. As any additional infiltration of leukocytes and macrophages into organotypic tissue injured in vitro is prohibited by the nature of the slice environment, it seems that the requisite actions of phagocytosis and scar formation are fully relegated to slice microglia and astrocytes following injury. This is a role that may or may not be shared with immune-derived cells in vivo, presumably dependent on the type of injury.

Particular attention should be paid to the absence of communication between neural tissue and the peripheral immune system. Astrocytes at the BBB and ependymal cells along ventricles and the choroids plexus help to initiate neuroimmune responses. It is clear that the CNS, and presumably OSCs, possess the full complement of the innate immune response (complement, scavenger receptors, etc.), but not the adaptive immune system (T and B cells, antibodies). The innate immune response allows for the removal of toxic debris, amyloid fibrils, and similar factors expressing "eat me" signals by brain glia (123). Of equal importance are the signals that distinguish "self," so-called "don't eat me" signals, as well as those that help to silence innate immunity and limit localized inflammation. Using these signals, exogenous cells and toxic debris are targeted for destruction by astrocytes and later cleared by brain glia without ancillary damage to proximal brain tissue (124). Some of these signals may be produced and recognized by, not only brain-resident glial cells and neurons, but also infiltrating cells which routinely survey the brain parenchyma. Alternatively, the release of these signals from brain glia may, in part, occur in response to interactions with systemic immune signals at sites lost in OSCs, such as capillaries or ventricles. Thus, slices may fail to fully manifest or recognize soluble signals following select injury paradigms that involve a peripheral immune cell component in vivo.

Neurogenesis

Recently, a large portion of the neurobiology literature has focused on the functional neurogenesis now accepted to occur throughout the lifetime of mammals. As maturing slices appear to develop in a manner approximating in vivo aging, and some recent progress has been purported in the establishment of cultures from adult animals, this culture model seems a likely candidate to study the nuances of the neurogenic mechanism. Neurogenesis continues throughout the in vitro lifespan of organotypic cultures. Using BrdU label colocalization with markers indicative of maturing neuronal phenotypes, researchers have been able to show that proliferating cells, from neurogenic zones identified in vivo, mature into neuronal phenotypes and integrate into the existing circuitry in vitro (see below).

The subgranular zone (SGZ) of the dentate gyrus and the subventricular zone (SVZ) along the lateral ventricle wall are sites of active neurogenesis in adults. In OSCs, the SGZ has been used to model neurogenesis using in vitro matured cultures from postnatal animals. Using the roller tube (25) and the interface method (125), the incorporation of BrdU into proliferative cells within this discreet region has been shown to occur in vitro. The cells subsequently mature along a neuronal lineage for extended periods of time. These regions remain responsive to neurogenic stimuli, for example FGF, after some time in vitro (126). In a study looking at neurogenesis at earlier in vitro timepoints (<14DIV), it was found that the most robust cellular proliferation occurs immediately after explantation and, overall, is less robust in the presence of serum (127).

Implantation studies of neural stem cells recovered from GFP-transgenic animals into 7DIV OSCs showed they differentiate into neurons (21%) and astrocytes (38%) over seven more DIV (128). It is difficult to discern with confidence whether these percentages differ from those produced in vivo. However, the number of new neurons approximates those witnessed in vivo, while the percentage of astrocytes is quite elevated, nearly doubling in vivo observations (129). Regardless, the distribution of these newly born neurons (NeuN/ BrdU colocalized) does closely mimic that of rodent neurons in vivo (25), arguing that the appropriate regional cues for proliferation and differentiation persist in vitro.

Complications Associated with Studying Neurogenesis in OSCs

Paralleling the explantation-induced increase in cellular death markers, the observation of a robust, transient elevation in cellular proliferation and neurogenesis should be considered, as this type of induction is widely recognized as a compensatory mechanism of neuronal cell loss in the adult animal (130). In vivo, and likely in OSCs as well, this proliferation typically exceeds the number of neurons necessary to integrate into available functional circuits. Thus, over subsequent weeks, there is a need to gradually reduce the number of newly generated cells that have failed to integrate. Such a process would be accelerated and amplified in OSCs, as there is still trauma-induced proliferation, but fewer remaining targets to which those new neurons can home and successfully integrate. The activation of mechanisms necessary to effect the clearance of these excess neurons remains poorly defined. In the adult at least, new neurons are activated more easily and have a lower threshold for LTP relative to pre-existing neurons; this phenotype may persist for several months in vivo [supported and summarized in (131)]. In addition, the long-term survival and establishment of mature, fully functional circuitry appears to be exquisitely sensitive to modification by enrichment, typically in the form of physical activity or experience (132,133). It has been shown that early stressors can cause life-long reductions to neurogenesis, which may be reversible with experience (134,135). OSCs lack the propensity for prototypical enrichment, so the longevity and completeness of neuronal integration in vitro, as well as new neurons' ability to change in response to experimental manipulations, remains suspect. To assuage some of these worries, researchers should be encouraged to provide direct observations of BrdU+ neuron contributions to long-lived plasticity in vitro.

In part, the presence of an abnormal amount of cellular proliferation in comparison to the developmental maturity of the cultures and coincident, potentially confounding, glial proliferation (120) reflects a compensatory response of OSCs to the neuronal loss induced during culture preparation. However, the potential advantages of this ongoing neurogenesis warrant further detailed investigation. To date, studies of OSC neurogenesis have relied heavily on implantation and cell tracking techniques, with negligible effort towards generating functional neuronal progenitors from in vitro cultured OSCs. This raises the question of whether these isolated cells would display functional phenotypes similar to those observed after isolation from the intact brain, as the immediate neural microenvironment has been shown to have considerable effects on progenitor cell function and integration in vivo (136). In short, studies of neurogenesis in OSCs should be considered with a critical eye.

INVESTIGATIVE STUDIES USING OSCS: NEURONAL TOXICITY

Perhaps most importantly and in clearest contrast to dissociated neurons, slice neurons routinely respond to a variety of CNS insults in a manner closely mimicking the regional specificity and selective vulnerability of neuronal subtypes following toxicant exposure in vivo. This toxicity often remains temporally matched, with a conserved mechanism of cell death. Similarly, the architecture of OSCs allows for assays of toxicants with an indirect mechanism of action, such as those requiring bioconversion or secondary signaling responses from other cell types prior to any observable neurotoxicity. Owing to organotypic setups maintaining much of the complexity of the in vivo system, it is unsurprising that the use of these cultures has already led to seminal discoveries in the fields of neurotoxicology and neurodevelopment.

Traumatic Injury

Using a physical device, a highly controlled, traumatic insult with reproducible injury is hard to achieve on a restrained animal. In vivo models of traumatic brain injury are comprised of multiple central and peripheral system components, including BBB damage and edema, production of oxygen free radicals by the blood vessels, and infiltrating macrophages or lymphocytes. Therefore, although the impact region following the initial mechanical insult may remain constant, a high amount of variability can be found in the subsequent components of the injury response. Efforts have been attempted to minimize some of the inherent variables and more clearly track the injury and recovery processes, such as open skull injury models and modeling of individual, secondary components, such as hypoxia. OSCs stand poised to help bridge the gap between in vivo injury and the responses of individual cells.

In OSCs, the absence of an intact vascular system and the lack of infiltrating immune cells allows for the selective examination of resident brain cell responses to physical injury. Two main physical injury models have been applied to organotypic cultures: namely, two-dimensional "stretch" (137,138) and "weight drop" (139,140) models. The former relies upon stretching the substrate upon which the organotypic preparation is cultured and the latter indicates a mechanical compression of the organotypic parenchyma, either atop axonal projections or intact neuronal fields. Typically, these models are used to examine the temporal spreading of cell death following injury using PI fluorescence. Recently, a flexible, microelectrode matrix was developed in order to record neuronal activity following trauma up to 14DIV, in an effort to explore trauma-induced functional deficits (141). The tools necessary to examine the complex features of this type of injury are still being developed, leaving this an infrequently examined, yet promising, area of OSC research.

In Vitro Lesions

Lesion models using slice cultures allow for a direct visualization of the sequelae to axonal transection within an intact tissue environment. Most commonly, fine blades or thin wire are used to sever synaptic projections, allowing for the observation of subsequent compensatory responses in organotypic tissues at sites representing either a loss of retrograde or anterograde signals. Oftentimes, hippocampus-entorhinal cortex preparations are used due to the ease of severing their well-described projections. The fundamental basis behind the use of this model is summarized in a paper from Stoppini et al. (142). The response of seven or 21DIV OSCs to lesioning encompassed both degenerative and regenerative processes, including functional synaptogenesis, activation of reactive and phagocytic glia, as well as the formation of a thin scar at the lesion site over subsequent days post-lesion. When the cultures were maintained in vitro for a longer duration, they became less plastic, exhibiting weaker signs of regeneration following injury. Slices transected immediately upon preparation also displayed a reduced ability to establish connections across the lesion zone. When connections were formed, there was a decreased density and fewer terminals. In cultures that failed to reestablish terminals distal to the cut, a delimited zone populated by non-neuronal cells was formed (12).

This method has been used to describe various aspects of the response to axonal transection. Studies observing labeled efferent projections, either retrogradely using tracers (143) or via co-culture studies with OSCs derived from transgenic mice expressing select fluorescent proteins (144), are useful in examining characteristics of axonal regrowth, such as an observed dependence on developmental age (143). In examinations of processes believed to be integral to recovery, pre-lesion modification by enzymatic treatment (145) or blocking antibody treatment (146,147), for example, have helped to describe proteins and receptors involved in circuit development. Modeling methods unique to the post-lesion situation include implantation of radial-like glia in the lesion site, which has been shown to increase the number of translesional axon projections (148). Modification of the excitatory potential of deafferented neurons through the use of neurotransmitter receptor antagonists, photolysis of caged glutamate at distal dendrites, and direct stimulation of residual synapses, pinpointed an effect at distal dendrites that is responsible for the hyperexcitability observed following deafferentation (149). The in vitro manipulation and real time observation of these immature cultures prior to, and shortly after, in vitro lesioning gives OSCs a unique advantage over in vivo experiments.

Studies of in vitro lesions on postnatally-derived slices maintained beyond two to three weeks in vitro have been relatively unsuccessful. This is most likely due to the developmental "age" of the tissue. It has been proposed that the maturation of non-neuronal cells within the lesion site is the causative factor behind this failed axon regeneration (148). Glia infiltrate into the lesion site to remove harmful cellular debris and limit secondary damage. However, their presence and maturation there also appears to impede local axonal recuperation.

Ischemia

OSCs have been used to experimentally examine the etiology of stroke through ischemic injury via hypoxia and hypoglycemia (a.k.a. OGD). This model is rapidly induced (typically <180 minutes) and results in a loss of ATP, release of excitatory amino acids, activation of glia, and, in the hippocampus, CA1 cellular injury by 24 hours, reproducing a similar in vivo sensitivity (150). As such, this model has been repeatedly used to examine the neuroprotective properties of a wide range of compounds, with select glutamate receptor antagonists serving as effective positive controls. Although most studies focus on the role of excitatory stimuli on the ischemia-induced cell death, there does also appear to be a strong inflammatory component to the toxicity, as microglia become activated prior to neuronal loss and inhibiting their activation can reduce the OGD-induced cellular damage (151); however, their exogenous addition after OGD is also neuroprotective (93). One of the main differences between ischemia in vitro and in vivo is the in vitro absence of delayed CA1 cell death observable in vivo. In addition, although the period of submersion in the modified medium solution is brief, the effects of this transient change in slice respiration are unknown. Thus, it is important and notable to recognize that multiple studies have been able to translate their in vitro results to the in vivo situation, i.e., the middle cerebral artery occlusion model, supporting the usefulness of the OGD model for studying neuroprotective compounds in slices.

Excitotoxicity

In OSCs, the toxic stimuli most commonly applied are compounds that are widespread in the neural environment and result in hyperexcitability of glutamatergic neurons, specifically glutamate, NMDA, and kainate. For the most part, these excitotoxins cause selective cell loss in a neuron subtype-specific and regionally-localized manner that mirrors in vivo exposure to the same chemical. Accordingly, many aspects of the specific mechanistic cell death pathways in these models have been mapped out in OSCs [e.g., the kainate cell death pathway in (152)]. The activation of glia in response to the increased depolarizations and neuron loss caused by these excitotoxins has also been temporally characterized.

There are notable differences in the responsivity to excitotoxicity between in vitro-maintained tissue and observed in vivo phenotypes. KA-induced toxicity in rats in vivo does not involve damage to the dentate gyrus; whereas, in organotypic culture, this toxicity extends to this resistant region (153). These types of inconsistencies may be due to differences in toxin distribution or regional metabolism between OSCs and in vivo. Similarly, although induced tonic-clonic seizure-like events in acutely-derived slices are responsive to standard antiepileptic drugs, mature hippocampal OSCs exhibit pharmacoresistance, presumably due to the presence of aberrant axonal connections (154).

OSCs are capable of generating complex profiles in response to excitotoxicity. The preconditioning-type protective effect of sublethal doses of neurotoxins has been replicated in OSCs, as is seen with NMDA and kainate excitotoxicity (155). Similar results have been noted with ischemia (156), albeit with mixed consistency (157), where sublethal ischemic preconditioning did not reduce neuronal damage from a subsequent, lethal ischemic injury. Importantly, as with many other features of OSCs, these protective effects are, at least in part, dependent upon the time in culture at exposure. Variability also seems to be dependent on the specific alterations made to media content (e.g., Mg^{2+} concentration or K^{+} channel blocker presence) and the age at explantation.

Compound-Related Neurotoxicity

An examination of several examples of toxicants explored using OSCs reveals that these preparations can yield informative, but somewhat unique or specialized, results. These toxicants range from simple amino acids encountered at the cellular level to complex molecules present at appreciable levels in the environment. In these experiments, OSCs, typically having been previously modified (e.g., derived from transgenic animals or stimulated with drug), are treated with the individual toxic compound in question and, some time later, assessed for overt neuronal cell loss. The endpoints in these types of experiments are often minimal: PI incorporation and LDH release into the media, in order to allow for rapid comparisons and multiple dose levels.

In a study of ethanol (EtOH) toxicity by Mooney et al. (158), OSCs derived from E16 rats were more sensitive to EtOH than P3-derived cultures. The cell death in E16-derived cultures was not responsive to nerve growth factor (NGF), whereas those cultures derived from P3 rats could be partially protected. These effects seemed to be regulated according to the neurons' migration pattern and differentiation state, both of which undergo considerable modifications during the early postnatal period, again highlighting the importance of age at explantation. Heavy metal toxicants that have been examined in brain OSCs include lead, iron, and tin (159–161). Slice cultures may also prove a suitable model for the exploration of other, less common, environmental pollutants. For example, the monohalomethane derivative S-methylcysteine electrophysiologically reduces GABA-induced currents and causes increased PI uptake in hippocampal slices (162). Similarly, PCB-induced oxidative stress causes GABAergic and dopaminergic neuron dysfunction (in the form of depleted levels of GABA and dopamine, DA) in co-cultured slices of rat striatum and ventral mesencephalon (163). One of the important caveats of the OSC system to be taken from these studies, and those like them, is that the method of toxicant application can have vast effects on the resultant levels and regional distribution of neuronal death. Implicit in this observation is the possibility that in vivo dosing and human exposures allow for a differential gradient of neurotoxic material to occur near specific neuronal networks in the brain; in slices, this gradient may be lost. Of course, in these studies, as well as those using excitotoxins, OSCs do maintain the overall, selective susceptibility of specific neuronal subtypes, once again arguing that the intrinsic properties of those neurons persist in vitro.

Ex Vivo OSC Maturation Following In Vivo Exposures

Aside from their well-recognized use as models of prenatal and postnatal neural development (164,165), these cultures can be used to examine real-time effects in vitro following an in vivo exposure. A limited number of studies have used this approach to examine the impact of in vivo manipulations on the establishment and maturation of the slice. Following prenatal exposure to LPS, the integration and connectivity between the substantia nigra (SN) and striatum were assessed in co-cultured organotypic preparations. Although DA neurons seemed to grow into the exposed striatum from the unexposed SN, these DA neurons were lost over longer durations in vitro. This indicates their inability to survive under the organotypic conditions, possibly due to some underlying, compromised function or acquired sensitivity (166). A similar use of this system has been applied to examining the effects of perinatal asphyxia on hippocampal neuron survival and proliferation, highlighting a potential compensatory effect of neurogenesis for neuron loss (167). This study emphasized decreased glial scar formation following asphyxia as having the potential to aggravate the exposure of these OSCs to toxic mediators from the surrounding environment.

It is thought that in vivo exposures prior to slice isolation directly affect the brain structure in question, revealing functional deficits in the resultant OSC. Alternatively, though, it is possible that the exposure could indirectly alter the appearance, viability, functionality, or longevity of the brain slice as a result of changes imparted during the generation or maturation of the culture. A relatively unexplored field of slice culture studies, it is unknown whether exposure-induced changes to the size and complexity of targeted brain regions are translated as phenotypes present within slices derived from these exposed regions. Similarly, if preculture exposures somehow impede the growth or spreading of the slice, changes in size have the potential to skew subsequent observations of cell number, cell density, and dose-target effects. Perhaps most likely of all, changes to regional glia in response to developmental exposures could be foreseen to have a robust effect on the remodeling capabilities of these cells, which are essential for the appropriate recovery of slice neurons following deafferentation. The consequences of such changes may be incorrectly attributed to direct effects of the developmental exposure, rather than alterations initiated during the culture preparation itself. These potential confounders to neurodevelopmental OSC research remain largely unexamined.

Models of Neurodegenerative Disease

OSCs have been used to model components of human disease (168,169). These models include stroke (via ischemia induced with OGD), epilepsy (induction of seizure-like events through artificial modification of extracellular salts, i.e., magnesium, calcium, and potassium), HIV neurotoxicity (through gp120 or Tat stimulation), as well as neurodegenerative diseases such as AD, Parkinson's disease (PD), and Huntington's disease (HD). Neurodegenerative diseases are most commonly modeled by directly applying proteins associated with disease pathology, using OSCs derived from transgenic mice, or overexpressing gene(s) of interest via in vitro transfer techniques, such as viral transduction, biolistics (170), lipofection or electroporation. More complex insults have been developed in OSCs to initiate neuronal degeneration similar to that of human diseases, but via an indirect mechanism of action. Many of these involve alterations to neuronal protein processing, as neuronal degeneration in AD, PD, and HD are all believed to involve a dysfunctional component of ubiquitination, lysosomal degradation, or proteasomal breakdown of misfolded and/ or aggregated proteins. One example of this is an interesting, alternative model of AD that exploits the experimental induction of lysosomal dysfunction by the chemical, chloroquine, to promote in situ accumulation of Aβ peptides (171).

Alzheimer's Disease

The progressive neurodegeneration witnessed in AD is thought to result from an accumulation of various toxic species of beta amyloid (Aβ) protein and a concomitant neuronal aggregation of hyperphosphorylated tau protein into destructive tangles. By far, the most common approach used to study components of AD in slices is the direct addition of various lengths and fibrillization or aggregation states of Aβ to mature OSCs. Typically, the overt cell death that results is later assessed following the addition of some protective factor. Recently in slices, Aβ was shown to cause an increase in neuronal complement C1q and microglial activation (172). Interestingly, neuronal C1q expression has been hypothesized to be involved in flagging damaged neurons and/ or synaptic sites for removal by microglia (or similar cells) to expedite recovery and remodeling (173,174). One of the most controversial questions with this type of research, similar to studies in dissociated cell culture, is whether the artificial aggregation of these peptides produces soluble and fibrillar species similar to those formed in the brain, at relevant concentrations and ratios. Additionally, it remains unknown whether any or all soluble species of Aβ remain closely associated with neurons for extended periods of time in vivo or are quickly and selectively cleared from the brain parenchyma, possibly through mechanisms absent in OSCs.

Recent advances in culturing OSCs from adult tissue and for extended periods of time in vitro may enable researchers to begin looking at combinations of variables that can contribute to neurodegenerative phenotypes in humans. For example, the ability to culture OSCs for longer periods of time in vitro, in this case supplementation with NGF allowed culture growth for 50 weeks, has revealed that slices begin to produce and release low levels of Aβ beginning after 12 weeks in vitro, which can be induced to aggregate (175). Researchers have had some success in culturing viable OSCs from 20- to 30-day old, "young adult," animals using lowered incubation temperatures and varied potassium concentrations, but these methods remain insufficiently characterized. Additionally, degeneration appears to persist and progress more rapidly in long-term cultures derived from older donors (176,177). Most recently, exposure of 10DIV hippocampal OSCs derived from long lifespan, adult mice (~15 months) that are resistant to oxidative stress and exhibit improved memory retention (Ames dwarf mouse), resulted in less toxicity from Aβ application than controls (178). Parelleling this is the observation of intranuclear, rod-like inclusions, in CA1 neurons of organotypic hippocampal slices aged over 14DIV (66). These species are found in neurons in adult animals and have been shown to decrease during neurodegenerative conditions such as AD, so it may be that naïve OSCs can be used to study general aging phenomena in vitro. Clearly, these approaches, once they are verified, are of interest across a wide range of neurobiological applications.

Huntington's Disease

HD is a neurodegenerative condition typified by the accumulation of N-terminal fragments of huntingtin protein possesseing an abnormal glutamine expansion within diseased neurons. The use of organotypic preparations from the R6/2 mouse model of HD allows for the in vitro development of polyglutamine aggregates, a hallmark of HD (179). In vivo, these aggregates begin to form in the CA1 beginning at three weeks of age, then the CA3 and DG by five weeks. In OSCs, this sequence of aggregation was sustained: CA1 aggregates by 14DIV and DG aggregates by three weeks in vitro. Induction of heat shock protein 70 (HSP70), by crossing these animals with transgenics overexpressing HSP70 or via chemical treatment, caused aggregate formation in OSCs to be delayed in a transient and DIV-dependent manner (180). Additional modeling of this disease in slice cultures has been achieved through transduction with pathological length huntingtin fragments (typically via biolistics) (Fig. 2), as well as the

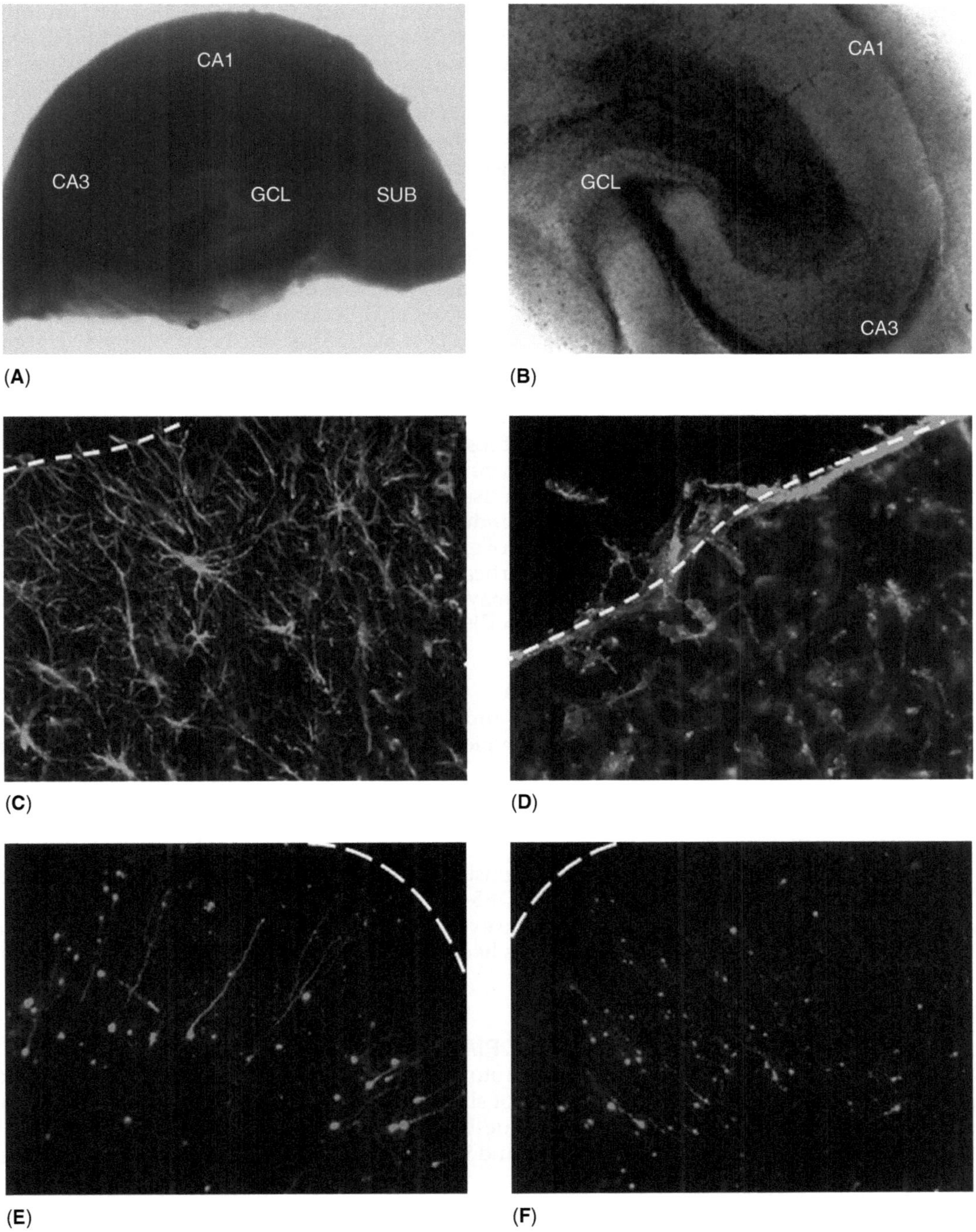

Figure 2 Select observations of interface slice cultures. (**A**) Freshly dissected hippocampal slice showing thickness of the sliced tissue and locations of the granule cell layer (GCL) of the dentate gyrus, CA subfields, and attached subiculum (SUB). (**B**) Transmission light microscopy of a 5DIV mouse hippocampal slice to show maintenance of hippocampal subfields in live organotypic tissue. (**C**) GFAP immunofluorescence of astrocytes at 3DIV in rat cortico-striatal slice cultures (the dashed line indicates the edge of the culture, where glia can often be found to migrate radially away from the slice proper). (**D**) Isolectin B4 (IB4) staining of microglia at 3DIV in OSCs. (**E**, **F**) Following biolistic transformation with YFP (**E**) or YFP coupled with mutant huntingtin containing a polyglutamine stretch of pathological length (**F**) immediately after explantation, 5DIV rat cortical slice cultures exhibit noticeable differences in the size and health of transformed neurons. *Abbreviations*: DIV, days in vitro; GFAP, glial fibrillary acidic protein; OSC, organotypic slice culture; YFP, yellow fluorescent protein.

exogenous addition of quinolinic acid or 3-nitroproprionic acid (3-NP) (181,182). Importantly, quinolinic acid is added to cultures at 3–5× the levels found endogenously and 3-NP was found to be considerably more toxic after glucose concentrations were artificially lowered to more physiological levels. These relatively simple models are currently being used to evaluate anti-inflammatory compounds and inhibitors of oxidative stress as neuroprotective drug targets. Perhaps the most appealing feature of these models for drug targeting and screening applications is their short experimental timecourse, typically less than 7DIV.

Parkinson's Disease

In vivo PD models, including injections with the complex I inhibitors, 1-methyl-4-phenyl-1,2,3,6-tetrahydropyridine (MPTP), and rotenone—both relatively acute insults—induce a selective loss of DA neurons in a manner reminiscent of DA neuronal loss in human disease. In an interesting application of the OSC system, Testa et al. (183) established a relatively chronic paradigm of DA neuron toxicity using midbrain slices treated with low dose rotenone for multiple weeks in vitro. To overcome the variability between slices, the authors transected the coronal slices and used one half as an internal control for their toxicant treatment. In this way, they showed an oxidative stress-dependent loss of tyrosine hydroxylase-positive neurons, as the neurotoxic effects were mitigated by concurrent treatment with the vitamin E compound, α-tocopherol. In an organotypic setup with tissues derived from three distinct brain regions, Neely et al. examined dopamine depletion-induced loss of dendritic spines on medium spiny neurons (184). The authors were able to uncover a selective sensitivity of morphologically thin spines that was reliant on the activity of cortical efferents. By modeling only select aspects of a disease phenotype, studies such as this may be capable of examining discreetly observed components of an otherwise overly complex disease phenotype.

Tumorigenic and Epileptic Human Tissue

Currently, methods are being developed and evaluated in order to examine the in vitro progression and responsiveness of organotypic cultures isolated from diseased human brain tissue. These specimens are typically isolated following elective surgery for invasive tumors or intractable epilepsy. The latter procedure can sometimes yield associated, overlying non-epileptogenic tissue. These methods bear further scrutiny for their use in studying epileptic and tumorous tissue, as well as the myriad potential applications of non-diseased adult tissue. Preliminary use of these cultures has shown considerable promise in regards to tissue architecture, neurogenesis, and viability measurements (185–187). The main drawback of these approaches is the lack of a true control tissue, in that the extra cortical tissue removed is juxtaposed to the resected epileptic cortex, for example, and is likely influenced by the diseased tissue's aberrant function.

CONSIDERATIONS FOR MODEL DEVELOPMENT WITH OSCS

OSCs provide a unique platform for neurotoxicological studies. The wealth of published literature regarding the distinct properties of slices in culture provides a significant neurobiological base upon which to direct specific questions regarding the neurotoxicity of exogenous compounds. Individual cellular properties and interactions can be examined within the framework of regional phenomena, an effect that cannot be replicated using dissociated cell cultures. On the other hand, the ability to visualize neurons in real time yields advantages over in vivo models. The fact that these cultures circumvent first pass metabolism and the BBB provides accessory support for their use in toxicity assessment. However, appropriate modeling is dependent upon the specific question being asked. While this seems a very basic factor, it is often lost in the translation of neurobiological models to the toxicological arena. In the development of OSCs as a model to examine the neurotoxic potential of any compound, one of the first considerations for expanded research is the inherent variability of the system, which needs to be taken into account and, where possible, controlled. Cultures should be generated from animals that are as phenotypically identical as can be achieved. The source, developmental stage, and physical characteristics of the slice need to be considered. For toxicological effects, evaluations of dose, temporal events, sampling requirements, and physiologically relevant endpoints are necessary. Similarly, the dosage and selection criteria need to reflect appropriate exposures, adequate sampling, and physiological, study-relevant measures of viability. In neurobiological studies, changes in individual cells can often be assayed independent of surrounding cells. In this manner, criteria are typically set for a "normal" cell and others are selected and then assayed according to that control cell. In toxicological studies, however, this type of approach can result in a selection bias, particularly when looking at the responses of individual cells within groups of differently responding cells. It is, thus, very important to consider contributions from juxtaposed cells within the slice that have the potential to be affected by the chosen experimental manipulation or exogenous compound, as one would in in vivo studies. There is also a growing need for stereotyped descriptions of the experimental setup, tissue processing, and visualization techniques in order to make accurate comparisons of drug treatment and toxicant exposure across studies.

Explantation Phenotype

A variable that is not often considered in OSCs is the phenotype of donor animals at the time of explantation. Because OSCs are routinely prepared from early postnatal rodents, sex and weight are seldom incorporated into these studies. The influence of these factors on in vitro responses is unclear. Depending upon the specific experimental question, the sex of the pups may prove to be an important variable for consideration. This would assuredly be an issue with slices obtained from sexually mature animals or if tissue is obtained from animals exposed to experimental manipulation in vivo.

Sex steroids such as estradiol and testosterone can have profound modulatory effects on the growth, migration, and differentiation of immature brain neurons. In vivo, females are less vulnerable to neurotrauma, ischemia, and drug-induced toxicity (188). Similarly, estrogen offers a level of neuroprotection in OSCs (189), which may be reflected in estrogen receptor activation. Though hormonal differences during early postnatal development are thought to be minimal, an intrinsic resilience of female hippocampal OSCs to OGD and NMDA has been demonstrated, alongside differences in neuroprotective compound efficacy (190). These studies demonstrate the need to consider the impact of any sex-specific effects and to modify the experimental design accordingly. One may want to include sex as an experimental factor or exclude it by collecting slices from only one sex. While rarely done, inclusion of sex as an experimental variable may serve to decrease variability within data that are known to be notoriously variable. This would be significant when examining a manipulation that would modify sex-specific hormonal factors. These considerations may also prove helpful when examining low-dose effects or test compounds with an ill-defined mechanism of action. While there is little data to support body weight as a significant confounder in OSC studies, it is well known to influence developmental ontogenic processes in vivo. Birth weight is a reflective indicator of nutrition and organotypic preparations are derived during a period of ongoing, active brain development, which is reliant upon an appropriate diet. Thus, incorporation of an exclusion factor based upon body weight, i.e., excluding runts of the litter, may also serve to decrease variability.

Developmental Stage

The rapidly dynamic development of the brain necessitates the need for accurate maintenance of the age at explantation (191). This should incorporate a consideration of the specific in vivo insult that is being modeled, as well as the specific endpoint under study. The optimal age for cultivation is highly dependent on the region from which the CNS tissue is derived and the developmental maturation timecourse for this specific region of interest. The optimal culture time is typically during an immature stage of the neuronal population. Once the neurons mature and make permanent synaptic contacts, culturing can become a significantly traumatic event for the tissue. This stress, combined with the deafferentation of the neurons, results in a considerable exacerbation of trauma-induced cellular death. Thalamic and spinal tissue is often prepared from fetal tissue (60), other brain regions from P0–P15. In the OSC field, it is fairly well accepted that cultures derived at early postnatal ages (P0–P3) have better neuronal survivability than those from later postnatal ages (P9–P15), but suffer from poor tissue organization, as many neurons are still in their migratory phase at explantation and may inappropriately integrate. As such, P5–P7 is often used as a compromise between these two, desirable features (192).

There are direct relationships between the animal age at the time of culture, the CNS developmental stage, the morphology and maturation state of brain glia, and the migratory ability of slice neurons. This consideration becomes critical in cross-study comparisons or when establishing injury models. One example is the observed difference in the expression of the presynaptic protein, synaptophysin, in hippocampal slices derived from animals at different postnatal ages. Using P6–P8 rats, a robust increase was noted by Buckby et al. from 0 to 21DIV, whereas Bahr et al. used P11–P13 rats and noticed no change in expression over DIV (91,171). Additionally, because slice preparations mature for some time in vitro in the presence of soluble and contact-dependent signals seldom encountered in the intact brain, the tendency to compare maturation state in vitro to a specific in vivo age should be approached with hesitancy. Such samples would be expected to have experienced different maturation signals in these disparate environments.

It is also important to use organotypic tissue that is developmentally age-appropriate for the specific in vivo insult being modeled. For example, the hippocampus, cerebellum, and cortex all have postnatally-developing or migrating neuronal populations. In the rodent hippocampus, 85% of dentate granule cells divide and differentiate postnatally, with 40% remaining immature beyond P7 and proliferation continuing into the third postnatal week; conversely, pyramidal cells are relatively established, although CA1 and CA3 mature on different postnatal timecourses, prior to birth (193–195). Thus, in vitro modeling of toxicants that elicit effects on specific neuronal nuclei, if done at an inappropriate developmental stage, can yield results that are impossible to replicate in vivo. Likewise, in the late-developing cerebellum, where much of the maturation occurs after P12 (196), differentiated Purkinje cells and synaptic connections are not formed to any appreciable extent until after the second postnatal week (197), with a corresponding differential in the development of individual cerebellar lobules. Thus, it may be possible to miss regional-specific toxicity or to induce toxicity in sensitive, still-maturing cell populations if the developmental age is not carefully controlled.

Considerations in Methodology

The different methods for culturing brain slices require that a decision between the roller tube and interface techniques be made at the study onset. Some differences between organotypic preparations grown in interface versus roller tube cultures have been noted, such as variations in neuronal migration patterns (96). As such, care is required to validate the appropriate slice methodology and prevent comparisons that would be confounded by preparation techniques. In addition to a consideration of the tissue region to be assessed, this decision must involve an analysis of the expected experimental endpoints to be used in the study.

The roller tube technique appears better suited for the establishment of extremely long term cultures, over several months. These cultures thin out to near single cell layer and, accordingly, are optimal for direct visualization techniques. Additionally, it is easy to identify and observe individual cells for dye loading of the entire neuronal structure. Roller tube cultures are grown upon a glass coverslip, making them highly amenable to immunostaining techniques. During the first week in vitro, the plasma clot has yet to break down, making most studies difficult or impossible during this time.

Interface cultures, which are somewhat easier to prepare than roller tube cultures, are preferred for acute or short-term cultures, particularly over the first two weeks in vitro. These thicker sections retain a better three-dimensional architecture, as they remain several cell layers thick after several weeks in vitro. This feature makes them better suited for studies of 3D groupings of neurons and neuronal networks. Also due to this, the tissue is less fragile and more resilient to traumatic labeling techniques such as biolistics and microinjection. In addition, the thicker tissue allows for easier biochemical analyses, including RNA and protein isolation.

A similar consideration should be paid to the matrix upon which these cultures are grown: glass for roller tube cultures and optically-clear, low protein binding, porous polymer membranes for interface cultures. Contact-dependent effects in the brain are regulated, to a large extent, by cues from the extracellular matrix. Processes such as migration and differentiation are particularly sensitive to these cues. Studies into alterations to these processes initiated as a direct result of slice incubation upon porous membranes or within a plasma clot, even temporarily, are nonexistent and difficult to design. In addition, a penetration of neuronal and glial processes through interface inserts, which increases over time in vitro (55,65), begs the question of what influence this may have on glial and neuronal activity or signaling processes in nearby cellular fields. Fortunately, these interfacing regions of the slice preparation are routinely barred from experimental endpoints. Caveats such as these should always be considered in relation to the specific neuronal population under study.

Physical Parameters

Several physical variables across organotypic setups can limit the generation of reproducible data. These include, but are not limited to, the plane of cut and thickness of the OSC, as well as the levels of specific media constituents, notably serum, and glucose. There are difficulties inherent in processing these tissues in an architecturally sound manner and in maintaining consistent visibility across cell populations of interest at the culture endpoint. In addition to morphologically selecting only the most structurally-intact and healthy looking sections of tissue for culture, as is routine for most experimenters, careful attention needs to be paid to minimizing any additional physical discrepancies across slices.

The thickness and size of the organotypic tissue directly influences the diffusion of media and gases in vitro. With increasing thickness, there is the possibility of developing mild necrosis in the center of the slice due to an incomplete or graded perfusion of the tissue by the requisite gases. With time in culture, considerable thinning out of sections occurs with a subsequent reorganization of neuronal fields over DIV. Such thinning continues beyond 7DIV (66). Thus, the changing slice thickness can alter the normal morphology of intact neuronal fields. For example, a widening of the dentate granule cell field is seen in hippocampal slices over time in culture (127). If the study design or the endpoints are dependent upon the distance between individual cells, changes to slice thickness or DIV can have significant, non-specific effects on the study, skewing distance-dependent observations.

A consistent plane of cut and maintenance of the approximate in vivo stereological positioning of the cultured brain region is also essential. The importance of maintaining septo-temporal positioning across samples has been previously described (127). This work demonstrated that the positioning across slices can influence cell number, cell density, BrdU+ cell counts, and the number of caspase-3+ apoptotic cells observed under routine experimental conditions. The angle at which a brain region is cut can significantly alter the presence of out-of-plane projections and the maturity of the tissue region of interest. For example, there is a gradient of developmental maturation in the hippocampus, with the septal portion being more mature than the temporal part in early postnatal stages (198). As such, particular consideration should be paid to these physical slice characteristics when attempting to use OSCs to model developmental exposures, as it remains unclear whether in vivo manipulations distort fundamental aspects of slices during their subsequent generation and maturation. In addition, the sectioning angle can change the proportion of white matter, myelin, or persisting elements of the ventricle. This can result in significant variability across slices owing to a differential in the population of non-neuronal cells, as regards proliferation, migration, and activity. All of these features appear to be elevated in the white matter regions of the slice and likely differ in other distinct substructures of the slice as well.

Serum

As in all culture models of the nervous system, the in vivo protective function of the BBB makes the presence of serum reflective of an atypical biological system. Serum contains various factors seldom encountered by brain cells behind an intact BBB in vivo, including various hormones and blood-derived factors, as well as a number of unknown or unidentified constituents. Serums vary across manufacturers and across lot numbers, differing even in the concentrations of trophic factors they contain. One of the most important components of serum to be considered in slice culture setups is endotoxin, a.k.a. LPS, a well-described contaminant known to stimulate microglia. Within OSCs, changes to serum-free media have recently unmasked a number of changes that are induced by, or an artifact of, serum (25,199,200). As of yet, the full range of differences in neural morphogenics and signaling between serum and serum-free conditions remains undetermined. For these reasons, if the in vivo condition that is being modeled with the slice culture does not involve damage to the BBB, as can occur in a stroke infarct for example, all efforts should be made to exclude serum from the media. If the insult being modeled involves an alteration in the permeability of the vascular system, then the presence of serum may be appropriate.

Specifically in the context of studies of neuroglia, the presence of serum can increase microglial proliferation, enhance the formation of the "scar"-like astrocyte barrier at the slice interface, and stimulate oligodendrocytes, resulting in inhibited axonal growth of GABAergic neurons (201). Of additional concern are the effects on neuronal morphology and signaling. There are significant, serum-dependent differences in dendritic spine development, as well as changes in neuronal responsiveness to growth factors and other signaling molecules (199). A final consideration in the use of serum-containing media is the potential alteration of cellular proliferation in zones of neurogenesis. In interface cultures, it was noted that there is an earlier and more rapid reconstruction of cellular layers in serum-free preparations, with these cultures achieving a higher cell number and density days earlier, alongside an elevated number of BrdU+ cells (127). Further, the concentration of serum in slice media can also influence viral infectivity and transfection, toxicant and drug effects, gene induction, and protein expression (157,200,the author's unpublished observations). It remains unclear exactly how the serum level in organotypic media will affect a specific endpoint and, as such, changes to this variable should be evaluated prior to establishing or altering toxicological or pharmacological models.

Glucose

The glucose concentration in the media of OSCs is also of key relevance for the evaluation of neurotoxicological endpoints. Typically, slice culture media has glucose concentrations in the vicinity of 25 mM, nearly 50× higher than the in vivo neuronal environment (202). This has been shown to be a determinant factor on the ability of toxicants to elicit effects on OSC neurons at doses with physiological relevance. Unfortunately, it is clear that lowering the glucose concentration in order to tighten this dose-dependence in vitro can initiate a significant impediment to natural neuronal metabolism processes and is capable of causing ischemic events. The effects of glucose on the non-neuronal cells within a slice have not been thoroughly examined; however, one would expect that similar protective and stimulatory properties would be observed.

Sampling

In reviewing many of the major studies utilizing OSCs, it is often difficult to determine the experimental unit. Organotypic "n" numbers can be expressed in vague terms, such as "samples" or "independent experiments." They can also be specified as samples from "independent animals," separate "inserts," or individual "slices." Quite often, slices from any one animal are used for multiple manipulations and endpoints, and may even be considered as a way to minimize variability, with the "animal" serving as its own control. However, as mentioned above, the plane of cut and the percentage of the slice that includes, for example, white matter tracts and ventricular elements, can significantly influence endpoints under study. While, technically, the animal may serve as its own control, the cut across the region of interest may alter the feasibility of such comparisons. An alternative approach has been described by Testa et al. (183). In this study, hemisected coronal sections were used as "within slice" internal controls for pharmacological manipulations. This approach may provide a strategy to identify low-level and specific effects.

If one considers an OSC experiment within the framework of a neuro-developmental study, the dam and litter would determine the n. One could extrapolate that comparisons using OSCs should be made across animals, rather than across slices prepared from the same pup or pool of pups. For scaled-up experiments, however, the use of multiple animals as a source of slices can be used, not only to minimize the slice-to-slice variability inherent in these preparations, but also to maintain a high degree of genetic heterogeneity across treatments. Ideally, slices should be sampled across animals within a litter and then compared to slices from animals derived from a separately-prepared, age-matched litter, with experimental manipulations carried out independent of one another. Unfortunately, this can generate additional problems with regards to the time frame for generating slices and conducting experiments, as would be a concern with acute slices for electrophysiology, for example.

Dose Selection, Delivery, and Metabolism

Within the whole organism, the metabolism and biotransformation of foreign compounds in the body significantly contributes to adverse effects at the target organ. Many compounds require metabolic activation, while others are rapidly broken down by detoxification systems. These effects can be directly adverse or can be minimized and compensated for by the surrounding environment en route to, or within, the target organ. In any case, this can have a significant impact on the translation of an in vitro effect to the in vivo environment. As in any culture system, first pass metabolism is circumvented in the OSC and thus, unless there is in situ metabolism, only effects of the parent compound can be examined. Thus, the pharmaco- and toxicokinetics of the compound under study, particularly whether the parent compound actually reaches the brain, should be known prior to the initiation of a study. Xenobiotic (in)activation systems are predominantly expressed in non-neuronal cells (203). For example, astrocytes metabolize the parent compound, MPTP, to the neurotoxic species, N-methyl-4-phenylpyridinium ion (MPP+), convert glutamate to glutamine, and sequester heavy metals from the neural environment (204–206). While these cells are present in OSCs, their activation state is significantly altered, which may influence their ability to perform normal metabolic functions.

Selection of dose levels for examination in vitro can become a major limiting factor for any study. If the neurotoxic potential of a chemical in vivo is dependent upon metabolic parameters, such as biotransformation, lipid-aqueous partitioning, and distribution, then the expression of toxicity in a culture system may be irrelevant when related to whole-animal studies. In short, the metabolism of compounds in vitro can vary considerably from that in vivo. Of particular concern in the context of slice studies are variations dependent on the in vivo age at explantation and the in vitro age of slice "development." To date, no glaring differences have been noted between in vivo parenchymal brain cells and slice cells in regard to overt metabolic activity. One example of this is the conserved temporal expression of epoxide hydroxylase and NADPH cytochrome P-450 reductase (207). Similar to this is the sensitivity of OSCs to MPTP, a toxicant ineffective in dissociated neuronal cultures because it requires astrocytic conversion (208). In this manner, slice preparations provide a unique system with which researchers can manipulate and precondition neurons, while maintaining at least some of the bioactivation characteristics present in vivo.

Organotypic cultures have been shown to exhibit an altered permeability and responsiveness to several diffusible signals released from active glia, including nitric oxide (NO) and free amino acids, such as glutamate (45,209). Similarly, penetration into the slice decreases over time in vitro, often proving an impediment to the examination of toxicological endpoints. Antibodies such as BrdU (127), drugs, and toxicants all exhibit penetration difficulties throughout the full depth of mature organotypic tissues. It remains unclear whether this altered permissiveness is the result of metabolic clearance, a loss or down-regulation of specific transporters present in vivo, the rapid formation of a barrier-type system of activated glia at the slice interface, or some as-yet-unidentified reason. In OSCs, the transport of various factors appears to be highly molecule-specific and heavily reliant upon the time in vitro that the OSC is exposed. The possibility of a differential in access or penetration of compounds, protective agents, and toxicants due to their specific chemical composition, such as the presence of a highly charged moiety, remains a critical concern in slice culture research.

Quite often, the ability to detect changes in OSCs is dependent upon a very high dose level as compared to in vivo exposure, adding complexity to translation of these data to human exposure. High-dose exposures in the absence of a very specific and relevant mechanistic endpoint can be difficult to interpret. Data obtained from isolated cell culture systems can be used as a starting point for dose selection; however, as would be expected, effective dose levels in OSCs are often significantly higher than those than those required for isolated cell cultures. Thus, it may also be important to thoroughly re-examine the temporal timecourse and re-establish the primary action of the applied toxicant within this closed system, particularly when examining toxicants with complex secondary responses. These extra steps may have a considerable effect on high throughput-type analyses.

Once the dose is established, the malleability of the slice environment requires evaluation. A distinct advantage of this system is that substances can be added or withdrawn from the culture medium, allowing for an analysis of sequential events. Techniques also exist that allow for the direct injection of a substance into an individual cell or discreet tissue region and examination of the resulting cellular responses. The amenability of slices to transformation using transfection, viral products, or gene gun-mediated biolistics decreases rapidly and considerably over DIV, with older cultures also exhibiting signs of toxicity from these types of experimental manipulations. Thus, these modifications are often carried out soon after explantation. While the localized delivery of a chemical can provide very interesting information with regards to an acute response, this is unlikely to occur in vivo. In addition, the injury induced by the direct application of compound or by injection is, by itself, capable of stimulating cells at the surface of the slice culture. Interface glia represent a population unique to the slice culture setup and one not present in the normal brain. They are highly active and migratory, having the potential to track along gradients of damage from their vigilant, sentry position at the slice periphery. These cells can have robust actions if allowed to penetrate into the relatively isolated slice proper, at least over the short term, with the potential to influence experimental endpoints. Thus, care should be taken during dosing to minimize disruption of the slice interface. One may consider efforts to maximize compound delivery while minimizing local injury during delivery of substances to the brain via either a direct injection or cannula (i.e., injection volume, duration, etc.). Further, as in any direct delivery to neural cells, whether a direct injection to the brain or to isolated cultured cells, the physicochemical properties of the chemical, such as solubility, pH, and binding to serum must be considered prior to any study design. Quite often, solubility issues require the use of dimethyl sulfoxide or ethanol in the vehicle. These substances are chemically useful as carriers, however, they can have adverse effects in and of themselves or synergistic effects when combined with the drug or compound of interest. In exquisitely complex systems such as slice cultures, the use of such compounds, while often necessary, can distort toxicological or pharmacological effects and raise questions when translating the findings to the in vivo situation.

The capacity to detoxify the OSC environment following injury is likely reduced or, at the least, is mechanistically distinct from the ability of the intact brain to remove unwanted factors. For example, it is unclear whether inflammatory molecules and cellular debris capable of enhancing toxicity paradigms persist in OSCs, or if the rapid removal of such items is one of the explanations behind the early and robust activation of microglia and the prolonged vigilance of slice astrocytes. The metabolic activity of slice astrocytes, particularly those compromising the glial barrier at the culture interface, and the lack of a vascular element for the removal of toxic byproducts, may also lead to an amplification of secondary effects. Although these caveats may complicate dose-effect relationships and data translation to specific in vivo models, the differences should remain minor so long as OSCs are recognized as closed systems during data interpretation. Conversely, the relative lack of systemic factors does make the use of OSCs quite attractive for neuroimmune-oriented examinations in which these cells and cell products are exogenously added.

Viability of OSCs

As a measure of quality control and selection criteria for the experimental use of individual slices, or when establishing dosage parameters, cell viability is an effective, initial assessment. Simple, nonspecific measures of membrane compromise and cellular damage that are most frequently used to assess OSC viability include PI uptake and LDH efflux. These methods are useful because they can be applied in intact preparations to estimate cell loss over DIV; however, both markers have limitations when they are employed as a physiological endpoint. While it is often reported that changes observed with these markers are indicative of the neuronal populations, the fact is that they are cellularly non-specific markers for cell membrane permeability, detecting changes in all cell types.

PI is typically added and measured following toxicant exposure. When PI uptake is used as a biological marker for toxicity in slice cultures, the high level of background fluorescence requires an estimation of the ratio of the experimental signal to background noise to determine changes in specific staining. This often results in the exclusion of data from slices exhibiting high basal staining intensity. This signal intensity can be influenced by the age of the slices in culture, slice thickness, media composition, and tissue autofluorescence. Finally, similar to other vital dyes, an assessment of PI requires in situ exposures. There is little data available to evaluate how the vital dye uptake and fluorescent imaging, in and of itself, may affect the function and integrity of cells, as well as the photo-stability of the compound.

Together, these non-linear manipulations result in data that should be viewed as sufficient for qualitative, not quantitative, assessments.

LDH efflux can provide a more quantitative assessment of membrane permeability; however, it too can be influenced by media composition and metabolism. An important consideration in the use of LDH is that its efflux is delayed compared to the actual deterioration of cellular morphology (210). This can result in an underestimation of cell membrane damage. An additional concern with regards to OSCs is that LDH is measured in the media and, depending on the culture conditions, may not reflect changes occurring within the protected neuronal layers of the organotypic tissue. It is impossible to discount the aberrant, interfacing region of the OSC from such endpoints. As such, direct measures of morphological changes indicative of the loss of select cell types, for example, the quantifiable loss of neuronal MAP2 staining, may be more appropriate. However, nearly any such marker of cell death would have similar problems dependent on the metabolic or cellular process measured, penetration and longevity of the marker, background signal, as well as the method of data collection.

Experimental Processing and Visualization

Compared to both dissociated cell cultures and tissues processed from *in vivo* experiments, the morphology of OSC tissue maintained in vitro for any significant length of time is of a poorer quality. As mentioned previously, roller tube cultures are well-suited for whole culture immunohistochemistry and maintain better visibility of their cellular fields. However, they possess a reduced three-dimensionality and cannot be used for long-term live cell imaging. Conversely, with interface cultures, particularly if the desire is to observe the cultures during early stages (<14DIV), cryosectioning is both necessary and appropriate. Due to their thin and extremely fragile nature, sectioning of organotypic tissue to yield morphologically acceptable slides is only accomplished through painstaking and time-consuming methods development. Without damaging the samples, attempts to remove OSCs from membranes, to cut OSCs directly using the membrane to mount the section to the chuck, or after attaching OSCs to larger tissue sections, have been largely unsuccessful. Rather, fixation, cryoprotection, and freezing within intact wells, using minimal agitation and carefully established protocols, and prior to removing the membrane, appears to work best. In addition, because, quickly after explantation, glial barriers have formed at the points of interface in these cultures, most protocols require that the sections immediately apposing either the membrane or air interface are excluded from any type of analysis.

One of the most attractive characteristics of OSCs is their potential for live cell tracking. Thus, it is worth mentioning some notable difficulties inherent to using this procedure in slices. Again, the glial barrier comes into consideration, as most cells of interest reside in the inner layers of the OSC. This necessitates the use of confocal microscopy. Following implantation, cells routinely and immediately migrate either into the deep tissue or to the interfacing areas of the culture. Unfortunately, as OSCs exhibit a large amount of autoflourescence, it becomes quite difficult to track, with any notable resolution, cells tagged with vital dyes or cell surface antibodies in vitro. An additional, complicating factor common to many transplant studies is that these donor implants and OSC recipients are infrequently genetically-matched, matched for developmental stage, or, perhaps more importantly, derived from similar brain regions. This can be expected to skew interactions of implanted cells with phenotypically-different, and more mature, OSC cells. To visualize individual cells, these studies are typically carried out using brightly-labeled, regionally specified cell populations, such as promoter-driven eGFP transgenics, so that these cells are easy to locate with a confocal scope, minimizing toxicity to the surrounding tissue during excitation. Technologies are evolving, as shown in the use of photo-convertible molecules to identify and track individual cells in non-transgenic, non-manipulated OSCs (211). Also, as many of the changes impeding these types of observations may be attributed to the glial barrier, some researchers have begun to test methods to minimize its formation. However, the isolation of these sensitive neurons may be necessary and beneficial, so the long-term effects of these manipulations on synapse maturation and neuronal morphology need to be examined in greater detail. These types of studies hold a great deal of future promise, not only for CNS transplant applications, but also for dissecting function-specific effects of individual cell populations on brain tissue, following both pathological insult and pharmacological manipulation.

CONCLUDING REMARKS

Slice culture preparations are poised for widespread use in neurobiological and toxicological studies of integrated cellular responses. Their ability to maintain intact brain architecture and neuronal connectivity within a platform that allows for easy manipulation and visibility sets OSCs apart from other in vitro models. Derived from early postnatal tissue, slice cultures allow for observations of neuronal network development in vitro. The presence of regionally conserved, supportive glia cells and the concomitant absence of a brain barrier system and peripheral metabolism makes them desirable for use in examinations of toxicant action on specific areas of intact brain tissue. Similarly, the absence of influence from vascular factors and systemic immune cells allows for slices to be used as a closed system to examine the responses of parenchymal brain tissue in isolation. Current techniques allow for in vitro observations which are only beginning to be applied to in vivo systems, including live cell tracking, coordinated manipulations with multiple pharmacological and DNA-based agents, and rapidly-induced degenerative processes reminiscent of human conditions. In all, the current impression derived from the slice culture literature is that of an easily adaptable system, well suited to assaying complex functions common to the in vivo brain.

Taken individually, however, the benefits of the slice setup each possess significant caveats. These caveats may prove prohibitive when attempting to translate toxicity assessment, drug effect, and neurobiological phenomena to in vivo responses. The OSC architecture is a reasonable approximation of the intact brain, but it is notable that the tissue thins and spreads apart over DIV, rests upon a non-native matrix material, and contains large regions of cellular and axonal loss that are cleared and replaced with less mature cellular components. Although neuronal connectivity and network development are maintained or reestablished in vitro, these processes are highly influenced by the loss of afferent and efferent connections outside of the cultured region, resulting in changes to synapse organization and cellular morphogenics. Similarly, as the in vitro development of organotypic neuronal networks occurs in the absence of experience-dependent learning and is likely to be aberrant, the inability to effectively culture adult tissue remains a key concern in slice culture research.

There are several aspects of working with OSCs that should be evaluated when considering their use as an experimental model. Manipulation and visualization within slices can be problematic and should be conducted with care, particularly at later DIV or if the region being assayed is large. As the techniques used in slices become more refined, attention needs also be paid to maintaining appropriate dose levels, developmental stage, and septo-temporal location, as each of these variables can have a considerable effect on experimental endpoints. Of specific concern with delivery of compounds or molecules of interest, small disturbances to the organotypic tissue can alter the responsiveness of highly active cells at the tissue interface. There are notable downsides to lacking peripheral metabolism and influence from systemic immune cells. Namely, bioconversion can only occur in situ and neuroimmune interactions, which may be important to the function of the intact brain, are lost in slices. Deafferentation and the lack of a BBB make it necessary for a scar-like cell layer to form at the culture interface and regulate the entry of foreign compounds or harmful factors into the slice proper, but this scar simultaneously limits drug and antibody penetration. Finally, the supportive microglia and astrocytes present within the slice act in a manner dissimilar to their in vivo counterparts, as they respond to vascular-derived factors and axonal transection during culture preparation. This "pre-conditioning" of glia may alter their reactions so that they do not accurately represent endogenous phenotypes.

The variables inherent to the organotypic preparation can be effectively minimized through careful study design and appropriately timed manipulations. However, translations from slices to in vivo need to include a consideration of the characteristics unique to these in vitro models. These aspects of OSCs, which are essentially artifacts of the in vitro culture situation, have the ability to add variability and shift experimental outputs, as well as complicate deductive interpretations to in vivo brain processes. It becomes a requisite aspect of establishing organotypic models for human conditions to consider the contribution of some or all of these facets individually. Efforts such as these help to ensure that the in vivo profile of the model in question is not significantly altered as a result of the preparation itself. Despite the presence of potential confounders across multiple aspects of slice biology, certain focused questions in neurobiology and toxicology can only be answered through the careful use of this unique and informative culture system.

REFERENCES

1. McIlwain H, Buchel L, Cheshire JD. The inorganic phosphate and phosphocreatine of brain especially during metabolism in vitro. Biochem J 1951; 48: 12–20.
2. McIlwain H, Buddle HL. Techniques in tissue metabolism. I. A mechanical chopper. Biochem J 1953; 53: 412–20.
3. Li CL, McIlwain H. Maintenance of resting membrane potentials in slices of mammalian cerebral cortex and other tissues in vitro. J Physiol 1957; 139: 178–90.
4. Gahwiler BH. Organotypic monolayer cultures of nervous tissue. J Neurosci Methods 1981; 4: 329–42.
5. Gahwiler BH. Slice cultures of cerebellar, hippocampal and hypothalamic tissue. Experientia 1984; 40: 235–43.
6. Stoppini L, Buchs PA, Muller D. A simple method for organotypic cultures of nervous tissue. J Neurosci Methods 1991; 37: 173–82.
7. Moser KV, Schmidt-Kastner R, Hinterhuber H, et al. Brain capillaries and cholinergic neurons persist in organotypic brain slices in the absence of blood flow. Eur J Neurosci 2003; 18: 85–94.
8. Frotscher M, Misgeld U, Nitsch C. Ultrastructure of mossy fiber endings in in vitro hippocampal slices. Exp Brain Res 1981; 41: 247–55.
9. Okada M, Sakaguchi T, Kawasaki K. Correlation between anti-ubiquitin immunoreactivity and region-specific neuronal death in N-methyl-D-aspartate-treated rat hippocampal organotypic cultures. Neurosci Res 1995; 22: 359–66.
10. Pozzo Miller LD, Mahanty NK, Connor JA, et al. Spontaneous pyramidal cell death in organotypic slice cultures from rat hippocampus is prevented by glutamate receptor antagonists. Neuroscience 1994; 63: 471–87.
11. Zimmer J, Gahwiler BH. Cellular and connective organization of slice cultures of the rat hippocampus and fascia dentata. J Comp Neurol 1984; 228: 432–46.
12. Zimmer J, Gahwiler BH. Growth of hippocampal mossy fibers: a lesion and coculture study of organotypic slice cultures. J Comp Neurol 1987; 264: 1–13.
13. Coltman BW, Earley EM, Shahar A, et al. Factors influencing mossy fiber collateral sprouting in organotypic slice cultures of neonatal mouse hippocampus. J Comp Neurol 1995; 362: 209–22.
14. Debanne D, Guerineau NC, Gahwiler BH, et al. Physiology and pharmacology of unitary synaptic connections between pairs of cells in areas CA3 and CA1 of rat hippocampal slice cultures. J Neurophysiol 1995; 73: 1282–94.
15. Gutierrez R, Heinemann U. Synaptic reorganization in explanted cultures of rat hippocampus. Brain Res 1999; 815: 304–16.
16. De Simoni A, Griesinger CB, Edwards FA. Development of rat CA1 neurones in acute versus organotypic slices: role of experience in synaptic morphology and activity. J Physiol 2003; 550: 135–47.
17. Lindroos MM, Soini SL, Kukko-Lukjanov TK, et al. Maturation of cultured hippocampal slices results in increased excitability in granule cells. Int J Dev Neurosci 2005; 23: 65–73.
18. Thompson SM, Gahwiler BH. Activity-dependent disinhibition. II. Effects of extracellular potassium, furosemide, and membrane potential on ECl- in hippocampal CA3 neurons. J Neurophysiol 1989; 61: 512–23.
19. Buckby LE, Mummery R, Crompton MR, et al. Comparison of neuroplastin and synaptic marker protein expression in acute and cultured organotypic hippocampal slices from rat. Brain Res Dev Brain Res 2004; 150: 1–7.
20. del Rio JA, Heimrich B, Soriano E, et al. Proliferation and differentiation of glial fibrillary acidic protein-immunoreactive glial cells in organotypic slice cultures of rat hippocampus. Neuroscience 1991; 43: 335–47.
21. Berger T, Frotscher M. Distribution and morphological characteristics of oligodendrocytes in the rat hippocampus in situ and in vitro: an immunocytochemical study with the monoclonal Rip antibody. J Neurocytol 1994; 23: 61–74.
22. Birgbauer E, Rao TS, Webb M. Lysolecithin induces demyelination in vitro in a cerebellar slice culture system. J Neurosci Res 2004; 78: 157–66.
23. Coltman BW, Ide CF. Temporal characterization of microglia, IL-1 beta-like immunoreactivity and astrocytes in the dentate gyrus of hippocampal organotypic slice cultures. Int J Dev Neurosci 1996; 14: 707–19.
24. Hailer NP, Jarhult JD, Nitsch R. Resting microglial cells in vitro: analysis of morphology and adhesion molecule expression in organotypic hippocampal slice cultures. Glia 1996; 18: 319–31.
25. Raineteau O, Rietschin L, Gradwohl G, et al. Neurogenesis in hippocampal slice cultures. Mol Cell Neurosci 2004; 26: 241–50.
26. Giulian D, Baker TJ, Shih LC, et al. Interleukin 1 of the central nervous system is produced by ameboid microglia. J Exp Med 1986; 164: 594–604.

27. Derouiche A, Heimrich B, Frotscher M. Loss of layer-specific astrocytic glutamine synthetase immunoreactivity in slice cultures of hippocampus. Eur J Neurosci 1993; 5: 122–7.
28. Schmidt-Kastner R, Humpel C. Nestin expression persists in astrocytes of organotypic slice cultures from rat cortex. Int J Dev Neurosci 2002; 20: 29–38.
29. Miyaguchi K. Ultrastructure of intermediate filaments of nestin- and vimentin-immunoreactive astrocytes in organotypic slice cultures of hippocampus. J Struct Biol 1997; 120: 61–8.
30. Davalos D, Grutzendler J, Yang G, et al. ATP mediates rapid microglial response to local brain injury in vivo. Nat Neurosci 2005; 8: 752–8.
31. Stence N, Waite M, Dailey ME. Dynamics of microglial activation: a confocal time-lapse analysis in hippocampal slices. Glia 2001; 33: 256–66.
32. Bernaudin M, Nouvelot A, MacKenzie ET, et al. Selective neuronal vulnerability and specific glial reactions in hippocampal and neocortical organotypic cultures submitted to ischemia. Exp Neurol 1998; 150: 30–9.
33. Huuskonen J, Suuronen T, Miettinen R, et al. A refined in vitro model to study inflammatory responses in organotypic membrane culture of postnatal rat hippocampal slices. J Neuroinflammation 2005; 2: 25.
34. Hailer NP, Bechmann I, Heizmann S, et al. Adhesion molecule expression on phagocytic microglial cells following anterograde degeneration of perforant path axons. Hippocampus 1997; 7: 341–9.
35. Kreutzberg GW. Microglia: a sensor for pathological events in the CNS. Trends Neurosci 1996; 19: 312–18.
36. Gall C, Rose G, Lynch G. Proliferative and migratory activity of glial cells in the partially deafferented hippocampus. J Comp Neurol 1979; 183: 539–49.
37. Brockhaus J, Moller T, Kettenmann H. Phagocytozing ameboid microglial cells studied in a mouse corpus callosum slice preparation. Glia 1996; 16: 81–90.
38. Petersen MA, Dailey ME. Diverse microglial motility behaviors during clearance of dead cells in hippocampal slices. Glia 2004; 46: 195–206.
39. Eyupoglu IY, Bechmann I, Nitsch R. Modification of microglia function protects from lesion-induced neuronal alterations and promotes sprouting in the hippocampus. FASEB J 2003; 17: 1110–11.
40. Savill J, Dransfield I, Gregory C, et al. A blast from the past: clearance of apoptotic cells regulates immune responses. Nat Rev Immunol 2002; 2: 965–75.
41. Grossmann R, Stence N, Carr J, et al. Juxtavascular microglia migrate along brain microvessels following activation during early postnatal development. Glia 2002; 37: 229–40.
42. Skibo GG, Nikonenko IR, Savchenko VL, et al. Microglia in organotypic hippocampal slice culture and effects of hypoxia: ultrastructure and lipocortin-1 immunoreactivity. Neuroscience 2000; 96: 427–38.
43. Morrison B, Eberwine JH, Meaney DF, et al. Traumatic injury induces differential expression of cell death genes in organotypic brain slice cultures determined by complementary DNA array hybridization. Neuroscience 2000; 96: 131–9.
44. Hickey WF. Basic principles of immunological surveillance of the normal central nervous system. Glia 2001; 36: 118–24.
45. Keynes RG, Duport S, Garthwaite J. Hippocampal neurons in organotypic slice culture are highly resistant to damage by endogenous and exogenous nitric oxide. Eur J Neurosci 2004; 19: 1163–73.
46. Bendfeldt K, Radojevic V, Kapfhammer J, et al. Basic fibroblast growth factor modulates density of blood vessels and preserves tight junctions in organotypic cortical cultures of mice: a new in vitro model of the blood-brain barrier. J Neurosci 2007; 27: 3260–7.
47. Baslow MH, Suckow RF, Gaynor K, et al. Brain damage results in down-regulation of N-acetylaspartate as a neuronal osmolyte. Neuromolecular Med 2003; 3: 95–104.
48. Popovich PG, Longbrake EE. Can the immune system be harnessed to repair the CNS? Nat Rev Neurosci 2008; 9: 481–93.
49. Hellstrom IC, Danik M, Luheshi GN, et al. Chronic LPS exposure produces changes in intrinsic membrane properties and a sustained IL-beta-dependent increase in GABAergic inhibition in hippocampal CA1 pyramidal neurons. Hippocampus 2005; 15: 656–64.
50. Ostergaard K, Finsen B, Zimmer J. Organotypic slice cultures of the rat striatum: an immunocytochemical, histochemical and in situ hybridization study of somatostatin, neuropeptide Y, nicotinamide adenine dinucleotide phosphate-diaphorase, and enkephalin. Exp Brain Res 1995; 103: 70–84.
51. Gahwiler BH, Capogna M, Debanne D, et al. Organotypic slice cultures: a technique has come of age. Trends Neurosci 1997; 20: 471–7.
52. Baker RE, Van Pelt J. Cocultured, but not isolated, cortical explants display normal dendritic development: a long-term quantitative study. Brain Res Dev Brain Res 1997; 98: 21–9.
53. Frotscher M, Gahwiler BH. Synaptic organization of intracellularly stained CA3 pyramidal neurons in slice cultures of rat hippocampus. Neuroscience 1988; 24: 541–51.

54. Frotscher M, Zafirov S, Heimrich B. Development of identified neuronal types and of specific synaptic connections in slice cultures of rat hippocampus. Prog Neurobiol 1995; 45: 143–64.
55. Buchs PA, Stoppini L, Muller D. Structural modifications associated with synaptic development in area CA1 of rat hippocampal organotypic cultures. Brain Res Dev Brain Res 1993; 71: 81–91.
56. Mohajerani MH, Cherubini E. Spontaneous recurrent network activity in organotypic rat hippocampal slices. Eur J Neurosci 2005; 22: 107–18.
57. Cherubini E, Gaiarsa JL, Ben-Ari Y. GABA: an excitatory transmitter in early postnatal life. Trends Neurosci 1991; 14: 515–19.
58. Bausch SB, McNamara JO. Synaptic connections from multiple subfields contribute to granule cell hyperexcitability in hippocampal slice cultures. J Neurophysiol 2000; 84: 2918–32.
59. Routbort MJ, Bausch SB, McNamara JO. Seizures, cell death, and mossy fiber sprouting in kainic acid-treated organotypic hippocampal cultures. Neuroscience 1999; 94: 755–65.
60. Bolz J, Novak N, Gotz M, et al. Formation of target-specific neuronal projections in organotypic slice cultures from rat visual cortex. Nature 1990; 346: 359–62.
61. Benediktsson AM, Schachtele SJ, Green SH, et al. Ballistic labeling and dynamic imaging of astrocytes in organotypic hippocampal slice cultures. J Neurosci Methods 2005; 141: 41–53.
62. Mateos JM, Luthi A, Savic N, et al. Synaptic modifications at the CA3-CA1 synapse after chronic AMPA receptor blockade in rat hippocampal slices. J Physiol 2007; 581: 129–38.
63. Martens U, Wree A. Distribution of [3H]MK-801, [3H]AMPA and [3H]Kainate binding sites in rat hippocampal long-term slice cultures isolated from external afferents. Anat Embryol (Berl) 2001; 203: 491–500.
64. Verkuyl JM, Matus A. Time-lapse imaging of dendritic spines in vitro. Nat Protoc 2006; 1: 2399–405.
65. Empson RM, Buckby LE, Kraus M, et al. The cell adhesion molecule neuroplastin-65 inhibits hippocampal long-term potentiation via a mitogen-activated protein kinase p38-dependent reduction in surface expression of GluR1-containing glutamate receptors. J Neurochem 2006; 99: 850–60.
66. Mielke JG, Comas T, Woulfe J, et al. Cytoskeletal, synaptic, and nuclear protein changes associated with rat interface organotypic hippocampal slice culture development. Brain Res Dev Brain Res 2005; 160: 275–86.
67. Beique JC, Andrade R. PSD-95 regulates synaptic transmission and plasticity in rat cerebral cortex. J Physiol 2003; 546: 859–67.
68. Nagerl UV, Kostinger G, Anderson JC, et al. Protracted synaptogenesis after activity-dependent spinogenesis in hippocampal neurons. J Neurosci 2007; 27: 8149–56.
69. Roelandse M, Welman A, Wagner U, et al. Focal motility determines the geometry of dendritic spines. Neuroscience 2003; 121: 39–49.
70. Nikonenko I, Jourdain P, Alberi S, et al. Activity-induced changes of spine morphology. Hippocampus 2002; 12: 585–91.
71. De Simoni A, Edwards FA. Pathway specificity of dendritic spine morphology in identified synapses onto rat hippocampal CA1 neurons in organotypic slices. Hippocampus 2006; 16: 1111–24.
72. Cullheim S, Thams S. The microglial networks of the brain and their role in neuronal network plasticity after lesion. Brain Res Rev 2007; 55: 89–96.
73. Ninkovic J, Gotz M. Signaling in adult neurogenesis: from stem cell niche to neuronal networks. Curr Opin Neurobiol 2007; 17: 338–44.
74. Ziv Y, Ron N, Butovsky O, et al. Immune cells contribute to the maintenance of neurogenesis and spatial learning abilities in adulthood. Nat Neurosci 2006; 9: 268–75.
75. Battista D, Ferrari CC, Gage FH, et al. Neurogenic niche modulation by activated microglia: transforming growth factor beta increases neurogenesis in the adult dentate gyrus. Eur J Neurosci 2006; 23: 83–93.
76. Chattopadhyaya B, Di Cristo G, Higashiyama H, et al. Experience and activity-dependent maturation of perisomatic GABAergic innervation in primary visual cortex during a postnatal critical period. J Neurosci 2004; 24: 9598–611.
77. Ruijter JM, Baker RE, De Jong BM, et al. Chronic blockade of bioelectric activity in neonatal rat cortex grown in vitro: morphological effects. Int J Dev Neurosci 1991; 9: 331–38.
78. Britto JM, Obata K, Yanagawa Y, et al. Migratory response of interneurons to different regions of the developing neocortex. Cereb Cortex 2006; 16(Suppl 1): 57–63.
79. Dayer AG, Jenny B, Potter G, et al. Recruiting new neurons from the subventricular zone to the rat postnatal cortex: an organotypic slice culture model. Eur J Neurosci 2008; 27: 1051–60.
80. Aloisi F. Immune function of microglia. Glia 2001; 36: 165–79.
81. Napoli I, Neumann H. Microglial clearance function in health and disease. Neuroscience 2009; 158: 1030–38.
82. Czapiga M, Colton CA. Function of microglia in organotypic slice cultures. J Neurosci Res 1999; 56: 644–51.

83. Boucsein C, Kettenmann H, Nolte C. Electrophysiological properties of microglial cells in normal and pathologic rat brain slices. Eur J Neurosci 2000; 12: 2049–58.
84. Boucsein C, Zacharias R, Farber K, et al. Purinergic receptors on microglial cells: functional expression in acute brain slices and modulation of microglial activation in vitro. Eur J Neurosci 2003; 17: 2267–76.
85. Haynes SE, Hollopeter G, Yang G, et al. The P2Y12 receptor regulates microglial activation by extracellular nucleotides. Nat Neurosci 2006; 9: 1512–19.
86. Wolf SA, Gimsa U, Bechmann I, et al. Differential expression of costimulatory molecules B7-1 and B7-2 on microglial cells induced by Th1 and Th2 cells in organotypic brain tissue. Glia 2001; 36: 414–20.
87. Bechmann I, Peter S, Beyer M, et al. Presence of B7–2 (CD86) and lack of B7–1 (CD80) on myelin phagocytosing MHC-II-positive rat microglia is associated with nondestructive immunity in vivo. FASEB J 2001; 15: 1086–8.
88. Neumann H, Boucraut J, Hahnel C, et al. Neuronal control of MHC class II inducibility in rat astrocytes and microglia. Eur J Neurosci 1996; 8: 2582–90.
89. Streit WJ, Graeber MB, Kreutzberg GW. Functional plasticity of microglia: a review. Glia 1988; 1: 301–7.
90. Eliason DA, Cohen SA, Baratta J, et al. Local proliferation of microglia cells in response to neocortical injury in vitro. Brain Res Dev Brain Res 2002; 137: 75–9.
91. Neumann J, Gunzer M, Gutzeit HO, et al. Microglia provide neuroprotection after ischemia. FASEB J 2006; 20: 714–16.
92. Moriguchi S, Mizoguchi Y, Tomimatsu Y, et al. Potentiation of NMDA receptor-mediated synaptic responses by microglia. Brain Res Mol Brain Res 2003; 119: 160–9.
93. Neumann J, Sauerzweig S, Ronicke R, et al. Microglia cells protect neurons by direct engulfment of invading neutrophil granulocytes: a new mechanism of CNS immune privilege. J Neurosci 2008; 28: 5965–75.
94. Volterra A, Meldolesi J. Astrocytes, from brain glue to communication elements: the revolution continues. Nat Rev Neurosci 2005; 6: 626–40.
95. Jacobsen CT, Miller RH. Control of astrocyte migration in the developing cerebral cortex. Dev Neurosci 2003; 25: 207–16.
96. Berglof E, Af Bjerken S, Stromberg I. Glial influence on nerve fiber formation from rat ventral mesencephalic organotypic tissue cultures. J Comp Neurol 2007; 501: 431–42.
97. Johansson S, Stromberg I. Guidance of dopaminergic neuritic growth by immature astrocytes in organotypic cultures of rat fetal ventral mesencephalon. J Comp Neurol 2002; 443: 237–49.
98. Jabaudon D, Shimamoto K, Yasuda-Kamatani Y, et al. Inhibition of uptake unmasks rapid extracellular turnover of glutamate of nonvesicular origin. Proc Natl Acad Sci U S A 1999; 96: 8733–8.
99. Haber M, Zhou L, Murai KK. Cooperative astrocyte and dendritic spine dynamics at hippocampal excitatory synapses. J Neurosci 2006; 26: 8881–91.
100. Blanc EM, Bruce-Keller AJ, Mattson MP. Astrocytic gap junctional communication decreases neuronal vulnerability to oxidative stress-induced disruption of Ca2+ homeostasis and cell death. J Neurochem 1998; 70: 958–70.
101. Seil FJ, Drake-Baumann R. Activity-dependent changes in "transplanted" cerebellar cultures. Exp Neurol 1996; 138: 327–37.
102. Rolls A, Shechter R, Schwartz M. The bright side of the glial scar in CNS repair. Nat Rev Neurosci 2009; 10: 235–41.
103. Muller CM. A role for glial cells in activity-dependent central nervous plasticity? Review and hypothesis. Int Rev Neurobiol 1992; 34: 215–81.
104. Hailer NP, Heppner FL, Haas D, et al. Astrocytic factors deactivate antigen presenting cells that invade the central nervous system. Brain Pathol 1998; 8: 459–74.
105. Bechmann I, Nitsch R. Involvement of non-neuronal cells in entorhinal-hippocampal reorganization following lesions. Ann N Y Acad Sci 2000; 911: 192–206.
106. Hailer NP, Wirjatijasa F, Roser N, et al. Astrocytic factors protect neuronal integrity and reduce microglial activation in an in vitro model of N-methyl-D-aspartate-induced excitotoxic injury in organotypic hippocampal slice cultures. Eur J Neurosci 2001; 14: 315–26.
107. Hailer NP, Vogt C, Korf HW, et al. Interleukin-1beta exacerbates and interleukin-1 receptor antagonist attenuates neuronal injury and microglial activation after excitotoxic damage in organotypic hippocampal slice cultures. Eur J Neurosci 2005; 21: 2347–60.
108. Dehghani F, Conrad A, Kohl A, et al. Clodronate inhibits the secretion of proinflammatory cytokines and NO by isolated microglial cells and reduces the number of proliferating glial cells in excitotoxically injured organotypic hippocampal slice cultures. Exp Neurol 2004; 189: 241–51.
109. Kohl A, Dehghani F, Korf HW, et al. The bisphosphonate clodronate depletes microglial cells in excitotoxically injured organotypic hippocampal slice cultures. Exp Neurol 2003; 181: 1–11.

110. Rappert A, Bechmann I, Pivneva T, et al. CXCR3-dependent microglial recruitment is essential for dendrite loss after brain lesion. J Neurosci 2004; 24: 8500–9.
111. Kurpius D, Nolley EP, Dailey ME. Purines induce directed migration and rapid homing of microglia to injured pyramidal neurons in developing hippocampus. Glia 2007; 55: 873–84.
112. Ide CF, Scripter JL, Coltman BW, et al. Cellular and molecular correlates to plasticity during recovery from injury in the developing mammalian brain. Prog Brain Res 1996; 108: 365–77.
113. Ho A, Blum M. Regulation of astroglial-derived dopaminergic neurotrophic factors by interleukin-1 beta in the striatum of young and middle-aged mice. Exp Neurol 1997; 148: 348–59.
114. DiProspero NA, Meiners S, Geller HM. Inflammatory cytokines interact to modulate extracellular matrix and astrocytic support of neurite outgrowth. Exp Neurol 1997; 148: 628–39.
115. Neumann H. Control of glial immune function by neurons. Glia 2001; 36: 191–9.
116. Wekerle H, Schwab M, Linington C, et al. Antigen presentation in the peripheral nervous system: Schwann cells present endogenous myelin autoantigens to lymphocytes. Eur J Immunol 1986; 16: 1551–7.
117. Iliev AI, Stringaris AK, Nau R, et al. Neuronal injury mediated via stimulation of microglial toll-like receptor-9 (TLR9). FASEB J 2004; 18: 412–14.
118. Ling C, Verbny YI, Banks MI, et al. In situ activation of antigen-specific CD8+ T cells in the presence of antigen in organotypic brain slices. J Immunol 2008; 180: 8393–9.
119. Wolf SA, Fisher J, Bechmann I, et al. Neuroprotection by T-cells depends on their subtype and activation state. J Neuroimmunol 2002; 133: 72–80.
120. Gimsa U, Peter SV, Lehmann K, et al. Axonal damage induced by invading T cells in organotypic central nervous system tissue in vitro: involvement of microglial cells. Brain Pathol 2000; 10: 365–77.
121. Gimsa U, Wolf SA, Haas D, et al. Th2 cells support intrinsic anti-inflammatory properties of the brain. J Neuroimmunol 2001; 119: 73–80.
122. Manoonkitiwongsa PS, Jackson-Friedman C, McMillan PJ, et al. Angiogenesis after stroke is correlated with increased numbers of macrophages: the clean-up hypothesis. J Cereb Blood Flow Metab 2001; 21: 1223–31.
123. Elward K, Gasque P. "Eat me" and "don't eat me" signals govern the innate immune response and tissue repair in the CNS: emphasis on the critical role of the complement system. Mol Immunol 2003; 40: 85–94.
124. Griffiths M, Neal JW, Gasque P. Innate immunity and protective neuroinflammation: new emphasis on the role of neuroimmune regulatory proteins. Int Rev Neurobiol 2007; 82: 29–55.
125. Kamada M, Li RY, Hashimoto M, et al. Intrinsic and spontaneous neurogenesis in the postnatal slice culture of rat hippocampus. Eur J Neurosci 2004; 20: 2499–508.
126. Chechneva O, Dinkel K, Schrader D, et al. Identification and characterization of two neurogenic zones in interface organotypic hippocampal slice cultures. Neuroscience 2005; 136: 343–55.
127. Sadgrove MP, Laskowski A, Gray WP. Examination of granule layer cell count, cell density, and single-pulse BrdU incorporation in rat organotypic hippocampal slice cultures with respect to culture medium, septotemporal position, and time in vitro. J Comp Neurol 2006; 497: 397–415.
128. Shetty AK, Rao MS, Hattiangady B. Behavior of hippocampal stem/progenitor cells following grafting into the injured aged hippocampus. J Neurosci Res 2008.
129. Pencea V, Bingaman KD, Wiegand SJ, et al. Infusion of brain-derived neurotrophic factor into the lateral ventricle of the adult rat leads to new neurons in the parenchyma of the striatum, septum, thalamus, and hypothalamus. J Neurosci 2001; 21: 6706–17.
130. Kuhn HG, Palmer TD, Fuchs E. Adult neurogenesis: a compensatory mechanism for neuronal damage. Eur Arch Psychiatry Clin Neurosci 2001; 251: 152–8.
131. Kee N, Teixeira CM, Wang AH, et al. Preferential incorporation of adult-generated granule cells into spatial memory networks in the dentate gyrus. Nat Neurosci 2007; 10: 355–62.
132. Hensch TK. Critical period regulation. Annu Rev Neurosci 2004; 27: 549–79.
133. van Praag H, Christie BR, Sejnowski TJ, et al. Running enhances neurogenesis, learning, and long-term potentiation in mice. Proc Natl Acad Sci U S A 1999; 96: 13427–31.
134. Lemaire V, Koehl M, Le Moal M, et al. Prenatal stress produces learning deficits associated with an inhibition of neurogenesis in the hippocampus. Proc Natl Acad Sci U S A 2000; 97: 11032–7.
135. King RS, DeBassio WA, Kemper TL, et al. Effects of prenatal protein malnutrition and acute postnatal stress on granule cell genesis in the fascia dentata of neonatal and juvenile rats. Brain Res Dev Brain Res 2004; 150: 9–15.
136. Monje ML, Mizumatsu S, Fike JR, et al. Irradiation induces neural precursor-cell dysfunction. Nat Med 2002; 8: 955–62.
137. Morrison B 3rd, Meaney DF, McIntosh TK. Mechanical characterization of an in vitro device designed to quantitatively injure living brain tissue. Ann Biomed Eng 1998; 26: 381–90.

138. Morrison B 3rd, Cater HL, Benham CD, et al. An in vitro model of traumatic brain injury utilising two-dimensional stretch of organotypic hippocampal slice cultures. J Neurosci Methods 2006; 150: 192–201.
139. Wallis RA, Panizzon KL. Felbamate neuroprotection against CA1 traumatic neuronal injury. Eur J Pharmacol 1995; 294: 475–82.
140. Adamchik Y, Frantseva MV, Weisspapir M, et al. Methods to induce primary and secondary traumatic damage in organotypic hippocampal slice cultures. Brain Res Brain Res Protoc 2000; 5: 153–8.
141. Yu Z, Tsay C, Lacour SP, et al. Stretchable microelectrode arrays—a tool for discovering mechanisms of functional deficits underlying traumatic brain injury and interfacing neurons with neuroprosthetics. Conf Proc IEEE Eng Med Biol Soc 2006; (Suppl): 6732–5.
142. Stoppini L, Buchs PA, Muller D. Lesion-induced neurite sprouting and synapse formation in hippocampal organotypic cultures. Neuroscience 1993; 57: 985–94.
143. Woodhams PL, Atkinson DJ. Regeneration of entorhino-dentate projections in organotypic slice cultures: mode of axonal regrowth and effects of growth factors. Exp Neurol 1996; 140: 68–78.
144. Hechler D, Nitsch R, Hendrix S. Green-fluorescent-protein-expressing mice as models for the study of axonal growth and regeneration in vitro. Brain Res Rev 2006; 52 : 160–9.
145. Muller D, Stoppini L, Wang C, et al. A role for polysialylated neural cell adhesion molecule in lesion-induced sprouting in hippocampal organotypic cultures. Neuroscience 1994; 61: 441–5.
146. Mingorance A, Fontana X, Sole M, et al. Regulation of Nogo and Nogo receptor during the development of the entorhino-hippocampal pathway and after adult hippocampal lesions. Mol Cell Neurosci 2004; 26: 34–49.
147. Hakkoum D, Stoppini L, Muller D. Interleukin-6 promotes sprouting and functional recovery in lesioned organotypic hippocampal slice cultures. J Neurochem 2007; 100: 747–57.
148. Hafidi A, Grumet M, Sanes DH. In vitro analysis of mechanisms underlying age-dependent failure of axon regeneration. J Comp Neurol 2004; 470: 80–92.
149. Cai X, Wei DS, Gallagher SE, et al. Hyperexcitability of distal dendrites in hippocampal pyramidal cells after chronic partial deafferentation. J Neurosci 2007; 27: 59–68.
150. Bonde C, Noraberg J, Noer H, et al. Ionotropic glutamate receptors and glutamate transporters are involved in necrotic neuronal cell death induced by oxygen-glucose deprivation of hippocampal slice cultures. Neuroscience 2005; 136: 779–94.
151. Chechneva O, Dinkel K, Cavaliere F, et al. Anti-inflammatory treatment in oxygen-glucose-deprived hippocampal slice cultures is neuroprotective and associated with reduced cell proliferation and intact neurogenesis. Neurobiol Dis 2006; 23: 247–59.
152. Liu W, Liu R, Chun JT, et al. Kainate excitotoxicity in organotypic hippocampal slice cultures: evidence for multiple apoptotic pathways. Brain Res 2001; 916: 239–48.
153. Bruce AJ, Sakhi S, Schreiber SS, et al. Development of kainic acid and N-methyl-D-aspartic acid toxicity in organotypic hippocampal cultures. Exp Neurol 1995; 132: 209–19.
154. Albus K, Wahab A, Heinemann U. Standard antiepileptic drugs fail to block epileptiform activity in rat organotypic hippocampal slice cultures. Br J Pharmacol 2008; 154: 709–24.
155. Best N, Sundstrom LE, Mitchell J, et al. Pre-exposure to subtoxic levels prevents kainic acid lesions in organotypic hippocampal slice cultures: effects of kainic acid on parvalbumin-immunoreactive neurons and expression of heat shock protein 72 following the induction of tolerance. Eur J Neurosci 1996; 8: 1209–19.
156. Xu GP, Dave KR, Vivero R, et al. Improvement in neuronal survival after ischemic preconditioning in hippocampal slice cultures. Brain Res 2002; 952: 153–8.
157. Pringle AK, Thomas SJ, Signorelli F, et al. Ischaemic pre-conditioning in organotypic hippocampal slice cultures is inversely correlated to the induction of the 72 kDa heat shock protein (HSP72). Brain Res 1999; 845: 152–64.
158. Mooney SM, Miller MW. Nerve growth factor neuroprotection of ethanol-induced neuronal death in rat cerebral cortex is age dependent. Neuroscience 2007; 149: 372–81.
159. Sui L, Ruan DY. Impairment of the Ca2+-permeable AMPA/kainate receptors by lead exposure in organotypic rat hippocampal slice cultures. Pharmacol Toxicol 2000; 87: 204–10.
160. Liu R, Liu W, Doctrow SR, et al. Iron toxicity in organotypic cultures of hippocampal slices: role of reactive oxygen species. J Neurochem 2003; 85: 492–502.
161. Noraberg J, Gramsbergen JB, Fonnum F, et al. Trimethyltin (TMT) neurotoxicity in organotypic rat hippocampal slice cultures. Brain Res 1998; 783: 305–15.
162. Pancetti F, Oyarce M, Aranda M, et al. S-methylcysteine may be a causal factor in monohalomethane neurotoxicity. Neurotoxicology 2004; 25: 817–23.
163. Lyng GD, Seegal RF. Polychlorinated biphenyl-induced oxidative stress in organotypic co-cultures: experimental dopamine depletion prevents reductions in GABA. Neurotoxicology 2008; 29: 301–8.

164. Siegenthaler JA, Miller MW. Ethanol disrupts cell cycle regulation in developing rat cortex interaction with transforming growth factor beta1. J Neurochem 2005; 95: 902–12.
165. Nikonenko I, Toni N, Moosmayer M, et al. Integrins are involved in synaptogenesis, cell spreading, and adhesion in the postnatal brain. Brain Res Dev Brain Res 2003; 140: 185–94.
166. Snyder-Keller A, Stark PF. Prenatal inflammatory effects on nigrostriatal development in organotypic cultures. Brain Res 2008; 1233: 160–7.
167. Morales P, Reyes P, Klawitter V, et al. Effects of perinatal asphyxia on cell proliferation and neuronal phenotype evaluated with organotypic hippocampal cultures. Neuroscience 2005; 135: 421–31.
168. Noraberg J. Organotypic brain slice cultures: an efficient and reliable method for neurotoxicological screening and mechanistic studies. Altern Lab Anim 2004; 32: 329–37.
169. Cho S, Wood A, Bowlby MR. Brain slices as models for neurodegenerative disease and screening platforms to identify novel therapeutics. Curr Neuropharmacol 2007; 5: 19–33.
170. Lo DC, McAllister AK, Katz LC. Neuronal transfection in brain slices using particle-mediated gene transfer. Neuron 1994; 13: 1263–8.
171. Bahr BA, Abai B, Gall CM, et al. Induction of beta-amyloid-containing polypeptides in hippocampus: evidence for a concomitant loss of synaptic proteins and interactions with an excitotoxin. Exp Neurol 1994; 129: 81–94.
172. Fan R, Tenner AJ. Complement C1q expression induced by Abeta in rat hippocampal organotypic slice cultures. Exp Neurol 2004; 185: 241–53.
173. Stevens B, Allen NJ, Vazquez LE, et al. The classical complement cascade mediates CNS synapse elimination. Cell 2007; 131: 1164–78.
174. Nauta AJ, Trouw LA, Daha MR, et al. Direct binding of C1q to apoptotic cells and cell blebs induces complement activation. Eur J Immunol 2002; 32: 1726–36.
175. Marksteiner J, Humpel C. Beta-amyloid expression, release and extracellular deposition in aged rat brain slices. Mol Psychiatry 2008; 13: 939–52.
176. Finley M, Fairman D, Liu D, et al. Functional validation of adult hippocampal organotypic cultures as an in vitro model of brain injury. Brain Res 2004; 1001: 125–32.
177. Xiang Z, Hrabetova S, Moskowitz SI, et al. Long-term maintenance of mature hippocampal slices in vitro. J Neurosci Methods 2000; 98: 145–54.
178. Schrag M, Sharma S, Brown-Borg H, et al. Hippocampus of Ames dwarf mice is resistant to beta-amyloid-induced tau hyperphosphorylation and changes in apoptosis-regulatory protein levels. Hippocampus 2008; 18: 239–44.
179. Smith DL, Portier R, Woodman B, et al. Inhibition of polyglutamine aggregation in R6/2 HD brain slices-complex dose-response profiles. Neurobiol Dis 2001; 8: 1017–26.
180. Hay DG, Sathasivam K, Tobaben S, et al. Progressive decrease in chaperone protein levels in a mouse model of Huntington's disease and induction of stress proteins as a therapeutic approach. Hum Mol Genet 2004; 13: 1389–405.
181. Whetsell WO, Jr, Schwarcz R. Prolonged exposure to submicromolar concentrations of quinolinic acid causes excitotoxic damage in organotypic cultures of rat corticostriatal system. Neurosci Lett 1989; 97: 271–5.
182. Storgaard J, Kornblit BT, Zimmer J, et al. 3-Nitropropionic acid neurotoxicity in organotypic striatal and corticostriatal slice cultures is dependent on glucose and glutamate. Exp Neurol 2000; 164: 227–35.
183. Testa CM, Sherer TB, Greenamyre JT. Rotenone induces oxidative stress and dopaminergic neuron damage in organotypic substantia nigra cultures. Brain Res Mol Brain Res 2005; 134: 109–18.
184. Neely MD, Schmidt DE, Deutch AY. Cortical regulation of dopamine depletion-induced dendritic spine loss in striatal medium spiny neurons. Neuroscience 2007; 149: 457–64.
185. Chaichana KL, Capilla-Gonzalez V, Gonzalez-Perez O, et al. Preservation of glial cytoarchitecture from ex vivo human tumor and non-tumor cerebral cortical explants: a human model to study neurological diseases. J Neurosci Methods 2007; 164: 261–70.
186. Gonzalez-Martinez JA, Bingaman WE, Toms SA, et al. Neurogenesis in the postnatal human epileptic brain. J Neurosurg 2007; 107: 628–35.
187. Jung S, Kim HW, Lee JH, et al. Brain tumor invasion model system using organotypic brain-slice culture as an alternative to in vivo model. J Cancer Res Clin Oncol 2002; 128: 469–76.
188. Wise PM, Dubal DB, Wilson ME, et al. Estradiol is a protective factor in the adult and aging brain: understanding of mechanisms derived from in vivo and in vitro studies. Brain Res Brain Res Rev 2001; 37: 313–19.
189. Wilson ME, Dubal DB, Wise PM. Estradiol protects against injury-induced cell death in cortical explant cultures: a role for estrogen receptors. Brain Res 2000; 873: 235–42.
190. Li H, Pin S, Zeng Z, et al. Sex differences in cell death. Ann Neurol 2005; 58: 317–21.
191. Linnemann D, Lyles JM, Bock E. A developmental study of the biosynthesis of the neural cell adhesion molecule. Dev Neurosci 1985; 7: 230–8.

192. Gahwiler BH, Thompson SM, Muller D. Preparation and maintenance of organotypic slice cultures of CNS tissue. Curr Protoc Neurosci 2001; Chapter 6: Unit 6.11.
193. Schlessinger AR, Cowan WM, Gottlieb DI. An autoradiographic study of the time of origin and the pattern of granule cell migration in the dentate gyrus of the rat. J Comp Neurol 1975; 159: 149–75.
194. Bayer SA. Neuron production in the hippocampus and olfactory bulb of the adult rat brain: addition or replacement? Ann N Y Acad Sci 1985; 457: 163–72.
195. Bayer SA, Altman J. Hippocampal development in the rat: cytogenesis and morphogenesis examined with autoradiography and low-level X-irradiation. J Comp Neurol 1974; 158: 55–79.
196. Eccles JC. Neurogenesis and morphogenesis in the cerebellar cortex. Proc Natl Acad Sci U S A 1970; 66: 294–301.
197. Altman J. Postnatal development of the cerebellar cortex in the rat. II. Phases in the maturation of Purkinje cells and of the molecular layer. J Comp Neurol 1972; 145: 399–463.
198. Cowan WM, Stanfield BB, Kishi K. The development of the dentate gyrus. Curr Top Dev Biol 1980; 15: 103–57.
199. Chapleau CA, Carlo ME, Larimore JL, et al. The actions of BDNF on dendritic spine density and morphology in organotypic slice cultures depend on the presence of serum in culture media. J Neurosci Methods 2008; 169: 182–90.
200. Kawasaki H, Kosugi I, Arai Y, et al. The amount of immature glial cells in organotypic brain slices determines the susceptibility to murine cytomegalovirus infection. Lab Invest 2002; 82: 1347–58.
201. Dahl-Jorgensen A, Ostergaard K, Pedersen EB, et al. Serum and CNTF stimulate oligodendroglia and reduce fiber outgrowth from striatal cultures. Exp Neurol 1999; 157: 88–95.
202. McNay EC, Gold PE. Extracellular glucose concentrations in the rat hippocampus measured by zero-net-flux: effects of microdialysis flow rate, strain, and age. J Neurochem 1999; 72: 785–90.
203. Lowndes HE, Beiswanger CM, Philbert MA, et al. Substrates for neural metabolism of xenobiotics in adult and developing brain. Neurotoxicology 1994; 15: 61–73.
204. Holtzman D, DeVries C, Nguyen H, et al. Maturation of resistance to lead encephalopathy: cellular and subcellular mechanisms. Neurotoxicology 1984; 5: 97–124.
205. Ransom BR, Kunis DM, Irwin I, et al. Astrocytes convert the Parkinsonism inducing neurotoxin, MPTP, to its active metabolite, MPP+. Neurosci Lett 1987; 75: 323–8.
206. Hamberger A, Nystrom B, Sellstrom A, et al. Amino acid transport in isolated neurons and glia. Adv Exp Med Biol 1976; 69: 221–36.
207. Teissier E, Fennrich S, Strazielle N, et al. Drug metabolism in in vitro organotypic and cellular models of mammalian central nervous system: activities of membrane-bound epoxide hydrolase and NADPH-cytochrome P-450 (c) reductase. Neurotoxicology 1998; 19: 347–55.
208. Mytilineou C, Cohen G. 1-methyl-4-phenyl-1,2,3,6-tetrahydropyridine destroys dopamine neurons in explants of rat embryo mesencephalon. Science 1984; 225: 529–31.
209. Garthwaite J. Cellular uptake disguises action of L-glutamate on N-methyl-D-aspartate receptors. With an appendix: diffusion of transported amino acids into brain slices. Br J Pharmacol 1985; 85: 297–307.
210. Fukuda T, Ackerman SJ, Reed CE, et al. Calcium ionophore A23187 calcium-dependent cytolytic degranulation in human eosinophils. J Immunol 1985; 135: 1349–56.
211. Mutoh T, Miyata T, Kashiwagi S, et al. Dynamic behavior of individual cells in developing organotypic brain slices revealed by the photoconvertable protein Kaede. Exp Neurol 2006; 200: 430–7.

10 Design-Based Stereology and Video Densitometry for Assessment of Neurotoxicological Damage

Peter R. Mouton

Byrd Alzheimer Center and Research Institute, University of South Florida College of Medicine, Tampa, Florida, and Stereology Resource Centers, Chester, Maryland, U.S.A.

Marcia N. Gordon

Byrd Alzheimer Institute and Department of Molecular Pharmacology and Physiology, School of Basic Biomedical Sciences, University of South Florida College of Medicine, Tampa, Florida, U.S.A.

INTRODUCTION

The Swiss physician Paracelsus, also known as Theophrastus Phillipus Auroleus Bombastus von Hohenheim (1493–1541), based his life work on a simple philosophy: "All things are poison and nothing is without poison; only the dose makes a thing a poison." In 1813 the Spanish physician Mathieu Orfila incorporated this principle into *Toxicologie Générale*, the first comprehensive review of the clinical and pathology effects of known poisons on the human body. Today, almost two centuries later, toxicologists in government agencies, academic institutions, and private industry continue to test important hypotheses about the toxic effects of chemical agents on the central and peripheral nervous systems of humans and experimental animals. Modern methods for quantitative and semi-quantitative analysis of tissue, with support from high-resolution microscopy, computer-based imaging, and immunohistochemistry, provide the foundation for conclusions about changes in the number of discrete cell populations (apoptosis, proliferation); neuritic fibers and blood vessels (angiogenesis, sprouting, degeneration); absorptive membranes surfaces; and volumes of cells and tissues (atrophy, hypertrophy). This chapter reviews quantitative (design-based stereology) and semi-quantitative (video densitometry) techniques in contemporary neurotoxicology for the assessment of neural damage and neuroprotective effects at the organ, tissue, cellular, and sub-cellular levels.

DESIGN-BASED STEREOLOGY

The term stereology derives from the Greek *stereos* for "the study of objects in three-dimensions." In contrast to older morphometric approaches based on Euclidean geometry ("assume a cell is a sphere"), modern stereology is based on the application of stochastic geometry and probability theory for assessment of biological parameters, and rejects faulty assumptions and inappropriate models in previous approaches. Like a well-designed survey or poll, design-based stereology uses unbiased sampling to randomize data collection across an identified population of tissue, with objective questions in the form of unbiased probes. The goal is to ensure that all objects and regions of interest in an anatomically defined reference space (e.g., hippocampus, locus coeruleus) share equal probabilities of being sampled and analyzed. When used in combination with good study design, as levels of sampling stringencies increase the findings from both approaches approach the expected value for the parameters of interest. As introduced in the following sections, intense focus on sampling design and the elimination of bias led to the terms, unbiased and design-based, for modern stereological approaches, the current *sine qua non* for reliable quantification of pathological and toxicological changes in tissue. This section examines key concepts of modern stereology (1), with particular attention on the quantification of first-order stereological parameters (number, N; length, L; surface area, S; and volume, V) for light microscopy-based neurotoxicology studies.

Accuracy and Precision

The first step toward appreciating the differences between design-based stereology and older biased (assumption- and model-based) approaches requires a clear separation between the terms accuracy and precision. In terms of biological morphometry, when a method avoids all known sources of bias, the approach is unbiased; as sampling increases the results progressively converge on the true central tendency, and individual data points cluster around that expected value (Fig. 1). With biased methods, i.e., techniques based on verifiable assumptions and faulty models, the data continue to cluster around a central point; however, that point is an unknown and unknowable distance away from the expected value. In contrast, for unbiased methods data cluster in a random, non-systematic pattern around the true value (center of targets in Fig. 1).

Thus, both biased and unbiased approach may generate data with low levels of variability (high precision) around a central value; however, only unbiased methods generate accurate data that cluster around the true value. For biased methods the central tendency of the data points is systematically displaced from the true value. On first glance a reasonable notion might be to develop *post-hoc* approaches to remove the systematic error after the results have been generated, i.e., to develop so-called "correction factors." In practice, however, problems arise with this idea. First, correction factors introduce further bias through additional assumptions and models ("assume all cells are 30% spherical"). Second, not all treatments affect tissue in known ways, and not all cells and tissues respond uniformly to the same treatments. The core obstacle to correction factors is that the systematic error arises not in the data, but rather in the methods used to generate the data. Any correction factor that applies Euclidean-based geometry to biological tissue is bound to fail because biological tissues are not classical Euclidean structures (spheres, cubes, tetrahedrons, trapezoids, etc.). To eliminate bias in the analysis of non-classically shaped biological features requires that all faulty assumptions and models in the methods are avoided.

Although these problems were known for many decades earlier, in 1961 scientists from a wide range of disciplines formed the International Society for Stereology, a scientific confederation of biologists, geologists, mathematicians, and material scientists with the explicit purpose to develop reliable methods for quantification of non-classically shaped objects based on their planar appearance on 2D sections. In time, during the past four decades these methods formed the foundation for modern stereology (1–6).

In the past decade neurotoxicologists have increasingly recognized the value of stereological approaches for experiments that require the quantification of morphometric changes in biological tissues. At the start of a neurotoxicological study, when the expected values for parameters of interest are unknown, the combination of unbiased sampling designs and unbiased geometric probes will serve to maximize the probability that data points will be accurate, and cluster around the true values for the parameter. The desired or optimal level of precision in the

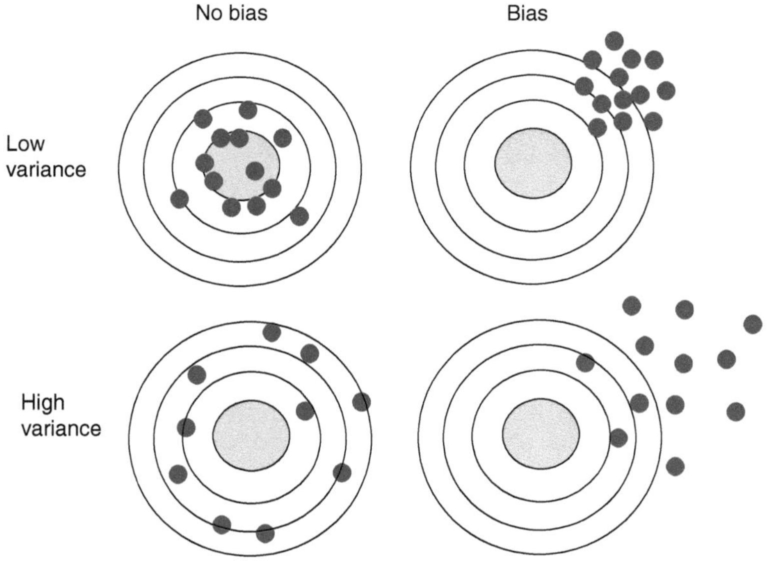

Figure 1 Comparison between bias and variance in morphometric studies. Center of targets indicate the expected or true value. Increased sampling reduces variance, independent of whether bias present in methods. *Source*: From Ref. 1.

results depends on a number of factors, including the magnitude of effects, within-sample and between-sample differences, and the specific scientific question at hand. That is, the scientist must ask: What level of differences do I consider scientifically interesting? If a significant difference of 25% for five to six animals per group is novel and important, then the level of sampling required to achieve these results is optimal. However, this level of sampling will likely be higher than that for a screening study of 100 animals designed to detect significant differences of 50% or more in the parameter of interest. Although further in-depth analysis of this issue is beyond our scope, in later sections of this review we will outline optimal strategies to achieve maximal sampling efficiency based on the observed variability in the data. In the next section, we review a prototypical application of stereology to neurotoxicology, a study designed to detect whether a known toxin, mercury, administered in low doses to neonates, leads to significant changes in the brains of a strain of mice highly susceptible to immune-based damage.

In July 2008, Berman et al. published a study in *Toxicological Sciences* entitled, "Low level neonatal thimerosal exposure: further evaluation of altered neurotoxic potential in SJL mice." For these studies investigators chose a strain of rodent, the SJL mouse, developed from three different sources of Swiss Webster mice in 1955 by James Lambert at The Jackson Laboratory (Bar Harbor, Maine). Thirty years later investigators at the National Cancer Institute (Frederick, MD) discovered that around one year of age SJL mice display a markedly higher incidence of reticulum cell sarcomas, a condition resembling Hodgkin's disease in humans, than mice from the founder strains. Other investigators subsequently reported that SJL mice have increased susceptibility to experimental autoimmune encephalomyelitis, making them an excellent model for multiple sclerosis research, and severe autoimmune reactions in response to organic mercury exposure (8). An earlier study using these mice in combination with qualitative, non-stereological approaches (9) reported morphological changes in hippocampal structures of neonates after mercury exposure to pregnant dams; however, the attempt by Berman et al. (7) to replicate those results using design-based stereology found no evidence of mercury damage to hippocampal structures. The most likely basis for these discrepant findings is in the methods use to sample tissue and quantify morphological endpoints, as outlined in the following section.

THE CORPUSCLE PROBLEM

An issue well known to microscopists for over 75 years (10), the corpuscle problem is a source of systematic bias that arises from the methods required to view microscopic structures. To assess the microscopic effects of potentially toxic compounds on biological tissues, toxicologists must cut tissue thin enough for light to pass through the sections. This tissue sectioning removes one dimension from the regions and objects in the resulting tissue sections: the 3D volumes of objects appear as 2D areas; 2D surface areas appear as 1D lines; 1D lengths appear as 0D points.

As shown in Figure 2, biological objects viewed under the microscope appear as 2D profiles, rather than 3D cells. Note that the number of 2D profiles ($n = 10$) in Figure 2 does not equal

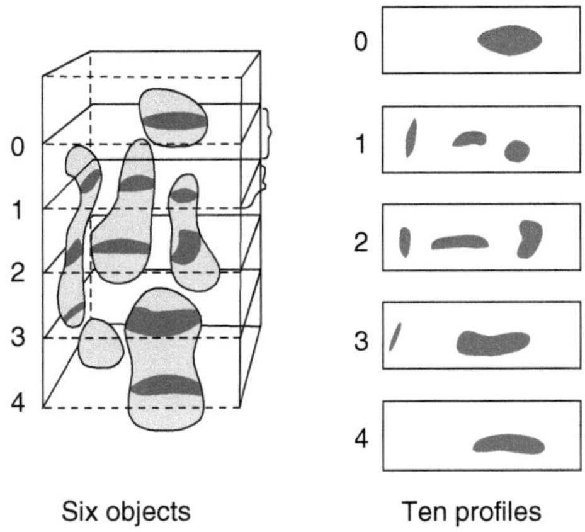

Figure 2 Corpuscle problem. Difference between numbers of profiles on 2D sections (*right*) versus true number of objects in 3D (*left*). *Source*: From Ref. 1.

the actual number of 3D objects in the tissue ($n = 6$); mathematically speaking, $N_A \neq N_V$. The systematic error (bias) arises because sectioning 3D objects with a 2D plane (knife blade) changes the sampling probability; instead of every object having the same chance of appearing on the section, larger objects, objects with more unusual shapes, and objects with their long axis perpendicular to the direction of sectioning have a higher probability to appear on the section, and therefore to be counted. Rather than simple semantics, the distinction between cells and profiles introduces bias for quantification of the number of biological objects in tissue.

Correction Factors

In the 1920s, the Swedish medical scientist S.D. Wicksell introduced a correction factor to remove the bias associated with the corpuscle problem (11). This formula, $N_v = N_A \times D$, where D = mean particle diameter, like all subsequent correction formulas, involves a function, $f(x)$, with an input value x obtained from measurements on sections, and the output value y from the equation $y = f(x)$, where x = the observed number of 2D profiles per unit area, N_A (mm^{-2} units); y = the actual number of 3D cells per unit volume, N_v (mm^{-3} units).

The problem is that a reliable estimate of D, the mean particle diameter, cannot be determined on sections cut through tissue. In response to perceived weaknesses in the Wicksell correction, a number of other investigators introduced similar factors into the scientific literature to overcome the corpuscle problem (Fig. 2).

All of the equations shown in Table 1 fail to "correct" the problem, either because the function is based on arbitrary, non-verifiable assumptions, or because it requires information that cannot be known with sufficient confidence, such as coefficient for shape (β), particle distribution (K), or the number of sections that contain the objects of interest (m). The problem with all correction factors is that they attempt to force arbitrary-shaped biological features into an equation designed for classically shaped Euclidean structures such as spheres or ellipsoids. Naturally occurring biological features rarely, if ever, exist in these classical shapes. If they did, the effects of diseases, aging, and treatment with toxic substances would change the sampling probability of treated groups in an unpredictable manner. Rather than correcting for the corpuscle problem, these correction factors simply substitute one biased value for another.

With these issues in mind, the purpose of design-based stereology is to answer the apparently simple and basic question: How should one accurately quantify the number, size, surface area, and length of biological objects present in 3D tissue, based on measurements made on 2D sections cut through the tissue, without assumptions or models that do not apply to biological objects? In contrast to earlier approaches, design-based stereology is based on a solid mathematical foundation of stochastic geometry and probability theory. Design-based stereology successfully addresses this issue through systematic-random sampling and geometric probes.

SYSTEMATIC-RANDOM SAMPLINE (SRS) OF DEFINED REFERENCE SPACES

As for all population-based studies, the process of sampling animals, individuals, or subjects for design-based stereology begins with randomization into control and treatment groups. Sampling within each case begins with the definition of a clear, anatomically well-defined reference space, the region of tissue containing the biological features of interest, which is cut serially and then subsampled into ~8–12 sections using systematic-random sampling (SRS). For

Table 1 Correction Factors Depend on Euclidean-based Assumption and Models that Do Not Apply to Non-classically Shaped Biological Features

Correction formula	Reference
$N_v = N_A/D$	Wicksell, 1925 (11); DeHoff and Rhines 1961 (12)
$N_v = (N_A \times T)/(T + 2r - 2k)$	Abercrombie, 1946 (13)
$N_v = [K(N_A)^{3/2}]/\beta(V_v)^{1/2}$	Weibel and Gomez, 1962 (14)
$N_v = (N_{A1} - N_{A2})/(T_1 - T_2)$	Ebbeson and Tang, 1965 (15)

Abbreviations: N_v, actual number per volume unit; N_A, number counted in area A; D, mean cell diameter; T, thickness of section; $2r$, diameter of structure; $2k$, smallest detectable part of structure; K, size distribution coefficient; β, shape coefficient.

example, assume a reference space sectioned into 200 sections at an instrument setting of 40 μm. Selecting every 20th section with a random start in the first 20 sections will yield ten sets of SRS sections, with $n = 10$ sections in each set. Each reference space in subsequent cases should be sampled with a different random start. This SRS approach ensures all regions of the reference spaces within each case have an equal probability of being analyzed.

UNBIASED GENOMETRIC PROBES

As described in detail below for the needle problem, a prototypical exercise in stereological probability, probes of known geometry (points, lines, planes, etc.) are used to quantify the probability that an intersection occurs between the probe and the features within a reference space. Knowing the probability of a probe-feature intersection leads to accurate (unbiased) estimates of the four first-order stereological parameters [volume (V), S, length (L), number (N)] and their variability. As mentioned above, with increased sampling these unbiased estimates progressively converge on the true or expected value. Because the procedure is based on probability, random placement of probes at more than ~100–200 locations is sufficient to correctly determine probability of intersection—like flipping a coin ~200 times to estimate the probability of heads. In a later section ("Do More, Less Well"), we will revisit the issue of how many animals, how many sections, and how many probe locations to analyze through the reference space to achieve maximum sampling efficiency.

For each of the first-order stereological parameters, the choice of the correct probe is guided by the number of dimensions in the parameters of interest, as shown below in Table 2.

For an unbiased method, the sum of dimensions in the parameter plus the sum of dimensions in the probe must equal at least 3 ($\text{dim}_{\text{parameter}} + \text{dim}_{\text{probe}} > 3$), where 3 equals the total number of dimensions in matter. For the example of profile counting shown in Figure 2, use of a single 2D section to quantify total number, a 0D parameter, yields a total of two dimensions ($2 + 0 = 2$), which is one less than the number required for an unbiased design. In this example, objects could, in effect, "hide" in the excluded dimension, and therefore not be counted. In this case, one could say that a 2D probe is biased for number.

NUMBER (*N*)

The first-order parameter N refers to the total number of discrete objects (e.g., cells) in a defined reference space. As mentioned previously (Fig. 2), the corpuscle problem described by Wicksell (11) illustrated the stereological bias associated with counting 2D profiles, rather than 3D objects. In 1984, the *Journal of Microscopy* published the first approach to overcome the corpuscle problem (16). This paper demonstrated a technique to quantify the true number of objects in a defined 3D volume of tissue, N_v, with no assumptions or models required about the biological features of interest. The disector principle is based on two sections to create a virtual 3D probe, the disector.

Geometrically, a 3D disector probe consists of an unbiased counting frame of known area, superimposed on one of two adjacent sections (a disector-pair) separated by a known distance (disector volume = frame area × disector height). Referring to Figure 3, the user counts the number of object profiles that appear on one section (reference section), but not on the other section (look-up section); this process provides an accurate count of number of object "tops" in the known disector volume. Since all objects, regardless of their size, shape, and orientation have only one top, the number of tops equals the number of objects in the disector volume. The older stereology literature often uses the term Q^-, which refers to the German term for cross-section, *querschnitt*, and $\sum Q^-$ as the number of objects counted (the sum of disappearing cross-section) within the disector volume. The disector principle uses unbiased counting frames

Table 2 Dimensions (dim) of Objects and Probes in 3D

Parameter (dim)	Probe (dim)	Total dim (parameter + probe)
Volume (3)	Point grid (0)	3
Area (2)	Line (1)	3
Length (1)	Plane (2)	3
Number (0)	Disector (3)	3

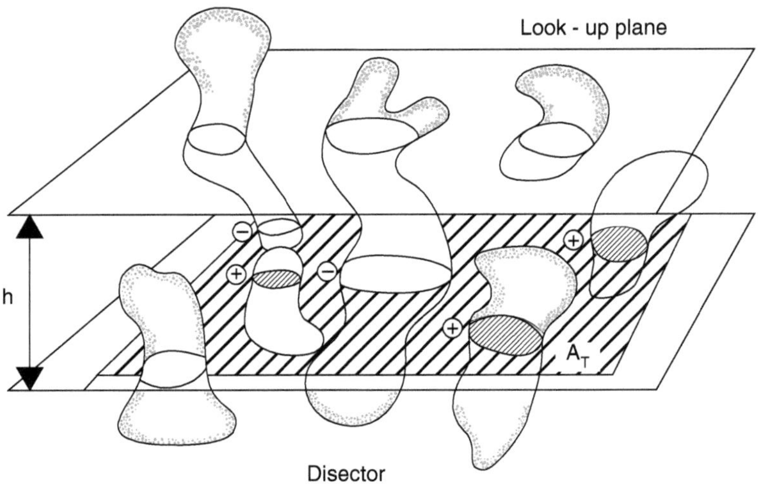

Figure 3 The disector principle. The use of two adjacent sections allows for quantification of the true number of objects per unit volume. *Source*: From Ref. 1.

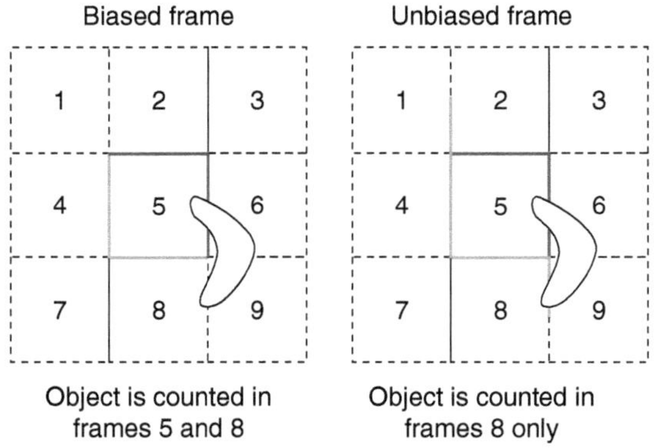

Figure 4 Comparison of biased versus unbiased counting frames. *Source*: From Ref. 1.

with inclusion and exclusion lines (Fig. 4) (17) to avoid size, shape, and orientation biases at the edges of the counting frame.

The disector probe is placed at multiple systematic-random sampled (SRS) locations through the tissue to generate a reliable estimate of the neuronal density, N_V, the total number of objects within a known volume of tissue, $\sum Q^- / \sum V$. Finally, the following formula converts density of objects, N_V, to total number of objects, N.

$$N = (\sum Q^- / \sum V) \times V_{ref}$$

where V_{ref} refers to the total volume of tissue containing the objects of interest (reference volume).

Because the original paper by Sterio (16) used two thin physical sections cut a known distance apart, the first application of the disector principle is termed the *physical disector*. In 1986, Gundersen expanded the disector principle from thin physical sections to parallel optical scanning planes through a thick section, the so-called optical disector (18). For thick sections the optical disector is far more efficient than the physical disector; however, in studies where optical scanning is not possible (e.g., thin sections from confocal or electron microscopy), the physical disector method provides the only unbiased approach to estimate number of biological features.

One year after publication of the disector principle (18), Canadian scientists used a gold-standard approach to test the disector principle, along with several others techniques currently in use for object counting (19). By embedding a known number of phantom objects in a matrix, and then sectioning the material and applying each of the methods to estimate the number of objects, this approach provided a direct test of bias inherent to each approach. Of the various methods evaluated, only estimates from the disector principle produced accurate results, without assumptions about the size or shape of objects. Note that the disector fulfills the requirement shown in Table 2 for a three dimensional probe to quantify a zero dimensional parameter (number); in this case the sum of dimensions in the probe and parameters equals three, the total number of dimensions in tissue.

In 1986, Gundersen introduced a straightforward method called the *fractionator* for scaling local density estimates (N_V, L_V, S_V, V_V) to total values for the reference space. The fractionator method involves using unbiased geometric probes to estimate the probability of intersections with biological features in a known fraction of the reference space (18). The *optical fractionator* combines the optical disector and fractionator concepts into the state-of-the-art approach for making reliable estimates of *N* objects, as shown by West and colleagues in the early 1990s (20).

THE REFERENCE TRAP: TOTAL VALUES VERSUS DENSITIES

First-order stereological parameters refer to absolute parameters (i.e., total *N*, total *L*, total *S*, total *V*). Prior to the advent of modern stereology, however, morphometric endpoints relied on ratio estimators for these parameters, such as number per unit volume (N_V), surface per unit volume (S_V), and length per unit volume (L_V). The problem with these estimators is that they assign a single value to two different quantities, the numerator and the denominator in the ratio, which vary independent of each other. The reality is that any ratio estimator like N_V varies from changes in the numerator alone, the denominator alone, or both. Across different treatment groups and within animals of a specific group, variation in one, both, or none of the values in the ratio can lead to confounding sources of variation in the results. Despite this ambiguity the majority of non-stereology studies interpret changes in ratio estimators only in terms of the numerator, and completely disregard possible effects of changes in the denominator on the ratio estimator. One explanation for this erroneous interpretation is that, historically speaking, biologists in the 1960s adopted existing stereology approaches from materials sciences and geology, disciplines that focus on materials that do not experience changes in the denominator (reference volume) during processing for quantification of stereological parameters. Since these relatively hard materials are resistant to shrinkage, changes in total values correlate directly with changes in density (total $N = N_V$). However, for biological tissue in which agonal changes, fixation, drying, dehydration, rehydration, etc. cause alterations in reference volumes in the denominator of ratio estimators, the situation is clearly different. For these applications ratio estimators such as number per unit area or volume can provide misleading results.

The advantage of total values over ratio estimators came into strong focus two decades ago in studies concerned with changes in numbers of brain cells (neurons) as a function of aging. At that time, the widely accepted dogma held that marked neuron loss in the neocortex and other defined brain regions begins around age 50, and continues unabated through old age (21–23). This theory gained wide acceptance in the peer-reviewed scientific literature and medical textbooks as the most logically compelling explanation for age-related reductions in motor skills and certain cognitive abilities. Rather than total numbers of neurons, these studies reported neuronal density, N_V, a ratio estimator quantified on a small sample of sections through anatomically important brain regions.

This "age-related neuron loss" dogma began to unravel with reports (24,25) that an inverse relationship exists between age and fixation-induced shrinkage of neural tissue; that is, following an equivalent period of standard fixation in aldehydes, aged neural tissue shrinks less than young tissue. This increased shrinkage of the reference volume leads to an increase in N_V in young brains compared to old brains, without any change in numbers of cells. Thus, age-related decreases in neurons reflects age-related differential in tissue shrinkage, rather than age-related neuronal loss. As a test of this issue, several studies in the 1990s used design-based stereological methods to quantify total neuron number in brain regions that were reported to undergo age-related neurodegeneration on the basis of densities (locus coeruleus, hippocampus, neocortex). These design-based stereology studies reported no evidence of age-related neuron loss (26–30), since quantification of total neuron number avoids reference volume shrinkage that may dramatically affect neuronal density (neuron number per unit tissue).

VOLUME (*V*)

The Greek mathematician and engineer Archimedes (287–212 BC) proposed the first non-Euclidean approach to quantify a first-order parameter. With this approach the true volume of an object, regardless of its shape, size, or orientation, could be determined from its water displacement. This approach is limited to solid, non-porous objects, however, and not easily applicable to biological tissue. For this purpose, stereologists in the 1960s discovered a preferred method on another page of mathematical history: the Cavalieri principle (31).

During the high Italian Renaissance, a student of Galileo Galilei named Bonaventura Cavalieri (1598–1647) published a technique for volume estimation in a paper entitled "Geometria Indivisibilis Continuorum Nova Quadam Ratione Promota" ("A certain method for the development of a new geometry of continuous indivisibles"). This methodology of Cavalieri (31), which contributed to the theoretical development of integral calculus by Leibniz and Newton (32), argued that all solid objects consist of "infinitely divisible" areas, as shown in Figure 5.

The formula for estimation of V in units of mm^3 by the Cavalieri method follows:

$$V = \sum A \times T$$

where $\sum A$ = sum of areas on cut surfaces (mm^2); T = distance between the cut surfaces (mm).

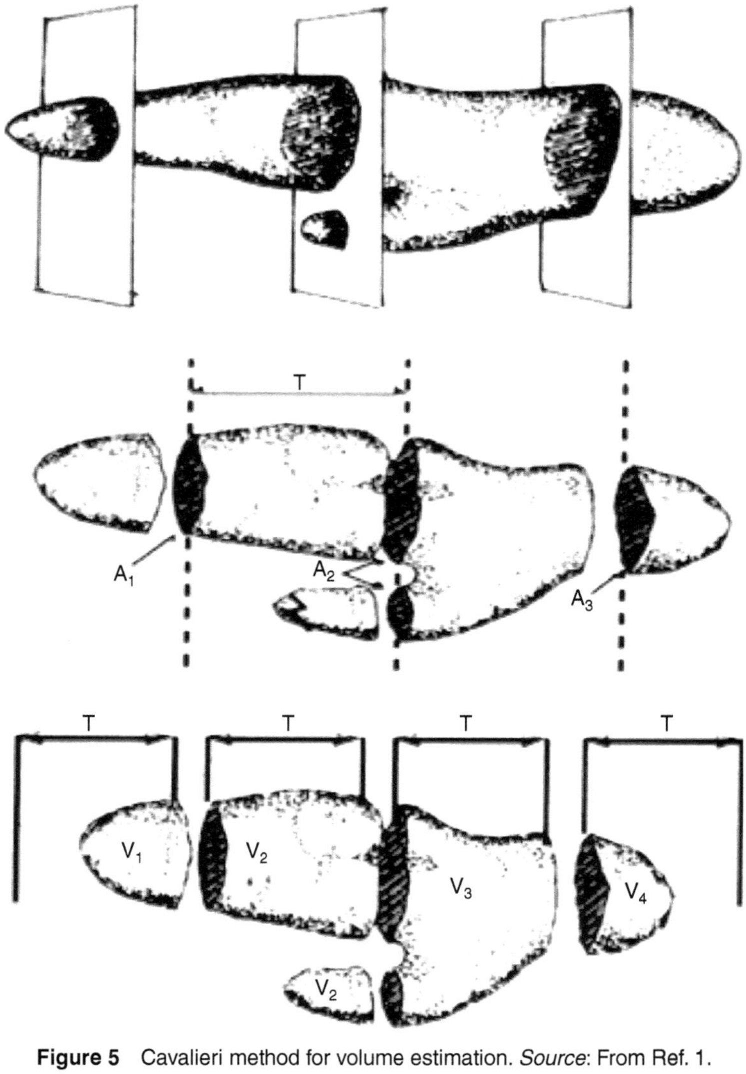

Figure 5 Cavalieri method for volume estimation. *Source*: From Ref. 1.

In the 1980s two prominent Danish stereologists, Professor Hans Juergen G. Gundersen and Dr. Eva Vedal Jensen, combined the Cavalieri principle with area estimation by point-counting (10); this approach allows for rapid, precise, and unbiased volume estimation of arbitrarily shaped 3D objects, based on the areas of their sectioned surfaces (Fig. 6). Validation of method requires sectioning of the objects into parallel planes, with the first plane at a random location through the object, and is highly correlated with volumes quantified by the Archimedes method of water displacement (33).

In this application, the sum of areas, $\sum A$ in mm^2, of the cut surface is quantified in an accurate and efficient manner by point counting:

$$\sum A = \sum P \times a(p)$$

where $\sum P$ = sum of points hitting the reference space on each section; $a(p)$ = area per point = distance in x direction (mm) × distance in y direction (mm).

Today computer-assisted and manual applications of the Cavalieri point-counting method provide the state-of-the-art methods for estimation of N and V in neural tissue from humans (28,33,34), monkeys (35,36), rats (37,38), and mice (7,39–42).

ALL VARIATION CONSIDERED

Variation in design-based stereology arises from within-individual (sampling error) and between-individual (biological variation) sources that together constitute the total observed variation in results, as quantified to the coefficient of variation (CV = SD/Mean). Within-individual variation results from the repeated application of estimator probes at systematic-random locations on sections cut in a systematic-random manner through the reference space. The extent of this repeated sampling through the reference space determines the sampling error, as quantified by the coefficient of error (CE). A relatively high CE value (<0.20) indicates a high level of error due to sampling in the results. Sampling error can be reduced by increasing the sampling stringency by increasing the number of sections from the minimum of ~6 to the maximum of ~12, and placement of increased numbers of probes on those sections from the minimum of ~75 to the maximum of ~200.

Biological variation (BV) that exists between individuals in a particular group reflects a variety of biological influences, including variation from evolution, genotype, gender environmental factors, aging, gender, treatment, etc. This source of variation diminishes as a result of sampling more individuals at random from the population. While simply adding more subjects (animals) in the hope of quickly reducing biological variation might seem effective, the cost of analyzing more individuals is high in terms of time, effort, and resources. Since all of the sources of variation can be estimated using design-based approaches, the most effective sampling designs

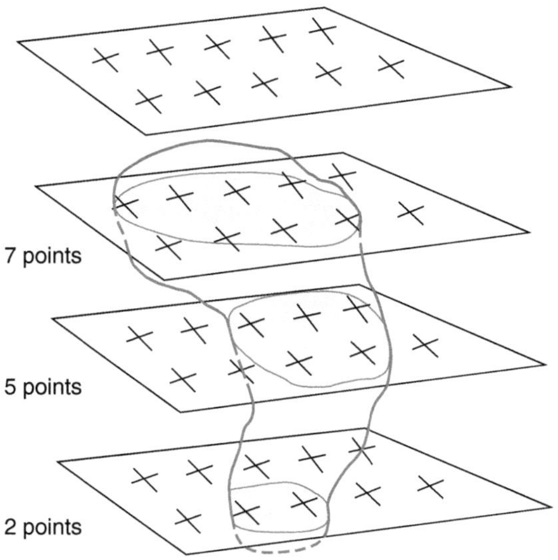

Figure 6 Combination of the Cavalieri Principle with point-counting for volume estimation. *Source*: From Ref. 1.

focus resources on the sources of error that contribute most to the variation in the results, and less effort on areas that contribute relatively less to the observed variation in the results.

The above discussion focuses primarily on *N* and *V*, two of the four first-order stereological parameters. In summary, design-based approaches discussed thus far emphasize the avoidance of known sources of bias, as summarized below:

- Unbiased Sampling: Definition of well-defined reference spaces based on unambiguous anatomical criteria, e.g., granule cell layer of the dentate gyrus (DG) and the pyramidal cell layer of Ammon's horn sector 1 (CA1). Within both of these references spaces, selection of 8 to 12 sections in a systematic-random manner will ensure that all subregions of the DG and CA1 have the same chance to be sampled and analyzed.
- Quantification of 3D volume rather than 2D area: Sectioning 3D tissue can generate planar sections with unusual appearances on qualitative examination, leading to increased chance for mistaken conclusions about possible volumetric effects. This source of bias can be avoided by quantitative evaluation of volumes using the Cavalieri point-counting method.
- Avoidance of the corpuscle problem: Quantification of possible changes in neuron numbers using the optical fractionator method, an efficient combination of the disector principle and the fractionator sampling scheme avoid bias arising from the corpuscle problem.
- Avoidance of the reference trap: The disector method, an application of the disector principle using a virtual 3D probe, counts cells of interest, without bias arising from size, shape, and orientation. Rather than reporting a ratio estimator such as "profiles per unit area," the optical fractionator quantifies total neuron number and thus avoids the reference trap.
- Avoidance of recognition errors: Counting at high magnification, in combination with Gundersen's unbiased 3D counting rules to avoid edge effects (13), for quantification of total neuron numbers.
- Low sampling error: High stringency of sampling within each brain, i.e., to a level necessary to achieve a mean CE between 5% and 10% for each group, will dramatically reduce statistically errors resulting from insufficient sampling. With a CE of 10% or less, variability from biological sources contributes the majority of variability to the results (>90%), thus increasing the validity of conclusions about significant differences or not, between controls and treated groups.

The next section reviews the remaining two first-order parameters, surface area (*S*) and length (*L*), of significant interest to neurotoxicological studies.

SURFACE AREA (*S*) AND LENGTH (*L*)

Design-based stereological approaches to quantify *S* and *L* require the recognition of three unique concepts: the needle problem, scale-dependence, and anisotropy.

The Needle Problem

In 1777, the French naturalist and mathematician Count George Leclerc Buffon (1707–1788) presented the "needle problem" to the Royal Academy of Sciences in Paris, France (43): What is the probability that a needle tossed at random over a parquet floor will intersect the lines on the floor (Figure 7)?

Buffon developed the needle problem as a mathematical exercise in probability theory. He correctly reasoned that if he knew one of two geometric quantities—either the length of a needle probe or the distance between the lines on the floor—he could calculate the unknown parameter by estimating the probability of an intersection between the two physical structures (39). As the exercise to estimate the true probability proceeded, calculation of the unknown quantity progressively converged on the true value. In this sense the needle problem represents the foundation for probability theory that underlies all design-based stereology methods—to estimate the probability that a geometric probe placed at random will intersect biological features in the tissue.

Scale Dependence

In the 1960s two groups of investigators, one in Europe (44) and the other in the United States (45), decided to test the inter-rater reliability of stereology based on the needle problem and

Figure 7 Buffon's needle problem. *Source*: From Ref. 1.

made a paradoxical discovery. Both investigators analyzed the same parameters in the same biological feature—the total surface area of endoplasmic reticulum in rat hepatocytes—and both groups used the same stereological method based on the needle problem; the only methodological difference between the studies was ultrastructural resolution: 80,000× and 12,000× by Weibel (44) and Loud (45), respectively. Surprisingly, the results from these studies were dramatically different: $6\,m^2/cm^3$ for the Loud study and $11\,m^2/cm^3$ for the study by Weibel et al. (44). These findings represented a paradox in the field of stereology for a decade. Then in 1977, at a conference organized in Paris to commemorate the 200th anniversary of the needle problem presentation by Buffon to the Academy of Sciences, Professor Benoit Mandelbrot, a Polish-born mathematician, showed that *S* and *L* for biological objects were *fractal dimensions*. According to Mandelbrot's theory of fractal sets (46,47), both the results by Weibel et al. (44) at the higher resolution and the results by Loud (45) at the lower resolution were correct; at higher resolution more structure is present. From the viewpoint of stereology comparison between studies, comparisons of data with estimates of *S* and *L* must be limited to data collected at the same or closely similar resolution. Note that this caveat does not apply to non-fractal dimensions, *N* and *V*.

Isotropic Intersections

As exemplified by the needle problem, design-based methods of stereology involve counting the number of intersections between a probe of known geometry and the unknown parameter associated with the biological feature. For *S* and *L*, the number of intersections is directly proportional to the surface area and length, *provided that the intersections occur in an isotropic manner*. Isotropy, meaning independent of direction, applies to *S* and *L* because the appearance of surface area of objects and length of linear features varies as a function of the angle of sectioning. The requirement for isotropic intersections between the biological feature and the geometric probe overcomes this inherent anisotropy of *S* and *L*. Note that this isotropic requirement can be met by randomization of *either* the probe *or* the biological feature. Prior to the computer age, this requirement for isotropic intersections during estimation of *S* and *L* meant spatial randomization of the tissue prior to sectioning (48). However, the availability of computer-generated virtual probes spawned approaches that randomize the orientation of the probe, as shown below for surface area (49) and length (50).

The use of computer-generated virtual probes, as illustrated in Figure 8, allows investigators to analyze tissue cut at any convenient orientation; and to quantify archival sections from previous studies.

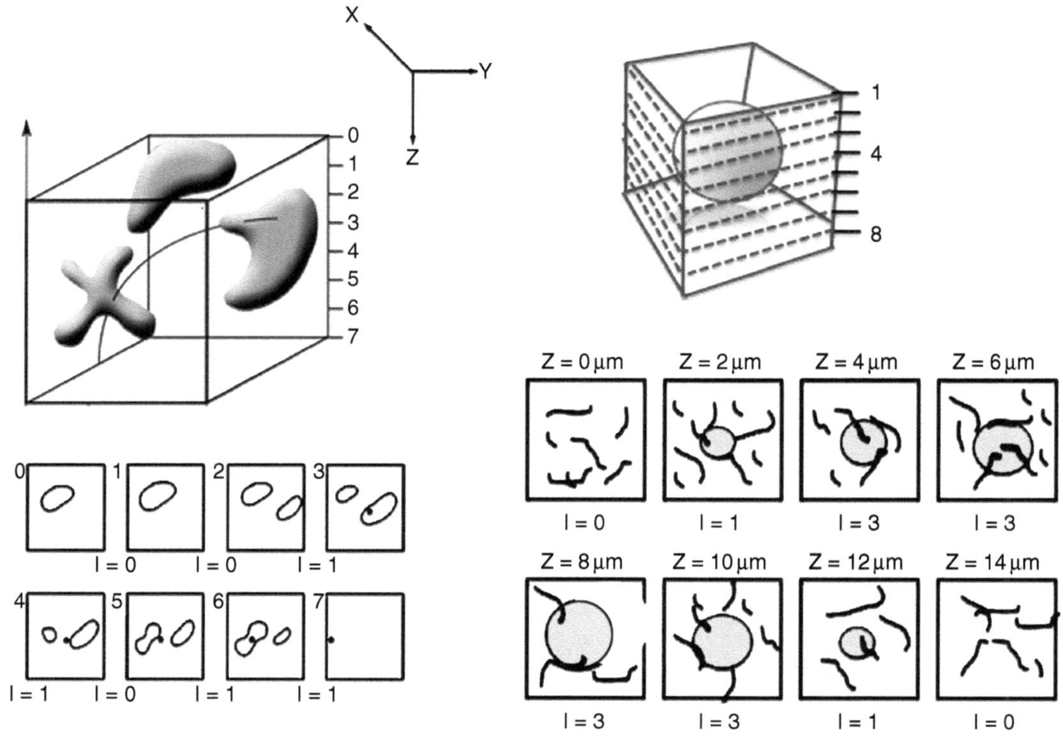

Virtual cycloid probe for surface area **Virtual sphere probe for length**

Figure 8 Virtual probes for length and surface area estimation. *Source*: From Ref. 1.

BIAS AND UNCERTAINTY

Despite the differences outlined above for design-based methods to estimate *N*, *V*, *S*, and *L*, all design-based stereological approach share a number of similarities that ensure the results will be unbiased for all practical purposes. For instance, all stereological methods involve the placement of geometric probes (estimators) at systematic-random locations through the tissue. Second, all approaches require an investigator to count and record the number of intersections between the biological feature of interest (cells, fibers, surface membrane, tissue volume) and the geometric probe (disector, point-grid, virtual cycloid, sphere probe). A third feature is the direct relationship between the number of intersections and the quantity of the parameter of interest, without further assumptions related to the biological objects of interest. The introduction of assumption, model, or correction factor that deviates from these procedures carries the potential to introduce systematic error (bias), and therefore could cause the data to vary from the true value.

In contrast to stereological bias, which cannot be measured, diminished, or removed, non-stereological bias refers to uncertainties. Potential sources of uncertainty include failure to carry out systematic-random sampling through the reference space, or uncertainty arising from insufficient magnification or poor staining that interferes with the recognition of intersections between stereological probes and biological features of interest. Although the error arising from these uncertainties, like the error from stereological bias, carries the same potential to cause results to deviate from the true value, in the case of uncertainties there is an important difference: Once identified, uncertainties can be eliminated through careful attention to the following factors in control of the investigator.

Reference Space

Identification of a *bounded* reference space, the volume of tissue containing the biological features of interest, is critical for estimation of all first-order stereological parameters. Error introduced by unclear anatomical boundaries can be reduced and eliminated through careful delineation between adjacent structures, optimization of tissue processing protocols, and consultation with anatomical atlases.

Microscopy

The optimal resolution for design-based stereology is the lowest magnification that allows for clear recognition of the biological feature of interest.

For the vast majority of neurotexicological studies, optical scanning through a thick section requires a thin focal plane afforded by high magnification objectives (60–100×) and high numerical apertures (n.a. 1.3–1.4).

Staining

Prior to the mid-to-late 20th century, histological processing for neurotoxicological studies depended on experts who required relatively little in terms of tissue preparation, typically a few thin sections cut at an instrument setting of 10 µm or less stained with water-soluble, basophilic dyes (hematoxylin, cresyl violet, giemsa) to reveal Nissl-positive structures. In these studies the evaluation of toxicity depended on the expert's experience related to the microscopic appearance of diseases or experimental models. Today, this burden has shifted from the expert to personnel responsible for the tissue processing and histological procedures that prepare tissue for data collection by trained technicians using computerized stereology systems. The goal of staining in stereological studies is to visualize biological features of interest in the tissue. As described in the following section on Computerized Video Densitometry, and standardized tissue processing and staining protocols will reduce uncertainty arising from histological sources.

DO MORE, LESS WELL

The sampling strategy is under the control of the investigator, with the goal to focus on sampling the source that contributes most heavily to the total observed variation in the study. The remarkable efficient of design-based stereology is that accurate and precise estimates of structural parameters can be estimated by counting a relatively small number of features (10). The total number of sampling locations to achieve optimal sampling ranges from about 100 locations for a relatively homogeneous distribution of features of the interest, to about 200 for features of interest distributed in a relatively non-homogeneous (heterogeneous) manner. Since variation between individuals in a particular group typically is greater than variation within a single individual, sampling more than ~200 locations through a reference space in any given individual approaches a point of diminishing return. As shown in Figure 9, estimates of total cell number in a defined brain region changes little by sampling more than ~200 cells in each individual.

Thus, the optimal strategy to make reliable estimates of *N*, *L*, *S*, and *V* is the sampling design that most efficiently reduces the total observed variation (CV) per unit of time spent analyzing tissue. Because design-based methods account for all sources of error in methods, variation in an estimate can be partitioned into the constituent sources of error using the following equation:

$$CV^2 = BV^2 + CE^2$$

where,

CV^2 = total observed variance
BV^2 = biological variance
CE^2 = error variance

Because CE and CV can be calculated from the results of a study, the third value (BV) can be obtained through rearrangement ($BV^2 = CV^2 - CE^2$). By identification of the source that contributes the majority of variation to the results, the investigator can design a sampling strategy that focuses effort on that level.

The work-flow for maximum efficiency begins with sampling the reference space into ~10 (8–12) systematic-random sections; second, to quantify the parameter of interest on about two to three individuals for each group using one of the commercially available computerized stereology systems (*Stereologer*, Stereology Resource Center, Chester, MD); and finally, to

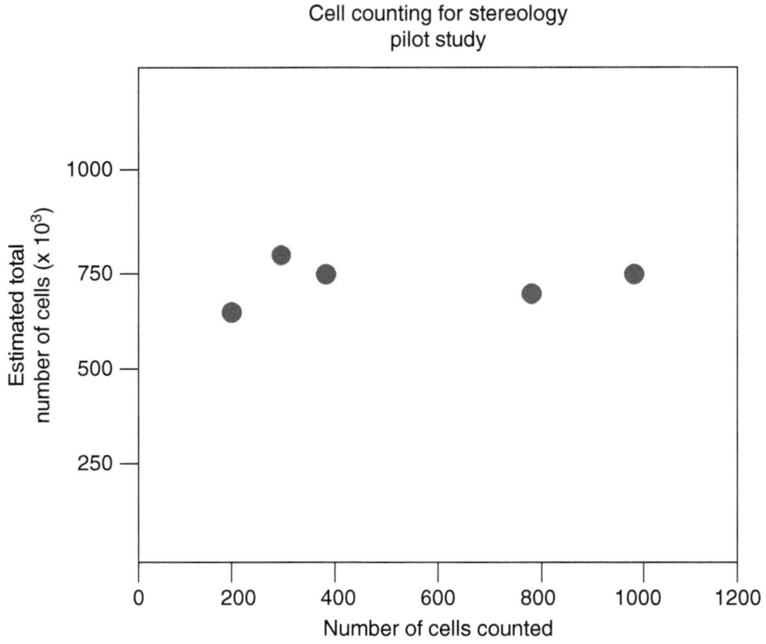

Figure 9 Do more, less well. Sampling greater than ~200 locations in the tissue results in low changes in accuracy of an estimate. *Source*: Courtesy of Dr. Kebreten F. Manaye.

calculate the values for CV and CE, and solve for BV. At this point, inspection of these results will guide further decisions. For example, suppose the initial analysis of n = 5 subjects per group in a study fails to reveal significant effect in the parameter of interest (number of cells in a defined brain region), but shows a trend toward a significant effect. In this case the investigator has the option to: (*i*) analyze more subjects; or (*ii*) analyze more locations within the sections from the existing subjects. Using the above equation, the results indicate that BV accounts for 20% of the observed variation while CE accounts for 80%. In this case the obvious choice is to apply a higher stringency of sampling on the existing sections, which is less expensive, faster, and least wasteful of resources. Once the CE is reduced to below 10%, one approaches a point of diminishing returns. If the results continue to show a trend in support of possible significant differences, then time and effort should be used to analyze more individuals in each group using the sampling strategy developed in the original small number of subjects. The statistical power required to reveal significant differences depends on the expected strength of the effect, but often $n \approx$ 10–12 subjects per group reveals a weak effect, $n \approx$ 7–8 subjects per group a moderate effect, and $n \approx 5$ subjects per group a strong effect.

One final note with regard to the selection of an optimal sampling stringency is the distribution of the biological objects or features within the reference space. For objects distributed in a relatively uniform, homogeneous manner, a lower sampling stringency, i.e., fewer probes placed at a small number of locations on fewer sections, can be expected to reduce the CE in an efficient manner. If the biological features exist in a relatively heterogeneous, clustered distribution within the reference space, however, a higher sampling stringency—more probes placed on more sections—will be required to reduce the CE. Given that the goal is to capture the majority of variation for a mean parameter estimate, excessive sampling within a single individual is far less efficient than using the same effort to sample a greater number of individual. This process of focusing work effort on the greatest source of variation has been termed, "Do More, Less Well," by Professor Ewald Weibel, one of the pioneers of modern stereology.

REVIEWER CONSIDERATIONS: DESIGN-BASED STEREOLOGY

In 1996 *The Journal of Comparative Neurology*, one of the most highly regarded journals in the neurosciences, published a review of design-based stereology (51). In an editorial statement printed with the review, the journal's Editor-In-Chief, Dr. Clifford Saper (52), shared his preference for studies that use state-of-the-art approaches of unbiased stereology, rather than model- and assumption-based methods. Furthermore, Professor Saper explained the journal's new policy to require design-based stereology, when appropriate, for quantification of morphological endpoints; otherwise, authors would be expected to justify their alternative methodology. This editorial statement sent a clear message to the international neuroscience community about the value of stereological techniques. Several other publications of peer-reviewed biomedical research established similar review criteria in the past decade, and this trend is expected to continue.

Although what qualifies as basic information varies among investigators and reviewers, a minimum set of information exists to communicate the application of stereology principles and practices in studies for publication in peer-review journals, as summarized below.

- Definition of a clear, unambiguous reference space
- Calculate and report volume of the reference space (V_{ref})
- Sampling approach (every 5th with a random start)
- What are the sampling fractions
- What was counting item (e.g., Chat-pos neuronal cell bodies, nucleoli, TH-positive fibers)
- Specific stereological probe (optical fractionator, rotator, etc.)
- How many animals (age, gender), how many sections, how many actual cells counted (mean cells counted per case ≈ 150 cells)
- Descriptive stats: group means, variance (SD, SEM), and CV, mean CE (extra credit for calculation of biological variance, BV).

Inclusion of the above minimal information for peer-reviewed publications similarly addresses several of the common problems (e.g., points 3, 6, and 8 below in **bold**) in grant applications for federal funding, based on information provided by the Center for Scientific Review from the National Institute of Health.

1. Lack of new or original ideas
2. Absence of acceptable scientific rationale
3. **Lack of experience in essential methodology**
4. Uncritical approach
5. Diffuse, superficial or unfocused research plan
6. **Lack of sufficient experimental detail**
7. Lack of knowledge of published relevant work
8. **Unrealistic large amount of work**
9. Uncertainty concerning future direction

With respect to both peer-review journals and applications for extramural funding, a pilot study in a small group ($n < 5$) of treated subjects and controls highly effective preliminary data for stereological studies. Data can be generated in support of statistical trends or differences, and thus support a hypothesis related to morphological changes. Furthermore, a pilot study will reveal information about time, labor, and material resources required to complete the proposed studies. Lastly, for grant application the pilot study addresses the concerns of reviewers that the investigator has the necessary resources and expertise to carry out the proposed studies.

Modern stereology provides design-based (unbiased) methods to sample and quantify first-order stereological parameters in neural tissue, including length of nerve fibers or the number and size of neurons in a particular brain region. However, neurological conditions or models of brain injury may also include damage or changes to tissue components apparent only by variations in staining intensity. Loss of connectivity between neurons (synaptic degeneration) or increased expression of the astroglia intermediate filament, glial fibrillary acidic protein (GFAP) involve changes in immunoreactive intensity, as assessed by computer-assisted video densitometry and reviewed in the following section.

COMPUTERIZED VIDEO DENSITOMETRY

Video densitometry provides a semi-quantitative approach to assess relative differences in staining intensity on tissue sections. The approach involves digitization of color video images, with separate sampling of light in the red, green, and blue frequency ranges, followed by measurement of intensity in terms of numeric values that correspond to red-green-blue (RGB) or hue-saturation-intensity (HSI). Hue refers to the basic color of an object, while saturation refers to color brilliancy. Two features can share the same color, with one being more intense, such as candy apple red versus candy apple pink. In the HSI approach, hue and saturation are plotted as a hexagon along the x and y axes, with intensity represented as a composite measure of overall brightness along the z axis. Since the DAB reaction product is the same hue under all conditions, experimental differences in antigen expression lead to changes color saturation and/or intensity, rather than the formation of new colors.

To define positively stained elements, an HSI identification window is assigned to segment (tag) pixels that are positively stained. For each primary antibody, a series of sections encompassing the range of reaction product intensities present in the experiment are selected as standards. The optimal window, a segmentation of the positively stained cellular components, without background on all standard sections, is determined and then held constant for the analysis of all sections in the experiment. Each image is digitized, corrected for heterogeneity in background illumination, and segmented by using the HIS window parameters. The value of constant microscope settings, a constant voltage source, and manually entered HSI segmentation window settings is to allow results from different analysis sessions to be combined for statistical analysis, without further normalization. To confirm that microscope and camera settings remain constant, the standard sections or a photographic density wedge is imaged before each session. Densitometry software sums the number of pixels within the HSI identification window positive for reaction product. Calibration of the densitometry system in true spatial units allows expression of reaction product area in units of μm^2 (see reference 49 for further details on the HSI method).

The intensity of the peroxidase reaction product is expressed in terms of the optical density (OD) for each pixel within the reaction product HSI window, and these values averaged together to produce the average reaction product OD per section. Alternatively, the fractional area occupied by peroxidase reaction product can be calculated by dividing reaction product area by the total area of the digitized image field, generating a measure (percent area) that is relatively independent of the magnification used or the field size examined. Note that OD values and fractional area values can vary independently for a given treatment. Increases in cell size or cell number, in the absence of changes in protein concentration, would increase fractional area without changing OD, while increased protein expression per cell would increase staining intensity without changing fractional area. If both parameters vary, a useful measure is total immunoreactivity, calculated as the product of fractional area and reaction product OD.

Video densitometry can also be carried out without segmentation in particular cases. For example, tyrosine hydroxylase immunoreactivity in the neostriatum is localized in terminals of the dopaminergic afferents. Also, at relatively low magnification where individual synaptic boutons cannot be discriminated, staining appears homogeneous and the OD of each pixel in a circumscribed region can be averaged into a measure of field OD. This parameter may be calibrated with a photographic density step wedge, a strip of photographic film with a defined gradient of OD.

STAINING INTENSITY

The same requirements exist for both design-based stereology and video densitometry analysis to minimize bias from biological variability and observer error; to randomize animals into different treatment groups; and to ensure data collection by investigators blind to the experimental treatments. Where the methodologies for stereology and video densitometry analysis diverge, however, is in the level of rigor required from staining protocols. Staining protocols for stereology studies must achieve a minimum threshold for visualization of the biological features of interest; that is, of sufficient intensity to support of a binary (yes/no) decision about whether the feature of interest intersects the geometric probe. For stereology, above the threshold for recognition of positive staining, differences in staining intensity do not introduce variability into the results. In contrast, variation in staining intensity is the primary outcome variable for video densitometry analysis.

To ensure the high level of staining consistency required for video densitometry, the neuroanatomist must approach the process of staining tissue like a neurochemist, with strict attention to all details of the tissue processing protocol—treating sections cut at exactly the same thickness with exactly the same reagents for exactly the same duration—to ensure absolute uniformity across groups, with the code broken only after the analysis is complete (49–52). The following section reviews the important considerations required to minimize variability in staining intensity arising from non-biological sources.

FIXATION

A necessary evil, fixation represents a trade-off between the need to maximize tissue preservation, while minimizing changes to tissue antigens. Fixation is required to preserve tissue morphology and retain epitopes on antigens for antibodies recognition. Of the numerous fixatives available, including acetone, methanol, and aldehydes (formaldehyde, paraformaeldehyde, and glutaraldehyde), several can be combined or mixed with other compounds. Acetone and methanol denature and precipitate proteins, while the aldehydes cross-link proteins in place to prevent them from leaching out of tissue. Chemically reactive aldehyde groups interact with primary amines, such as lysine residues in proteins, and thiols to form cross-links of specific spatial characteristics. In this way, fixation alters the structure of tissue antigens, which can influence the binding of primary antibody to specific tissue antigens. Certain antibodies can fail to recognize their cognate antigen after incubation in some fixatives, while "neutralizing" antibodies recognize native protein confirmations, but fail to recognize their cognate antigen after fixation. Although acetone can be used to fix cells in culture, the typical disposable plastic-ware for cell cultures is composed of polystyrene that dissolves after contact with acetone and turns optically opaque. Glutaraldehyde, the required fixation for electron microscopy, suffers from a short shelf half-life, as well as being hazardous and expensive. As a result of these considerations, 10% formaldehyde and its 4% paraformaldehyde derivative have become the preferred choices for fixation of proteins in the histology laboratory.

Formalin, a commercial preparation of formaldehyde, contains 37% formaldehyde in distilled water on a weight:weight basis. Solutions of formaldehyde, however, carry a number of important caveats for use in conjunction with densitometry-based measurements. Formaldehyde molecules tend to spontaneously condense into long-chain polymers, a reaction that varies as a function of time, temperature, and pressure. This polymerization process reduces the fixation capability of formaldehyde but can be inhibited by alcohols; as a result, many commercial preparations typically contain 10% methanol. A relatively poor fixative, methanol is strongly hydrophobic, which leads to greater tissue shrinkage than aldehydes. Therefore, methanol-containing solutions of formaldehyde can impact antigen:antibody reactions in a manner that ultimately determines the signal strength of densitometry data. Furthermore, formaldehyde oxidizes to formic acid, leading to a change in the pH of the stock solution. However, because the formic acid concentration varies with time and storage conditions, this effect is low relative to the formaldehyde concentration and, therefore, can be prevented by dilution of formalin in buffers.

Fixation Preparation

Paraformaldehyde powder requires heat and the addition of base to dissolve into solution, and the use of a heating block and a magnetic stirrer minimizes polymerization. For a 4% solution of paraformaldehyde, which should be prepared on the day of use, 4 g of paraformaldehyde powder plus 40 ml of high-quality (distilled or nanopure) water is mixed with two drops of 6N NaOH and placed on a heat block with stirring at 60°C for five minutes. Fifty ml of cold (refrigerated) 2 × Sorensen's phosphate buffer is added and the solution allowed to reach room temperature. The pH is then adjusted to 7.4 with HCl and the solution brought to a final volume of 100 ml with water.

Fixation Duration

Another critical variable for standard tissue processing is the duration of fixation. Formaldehyde, with its very small molecular weight, penetrates tissue rapidly, while cross-links formed far more slowly. Older histopathology protocols placed various sized brains in formalin solution for varying durations, sometimes for extremely prolonged periods, even years. Fox et al. (53) suggested 24-hour fixation in formaldehyde to completely fix even small tissue samples. For standard immunostaining, variations in extent of fixation may affect antigen:antibody binding, and most labs have shortened the duration of fixation to a minimum to retain antigenicity to the greatest extent possible.

TISSUE PREPARATION

Buffering

Buffers maintain pH during all phases of the section storage and staining procedures. Phosphate buffers offer strong buffering capability at physiological pH, while an isotonic buffer prevents osmotic shrinkage or swelling of sections and cells in culture. A useful phosphate-buffered saline is composed of 50 to 100 mM phosphate salts plus 0.9% sodium chloride (NaCl). The addition of calcium and magnesium improves cell retention during the staining process, making Dulbecco's phosphate-buffered saline (DPBS) a convenient buffer. This buffer is isotonic, develops a pH of 7.4 when prepared without acid or base, and contains calcium and magnesium. However, phosphates must be in solution prior to the addition of calcium and magnesium to avoid the formation of insoluble precipitates. Addition of calcium and magnesium salts slowly with strong stirring, or addition of concentrated stock solutions of calcium and magnesium to generate the appropriate final desired concentration, will prevent the formation of these precipitates.

Cryoprotection

Freezing a fixed neural tissue causes the formation of large ice crystals within cells and in the interstitial space. With thawing, these ice crystals expand and punch holes in cell membranes, leading to visible damage and holes within stained sections. Histologists typically use either one-day or three-day methods to prevent the formation of ice crystals using sucrose as cryoprotectant. For the three-day approach, the tissue is saturated with sucrose by immersing the brain in a series of increasing concentrations from 10% to 30% (weight:volume), beginning with 10% overnight, followed by overnight incubations in 20% and 30% sucrose solutions. Initially the brain floats in the 10% sucrose solution; however, the brain gradually sinks as the osmotic pressure of the brain and surrounding solution equilibrate. Some investigators obtain satisfactory results with the one-day approach that uses a single overnight incubation in 30% sucrose. After cryoprotection, the brain is placed in a shallow container in 20 to 30 ml isopentane (2-methyl butane) chilled by dry ice powder. Once frozen, the brain is transferred with forceps into a storage vial and quickly placed into the freezer.

Tissue Sectioning

Sliding microtomes quickly section tissue in a systematic, uniform manner at instrument settings ranging from 25 to 50 micron. For stereology studies the choice of section thickness depends on the parameter of interest, with thicker sections required for studies with 3D probes. The choice of sectioning in the coronal, sagittal, or horizontal planes depends on a number of factors related to the identification of defined anatomical structures, and generally does not impact the collection of data for either stereology or densitometry. Typically the most efficient choice is the sectioning plane that includes the majority of neural structures of interest.

Typically, sections are collected into 12- or 24-well tissue culture dishes containing 2 ml of phosphate-buffered saline, with 100 mM sodium azide to prevent the growth of bacteria, mold, or yeast during protracted storage. After sectioning, tissue sections can be stored in the cold for long periods of time (months) without loss of immunostaining. Two 24-well culture dishes side by side at the microtome can receive all sections from an entire mouse brain in the horizontal plane (approximately 1 cm). The first section is placed into column 1, row 1 of the first dish, with subsequent sections placed in order into following columns of row 1: sections 1 to 6 in the first culture dish, 7 to 12 in row 1 of the second dish; section 13 in dish 1, column 1, row 2; and sections 13 to 24 continue in row 2 of both dishes. Section 25 is placed into dish 1, column 1, row 3, and section 37 is placed into dish 1, column 1, row 4. After completing section 48, 1 section will be present in each well of both culture dishes. This approach maintains the spatial orientation of the brain allowing the investigator to track sections in order. Section 49 is placed in the same well as section 1, and sections 50 to 96 continue in order as described above, a process that repeats until the brain is completely sectioned with four sections per well. This distance between sections allows sections within a given well to be easily distinguished by eye using anatomic landmarks, with reference to an atlas if needed. Each of the twelve columns contains 16 sections separated by 300 μm (12 × 25 μm), which can be operationally defined as one set of sections. For each experimental animal, selection of columns can be randomized by rolling dice or with a random number generator. Thus, sample 1 could be column 4, followed by sample 2 as column 7. Sections from each column are stained for a particular marker. If fewer sections are needed, it is convenient to use one-half set of sections by choosing either wells 1 and 3 or 2 and 4 of a given column, resulting in 8 sections, each separated by 600 μm. This section spacing is appropriate for spanning the large reference space of large brain structures, such as hippocampus and cerebral cortex, in rodent brains. Smaller structures or nuclei will require adjustment of collection parameters to allow selection of sections closer together to maintain appropriate systematic-random sampling as described above. Note that the above protocol for sectioning could be used for either stereology or densitometry, since the key requirement to maintain the uniform separation and organization of sections is met. Most important, the experimental and sampling protocol needs to be established prior to sectioning to ensure the capability to conduct an unbiased analysis.

IMMUNOSTAINING

Free-Floating Sections

For both densitometry and stereology studies that analyze stained sections across experimental animals and conditions, all samples should be stained simultaneously in the same assay with the same solutions and reagents. Staining free-floating sections allows penetration of antibodies and other reagents from both of the cut surfaces.

Slide-Mounted Sections

Staining sections mounted on glass on slides prior to immunostaining limits penetration to a single surface of tissue sections. This limitation may require higher antibody titers and longer incubation times to produce equivalent staining intensity (Fig. 10). However, the section thickness is usually less than for free-floating sections. Slides are typically placed horizontally on racks in a humid chamber and covered with antibody solutions, and moved as needed for wash and color development steps.

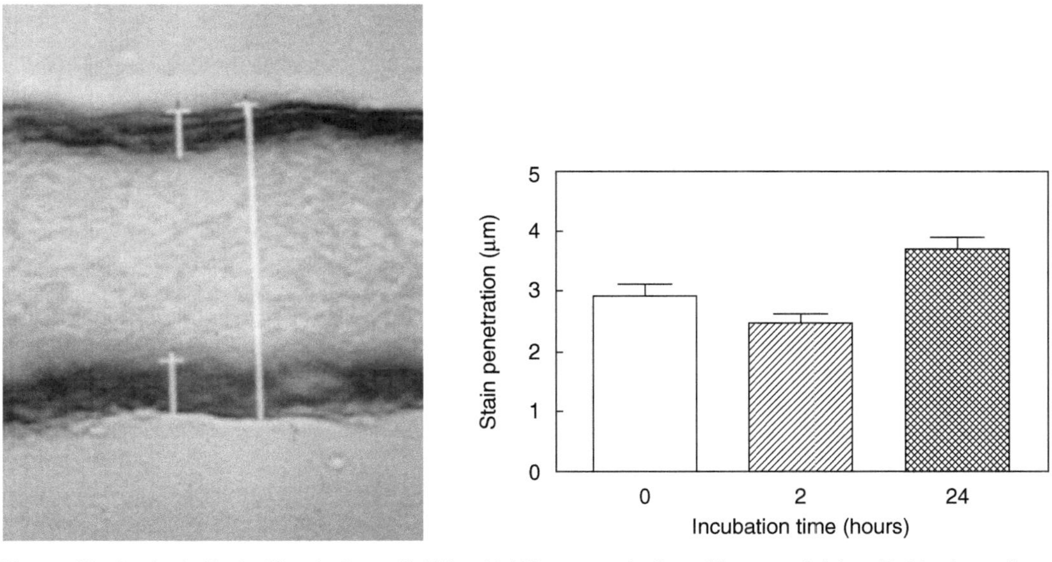

Figure 10 Lack of effect of incubation with Triton X-100 on penetration of immunostaining. Rat brain sections were incubated in 0.2% Triton X-100 in phosphate buffer for varying durations. Sections were then embedded in OCT compound, rotated 90°, and resectioned. Morphometry was used to measure the apparent distance immunostain penetrated from the cut surface of each section. (*Left*) Screen image showing immunostained cross section with the total section thickness, and stain penetration depth indicated by white lines. (*Right*) Stain penetration was measured at five positions along each cross section and nine cross sections were measured per condition. Data represents mean ± SEM.

FEATURE VISUALIZATION

For video densitometry and stereology, the peroxidase-diaminobenzidine (DAB) visualization system provides the most effective, uniform, and long-lasting results. Fluorescent immunostaining is prohibitively difficult to quantify by video densitometry and results in photo bleaching of the fluorochrome in an inconsistent manner across specimens. The peroxidase-DAB system contains toxic reagents that necessitate the use of a ventilated hood. The high specific activity of peroxidase yields significant reaction product in a short time period and DAB reveals crisp cellular detail. Inclusion of various salts can be added to modify color in the final reaction product. The next section discusses the details of DAB colorization.

TISSUE TREATMENT

Inhibition of Endogenous Peroxidase

One drawback to the use of DAB is that some tissues contain sufficient endogenous peroxidase to generate substantial color in the absence of positive immunostaining signal. In the brain, the majority of this endogenous peroxidase exists in red blood cells (RBCs), where it serves to detoxify peroxides produced in their high oxygen tensions. Perfusion of the brain minimizes the presence of RBCs, but is rarely 100% effective. After immunostaining using the peroxides-DAB system, RBCs will be a deep brown color due to their inherent peroxidase activity. Endogenous peroxidase activity can be quenched by incubation in a solution containing methanol and hydrogen peroxide. As the peroxidase enzyme binds to the hydrogen peroxide to reduce it, methanol in the solution precipitates and permanently inactivates the enzyme. For brain tissue, incubation in a solution containing 10% methanol (volume:volume) plus 3% hydrogen peroxide (H_2O_2) in DPBS for 30 minutes completely quenches endogenous peroxidase activity. For some tissues with very high endogenous peroxidase activity such as liver, this solution generates too much enzyme activity. The products of enzymatic activity are water and oxygen. Production of excessive oxygen is visible as bubbling at the section surface or in the solution, and may be so vigorous that it physically damages the section, producing holes and shredded tissue. To completely remove all methanol and hydrogen peroxide, sections can be washed thoroughly in at least three changes of buffer (5 min each). Note that all enzyme-linked methods can be affected by endogenous enzyme activity in tissues; thus, blockade of artifactual color development from other endogenouse enzymes may not be as simple and robust as that peroxidase.

Blocking

"Blocking" steps use reagents designed to reduce nonspecific signal generation. Typically, this step is carried out with 1–4% serum from the species in which the secondary antibody was generated. Serum contains relatively high concentrations of IgG, albumin, and globulins that mask nonspecific binding sites for the secondary antibody, and some investigators use commercially available solutions of bovine serum albumin. Milk-based blocking reagents contain significant concentrations of biotin, and therefore should be avoided in avidin-biotin complex (ABC) staining methods. Another useful blocking agent is 100 mM lysine. As discussed above, formaldehyde fixation generates chemical cross-links between lysine residues of proteins. If two lysine residues are not available in the necessary spatial configuration, a reactive formaldehyde group remains on the fixed tissue. Subsequent addition of proteins in solution, such as antibodies, can lead to undesired cross-linking to the tissue, and an undesirable increase in background staining. This problem occurs in techniques when formaldehyde fixation immediately follows staining reactions.

Immunostaining

Kinetics of antibody binding to tissue sections, at concentrations typically used for immunostaining procedures, confirms maximal binding during a two-hour incubation at room temperature. Because antibodies bind to antigens at two sites, the off-rate for antibodies from antibody:antigen complexes is essentially nonexistent, with virtually no antibody dissociating over 72 hours. As a reuslt, tissue incubations with antibodies require sufficient duration to bind for maximal signal, with sections typically held in primary antibody solutions overnight.

When the primary antibody is conjugated directly to a reporter molecule, binding affinity is reduced and staining less sensitive than other methods. Alternatively, a secondary antibody conjugated to a reporter molecule binds to the primary antibody. The secondary antibody is generated by injecting a large animal (horse or goat) with the IgG from the animal that produced the primary antibody (mouse or rabbit). Several commercially available kits contain secondary antibodies and color development reagents. For peroxidase-linked methods, these products fall into two levels of amplification. In the first case, secondary antibodies directly conjugated to peroxidase, with one peroxidase molecule bound to each antigen. Second, secondary antibodies can be conjugated to biotin, with tertiary reagents composed of avidin-peroxidase or streptavidin-peroxidase conjugates. Biotin has an extremely high affinity for binding to avidin. Because avidin contains four potential binding sites for biotin, the ABC technique results in formation of large complexes containing numerous avidin, biotin, and peroxidase molecules, resulting in greater sensitivity and amplified color development. On the other hand, this technique may not be ideal for quantification by image analysis because the extremely large molecular size of the avidin-biotin-peroxidase complex constrains antibody binding to closely spaced antigens (50). A typical protocol is to incubate floating sections with primary antibody overnight at 4°C, followed by sequential incubations in the biotinylated secondary antibody for two hours, then in streptavidin-peroxidase for one hour.

The enzyme peroxidase reduces hydrogen peroxide to water and oxygen. When the reaction occurs in the presence of DAB, the DAB is oxidized, causing it to change color and precipitate. Peroxidase has an extremely rapid specific activity, and its action quickly generates precipitates of oxidized DAB. This rapidly quenches further color development. A pre-incubation of the tissue with 1.4 mM DAB (but in the absence of hydrogen peroxide) ensures that DAB is available in the vicinity of the peroxidase enzyme. Color development reactions are then initiated by replacement of the incubation buffer to one containing 1.4 mM DAB plus 0.03% hydrogen peroxide. The color development reaction should be carefully timed, and performed for exactly the same duration for every specimen. Each assay should be balanced, so that specimens from each experimental group are included.

Other sections should be processed to assess nonspecific reaction product formation. These sections should be processed through all steps of the staining procedure, except they should not be exposed to primary antibody. This may be accomplished by either omitting the primary antibody, using "normal" or pre-immune serum in place of primary antibody, or by using immunoprecipitated serum. Preincubating antisera with saturating concentrations of the appropriate cognate antigen should reduce or remove specific binding.

COLLECTION OF DIGITAL IMAGES

Magnification should be sufficient to reveal anatomical structures, but also low enough to sample immunostaining. For cellular level resolution in brain, 100× or 200× work well. Imaging at a magnification that is too high results in small numbers of stained structures per field, and very large variability between fields. Appropriate Kohler illumination should be established, but then the position of the aperture diaphragm needs to be controlled. Even minor differences or changes in the position of the aperture diaphragm opening knob can result in substantial differences in the background illumination of individual specimens. Both the microscope and the video camera should be attached to a voltage regulator to prevent spurious changes in illumination or exposure caused by fluctuations in incoming line voltage. All video camera options should be set to predetermined manual exposure controls so that images are not modified by automatic compensations in the camera function for images of different overall brightness, color, or apparent contrast. These manual control settings should be held constant for imaging all specimens within an experiment. With computer-driven stereology systems, the operator identifies the boundaries of the brain region of interest by "mousing" around it, then the software selects systematic-random fields for analysis at high magnification. This capability is not a feature of most video densitometry programs, so it is up to the investigator to sample the desired space. Instead, the operator samples at low magnification (1–4×) to identify the brain region of interest. Selection of the final fields for analysis is randomized in the brain region of interest. All specimens should be imaged in a single sitting, whenever possible, with measurement of experimental and control specimens intermixed and blind to the operator. Group differences on the order of 30–50% can be successfully obtained with a sampling of 8 to 12 microscopic fields. Differences less than 30% may require additional sampling; however, they are often within the normal variance of the tissue.

REVIEWER CONSIDERATIONS: VIDEO DENSITOMETRY

Video densitometry has been used extensively to assess changes in protein expression due to aging, brain injury, transgene status, and other perturbations in biological tissues. Although measurements by video densitometry can be comparable across sections and perhaps across antibodies, it is important to emphasize that this approach is semi-quantitative and only applicable to relative comparisons. The slope of the relationship between antigen density and OD or total reactivity is unknown, and there is neither a standard curve nor an acceptable way to generate one. Even if a method for generating a standard curve did exist, tissue processing modifies epitopes and affects antibody-antigen affinity, thus preventing absolute quantification. Furthermore, the OD for a total immunoreactivity value that corresponds to zero antigen density is unknown, and the presence of some background staining predicts that zero antigen would correspond to a value greater than zero total reactivity. Thus, despite efforts to tightly controlled assay and measurement conditions, the semi-quantitative nature of the relationship between total immunoreactivity and antigen density predicts that fold changes between treatments will likely vary from those determined by quantitative assay procedures using standard curves.

With these caveats, there are critical steps involved in obtaining accurate data for immunoreactivity that also come into consideration when attempting to determine co-localization of proteins. The video camera image is collected and digitized by the frame grabber of the analysis system, and stored in the computer. In order to measure the amount of staining on a given section, it is necessary to identify what qualifies as positive stain to the computerized video densitometry system. Although it is possible to perform video densitometry in a black and white (sometimes called grey scale) mode, it is increasingly common to employ color video. It is necessary to represent the color of images as numeric values for computerized processing. There are several models available to represent color as an x, y, z grid, but the most commonly using in image analysis systems are RGB and HSI. Video cameras operate in RGB mode, with separate sampling of light in the red frequency range, the green range, and the blue range. Color may be numerically represented by the amount or brightness of red light, green light, or blue light from the image. The RGB color space may be imagined as a cube with each color arrayed on one axis. Each point in this cube would then represent a unique color. RGB allows the maximal number of colors to be defined.

An alternative mode of representing color space is HSI. Hue refers to the basic color of an object, while saturation refers to how brilliant that color is. Two items can be the same color, with one being more intense: candy apple red versus pink. In the HSI model, hue and saturation are plotted as a hexagon along the x and y axes. This color model is familiar to Microsoft Office™ users when selecting colors for fonts and other displays. Intensity is a composite measure of overall brightness, and is plotted along the z axis. The HSI model is superior for densitometric analysis. The DAB reaction product is the same hue under all conditions, but experimental differences in antigen expression should result in changes in saturation of that color and/or intensity, not in the formation of new and unique colors. These are precisely the primary fundamentals of discriminating positive staining used by the HSI model.

To define positively stained elements, an HSI identification window is assigned to segment (tag) pixels that are positively stained. For each primary antibody, a series of sections encompassing the range of reaction product intensities present in the experiment should be selected for use as standards. The optimal window, segmenting the positively stained cellular components without including background on all standard sections, should be determined and then held constant for analysis of all sections in the experiment. By keeping the microscope settings constant, by using a constant voltage source and by entering the HSI segmentation window settings manually, data from different analysis sessions should be sufficiently comparable that they could be combined for statistical analysis without normalization. To confirm that microscope and camera settings remain constant, the standard sections or a photographic density wedge may be imaged before each session. Details on the HSI method and other applications can be found elsewhere (54).

Each image is digitized, corrected for heterogeneity in background illumination, and segmented by using the HIS window parameters. The computer then sums the number of pixels within the HSI identification window (positive for reaction product). Calibration of the densitometry system in true spatial units allows expression of reaction product area in units of μm^2.

Alternatively, the fractional area occupied by peroxidase reaction product may be calculated by dividing reaction product area by the total area of the digitized image field, generating a measure (percent area) that is relatively independent of the magnification used or the field size examined. The intensity of the peroxidase reaction product may be determined by measuring the OD of each pixel within the reaction product HSI window, and averaging these values together to produce the average reaction product OD. Theoretically, either of these values could vary independently of the other for a given treatment. For example, increases in cell size or number but not in protein concentration would increase fractional area, but may leave reaction product OD unaffected. Alternatively, increased expression per cell may increase intensity independently of fractional area. Finally, "total immunoreactivity" may be calculated as the product of fractional area and reaction product OD, and is a useful measure if both parameters change. "Integrated OD" represents an analogous (although not necessarily identical) measurement.

It is also possible to perform video densitometry without segmentation. For example, tyrosine hydroxylase immunoreactivity in the neostriatum is localized in terminals of the dopaminergic afferents. At relatively low magnification, individual synaptic boutons cannot be discriminated, and staining appears homogeneous. The OD of each pixel in a circumscribed region may be averaged to produce a measurement of field OD. This parameter may be calibrated with a photographic density step wedge, a strip of photographic film with a defined gradient of OD.

Publications are available that may be especially useful for developing methodology (55–58). Failure to employ the types of precautions and care described above will introduce variability between sections that will compromise the ability to discriminate changes in marker expression.

SUMMARY

Taken together, design-based stereology and video densitometry provide powerful methods to assess toxicological effects at the regional, cellular, and subcellular levels in biological tissue. Like a well-designed survey, design-based stereology uses a combination of unbiased sampling to identify which element of the population to analyze, and unbiased probes to make accurate and precise estimates about parameters of interest from that element. Carrying out each step of these processes in an unbiased manner ensure the systematic error is eliminated, or at least minimized to a negligible level. When applied to test whether toxins lead to neurotoxicological effects on specific proteins, design-based stereology ensures that estimates stereological parameters, including numbers of cells, lengths of fibers and blood vessels, surface areas of objects, and volumes of objects and regions, without assumptions related to the geometry of the biological features of interest. Video densitometry uses a semi-quantitative approach, the relative transmission of light through tissue sections, to test the same hypotheses. For both cases the elimination of known sources of error, combined with well-planned sampling and consistent tissue processing, allow the investigator to draw strong and well-supported conclusions about the biological effects of potentially toxic compounds.

REFERENCES

1. Mouton PR. Principles and Practices of Unbiased Stereology: An Introduction for Bioscientists. Baltimore: The Johns Hopkins University Press, 2002.
2. Howell K, Hopkins N, McLoughlin P. Combined confocal microscopy and stereology: a highly efficient and unbiased approach to quantitative structural measurement in tissues. Exp Physiol 2002; 87: 747–56.
3. Hilliard JE, Lawson LR. Stereology and Stochastic Geometry (Computational Imaging and Vision Series), 1st edn. London: Springer, 2003.
4. Baddelay A, Jensen EV. Stereology for Statisticians. In: Monographs on Statistics and Applied Probability, 1st edn. Boca Raton: Chapman & Hall/CRC, 2004.
5. Howard CV, Reed MG. Unbiased Stereology: Three-Dimensional Measurement in Microscopy (Advanced Methods), 2nd edn. London: Taylor & Francis, 2005.
6. Nyengaard JR, Gundersen HJ. Direct and efficient stereological estimation of total cell quantities using electron microscopy. J Microsc 2006; 222: 182–7.
7. Berman RF, Pessah IN, Mouton PR, et al. Low level neonatal thimerosal exposure: further evaluation of altered neurotoxic potential in SJL mice. Toxicol Sci 2008; 101: 294–309.

8. Havarinasab S, Haggqvist B, Bjorn E, et al. Immunosuppressive and autoimmune effects of thimerosal in mice. Toxicol Appl Pharmacol 2005; 204: 109–21.
9. Hornig M, Chian D, Lipkin WI. Neurotoxic effects of postnatal thimerosal are mouse strain dependent. Mol Psychiatry 2004; 9: 833–45.
10. Gundersen HJG, Jensen EB. The efficiency of systematic sampling in stereology and its prediction. J Microsc 1987; 147: 229–63.
11. Wicksell SD. The corpuscle problem. A mathematical study of a biometric problem. Biometrika 1925; 17: 84–99.
12. DeHoff JF, Rhines FN. Determination of the number of particles per unit volume from measurements made on random plane sections: the general cylinder and ellipsoid. Trans Mettlur Soc AIME 1961; 221: 975–82.
13. Abercrombie M. Estimation of nuclear population from microtome sections. Anat Rec 1946; 94: 239–47.
14. Weibel ER, Gomex DM. A principle for counting tissue structures on random sections. J Appl Physiol 1962; 17: 343–8.
15. Ebbeson SOE, Tand DA. Method for estimation of the number of cells in histological sections. J Royal Microsc Soc 1965; 84: 449–64.
16. Sterio DC. The unbiased estimation of number and sizes of arbitrary particles using the disector. J Microsc 1984; 134: 127–36.
17. Gundersen HJG. Notes on the estimation of the numerical density of arbitrary profiles: the edge effect. J Microsc 1977; 111: 219–23.
18. Gundersen HJG. Stereology of arbitrary particles. A review of number and size estimators and the presentation of some new ones, in memory of William R. Thompson. J Microsc 1986; 143: 3–45.
19. Colonnier M, Beaulieu C. Counting of synaptic disks in the cerebral cortex. J Comp Neurol 1985; 231: 175–9.
20. West MJ, Slomianka L, Gundersen HJG. Unbiased stereological estimation of the total number of neurons in the subdivisions of the rat hippocampus using the optical fractionator. Anat Rec 1991; 231: 482–97.
21. Vijayanshankar N, Brody H. A study of aging in the human abducens nucleus. J Comp Neurol 1977; 1: 433–8.
22. Vijayanshankar N, Brody H. Aging in the human brain stem. A study of the nucleus of the trochlear nerve. Acta Anat 1977; 99: 169–72.
23. Vijayanshankar N, Brody H. A quantitative study of the pigmented neurons in the nuclei locus coeruleus and subcoeruleus in man as related to aging. J Neuropathol Exp Neurol 1979; 38: 490–7.
24. Haug H, Kuhl S, Mecke E, et al. The significance of morphometric procedures in the investigation of age changes in cytoarchitectonic structure of human brain. J Hirnforsch 1984; 25: 353–74.
25. Dam AM. Brain shrinkage during histological procedures. J Hirnforsch 1979; 20: 115–19.
26. West MJ. New stereological methods for counting neurons. Neurobiol Aging 1993; 14: 275–85.
27. West MJ. Stereological methods for estimating the total number of neurons and synapses: issues of precision and bias. Trends Neurosci 1999; 22: 51–61.
28. Mouton PR, Pakkenberg B, Gundersen HJG, et al. Absolute number and size of pigmented locus coeruleus neurons in the brains of young and aged individuals. J Chem Neuroanat 1994; 7: 185–90.
29. Pakkenberg B, Gundersen HJG. Neocortical neuron number in humans: effects of sex and age. J Comp Neurol 1997; 384: 312–20.
30. Long JM, Mouton PR, Jucker M, et al. What counts in brain aging? Design-based stereological analysis of cell number. J Gerontology 1999; 54A: B407–17.
31. Cavalieri B. Geometria Indivisilibus Continuorum. Typis Clementis Bonoiae. Reprinted 1966 as Geometris degli Indivisibili. Torinese, Torino: Union Tipografico-Editrice, 1635.
32. Cajori F. The history of notations of the calculus. Annals of Mathematics 1923; 25: 1–46.
33. Mouton PR, Martin LJ, Calhoun ME, et al. Cognitive decline strongly correlates with cortical atrophy in Alzheimer's dementia. Neurobiol Aging 1998; 19: 371–7.
34. Subbiah P, Mouton PR, Fedor H, McArthur J, Glass JD. Stereological analysis of cerebral atrophy in human immunodeficiency virus-associated dementia. J Neuropath Exp Neurol 1996; 55: 1032–7.
35. Marcario JK, Manaye KF, Mouton PR, et al. Severe subcortical degeneration in macaques infected with neurovirulent simian immunodeficiency virus. J Neurovirol 2004; 10: 1–13.
36. Mouton PR, Price DL, Walker LC. Empirical assessment of total synapse number in primate neocortex. J Neurosci Methods 1997; 75: 119–26.
37. Perry TA, Weerasuriya A, Mouton PR, et al. Pyridoxine-induced toxicity in rats: a stereological quantification of the sensory neuropathy. Exp Neurol 2004; 190: 133–44.

38. Perry TA, Holloway HW, Mouton PR, et al. Evidence of GLP-1-mediated neuroprotection in an animal model of pyridoxine-induced peripheral sensory neuropathy. Exp Neurol 2007; 203: 293–301.
39. Lei D-L, Long JM, Hengemihle J, et al. Effects of estrogen and raloxifene on neuroglia number and morphology in the hippocampus of aged female mice. Neuroscience 2003; 121: 659–66.
40. Long JM, Kalehua AN, Muth NJ, et al. Stereological analysis of astrocyte and microglia in aging mouse hippocampus. Neurobiol Aging 1999; 19: 497–503.
41. Mouton PR, Long JM, Lei D-L, et al. Age and gender effects on microglia and astrocyte number in brains of mice. Brain Res 2002; 956: 30–5.
42. O'Neil JN, Mouton PR, Tizabi Y, et al. Catecholaminergic neuron number in locus coeruleus of aged female dtg APP/PS1 Mice. J Chem Neuroanat 2007; 34: 102–7.
43. Buffon GL. Compte de. Essai d'arithmetique morale. Suplement al Historie Naturelle, Vol. 4. Paris: Impremarie Royale, 1777.
44. Weibel ER, Stäubli W, Gnägi HR, Hess FA. Correlated morphometric and biochemical studies on the liver cell. I. Morphometric model, stereologic methods, and normal morphometric data for rat liver. J Cell Biol 1969; 42: 68–91.
45. Loud AV. A quantitative stereological description of the ultrastructure of normal rat liver parenchymal cells. J Cell Biol 1968; 37: 27–46.
46. Mandelbrot BB. Form, Chance, and Dimension. New York: W.H. Freeman & Co., 1977.
47. Paumgartner D, Losa G, Weibel ER. Resolution effect on the stereological estimation of surface and volume and its interception in terms of fractal dimension. J Microsc 1981; 121: 51–63.
48. Baddeley A, Gundersen HJG, Cruz-Orive L. Estimation of surface area from vertical sections. J Micros 1986; 142: 259–76.
49. Gokhale AM, Evans RA, Mouton PR. Design-based estimation of surface area in thick tissue sections of arbitrary orientation using virtual cycloids. J Microsc 2004; 216: 25–31.
50. Mouton PR, Gokhale AM, Ward NL, et al. Stereological length estimation using spherical probes. J. Microsc 2002b; 206: 54–64.
51. Coggeshall R, Lekan H. Methods for determining numbers of cells and synapses: a case for more uniform standards of review. J Comp Neurol 1996; 364: 6–15.
52. Saper CB. Any way you cut it: a new journal policy for the use of unbiased counting methods. J Comp Neurol 1996; 354: 5.
53. Fox CH, Johnson FB, Whiting J, et al. Formaldehyde fixation. J Histochem Cytochem 1985; 33: 845–53.
54. Defigueiredo RJP, Cummings BJ, Mundkur PY, et al. Color image analysis in neuroanatomical research: application to senile plaque subtype quantification in Alzheimer's disease. Neurobiol Aging 1995; 16: 211–23.
55. Gordon MN, Kumar S, Espinosa de los Monteros A, et al. Ontogeny of glycerol phosphate dehydrogenase-positive oligodendrocytes in rat brain. Impaired differentiation of oligodendrocytes in the myelin deficient mutant rat. Int J Dev Neurosci 1992; 12: 243–53.
56. Gordon MN, Schreier WA, Ou X, et al. Exaggerated astrocyte reactivity after nigrostriatal deafferentation in the aged rat. J Comp Neurol 1997; 388: 106–19.
57. Gordon MN, Holcomb LA, Jantzen PT, et al. Time course of the development of Alzheimer-like pathology in the doubly transgenic PS1+APP mouse. Exp Neurol 2002; 173: 183–95.
58. Wilcock DM, Gordon MN, Morgan D. Quantification of cerebral amyloid angiopathy and paenchymal amyloid plaques with Congo red histochemical stain. Nat Protoc 2006; 1: 1591–5.

11 Effect of Neurotoxins on Brain Tissue and Function Revealed Through In Vivo MR Imaging and Spectroscopy

Margaret J. Rosenbloom

Department of Psychiatry and Behavioral Sciences, Stanford University School of Medicine, Stanford, and Neuroscience Program, SRI International, Menlo Park, California, U.S.A.

Edith V. Sullivan

Department of Psychiatry and Behavioral Sciences, Stanford University School of Medicine, Stanford, California, U.S.A.

Adolf Pfefferbaum

Department of Psychiatry and Behavioral Sciences, Stanford University School of Medicine, Stanford, and Neuroscience Program, SRI International, Menlo Park, California, U.S.A.

INTRODUCTION

The ten years since the second issue of Neurotoxicology was published have brought significant advances in neuroimaging techniques. However, the potential of in vivo neuroimaging to characterize and quantify deleterious effects of neurotoxin exposure on the brain and identify mechanisms for their effects remains to be fully realized. In this chapter we focus primarily on in vivo magnetic resonance imaging (MRI) that yields several imaging formats. Conventional structural MRI images are amenable to sensitive and reliable measurements of macrostructure—the size, shape, and some aspects of tissue quality of neuroanatomical structures, such as the basal ganglia, cerebellum, and hippocampus, which are often sites of neurotoxic damage. Structural images also reveal areas where the signal intensity is higher or lower than adjacent tissue within a structure, suggesting some form of pathology. diffusion tensor imaging (DTI) enables detection of microstructural deficits in white matter tracts, a feature of brain architecture particularly vulnerable to certain neurotoxins. Magnetic resonance spectroscopy (MRS) provides the means to assess a range of brain metabolites that can reflect neurotoxic effects on cell bodies, axonal constituents, and glia. Functional magnetic resonance imaging (fMRI) provides a proxy measure of localized changes in blood volume during brain activity and can reflect reorganization of functioning in response to neurotoxic damage. Other imaging techniques, such as positron emission tomography (PET) and single photon emission computerized tomography (SPECT), have also been used to delineate physiological, gene expression, and biochemical or cellular mechanisms for neurotoxic effects. The reader is referred elsewhere, however, for studies using these techniques and particularly for emerging applications of micro-PET scanning, which provides higher resolution data but is primarily confined to research settings and with animal models of neurotoxicity (1,2). In all studies, the contribution of nonspecific variables, such as age, gender, poor nutrition, seizure disorder, high blood pressure, or psychopathology to observed neurotoxic pathology remains a significant challenge. Several new studies have provided evidence linking a history of toxic exposure, behavioral deficits, and macroscopic, microscopic, or functional brain compromise.

IN VIVO NEUROIMAGING TECHNIQUES

MRI: Morphology and Intensity Differences

Magnetic Resonance brain imaging employs electromagnetic radiofrequency pulses to perturb atomic nuclei throughout the brain, appropriately tuned radiofrequency coils to detect the radio signals given off as atomic nuclei return to equilibrium, and magnetic field gradients applied in the three orthogonal axes to localize the signals. One of the strengths of MRI is that investigators can manipulate the amount of contrast between different types of biological tissue by varying the time between radiofrequency pulses (TR) and the echo time (TE) at which the signal is acquired. Gray matter, white matter, and CSF each produce somewhat different MRI signal intensities based on their biological properties, especially the amount of unbound water and the local environment in each. Conventional structural MRI takes advantage of these differences [gray matter is about 80% water, white matter is about 70% water, and cerebrospinal fluid (CSF) is essentially 100% water] to reveal the size, shape, and tissue composition (gray vs. white matter) of brain cortical and subcortical regions, and the size of the fluid-filled spaces (ventricles and cortical sulci) as well as of specific neuroanatomic structures such as the cerebellum, corpus callosum, hippocampus, and basal ganglia. The differing water composition of different brain tissues means that certain combinations of TR and TE times (e.g., a short TR (200–400 msec) short TE (20 msec) T1-weighted sequence) enhance the contrast between white and gray matter and bring the complex structures of the brain into focus, while other combinations (e.g., a long TR (2–3 sec), long TE (80 msec) T2-weighted sequence) will enhance CSF (CSF has a long T2) but reduce the contrast between tissue types. For brain imaging to assess global changes in volumes of white matter, gray matter, and CSF, the choice of T1 versus T2 weighted images presents a tradeoff between obtaining high contrast between gray matter and white matter and obtaining high contrast between CSF and brain parenchyma. An optimal study sequence for analysis focusing on neurotoxic effects in the morphometric characteristics of specific brain regions and structures includes both T1- and T2-weighted images. Many techniques have been developed to measure the size and shape of brain regions and structures with volumetric measurements replacing the earlier emphasis on area or linear measurement (3,4).

Structural MRI also reveals changes in signal intensity in focal regions. The basal ganglia nuclei, targeted by certain neurotoxins (5), are particularly prone to manifest signal abnormalities that can be enhanced for analysis by manipulating image acquisition parameters, even though the mechanisms by which specific toxins differentially affect signal intensities on T1 versus T2-weighted images are still not fully understood. Observed signal intensity differences (hyper- vs. hypo-) may be due merely to differences in acquisition sequence or, for a given acquisition weighting, may reflect different physiological mechanisms and processes. For example, methanol intoxication produces edematous changes secondary to hemorrhagic changes in the putamen. These lesions appear hyperintense on T2-weighted images, but hypo-intense on T1-weighted images (6). By contrast, manganese toxicity, whether induced by exogenous exposure (7) or indirectly from chronic liver disease(8), is manifest by hyper-intensity of the globuspallidus on T1-weighted images (Fig. 1), while iron accumulation, either as a normal developmental process (9,10) or associated with certain neurological diseases (11,12), produces hypo-intensity of the striatum on T2-weighted images (13). Exposure to the industrial solvent toluene has also been associated with hypo-intensity on T2 but not T1 weighted images (14). The T2-weighted image hypointensity may well be attributable to actual presence of toluene in the lipid-rich cell membranes of this region (15). That is, T2-weighting apparently has greater sensitivity than T1-weighting to membrane lipid packing and its interaction with water content (16).

Although some quantification approach is almost universally adopted for analyzing brain morphometry, quantification of signal intensity abnormality, especially in a clinical setting, typically employs a visual ranking scheme. This approach can be problematic given poor standardization of absolute signal intensity from scan to scan, but experts can make the necessary adjustment for inter scan variations to rank overall intensity on a simple scale (e.g., 0 = no signal; 1 = minimal; 2 = marked) (17). A preferable approach is to measure an absolute value such as T1 relaxation time in a target region but longer scanning time and additional processing are needed to generate quantitative T1 assessments. A common approach in the neurotoxicology literature is to express the intensity value for the target region, frequently the globuspallidus, as a ratio of signal intensity of subcortical white matter (assumed to be relatively immune from toxic accumulation), e.g., the Pallidal Index (PI) (18). The assumption that certain brain regions are in fact immune to toxic accumulation and can be used as a neutral denominator in the computation of any index has more recently been questioned by studies of non-human primates in which accumulations of manganese were detected not only in the basal ganglia, but widely throughout the brain, including frontal white matter (19,20). Recently, Choi and colleagues (21) compared quantitative pallidal T1 measurements with subjective PI and intensity ratings in a sample of manganese workers with well-documented cumulative lifetime exposure as well as current manganese blood levels. They found associations between PI and T1 only at higher exposure levels, suggesting a lack of sensitivity for the PI approach to detecting lower toxicity accumulations.

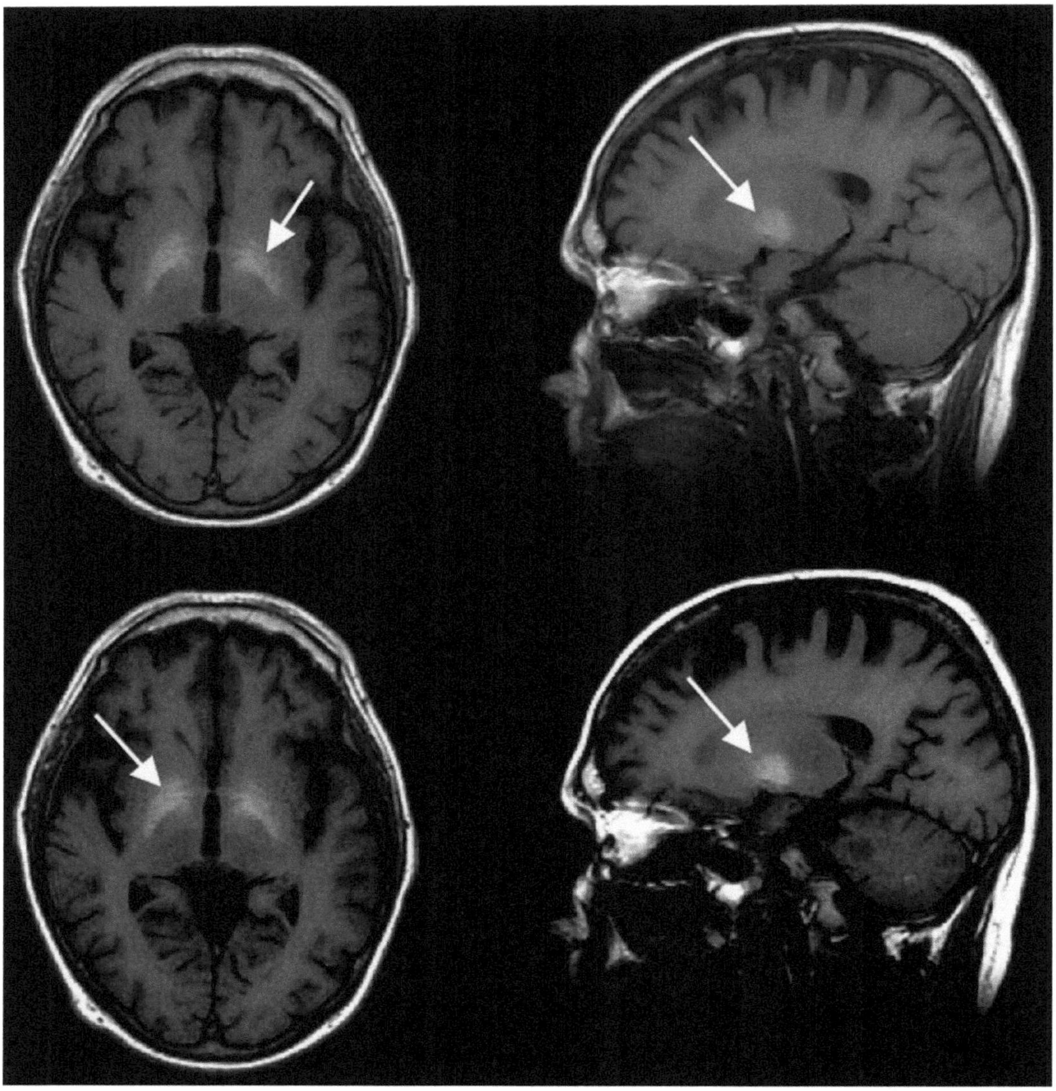

Figure 1 Hyperintensity of the globuspallidus (arrows) on a T1-weighted series (axial—*left* and sagittal—*right*) from a patient with alcohol-related cirrhosis of the liver.

DTI

Although revealing of neuroanatomical macrostructure, detailed conventional MRI does not clearly reflect the microstructure of brain tissue and its components, such as axons, microtubules, and myelin or the physical characteristics of white matter fibers, which vary widely in length, diameter, and myelination and by region (22). In contrast to conventional structural MRI, DTI measures the diffusion of water molecules within brain cells and in extracellular spaces. In white matter, diffusion is highly restricted, has a preferential orientation along the linear framework of the fiber, and is considered anisotropic. By contrast, unrestricted diffusion, as occurs in ventricular space, has no preferential orientation and is isotropic. Anisotropy is calculated on a within-voxel basis and is commonly expressed as a percent, that is, fractional anisotropy (FA) (23), ranging from 0 for CSF-filled spaces to 1 for perfectly organized, parallel bands of white matter. Diffusivity, expressed as apparent diffusion coefficient (ADC) or mean diffusivity (MD), provides a measure of the amount of water motility (independent of orientation) in a voxel and is commonly but not necessarily negatively correlated with FA within white matter samples (24,25). Figure 2 shows images of the brain obtained with DTI in which the white matter skeleton is emphasized. Aging, disease, or physical trauma can perturb the axon's cytoskeleton and myelin microstructure (26,27), and this disruption can be detected with DTI (28,29). Detailed reviews on DTI methods can be found elsewhere (25,30,32). As a relatively new technology, DTI quantification of white matter integrity has only recently been applied to the field of neurotoxicology, particularly the effects of pesticides such as sarin (33,34), presented in more detail below. DTI would be particularly appropriate for studies of effects of certain metals and inhalants for which white matter lesions of various sorts have been observed.

MRS

Structural MRI and DTI each reflect different aspects of water protons in the brain. By contrast, water (1H) MRS can detect other chemical substances by examining the chemical shift of their signal relative to the water signal and has the potential to detect neurotoxic effects of metals, solvents, pesticides, and drugs of abuse (30). Metabolite signals detected by MRS reflect regional biochemical health and yield biomarkers of disease and neurotoxic damage. Unlike MRI and DTI that each can yield high resolution images in two- and three-dimensions of the brain based on voxels as small as 1 mm^3, MRS typically requires larger voxels (2–12 cm^3) to detect metabolite signals (30). Spectroscopic values thus are often reported from discrete regions of interest (ROI) located in target brain structures such as the basal ganglia, frontal lobe, or cerebellum.

Among the spectroscopically-visible metabolites are *N*-acetyl aspartate (NAA), a marker of living, mature neurons and thus a potential indicator of neuronal and axonal integrity; myo-Inositol (mI), a marker of glial activity and gliosis; choline (Cho), reflecting cell membrane turnover; and creatine+phosphocreatine (Cr), a complex composite that reflects high-energy phosphate metabolism. Figure 3 shows an example of the spectrum encompassing these metabolites based on a voxel placed in the basal ganglia. For reasons related to the difficulty with early technology in obtaining absolute values of these low signal metabolites, many reports express NAA, mI, and cho as a ratio of Cr. Thus reduced NAA/Cr is interpreted as an indicator of reduced neuronal integrity, increased mI/Cr as an indicator of gliosis, and Cho/Cr as an indicator of enhanced cell membrane activity—all markers that can supplement MRI's gross assessment of reduced white or gray matter volume and also DTI's assessment of white matter integrity. The ratio approach has yielded useful data but is limited by its assumption that creatine is stable regardless of age, disease, or brain region. In fact, creatine concentration increases with age (35,36) and disease, including Alzheimer's disease (37) and alcoholism (38), varies by region of the brain (39), and is a significant source of metabolite measurement error (35). Thus measurement of absolute values is preferred.

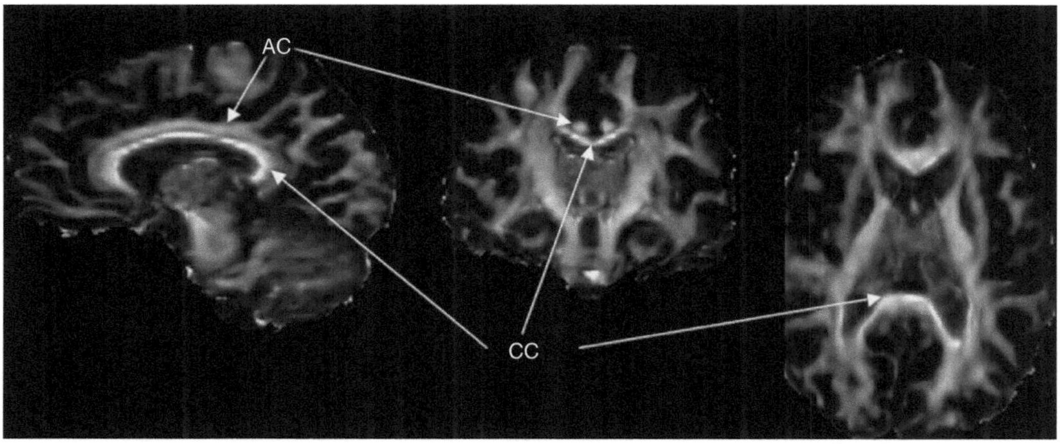

Figure 2 Sagittal, coronal, and axial fractional anisotropy (FA) images calculated from diffusion tensor imaging acquired in a 24-year-old woman. *Abbreviations*: AC, anterior cingulate bundle; CC, corpus callosum.

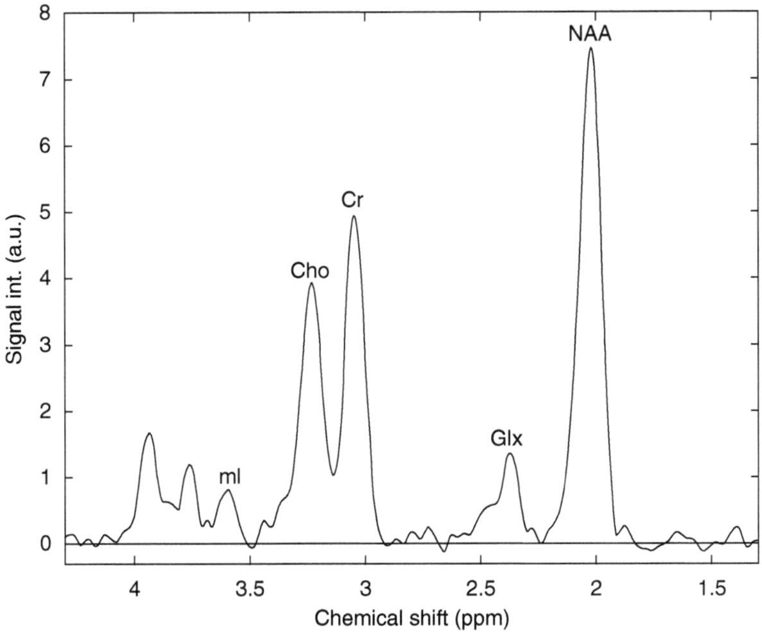

Figure 3 Magnetic resonance spectrum of a single voxel placed in the basal ganglia of a 20-year-old man. *Abbreviations*: Cho, choline; Cr, creatine+phosphocreatine; Glx, glutamate+glutamine; mI, myoInositol; NAA, *N*-acetyl aspartate.

Glutamate is another substance of considerable relevance for investigating excitotoxic and inflammatory mechanisms resulting from neurotoxic exposure (40,41). Its low signal to noise ratio and proximity to related resonances require a specialized acquisition sequence (42), and while a recent study has demonstrated the feasibility of using MRS to detect functionally relevant age-related differences in glutamate in the basal ganglia in healthy individuals (39), it has not yet been applied to assess effects of neurotoxins. By contrast, investigators are increasingly using proton MRS measurements of NAA, mI, and Cho to assess localized impact of neurotoxic exposure on neuronal integrity, gliosis, or cell membrane activity as described in more detail below.

fMRI

Functional MRI (fMRI) exploits the MRI-visible signal contrast between oxygenated (higher signal) and deoxygenated (lower signal) hemoglobin as it flows through small blood vessels in given brain regions. More activity increases the ratio of oxygenated to deoxygenated hemoglobin in the blood, enhancing the signal. This blood oxygen level dependent (BOLD) contrast mechanism does not directly measure blood flow or neuronal activity but rather the small, rapid changes in its paramagnetic properties that can be imaged by rapid sampling over the spatial domain (43). Changes in levels of originated hemoglobin in blood vessels, the hemodynamic response that occur in response to experimental manipulations, affect the local homogeneity of an MR signal. The BOLD effect is localized by measuring the difference between oxygenation time-locked to a specific cognitive or motor operation to that occurring during a rest period or another activity. The regions of the brain showing greatest difference between active and contrast conditions are believed to be those most involved in performing the operation under investigation (44,45). Further technical details about fMRI can be found in specialized reviews (30,46,47). Since first developed in the early 1990s, fMRI is now widely used to assess mental activities ranging from simple visual processing (48), inhibitory processes (49), expectations (50), memory (51), craving (52), and economic decision making (53). It is now also beginning to be applied in the field of neurotoxicology, for example to assess developmental cognitive abnormalities resulting from early childhood toxic exposure (54), or to study craving for drugs or the sequelae of drug abuse (55–57).

TYPES OF TOXICS AND SPECIFIC NEUROTOXIC MANIFESTATIONS

The spectrum of toxic substances known to cross the blood brain barrier and damage various parts of the brain, either directly, or secondary to hypoxic, ischemic, cancer-inducing, or other untoward events, is vast. Here we present the neuroradiological contribution to the study of neurotoxicity from the perspective of type of toxin rather than region of brain or type of brain cell affected. Neuroimaging studies of the neurotoxic effects of different chemicals include investigations of both acute (accidental or self-inflicted) and chronic (abusive or occupational) exposures. The chemical agent defines the type of damage produced—the quantity, frequency, and pattern of exposure defines the degree of damage—and the reversibility of any change depends upon all of these parameters. Unfortunately, clinical studies of chemical exposures are frequently hampered by limited or unreliable information on the specific agents involved as well as the quantity, frequency, and pattern of exposure. Neuroradiological techniques and methodological approaches used also vary considerably across this literature, testament to the enormous difficulty of performing clinical studies of neurotoxic exposure and the serendipitous nature of many observations. The clinical researcher is not at liberty to administer neurotoxins according to defined doses and schedules. Furthermore, the objects of exposure, unlike experimental animals, come in all ages, have experienced varying life-time exposures to the toxins under investigation, and are heterogeneous with regard to numerous other variables which also affect brain structure and function.

Metals

There are over 40 metallic elements in the environment. Some such as copper, iron, and zinc are essential to life, while others such as mercury, lead, aluminum, and cadmium are not, and indeed can induce neurological dysfunction even with low level chronic exposure. Many of these metals can take different forms (vapors, salts, organic, or inorganic compounds) and can be absorbed by inhalation, cutaneous exposure, or ingestion. Whereas postmortem neuropathological techniques can assess localized metal concentration in brain tissue or other microscopic manifestations of damage, metal neurotoxicity must either alter water proton concentration or mobility to affect signal intensities or produce gross changes in gray or white matter volume to be visible in vivo on conventional structural MRI. Over the past ten years, MR spectroscopy and MR spectroscopic imaging have been added to the armamentarium of in vivo neuroimaging techniques to characterize levels of various metabolites reflecting aspects of neuronal and glial functioning in local brain areas.

Manganese toxicity resulting from iatrogenic or occupational exposures has been widely studied using in vivo neuroimaging (58–60). Manganese preferentially accumulates in the globuspallidus (61) and its paramagnetic properties enhance the intensity of the signal seen on T1-weighted images by shortening the proton T1 relaxation time. Thus T1-weighted images are particularly useful for identifying areas of hyperintensity in basal ganglia structures in patients with iatrogenically, occupationally, or endogenously induced manganism, including children receiving long-term parenteral nutrition, which typically is manganese enriched (7,62), patient with hepatic encephalopathy (18,63), or factory workers with manganese exposure (64,65). These hyperintensities have been attributed to the accumulation of manganese in the globuspallidus, a finding consistent with manganese blood levels (7,18,21,63) and postmortem manganese concentrations in caudate nucleus and globuspallidus (18,66). Hyperintensities have also been related to upper (65) and lower (67) motor function in occupationally exposed workers, consistent with the known role of basal ganglia in control of movement.

Recently, spectroscopic studies of manganese toxicity have been undertaken to determine whether regions of hyperintensity demonstrate spectroscopic evidence of neuronal damage. One study (67) found no group differences in basal ganglia NAA/Cr between exposed and unexposed workers. However, when participants were segregated on the basis of smoking status, a strong relationship emerged between NAA/Cr and cumulative manganese exposure in the non-smokers. The authors speculate that the lack of association in smokers may indicate a protective effect of nicotine, either through its antioxidative properties or its neuroprotective effect on dopaminergic neurons.

Iatrogenic *aluminum* toxicity, such as in dialysis encephalopathy resulting from aluminum containing phosphate binding gels, or even the high aluminum content of dialysis water, is associated with elevated serum aluminum levels, impaired cognitive function, and increased concentration of aluminum in cortical tissue at autopsy (68). The aluminum lesion is less readily MRI-visible than other metals, consisting most probably of subtle decrease in tissue volume, rather than tissue quality alteration, which would produce an abnormal signal on MRI. Thus, in vivo neuroimaging observations of aluminum neurotoxicity are less commonly reported than behavioral observations, particularly in industrial workers. A meta-analysis (69) found some evidence that urinary aluminum concentrations below 135 mg/L can have an impact on cognitive performance but notes the need to pay more attention to confounding variables. A well designed follow-up study, however, found no support for cognitive deterioration over a four-year period in aluminum-exposed workers (70).

Inhaled *mercury* vapor is readily absorbed and retained by the brain. Animal models and autopsy data demonstrate distribution to occipital and parietal gray matter, brainstem, and cerebellar cortices (71,72). The neurotoxic effects of mercury have been long recognized. For example, Mad Hatters Disease was an occupational hazard for 18th and 19th century hat makers due to the use of a mercury solution to turn fur into felt. Neuroimaging technology played an important role in illustrating the neurotoxic effects of the organomercury compound methylmercury following large-scale outbreaks of poisoning from consumption of methylmercury contaminated fish in Japan in the 1950s and 1960. The slow diffusion of methylmercury into the brain accounts for the delay in the manifestation of neurological signs of poisoning to several weeks after exposure. Postmortem and in vivo CT (73) and MRI (74,75) studies have demonstrated predictable associations between tissue atrophy in the calcarine cortex and cerebellum and prominent symptoms of mercury poisoning, constriction of visual fields and ataxia. These studies of people exposed to high doses of methyl mercury 10 to 20 years earlier indicate that much of the damage is irreversible (76,77). More recently, neurobehavioral studies of the effects of low levels of methyl mercury exposure through consumption of contaminated fish in adults show impaired psychomotor performance and response inhibition (78,79). Neurodevelopmental effects for children, born to mothers with high blood levels, are an increasing source of concern, although a link between prenatal methylmercury exposure and developmental outcome has not been clearly established (80). Mechanisms for neurodevelopmental effects using in vitro and animal models have been recently reviewed (81), but no further applications of in vivo neuroimaging techniques to delineate effects or elucidate mechanisms for methyl mercury toxicity in the brain appear to have been reported.

Compared with mercury, exposure to *lead*, either organic or inorganic, is a more widespread phenomenon, and one that has prompted several in vivo brain imaging studies as well as many investigations of cognitive effects, particularly in children. Lead usage increased progressively with industrialization and rose dramatically with the introduction of lead-acid batteries and leaded fuel for automobiles in the 20th century. The predominant use of lead is now in batteries and, to a lesser extent, in construction materials, including paint, and lead-based chemicals. The use of lead in pipes, paints, and gasoline has been or is being phased out in many countries, but environmental and occupational exposure remain significant public health concerns.

Bleecker (82) reports an association, after covarying for age, education, cholesterol, between lifetime lead exposure of 61 lead smelter workers under the age of 50, assessed from work history blood levels as well as current lead bone levels, with presence or absence of white matter hyperintensities, assessed by a neuroradiologist from T2 weighted MR images. The relatively young age of these workers, before normal aging hyperintensities typically appear, adds significance to these findings. (Figure 4 shows an example of hyperintensities on a T2-weighted image in an otherwise healthy elderly woman.) Furthermore, both bone lead and white matter hyperintensities were associated with performance on the Grooved Pegboard task, which assessed upper motor speed, finger dexterity, and eye-hand coordination. A series of reports on more than 500 former (average time since last exposure = 18 yrs) organo-lead manufacturing workers also focused on associations between lead exposure, cognitive performance, and the brain. In these studies, both white matter lesions and volumetric measures of different brain areas were the measure of neurotoxicity (83–85). After adjusting for age, height, education, and smoking status, lead exposure—extrapolated from current bone lead level back to time of last exposure to reflect a peak lead level—was associated with incidence and severity of white matter lesions, as well as, smaller volumes of global white and gray matter with strongest effects for frontal gray and parietal white volume (84). Several domains of concurrent cognitive function of these workers were also associated with peak lead levels (86) and further

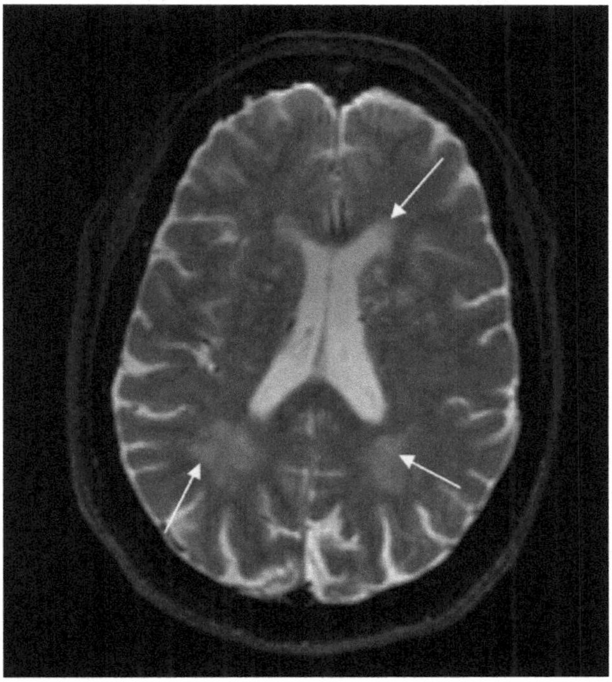

Figure 4 Axial late echo fast spin echo (TR/TE = 7850/96 msec) image from a 77-year-old woman volunteer with numerous white matter hyperintensities (arrows).

analysis (85) using both anatomically—and statistically—defined regions of interest revealed a number of associations between MRI measured brain volumes and specific cognitive domains. Path analyses (85) revealed associations between visuospatial construction and frontal gray matter, executive function and gray matter, and eye-hand coordination and white matter volumes that supported the hypothesis that the effects of lead exposure on these cognitive function were mediated by changes in regional brain volumes.

Further evidence suggestive that the association between chronic lead exposure and cognitive deficit is mediated by neuronal loss comes from an MRS study of a pair of monozygotic twins, both of whom were occupationally exposed to lead. While both twins scored below norms on memory and executive function tests, the twin with a history of higher lead exposure, due to a different job classification, not only scored worse than his brother but also had lower NAA/Cr ratios in hippocampus and frontal lobes (87). MRS has also been used to assess effects of chronic low-level lead accumulation in a healthy elderly sample without known occupational lead exposures—the Veterans Administration Normative Aging Cohort (88). Participants with the highest and lowest quintiles of patella bone lead were compared. Unlike the twin case study that found a difference in NAA/Cr between the higher and lower lead exposed twin (87), this population based study found differences between high and lower lead-exposed participants in creatine referenced myoInositol (mI), which is a marker for glial activity. Thus, evidence for greater mI/Cr in higher than lower lead-exposed individuals is consistent with animal and in vitro studies showing lead accumulation in glia, and may reflect the lingering effects of chronic low levels of environmental lead exposure.

Exposure of children to lead is an increasingly pressing public health concern. Despite a significant literature on cognitive developmental effects (89–91), no study has yet been published using MRI to trace morphometric effects in children exposed from an early age to lead. Two studies have used MRS to assess these effects in exposed children. In one study, six children living near a lead recycling plant in China were compared with six children from an area with no known source of pollution (92) using IQ testing and MRS. The lead-exposed children had a mean blood level seven times that in the unexposed children, lower total and verbal IQ scores, and reduced bilateral levels of NAA, Cr, and Cho in the frontal lobes and hippocampus. Another study also found lower levels of frontal gray matter NAA/Cr in lead-exposed children than their lead-free siblings (93). Interestingly, current lead blood levels in the exposed children were not associated with NAA/Cr, suggesting that current blood levels are not an effective measure of cumulative exposure.

Functional MRI (fMRI has been used to assess the impact at young adulthood of early childhood lead exposure on brain activation during performance of a language task. Yuan et al. (54) documented an inverse association between childhood lead levels and brain activation in two regions traditionally associated with semantic language function, the left frontal and middle temporal gyri. In addition, there was a positive association between language function and activation in a right hemisphere (non-dominant for language) region homologous to Wernicke's area in the left parietal cortex. The authors interpret this right hemisphere recruitment to perform a language task as evidence of compensatory activation to overcome earlier lead-induced deficits.

Organic Solvents

The clinical recognition and diagnosis of occupational toxic encephalopathy from solvent exposure was deemed complex and controversial 20 years ago (94) and continues to challenge clinicians (95). The wide range of available industrial solvents, varying mechanisms of action, and difficulties in identifying and measuring exposure levels present serious challenges for characterization and quantification of the input side of the function. Furthermore, the wide range of neurological, neuropsychological, and subjective complaints in chronically exposed workers make it hard to relate specific toxics to specific endpoints. The range of symptoms seen in toxic encephalopathy include fatigue, headaches, memory deficits, depression, anxiety, ataxia, tremors, dizziness, and insomnia, and suggests the involvement of a number of different brain regions. Given the diffuseness and non-specificity of the observed effects, it is particularly important that rigorous methodological controls for sources of non-specific variance be applied if in vivo neuroradiological techniques are to make a useful contribution to the assessment of solvent encephalopathy. Of particular relevance here is age, which will tend to be associated with years of exposure if the test population consists of career workers with occupational exposure to solvents. In addition, other risk factors such as head injuries, hypertension, and alcohol abuse, each of which can independently contribute to the observed structural brain changes need to be identified and either excluded or their contribution to brain structure statistically considered.

Neuroimaging has been widely used to complement neurological and neuropsychological assessments of both acute and chronic solvent neurotoxicity, but almost all published assessments have relied on subjective ratings of hypo- or hyper- intensities or of atrophy rather than quantitative assessments of tissue or CSF-filled volume or DTI measures of white matter integrity. Two recent reviews (96,97) assessed the contribution of in vivo brain imaging assessments in documenting long-term nervous system damage in workers exposed to a variety of industrial solvents as well as in inhalant abusers. These reviews conclude that evidence for any causal relationship is limited by methodological flaws in study design. Unlike toxic metals, for which cumulative exposure measures are available from bone tests, solvents do not leave a cumulative trace. Thus, studies linking brain changes to solvent exposure rely on estimates of duration, intensity, or frequency of exposure or use (98,99) which are significantly less satisfactory estimates of overall exposure than objective biological measures. Furthermore, most of the literature is devoted to case studies rather than controlled comparisons and relies on subjective assessments of abnormality, e.g., atrophy or anomalous signal intensity. Abnormalities reported include focal white matter hyperintensities on T2-weighted images in cerebrum, brain stem, and cerebellum; diffuse white matter changes; cortical, hippocampal and cerebellar atrophy; and thinning of the corpus callosum (97). Dose-dependent relationships have been demonstrated between duration and/or quantity of exposure to solvents and subjectively assessed abnormalities in solvent abusers (98,100,101). Volumes of the entire extent and anterior segment (genu) of the corpus callosum, measured with MRI in railroad workers with an average 24 years of occupational exposure and cognitive complaints (99),were less than in unexposed controls, and correlated with a ranking of severity of exposure. These abnormalities were detected at an average of eight years since last exposure, supporting earlier reports that changes may be irreversible (102), but systematic longitudinal assessments have not been performed.

MR spectroscopic imaging is beginning to be applied to assess effects of solvent exposure and characterize mechanisms, although the results to date are inconsistent, probably because of different sample characteristics as well as measurement from different brain regions. One study compared people who had been abusing toluene on average for four years with non-abusing controls and showed lower NAA/Cr and higher mI/Cr ratios in cerebellum and centrumsemiovale a region of white matter, but not in thalamus, a neuron-rich gray matter structure, that correlated with duration of abuse (103). Cho/Cr did not differentiate the groups. Negative findings for NAA/Cr from the thalamus and for Cho/Cr were interpreted as limited support for neuronal death or demyelination caused by toluene, while positive findings for NAA/Cr and Cho/Cr were interpreted as evidence of axonal depletion and gliosis. Different results were reported in a study comparing shoemakers exposed to toluene on average for 15 years with unexposed controls (104); no effects for NAA/Cr from thalamus, basal ganglia, and parietal white matter, but greater Cho/Cr in these regions. Cho/Cr from the basal ganglia correlated with duration of exposure, and these results were interpreted as indicating demyelination as a mechanism of toxicity. The vulnerability of white matter to solvents (96) makes this form of neurotoxicity particularly amenable to examination using DTI. To date, however, no such studies have been reported.

Pesticides

Unlike solvents, which are commonly abused for their euphorogenic properties, pesticide exposure is generally accidental or occupationally related. Neuroimaging has not yet played a prominent role in the diagnosis or assessment of various pesticide-related neurological syndromes. For organophosphates (including parathion and malathion), the most salient neurological symptoms involve the peripheral rather than the central nervous system (105). Organophosphate poisoning is associated with an acute cholinergic syndrome involving the inhibition of acetylcholinesterase, which causes muscle fiber necrosis. This may be followed by what has been termed organophosphate-induced delayed neuropathy, which appears weeks after apparent resolution of acute symptoms and can range from mild neuropathy to complete lower limb paralysis, persisting for up to six months (105), but is not readily detected by brain imaging.

Sarin and cyclosarin are two forms of organophosphate developed in World War II as chemical weapons that have since been classified by the United Nations as weapons of mass destruction. Neuroimaging studies of the delayed effects of sarin have been performed on victims of the Tokyo subway sarin attack in 1995 (33,34), and U.S. veterans serving in the 1991 Iraq war (106). With high exposure, sarin can kill rapidly through the loss of muscle function. At lower doses, victims recover from acute symptoms but suffer life-long medical and neurological symptoms. Of the 582 identified victims of the Tokyo subway attack, 29 received an MRI approximately five years later. A demographically matched group of controls was also tested. Voxel based morphometry revealed smaller gray volumes in the right insular, temporal cortices, and left hippocampus of the victims than controls.

Pesticide exposure is recognized as one factor in the multifaceted Gulf War Veterans Illness developing in later years among troops serving in the Gulf War of 1991 (107). Two recent studies have examined the delayed effects of sarin and cyclosarin on cognitive performance (108) and volumes of gross neuroanatomic structures seen with MRI (106) in a cohort of veterans who were exposed to different levels of these toxins when they detonated a storage dump. Cognitive assessment occurred 3.5 to 5 years after exposure in 141 veterans, 26 of whom subsequently completed an MR scan protocol 8 to 10 years after exposure. Cumulative exposure levels in mg min/m^3 during the three days of contact were estimated from military records for the units in which each veteran served. MR scanning yielded volumetric measures of overall white matter, gray matter, and CSF adjusted for intracranial volume. Analyses comparing exposed to unexposed veterans with age and current post-traumatic stress disorder (PTSD) as covariates found no group differences. However, analysis of relationship between brain variables and exposure level, after controlling for age and PTSD symptoms, revealed significant effects: The more the exposure, the smaller the white matter volume and larger the ventricles. Organochlorine insecticides (such as DDT, lindane, and dieldrin) are associated with a range of effects, including convulsions, ataxia, and tremor (109), but neuroimaging studies have not reported on CNS deficits associated with DDT exposure in humans.

CONCLUSIONS

In vivo neuroimaging has much to offer in the assessment of central nervous system effects of neurotoxins, with recent developments in MRS and DTI opening up new avenues, not only for identifying outcomes, but also for exploring mechanisms of neurotoxicity and neuroprotection. Applications for neuroimaging range from individual case reports of acute or chronic neurotoxin exposure to stringently designed research investigations comparing samples with well-documented and biologically measurable neurotoxic exposure to demographically matched samples without such exposure. The field is limited, however, more by the inherent difficulties of the naturalistic model—retrospective study of chronic, cumulative effects of poorly defined mixes of toxins rather than prospectively planned experiments of a single toxin (110) —than by technical limitations in imaging capacity. There is also a notable lack of studies that exploit the ability of MRI to track change over time. It is, of course, much less feasible to plan a prospective study of cumulative neurotoxic effects in humans than to monitor rates of recovery after neurotoxic exposure stops, but an effort to use MRI and DTI to evaluate burgeoning concerns over neurodevelopmental risks from environmental toxic exposure would be particularly promising.

Neuroimaging studies presented in this review have documented in vivo brain manifestations of neurotoxic exposure such as abnormalities in size of brain structures, signal intensity reflecting tissue quality, particularly in the basal ganglia, spectroscopically visible metabolites reflecting neural integrity, and indicators of white matter integrity. The location and pattern of some of these signs can pinpoint brain mechanisms or lesions responsible for behavioral deficits. If scanning is repeated over time, successive images can demonstrate the build-up and resolution of neurotoxic exposure. As many observed abnormalities may be caused by other factors, neuroimaging alone cannot be diagnostic for a particular form of toxic exposure. To date, the power of neuroimaging, particularly longitudinal study with MRS and DTI, has not been fully harnessed to document neurotoxicity, especially the pervasive subtle changes that accrue from inadvertent low level chronic exposure to environmental toxins including pesticides and metals.

REFERENCES

1. Aschner M, Guilarte TR, Schneider JS, et al. Manganese: recent advances in understanding its transport and neurotoxicity. Toxicol Appl Pharmacol 2007; 221: 131–47.
2. Walker RC, Purnell GL, Jones-Jackson LB, et al. Introduction to PET imaging with emphasis on biomedical research. Neurotoxicology 2004; 25: 533–42.
3. Thompson PM, Hayashi KM, Dutton RA, et al. Tracking Alzheimer's disease. Ann N Y Acad Sci 2007; 1097: 183–214.
4. Hoptman MJ, Gunning-Dixon FM, Murphy CF, et al. Structural neuroimaging research methods in geriatric depression. Am J Geriatr Psychiatry 2006; 14: 812–22.
5. Albin RL. Basal ganglia neurotoxins.Neurol Clin 2000; 18: 665–80.
6. Hantson P, Duprez T, Mahieu P. Neurotoxicity to the basal ganglia shown by magnetic resonance imaging (MRT) following poisoning by methanol and other substances. Clin Toxicol 1997; 35: 151–61.
7. Fell JM, Reynolds AP, Meadows N, et al. Manganese toxicity in children receiving long-term parenteral nutrition. Lancet 1996; 347: 1218–21.
8. Butterworth RF, Spahr L, Fontaine S, et al. Manganese toxicity, dopaminergic dysfunction and hepatic encephalopathy. Metab Brain Dis 1995; 10: 259–67.
9. Bartzokis G, Tishler TA, Lu PH, et al. Brain ferritin iron may influence age- and gender-related risks of neurodegeneration. Neurobiol Aging 2007; 28: 414–23.
10. Pfefferbaum A, Adalsteinsson E, Rohlfing T, et al. Diffusion tensor imaging of deep gray matter brain structures: effects of age and iron concentration. Neurobiol Aging 2008, in press, Epub May 28.
11. Michaeli S, Oz G, Sorce DJ, et al. Assessment of brain iron and neuronal integrity in patients with Parkinson's disease using novel MRI contrasts. Mov Disord 2007; 22: 334–40.
12. Bartzokis G, Lu PH, Tishler TA, et al. Myelin breakdown and iron changes in Huntington's disease: pathogenesis and treatment implications. Neurochem Res 2007; 32: 1655–64.
13. Pujol J, Junque C, Vendrell P, et al. Biological significance of iron-related magnetic resonance imaging changes in the brain. Arch Neurol 1992; 49: 711–17.
14. Thuomas KA, Moller C, Odkvist LM, et al. MR imaging in solvent-induced chronic toxic encephalopathy. Acta Radiol 1996; 37: 177–9.
15. Unger E, Alexander A, Fritz T, et al. Toluene abuse: physical basis for hypointensity of the basal ganglia on T2-weighted MR images. Radiology 1994; 193: 473–6.
16. Lee J, Barry J. Influence of membrane lipid packing on T2-weighted magnetic resonance images: study of relaxation parameters in model membrane systems. Magn Reson Med 1996; 36: 420–6.

17. Thuluvath PJ, Edwin D, Yue NC, et al. Increased signals seen in globuspallidus in T1-weighted mangetic resonance imaging in cirrhotics are not suggestive of chronic hepatic encephalopathy. Hepatology 1995; 21: 440–2.
18. Krieger D, Krieger S, Jansen O, et al. Manganese and chronic hepatic encephalopathy. Lancet 1995; 346: 270–4.
19. Guilarte TR, McGlothan JL, Degaonkar M, et al. Evidence for cortical dysfunction and widespread manganese accumulation in the nonhuman primate brain following chronic manganese exposure: a 1H-MRS and MRI study. Toxicol Sci 2006; 94: 351–8.
20. Dorman DC, Struve MF, Wong BA, et al. Correlation of brain magnetic resonance imaging changes with pallidal manganese concentrations in rhesus monkeys following subchronic manganese inhalation. Toxicol Sci 2006; 92: 219–27.
21. Choi DS, Kim EA, Cheong HK, et al. Evaluation of MR signal index for the assessment of occupational manganese exposure of welders by measurement of local proton T1 relaxation time. Neurotoxicology 2007; 28: 284–9.
22. Waxman SG, Kocsis JD, Stys PK. The Axon: Structure, Function and Pathophysiology. New York: Oxford University Press, 1995.
23. Pierpaoli C, Basser PJ. Towards a quantitative assessment of diffusion anisotropy. Magn Reson Med 1996; 36: 893–906.
24. Pfefferbaum A, Sullivan EV. Increased brain white matter diffusivity in normal adult aging: relationship to anisotropy and partial voluming. Magn Reson Med 2003; 49: 953–61.
25. Rosenbloom MJ, Sullivan EV, Pfefferbaum A. Using magnetic resonance imaging and diffusion tensor imaging to assess brain damage in alcoholics. Alcohol Res Health 2003; 27: 146–52.
26. Arfanakis K, Cordes D, Haughton VM, et al. Independent component analysis applied to diffusion tensor MRI. Magn Reson Med 2002; 47: 354–63.
27. Peters A, Sethares C. Aging and the myelinated fibers in prefrontal cortex and corpus callosum of the monkey. J Comp Neurol 2002; 442: 277–91.
28. Pfefferbaum A, Sullivan EV. Diffusion MR imaging in neuropsychiatry and aging. In: Gillard J, Waldman A, Barker P, eds. Clinical MR Neuroimaging: Diffusion, Perfusion and Spectroscopy. Cambridge: Cambridge University Press, 2005: 558–78.
29. Sullivan EV, Adalsteinsson E, Pfefferbaum A. Selective age-related degradation of anterior callosal fiber bundles quantified in vivo with fiber tracking. Cereb Cortex 2006; 16: 1030–9.
30. Adalsteinsson E, Sullivan EV, Pfefferbaum A. Biochemical, functional and microstructural magnetic resonance imaging (MRI). In: Liu Y, Lovinger DM, eds. Methods in Alcohol-Related Neuroscience Research. Boca Raton, FL: CRC Press, 2002: 345–72.
31. Le Bihan D. Looking into the functional architecture of the brain with diffusion MRI. Nat Rev Neurosci 2003; 4: 469–80.
32. Horsfield MA, Jones DK. Applications of diffusion-weighted and diffusion tensor MRI to white matter diseases—a review. NMR Biomed 2002; 15: 570–7.
33. Yamasue H, Abe O, Kasai K, et al. Human brain structural change related to acute single exposure to sarin. Ann Neurol 2007; 61: 37–46.
34. Abe O, Yamasue H, Kasai K, et al. Voxel-based diffusion tensor analysis reveals aberrant anterior cingulum integrity in posttraumatic stress disorder due to terrorism. Psychiatry Res 2006; 146: 231–42.
35. Li BS, Wang H, Gonen O. Metabolite ratios to assumed stable creatine level may confound the quantification of proton brain MR spectroscopy. Magn Reson Imaging 2003; 21: 923–8.
36. Pfefferbaum A, Adalsteinsson E, Spielman D, et al. In vivo spectroscopic quantification of the N-acetyl moiety, creatine and choline from large volumes of brain gray and white matter: effects of normal aging. Magn Reson Med 1999; 41: 276–84.
37. Pfefferbaum A, Adalsteinsson E, Spielman D, et al. In vivo brain concentrations of N-acetyl compounds, creatine and choline in Alzheimer's disease. Arch Gen Psychiatry 1999; 56: 185–92.
38. Pfefferbaum A, Adalsteinsson E, Sullivan EV. Cortical NAA deficits in HIV infection without dementia: influence of alcoholism comorbidity. Neuropsychopharmacology 2005; 30: 1392–9.
39. Zahr NM, Mayer D, Pfefferbaum A, et al. Low striatal glutamate levels underlie cognitive decline in the elderly: evidence from in vivo molecular spectroscopy. Cereb Cortex 2008; 18: 2241–50.
40. Segura Aguilar J, Kostrzewa RM. Neurotoxins and neurotoxic species implicated in neurodegeneration. Neurotox Res 2004; 6: 615–30.
41. Aschner M, Syversen T, Souza DO, et al. Involvement of glutamate and reactive oxygen species in methylmercury neurotoxicity. Braz J Med Biol Res 2007; 40: 285–91.
42. Mayer D, Spielman DM. Detection of glutamate in the human brain at 3 T using optimized constant time point resolved spectroscopy. Magn Reson Med 2005; 54: 439–42.
43. Logothetis NK, Pfeuffer J. On the nature of the BOLD fMRI contrast mechanism. Magn Reson Imaging 2004; 22: 1517–31.

44. Hennig J, Speck O, Koch MA, et al. Functional magnetic resonance imaging: a review of methodological aspects and clinical applications. J Magn Reson Imaging 2003; 18: 1–15.
45. Toma K, Nakai T. Functional MRI in human motor control studies and clinical applications. Magn Reson Med Sci 2002; 1: 109–20.
46. Buckner RL, Logan JM. Functional neuroimaging methods: PET and fMRI. In: Cabeza R, Kingstone A, eds. Handbook of Functional Neuroimaging of Cognition. Cambridge, MA: The MIT Press, 2001: 27–48.
47. Buxton RB. Introduction to Functional Magnetic Resonance Imaging: Principles & Techniques. Cambridge, UK: Cambridge University Press, 2002.
48. Hermann D, Smolka MN, Klein S, et al. Reduced fMRI activation of an occipital area in recently detoxified alcohol-dependent patients in a visual and acoustic stimulation paradigm. Addict Biol 2007; 12: 117–21.
49. Tapert SF, Schweinsburg AD, Drummond SP, et al. Functional MRI of inhibitory processing in abstinent adolescent marijuana users. Psychopharmacology (Berl) 2007; 194: 173–83.
50. Kufahl P, Li Z, Risinger R, et al. Expectation modulates human brain responses to acute cocaine: a functional magnetic resonance imaging study. Biol Psychiatry 2008; 63: 222–30.
51. Hayter AL, Langdon DW, Ramnani N. Cerebellar contributions to working memory. Neuroimage 2007; 36: 943–54.
52. Myrick H, Anton RF, Li X, et al. Differential brain activity in alcoholics and social drinkers to alcohol cues: relationship to craving. Neuropsychopharmacology 2004; 29: 393–402.
53. Peterson RL. The neuroscience of investing: fMRI of the reward system. Brain Res Bull 2005; 67: 391–17.
54. Yuan W, Holland SK, Cecil KM, et al. The impact of early childhood lead exposure on brain organization: a functional magnetic resonance imaging study of language function. Pediatrics 2006; 118: 971–7.
55. Magalhaes AC. Functional magnetic resonance and spectroscopy in drug and substance abuse. Top Magn Reson Imaging 2005; 16: 247–51.
56. Gruber SA, Silveri MM, Yurgelun-Todd DA. Neuropsychological consequences of opiate use. Neuropsychol Rev 2007; 17: 299–315.
57. Gonzalez R. Acute and non-acute effects of cannabis on brain functioning and neuropsychological performance. Neuropsychol Rev 2007; 17: 347–61.
58. Fitsanakis VA, Au C, Erikson KM, et al. The effects of manganese on glutamate, dopamine and gamma-aminobutyric acid regulation. Neurochem Int 2006; 48: 426–33.
59. Fitsanakis VA, Zhang N, Avison MJ, et al. The use of magnetic resonance imaging (MRI) in the study of manganese neurotoxicity. Neurotoxicology 2006; 27: 798–806.
60. Kim Y. Neuroimaging in manganism. Neurotoxicology 2006; 27: 369–72.
61. Newland MC, Ceckler TL, Kordower JH, et al. Visualizing manganese in the primate basal ganglia with magnetic resonance imaging. Exp Neurol 1989; 106: 251–8.
62. Quaghebeur G, Taylor W, Kingsley D, et al. MRI in children receiving total parenteral nutrition. Neuroradiology 1996; 38: 680–3.
63. Spahr L, Butterworth R, Fontaine S, et al. Increased blood manganese in cirrhotic patients: relationship to pallidal magnetic resonance signal hyperintensity and neurological symptom. Hepatology 1996; 24: 1116–20.
64. Jiang Y, Zheng W, Long L, et al. Brain magnetic resonance imaging and manganese concentrations in red blood cells of smelting workers: search for biomarkers of manganese exposure. Neurotoxicology 2007; 28: 126–35.
65. Shin YC, Kim E, Cheong HK, et al. High signal intensity on magnetic resonance imaging as a predictor of neurobehavioral performance of workers exposed to manganese. Neurotoxicology 2007; 28: 257–62.
66. Yamada M, Ohno S, Okayasu I, et al. Chronic manganese poisoning: a neuropathological study with determination of manganese distribution in the brain. Acta Neuropathol (Berl) 1986; 70: 273–8.
67. Kim EA, Cheong HK, Choi DS, et al. Effect of occupational manganese exposure on the central nervous system of welders: 1H magnetic resonance spectroscopy and MRI findings. Neurotoxicology 2007; 28: 276–83.
68. Alfrey AC, LeGendre GR, Kheany WD. The dialysis encephalopathy syndrome: possible aluminum intoxication. N Engl J Med 1976; 294: 184–8.
69. Meyer-Baron M, Schaper M, Knapp G, et al. Occupational aluminum exposure: evidence in support of its neurobehavioral impact. Neurotoxicology 2007; 28: 1068–78.
70. Kiesswetter E, Schaper M, Buchta M, et al. Longitudinal study on potential neurotoxic effects of aluminium: I. Assessment of exposure and neurobehavioural performance of Al welders in the train and truck construction industry over 4 years. Int Arch Occup Environ Health 2007; 81: 41–67.
71. Takahata H, Hayashi H, Watanabe B, et al. Accumulation of mercury in the brain of two autopsy cases with chronic inorganic mercury poisoning. Folia Psychiatr Neurol Jpn 1970; 24: 59–69.

72. Berlin M, Carlsen J, Norseth T. Dose-dependence of methylmercury metabolism. A study of distribution, biotransformation, and excretion in the squirrel monkey. Arch Environ Health 1975; 30: 307–13.
73. Matsumoto S, Okajima T, Inayoshi S, et al. Minamata disease demonstrated by computed tomography. Neuroradiology 1988; 30: 42–6.
74. Korogi Y, Takahashi M, Shinzato J, et al. MR findings in seven patients with organic mercury poisoning (Minamata disease). Am J Neuroradiol 1994; 15: 1575–8.
75. Korogi Y, Takahashi M, Hirai T, et al. Representation of the visual field in the striate cortex: comparison of MR findings with visual field deficits in organic mercury poisoning. Am J Neuroradiol 1997; 18: 1127–30.
76. Korogi Y, Takahashi M, Okajima T, et al. MR findings of Minamata disease—organic mercury poisoning. J Magn Reson Imaging 1998; 8: 308–16.
77. Ekino S, Susa M, Ninomiya T, et al. Minamata disease revisited: an update on the acute and chronic manifestations of methyl mercury poisoning. J Neurol Sci 2007; 262: 131–44.
78. Carta P, Flore C, Alinovi R, et al. Sub-clinical neurobehavioral abnormalities associated with low level of mercury exposure through fish consumption. Neurotoxicology 2003; 24: 617–23.
79. Yokoo EM, Valente JG, Grattan L, et al. Low level methylmercury exposure affects neuropsychological function in adults. Environ Health 2003; 2: 8.
80. Spurgeon A. Prenatal methylmercury exposure and developmental outcomes: review of the evidence and discussion of future directions. Environ Health Perspect 2006; 114: 307–12.
81. Johansson C, Castoldi AF, Onishchenko N, et al. Neurobehavioural and molecular changes induced by methylmercury exposure during development. Neurotox Res 2007; 11: 241–60.
82. Bleecker ML, Ford DP, Vaughan CG, et al. The association of lead exposure and motor performance mediated by cerebral white matter change. Neurotoxicology 2007; 28: 318–23.
83. Stewart WF, Schwartz BS. Effects of lead on the adult brain: a 15-year exploration. Am J Ind Med 2007; 50: 729–39.
84. Stewart WF, Schwartz BS, Davatzikos C, et al. Past adult lead exposure is linked to neurodegeneration measured by brain MRI. Neurology 2006; 66: 1476–84.
85. Schwartz BS, Chen S, Caffo B, et al. Relations of brain volumes with cognitive function in males 45 years and older with past lead exposure. Neuroimage 2007; 37: 633–41.
86. Stewart WF, Schwartz BS, Simon D, et al. Neurobehavioral function and tibial and chelatable lead levels in 543 former organolead workers. Neurology 1999; 52: 1610–17.
87. Weisskopf MG, Hu H, Mulkern RV, et al. Cognitive deficits and magnetic resonance spectroscopy in adult monozygotic twins with lead poisoning. Environ Health Perspect 2004; 112: 620–5.
88. Weisskopf MG, Hu H, Sparrow D, et al. Proton magnetic resonance spectroscopic evidence of glial effects of cumulative lead exposure in the adult human hippocampus. Environ Health Perspect 2007; 115: 519–23.
89. Canfield RL, Gendle MH, Cory-Slechta DA. Impaired neuropsychological functioning in lead-exposed children. Dev Neuropsychol 2004; 26: 513–40.
90. Cory-Slechta DA. Lead-induced impairments in complex cognitive function: offerings from experimental studies. Child Neuropsychol 2003; 9: 54–75.
91. Jusko TA, Henderson CR, Lanphear BP, et al. Blood lead concentrations < 10 mug/dL and child intelligence at 6 years of age. Environ Health Perspect 2008; 116: 243–8.
92. Meng XM, Zhu DM, Ruan DY, et al. Effects of chronic lead exposure on 1H MRS of hippocampus and frontal lobes in children. Neurology 2005; 64: 1644–7.
93. Trope I, Lopez-Villegas D, Cecil KM, et al. Exposure to lead appears to selectively alter metabolism of cortical gray matter. Pediatrics 2001; 107: 1437–42.
94. Schaumburg HH, Spencer PS. Recognizing neurotoxic disease. Neurology 1987; 37: 276–8.
95. Dick FD. Solvent neurotoxicity. Occup Environ Med 2006; 63: 221–6, 179.
96. Filley CM, Halliday W, Kleinschmidt-DeMasters BK. The effects of toluene on the central nervous system. J Neuropathol Exp Neurol 2004; 63: 1–12.
97. Ridgway P, Nixon TE, Leach JP. Occupational exposure to organic solvents and long-term nervous system damage detectable by brain imaging, neurophysiology or histopathology. Food Chem Toxicol 2003; 41: 153–87.
98. Rosenberg NL, Grigsby J, Dreisbach J, et al. Neuropsychologic impairment and MRI abnormalities associated with chronic solvent abuse. J Toxicol 2002; 40: 21–34.
99. Haut MW, Kuwabara H, Ducatman AM, et al. Corpus callosum volume in railroad workers with chronic exposure to solvents. J Occup Environ Med 2006; 48: 615–24.
100. Yamanouchi N, Okada S, Kodama K, et al. White matter changes caused by chronic solvent abuse. Am J Neuroradiol 1995; 16: 1643–9.
101. Aydin K, Sencer S, Demir T, et al. Cranial MR findings in chronic toluene abuse by inhalation. Am J Neuroradiol 2002; 23: 1173–9.

102. Rosenberg NL, Kleinschmidt-DeMasters KK, Davis KA, et al. Toluene abuse causes diffuse central nervous system white matter changes. Ann Neurol 1988; 23: 611–14.
103. Aydin K, Sencer S, Ogel K, et al. Single-voxel proton MR spectroscopy in toluene abuse. Magn Reson Imaging 2003; 21: 777–85.
104. Alkan A, Kutlu R, Hallac T, et al. Occupational prolonged organic solvent exposure in shoemakers: brain MR spectroscopy findings. Magn Reson Imaging 2004; 22: 707–13.
105. Abou-Donia MB. Organophosphorus pesticides. In: Chang LW, Dyer RS, eds. Handbook of Neurotoxicology. New York: Marcel Dekker, Inc., 1995: 419–73.
106. Heaton KJ, Palumbo CL, Proctor SP, et al. Quantitative magnetic resonance brain imaging in US army veterans of the 1991 Gulf War potentially exposed to sarin and cyclosarin. Neurotoxicology 2007; 28: 761–9.
107. Brown M. Toxicological assessments of Gulf War veterans. Philos Trans R Soc Lond B Biol Sci 2006; 361: 649–79.
108. Proctor SP, Heaton KJ, Heeren T, et al. Effects of sarin and cyclosarin exposure during the 1991 Gulf War on neurobehavioral functioning in US army veterans. Neurotoxicology 2006; 27: 931–9.
109. Wooley DE. Organochloride insecticides: neurotoxicity and mechanisms of action. In: Chang LW, Dyer RS, eds. Handbook of Neurotoxicology. New York: Marcel Dekker, Inc., 1995: 475–510.
110. Cory-Slechta DA. Studying toxicants as single chemicals: does this strategy adequately identify neurotoxic risk? Neurotoxicology 2005; 26: 491–510.

12 Principles of Behavioral Phenotyping of Neonatal and Adult Mice

Susan H. Brunssen

Division of Family Health, School of Nursing, and Neurodevelopmental Disorders Research Center, University of North Carolina, Chapel Hill, North Carolina, U.S.A.

Sheryl S. Moy

Neurodevelopmental Disorders Research Center and Department of Psychiatry, University of North Carolina, Chapel Hill, North Carolina, U.S.A.

Grace E. Kissling

National Institute of Environmental Health Sciences, National Institutes of Health, Research Triangle Park, North Carolina, U.S.A.

INTRODUCTION

Behavioral phenotyping is used to evaluate the functional consequences of exposure to chemicals, drugs, environmental contaminants or nutrients on the health, as well as on the reflex, sensory, motor, social, and cognitive repertoire of mice. Most frequently, studies query the effects of acute exposures on the behavior of adult animals, either immediately or across aging. Developmental neurotoxicity studies typically expose the embryo and/or fetus via the dam in utero, followed by continued exposure through lactation until weaning. Behavioral effects in the offspring often are measured only after sexual maturity, beginning at about 60 days of age, or in young and mature adults. In each of these cases, adult test batteries are employed. However, where the questions relate to the effect of substance exposures on the ontogeny of behavioral development, testing during the neonatal and pre-pubescent periods is necessary. Test strategies for neonatal mice must address a rapidly changing behavioral repertoire as well as effects of the birth and rearing litter.

This chapter discusses the purposes and considerations in undertaking behavioral testing in mice, both adult and neonatal. An overview of strategies for behavioral neurotoxicity screening and more explicit tests of adult mice is presented, organized by functional domains of behavior. Strategies are then discussed for testing neonatal mice, similarly organized by functional domains. Finally, some major issues in design, analysis, and interpretation of behavioral phenotyping studies are reviewed.

CONSIDERATIONS FOR BEHAVIORAL PHENOTYPING OF MICE

Strain Selection

Many studies in neurotoxicology are conducted with inbred mouse strains or outbred mouse stocks obtained from commercial suppliers. The use of inbred mouse strains allows the evaluation of neurotoxin effects on multiple individual animals with identical genotypes. Thus, the significant variability present in genetic diversity of outbred stocks can be controlled. At the same time, careful consideration must be given to the choice of inbred mouse strain for behavioral experiments (1–5). For example, the A/J mouse strain is characterized by a mutation of the dysferlin gene (6), leading to problematic issues with progressive muscle weakness, similar to muscular dystrophy in humans. A/J mice may have poor performance in tasks requiring motoric exertion, such as the rotarod (7) or Morris water maze (8–10). Marked differences exist in general activity levels in different strains of mice (4,8,11). Many inbred lines have heritable hearing loss (12,13) or blindness (14), which confounds performance in behavioral tests dependent upon auditory function or visual ability. Social behavior, including preference for affiliation or levels of aggression, varies across mouse strains (8,15–17). As with humans, genotype in mice can confer different degrees of resistance to the detrimental consequences of neurotoxin exposure (18–21). Chemical testing across multiple mouse strains can provide information on heritable factors in toxin susceptibility (22,23), while the heterogeneity of outbred stocks can be advantageous for high-resolution complex trait mapping, in order to identify specific susceptibility loci.

CNS Maturation and Timing of Assessments

Cross species comparisons between rodent and human brain development should be made relative to specific brain structures and cell types. In several important dimensions of forebrain and cerebellar development, rats and mice are born at a stage of maturation analogous to human mid-gestation (20 weeks post conception) (24). By the end of the first postnatal week, murine forebrain development is equivalent to about 80% human gestation (32 wks), especially related to oligodendrocyte maturation (25,26). The end of the second postnatal week in mice approximates human term birth to three months; the end of the third week corresponds to approximately two years of human age. Sexual maturity in mice is achieved by about six weeks of age, while human pubescence has an average onset at 12 to 15 years, varying by gender. Thus, assays of neurobehavioral functions in neonatal mice [postnatal days (PND) 0–28] are conducted within the context of a rapidly changing profile of emergent, maturing (and sometimes disappearing) reflex responses, and sensory, motor, and physical development.

These changing functional capacities reflect the underlying neurochemical and structural ontogeny. For example, cerebellar organization is complete by PND14 while differentiation and migration of neurons in hippocampal and striatal regions is ongoing into the fourth postnatal week (27–29). Synaptogenesis and gliogenesis continue throughout the neonatal period (29). Myelination of cortical and sub-cortical axons begins during the third week and progresses into adulthood, coincident with the onset of coordinated motor capacities. Biochemical differences in ionic concentration gradients and early GABA receptor subtype expression alter the cellular signaling environment of the cerebrum over the first two weeks, such that neurotransmitter signals that would be inhibitory in the adult brain are excitatory during this period (30) and may be reflected in the more global responses to stimulation seen in young neonatal mice.

The second postnatal week may be a particularly critical period for the effects of insults, toxins, and environmental stressors on the development of functional pathways underlying later behavioral phenotypes. Exposure to emotional stressors alters the later behavioral response to emotional stressors in adolescence (31); the hypothalamic circuits regulating food intake and glucose homeostasis can be modulated via alterations in systemic leptin expression (32), and exposure to nicotine, polychlorinated biphenyls and polybrominated diphenylethers (flame retardants) permanently affect cholinergic nicotinic receptor density, affecting behavior in adult mice (33).

The mouse is an altricial species which has implications for handling neonates and interpretation of behavioral testing. Pups are born weighing less than 1 gm and are dependent on the litter and the dam for tactile regulation of respiration and digestion, and for control of temperature and humidity. Tactile and olfactory senses develop rapidly during the first week, while audition and vision lag into the third postnatal week. Independent locomotion precedes the onset of vision and hearing. Thus, the neonatal mouse motorically explores its environment guided principally by tactile and olfactory cues. In contrast, the precocious human infant relies principally on vision, hearing, and midline tactile inputs at birth, long before developing the motor skills needed for independent exploration. These species differences in the temporal order of emerging abilities must be considered when extrapolating from effects identified in the neonatal mouse to potential effects on human development.

Target Processes and Their Neutral Substrates

Knowledge of the neural substrate of the behavior, as well as the usual phenotype and ontogeny in the particular strain and gender, informs the selection and timing of tests. Research on the neural circuitry underlying fear conditioning indicates that the hippocampus is important for forming the memory of environmental context, while the amygdala, nucleus accumbens, and other forebrain regions are important to memory for stimuli associated with aversive events (34–37). Spatial learning in the Morris water maze is dependent upon intact hippocampal function (37,38). In the neonate, nicotinic receptors in the hippocampus and cerebral cortex mediate impaired learning following exposure to methyl mercury (MeHg) (39). Learning in passive avoidance procedures is dependent on maturation of central cholinergic pathways in the second postnatal week (40–43). BALB/c mice evidence significant maturation in escape latency on the Morris water maze hidden platform task about PND23, regardless of when training is initiated (44). Thus, to test for behavioral teratogenesis mediated by hippocampal damage, Wagner et al. (44) tested neonatal BALB/c mice only in a narrow bracket around this age. Such timing may need to be shifted when designing a study using other strains. In studies involving genetic modifications, knowledge of both the mature capabilities and the ontogeny of the behaviors of interest in the background strain(s) should be documented before testing modified animals.

MULTI-DOMAIN BEHAVIORAL PHENOTYPING IN ADULT MICE

Behavioral Text Batteries

The chemical target for neurotoxicologic evaluation in rodents can be a drug or other substance with either a known detrimental impact on humans or one with unknown consequences of exposure. In the former cases, the rat or mouse model of toxin exposure is useful for identifying the neurologic alterations underlying effects on humans, and for testing possible therapeutic strategies to reverse toxic symptoms. In the latter case, rodents provide a model for rapid screening. The National Research Council has advocated a tiered strategy for evaluating the neurotoxic potential of environmental chemicals (45). This approach includes an initial screen in the first tier to assess compounds for hazardous effects. Chemicals found to have neurotoxic properties are then investigated for dose-response functions in second tier testing, and for the mechanism of action in third-tier testing. Since the identification of neurotoxins is dependent on tests conducted in the first tier, extensive efforts have been made to establish neurobehavioral testing protocols for rats that are both sensitive and reliable indicators of toxic effects (46–49).

A similar impetus to develop sensitive and reliable testing batteries in mice comes from behavioral genetics. Comprehensive sets of phenotyping assays have been proposed to detect the effects of gene mutation, overexpression, or deletion, especially in mouse models of neuropsychiatric disorders (3–5,50,51). Mouse test batteries usually include both observational and automated methods for response measurement and can be used to assess function across multiple domains of behavior in a first tier screen.

Preliminary evaluation of behavior in the home cage or in a novel cage can indicate whether mice have overt changes or impairments. Subjects are rated for the presence of abnormal responses, such as gait disturbance, tremors, or seizures, and for signs of general malaise or ill health, including poor coat condition or a hunched posture. Initial information on motor dysfunction or other signs of neurological impairment is important for deciding on the feasibility and utility of further behavioral testing, and for interpreting the results of any subsequent procedures. Preliminary assessments of motor function may be especially important in studies of long-term effects of exposure to a neurotoxin such as MeHg (52).

One issue with the use of a comprehensive phenotyping battery is that the experience of being tested in one procedure can alter performance in tests conducted later in the sequence (53). This may be especially problematic for free-exploration procedures to assess levels of activity and anxiety-like responses. Scheduling the least stressful tests near the beginning of the battery and the more aversive or invasive procedures at the end of the sequence can help control for the effects of testing order. Also, inserting one or more days between tests gives mice a period of recovery from different procedures, while still allowing several assays to be completed within three to four weeks (54). Additional weeks may be necessary for tests requiring learning acquisition, as in the Morris water maze procedure.

Observational Screen for Behavior

Mouse testing batteries frequently begin with a primary screen for health and neurologic function (3,55). Often, preliminary assessments of appearance, reflexes, and overt behavioral alterations are based on standard methods developed to detect drug effects in mice (56). A typical screen includes indices of coat condition, including any piloerection (hair standing erect) or barbering of fur or whiskers; condition of the eye, including lacrimation, ptosis (drooping of the upper eyelid), palperbral closure (squinting or closure of the eyes), or exopthalmas (bulging of the eyeball); and body position, including tail and pelvic elevation. In particular, piloerection or ruffled fur, discharge from nose or eyes, a hunched posture, reduced body weight, and hypoactivity are signs of exposure to toxins (57,58), including bacterial endotoxins (59,60). Like bacterial endotoxins, exposure to different types of neurotoxins, including mercury (61,62), 1-methyl-4-phenyl-1,2,3,6-tetrahydropyridine (MPTP) (60), and trimethyltin hydroxide (TMT) (64), has been shown to alter cytokine levels in mice. The cascade of physiological changes mediated by proinflammatory cytokines in brain (65,66) is associated with a constellation of symptoms referred to as "sickness behavior," which is a set of adaptive responses to infection or inflammation (65,67) that can be quantified by the observational screen.

Further observational measures include simple indices of sensory and motor function. The eye-blink reflex can be tested by approaching the eye with a cotton swab, or by gently touching the area of the inner eye where the upper and lower eyelids meet (the medial canthus). A similar light touch to the ear can be used to test the pinna reflex (movement of the ear or head). A touch to the whiskers should induce an orienting response. Deficiencies observed in these tests may be due to alterations in sensory or motor functions of the cranial nerves (68). Observation of the startle response to a sudden noise, such as the sound from a metal clicker, can be used as a preliminary assessment of auditory function and general reactivity. Overt changes in motor ability can be identified by a wire-hang test, in which mice hang onto the underside of a wire grid (such as a cage lid) for one minute.

A potential confound of these screening measures is that mice may freeze in response to anxiogenic conditions, such as exposure to a novel laboratory environment, and fail to demonstrate normal reactions during the reflex tests. Thus, it is important to record whether animals are showing exploration during the screen, including sniffing behavior, to aid in interpreting any deficiencies. Noting the occurrence of vocalization during the screen, reactivity to a brief period of tail suspension, and presence of piloerection (69) may also indicate a state of stress or fear in mice.

Tests of Motor Function

Motor function can be assessed by a number of tasks which can reflect different aspects of motor performance, such as muscle strength, motor coordination, strength endurance and fatigue, and fine motor skills.

Rotarod and Other Tasks

A commonly used measure of motor coordination is latency to fall from the turning barrel of a rotarod. Acquisition of motor skills and practice effects can be assessed with repeated rotarod trials during a single test session or over days (70). Task difficulty can be altered by changing the rate of barrel rotation, accelerating the rate of rotation within a trial, reversing the rotation direction, or even adding a rocking motion to the barrel. Since the rotarod test can be repeated over weeks or months, it has been used to provide information on symptom onset and the time course of development of motor deficiencies in mouse models to evaluate progressive impairments of neurodegenerative syndromes, such as amyotrophic lateral sclerosis (ALS) (71–73) and Huntington's disease (74), and in studies of chronic neurotoxin treatment (75). One issue with the rotarod test is that some mice will cling to the barrel and rotate back to the top, rather than falling off (70,76,77). Making the surface of the barrel smoother may prevent this clinging but can also lead to lower latencies to fall.

Progressive changes in motor function can also be measured using balance beam or dowel tasks. Carter, et al. (74) used beams with different widths and shapes (square or round) to track the emergence and progression of motor impairment in *R6/2* transgenic mice, a model for Huntington's disease. Measures were taken of latency to traverse each beam and the number of times the hind feet slipped off the beam. A Plexiglas beam with a tapering width has been used to evaluate mouse models of Parkinson's disease (78,79). In this test, the end of the beam leads to the mouse's home cage, in order to provide motivation for traversing the length of the beam. Latency to fall from a wooden dowel has also been used to investigate strain differences in the motor effects of ethanol (18). In each of these studies, additional tests provided complementary information on motor performance, including swimming ability (74), footprint patterns in a test of gait abnormalities (74), inverted grid or screen tests (18,79), and forepaw grip strength measured with a strain gauge (18).

Locomotion and Activity

Exploration in an open field is a commonly used test for drug and toxin effects (68), as well as evaluation of genetic alterations in mice. Automated testing systems, based on photobeam breaks or image tracking, allow multiple measures to be taken over extended periods. Often mice will show a higher level of activity in the first intervals of a test, with decreasing exploration across time. Failure to demonstrate this pattern may indicate a deficit in habituation, a fundamental form of adaptive learning in a novel environment. Impaired habituation in mice has been reported following the disruption of glutamatergic neurotransmission by pharmacological challenge (80) or by genetic mutation (81). Habituation deficits induced by treatment with MeHg can be exacerbated by co-exposure to flame retardant, polychlorinated diphenyl ether (PBDE 99) (39). Another measure that can be taken during the open field test is number of rearing movements, an index of vertical activity. The act of rearing is more physically demanding than ambulation, and it may be a more sensitive measure of motor impairment than horizontal locomotion (39). Non-ambulatory fine movements (repeated breaking of the same set of photobeams) can provide information on when mice are not locomoting but are still showing activity. Automated testing can be supplemented by direct observation or video coding to determine whether group differences in fine movements are associated with altered rates of sniffing, grooming, or other behaviors, or with the emergence of abnormal stereotypy, such as the repetitive head-twitch and head-weaving induced by phencyclidine (82).

Tests for Sensory Function

Visual Ability

Deficiencies in vision can cause poor performance on spatial learning tasks or other procedures involving visual cues or discrimination. Several commonly used inbred mouse strains carry *Pde6b*^{rd1}, a gene for retinal degeneration that leads to blindness around the age of weaning (14). Primary screening batteries include the visual placing task, in which a mouse is suspended by its tail and slowly lowered toward a visible surface (4,56). Visual ability is measured by whether the animal extends its forepaws toward the surface before or after the whiskers make contact. Visual ability can also be evaluated using a visual cliff task (84). Mice are placed in a chamber with a clear bottom, half of which extends over the floor. Mice with intact visual ability prefer the side of the chamber with the visually solid floor (the "safe" side) over the side with the visually absent floor (the "cliff" side). Another version of the task is a "step-down" visual cliff. A raised bar is placed across the chamber demarcating the safe side and the cliff side. Each mouse is placed on the bar, and the particular side where the subject first steps down is recorded. Mice can be given more than one trial, and the percent of trials when the safe side was chosen can be determined. Performance in the visual cliff task can be confounded by low exploration or a failure to step off the bar.

The cued visible platform task on the Morris water maze provides a method of evaluating visual ability that is less affected by general activity levels. In this test, mice are placed in a circular pool of water containing a clearly marked escape platform. Over several trials, the mice learn to swim directly to the cued platform. An advantage of this procedure is that most mice have an intrinsic motivation to escape from the water; however, the task is not appropriate to test animals with poor swimming ability.

Acoustic Startle Test

Like vision, hearing function varies across inbred mouse strains (13,85,86). One way to assess auditory ability (hearing threshold and habituation) is measurement of the startle response following an acoustic stimulus (87). This test is based on the whole-body flinch, or startle, that reflexively occurs after presentation of a sudden noise [for example, a 40-msec burst of white noise, from 110 to 120 decibels (dBs)]. Automated systems quantify the time (in msec) and the force displacement (relative amplitude) at the peak of the startle movement. Repeated presentation of an acoustic stimulus can provide information on habituation to auditory events in the environment, an adaptive behavioral response.

Sensorimotor gating, another adaptive process for responding to environmental events, can be evaluated by measuring prepulse inhibition of the startle response. In this procedure, presentation of a weaker sound (the prepulse) immediately before the startle-inducing acoustic stimulus suppresses the startle response. While the startle response tests the function of the eighth cranial nerve and its reflex arc through the brainstem and direct motor pathways, modulation by prepulse inhibition involves the medial prefrontal cortex, hippocampus, amygdala, and other brain structures important in higher-order cognitive and motivational processes (88,89). Deficits in the ability to modify, or gate, startle responses have been observed in several neuropsychiatric disorders, including schizophrenia (90,91) and autism (92). In mice, prepulse inhibition can be impaired by drugs with psychotomimetic effects, such as amphetamine and phencyclidine (83,85,93).

Olfactory Function

Olfactory ability can be assessed in mice using a simple buried food test (8,94). A piece of cereal or chocolate chip is placed beneath the bedding on the floor of a cage, and latency to find the hidden treat is measured. The food for the test is introduced to the mice in the home cage two or three days before the assay, in order to control for neophobia and to confirm that the treat is palatable to the mice. Deficits observed in this test may indicate impaired olfactory function or may be due to a lack of motivation. Depriving the mice of food for 16 to 24 hours before the test can increase motivation to find the buried item. The test of olfactory ability may be particularly relevant for interpreting changes in social behavior, since olfactory information is a key component of social interaction in mice (95–97).

Tactile and Pain Sensitivity

A simple test of tactile sensitivity is the tape removal test (98). In this assay, a small piece of adhesive tape is placed on the mouse, and measures are taken of latency to sniff, lick, or otherwise respond to the tape (the sensory score) and latency to remove the tape (the motor score). A more systematic method of determining sensitivity to non-noxious stimuli is the use of von Frey monofilaments to determine the threshold for paw withdrawal after mechanical stimulation. In this procedure, a series of monofilament hairs or wires are applied to the plantar surface of the paw; the monofilaments are carefully calibrated to bend at a specific amount of force, allowing a quantitative measure of tactile sensitivity. In an automated form of the test, a mechanized monofilament is positioned underneath the hindpaw of an unrestrained mouse. Measures are taken of grams of pressure (or force) leading to paw withdrawal. The von Frey assay has been used to demonstrate progressive neuropathy following streptozotocin treatment in mice (99). Nociceptive responses can also be evaluated by the hot plate analgesia test, which uses thermal stimulation. Mice are placed on a hot plate heated to 52–55°C and monitored for any signs of discomfort, such as licking or shaking a hindpaw, vocalization, or jumping. Measures are taken of latency to respond. The trial length is limited to 20 to 30 seconds to prevent tissue damage. The Hargreaves or Plantar test also uses thermal stimulation to evaluate nociceptive threshold. In this assay, a radiant light is focused on the paw, and latency to paw withdrawal is recorded (100). A battery of tests for nociception may be used for comprehensive assessment of pain sensitivity (101).

Anxiety-like and Depression-like Behavior

Most standard assays for anxiety-like behavior in mice are exploration tasks in which mice are given the choice between remaining in a "safe" location and exploring a novel environment. In the elevated plus maze test, mice can choose to remain in two walled, or closed, arms, or explore two open arms (102). Subjects with higher levels of anxiety or stress will spend less time than control mice in the open arms. Anxiety-like behavior on the elevated plus maze can be induced by short-term lead exposure (103) or administration of the endotoxin lipopolysaccharide (LPS) (60). In the light/dark preference task, mice are given the choice between spending time in a dark section of the test environment and exploring a lighted section (104). One study has shown that LPS decreases the amount of time spent in the light compartment without altering latency to enter the lighted side (60). In the open field test, the amount of time spent in the center of the chamber rather than near the walls of the chamber has been used as an index of anxiety. One drawback to all of these assays is that measures of anxiety-like behavior may be confounded by differences in overall activity levels. For example, LPS can cause marked decreases in locomotor activity (59,105), which may be related to the sickness behavior induced by the endotoxin (60). Deficits in activity and general exploration may be the cause of the decreased open arm time in the elevated plus maze and center time in the open field observed in LPS-treated mice (106).

Two behavioral aspects of human depression, loss of volition and anhedonia, can be modeled in mice. The tail suspension test and the forced swim test evaluate the amount of struggling or the number of escape attempts made by an animal in an aversive situation. The underlying premise is that immobility in the test reflects the loss of volition observed in human depression. In the tail suspension test, the mouse is attached to a hook or bar by a piece of tape or by a clasp around the tail. The amount of time the mouse is immobile can be measured by scoring a video record of the test, or by a strain gauge connected to an automated measurement system. In the forced swim test, the mouse is placed in a beaker of water, and the time spent swimming versus floating in an immobile state are compared. The sucrose preference test is a model for anhedonia. The mouse is given a free choice between plain water and a sucrose solution in the home cage. Mice that consume less sucrose may be demonstrating anhedonia.

As with the anxiety tests, performance in these assays can be confounded by alterations in general activity levels or by neurotoxin-induced sickness behavior. With some compounds, the optimal time for conducting tests of anxiety-like or depression-like behavior may be after the animal has recovered from the acute effects of a test agent on indices of activity or general health. One study (107) found that LPS increased the duration of immobility in the tail suspension and forced swim tests 24 hours post-treatment, a time point when no significant effects of LPS on locomotor activity were observed. The same study also found that LPS reduced sucrose consumption one and two days following treatment, when levels of drinking and eating had returned to normal. These persistent effects of LPS were linked to increased FosB/DeltaFosB gene expression in specific brain regions, including the amygdala, hippocampus, and hypothalamus. Depression-like behavior in the forced swim test, without concomitant changes in locomotor activity, has also been reported in both young and adult mice exposed to MeHg early in development (52). In contrast, immobility following MPTP exposure has been attributed to motor impairment rather than to depression-like behavior (108).

Social Behavior Tests

Assays of social behavior focus on many different aspects of mouse interaction including aggressive or sexual responses, approach or avoidance, juvenile play, and maternal behavior. In most procedures, mice are placed together in a home cage or in a novel environment, and records are made of behaviors during the test, such as social investigation, escape attempts, and agonistic responses (aggressive grooming, offensive/defensive postures, or fighting). In the resident-intruder paradigm, one mouse (the resident) is singly housed for a period of time, and then another mouse (the intruder) is introduced into the home cage. For drug testing, the resident mouse receives treatment with either the vehicle or the compound of interest. Intruders are group-housed, untreated mice. The use of juvenile mice as intruders can provide a size difference to allow discrimination between subjects during behavioral coding, and can also increase the tendency toward social exploration, rather than attacks or other agonistic responses. Studies using social interaction tests have shown both increases in social behavior (i.e., exposure to chlorpyrifos, an organophosphate pesticide, depending on type of response, sex, or time of exposure) (109–111), as well as suppression of social investigation (i.e., following treatment with LPS) (112).

Social recognition can be assessed using a habituation-dishabituation procedure (113). A subject mouse is first given repeated presentations of an unfamiliar stimulus mouse. Over several trials, the amount of social investigation decreases as the subject mouse becomes more familiar with the stimulus mouse. If a novel stimulus mouse is then introduced, the level of social investigation increases, indicating that the subject mouse has recognized that a new animal is now present. Failure to demonstrate this dishabituation effect may be interpreted as a deficit in social recognition, possibly attributable to an impairment of olfactory function (114). Non-investigatory social interactions, such as fighting, can be reduced in the test by using ovarectomized females as stimulus mice.

Tests for Learning and Memory

Operant Conditioning

Much of the behavioral work of neurotoxicology in rat models has involved operant conditioning using a lever-press response with different schedules of reinforcement. It is notable, however, that operant conditioning is often more challenging in mice than in rats. The studies may require weeks or even months of training. Difficulties in shaping mice to manipulate levers has led to the use of other types of operant responses, such as nose pokes detected by photobeam breaks (115) or turns of a running wheel (116).

Avoidance and Fear Conditioning

Learning to avoid aversive stimuli is a fundamental adaptive process, generally marked by rapid acquisition and persistent memory. Several procedures have been developed to assess avoidance learning in mice, including variants of passive and active avoidance, which are described briefly here.

The passive avoidance test is conducted in a chamber with two sides, one lighted and one dark. The subject mouse is placed in the lighted side of the chamber and generally quickly moves to the dark side, where it receives a mild foot shock. Twenty-four hours later, which is sufficient time for memory consolidation in the hippocampus, the mouse is returned to the lighted side of the chamber and latency to enter the dark side is recorded. Learning is demonstrated by a longer latency to enter the dark area, which is now associated with the aversive event. This test can also be run in a chamber containing a small wooden or rubber platform. In this case, the mouse is placed on the platform and receives a foot shock when it steps down onto the floor of the chamber. Exposure to carbon monoxide (CO) gas can lead to lower step-down latencies during the retention test (117).

With active avoidance, mice learn to move to a safe location during a test, rather than passively avoiding an aversive location. For example, mice learn to step up onto a platform in a chamber, and remain on the platform for the duration of a test session in order to avoid a foot shock. Intoxication with the heavy metal trimethyltin (TMT) can induce deficits in the avoidance step-up test (118). With both neurotoxins, CO gas and TMT, learning deficits have been associated with damage to hippocampal regions.

The fear conditioning procedure uses Pavlovian classical conditioning to demonstrate a learned association between an aversive event such as a foot shock (the unconditioned stimulus) and the environmental context or specific environmental cues (the conditioned stimulus). In this test, the conditioned response is immobility or freezing, a species-specific defense reaction to a threatening situation. The procedure typically has three components: a training trial on Day 1, a test for contextual learning on Day 2, and a subsequent test for cue learning on Day 2 or 3. For the training trial, a mouse is placed in a chamber and exposed to a foot shock which is preceded by a tone and/or light cue. Twenty-four hours later, the mouse is placed back in the chamber, and measures of freezing are taken, either by observational scoring or by automated detection of immobility (119). This test for context-dependent learning is followed by a second test for cue-dependent learning, in order to determine whether the mouse has learned to associate the light or tone presentation with the aversive stimulus. In the cue test, the testing context or environment is changed from the original context. For example, the shape of the chamber may be altered, the floor may be made from a different material, the walls may be darker or lighter, and a different odor may be introduced. The mouse is placed in this novel chamber, and presented with the cue (the tone and/or light). The onset of freezing during cue presentation is evidence that the mouse has formed an association between the previously neutral cue stimuli and the aversive footshock.

Extensive research has been directed toward identifying the neurocircuitry that mediates fear conditioning in mice (34–37). There is strong evidence that the hippocampus is important for forming the memory of environmental context, while the amygdala, nucleus accumbens, and other brain regions have roles in the acquisition and retention of memory for stimuli associated with aversive events. Neurotoxin-induced changes in fear conditioning may be due to effects on these forebrain structures. For example, researchers have reported that enhancement of fear conditioning by nicotine (120) or disruption by ethanol (121,122) may be mediated by drug-induced alterations in hippocampal function. One advantage of the fear conditioning procedure is that compounds can be administered before the training trial, and learning can be measured one or two days later. Wood et al. (123) have shown that treatment with a low dose of cocaine before the training trial can improve both context- and cue-dependent learning, while a moderate dose can disrupt both forms of fear conditioning. These performance-enhancing or disruptive effects were observed 24 hours following cocaine administration, after the time when stimulant effects on activity may confound the findings.

Spatial Learning Tests

Circular Mazes

The Morris water maze procedure is a standard test for evaluating rodent spatial learning that relies on intrinsic motivation to leave the water (38). The maze consists of a circular pool of water with a slightly submerged escape platform in one quadrant of the maze. Mice are given several trials a day to learn the position of the platform, using visual cues placed around the room. Following acquisition, the mice are given a probe trial with the escape platform removed. Spatial learning is demonstrated by a significant preference for swimming in the quadrant where the platform had previously been located. Subjects that fail to learn the task can be evaluated for visual impairment with the visual cue procedure (described under "Tests for Sensory Function"). Some mouse strains, such as C58/J, are very poor swimmers in the maze (8,10). Prenatal exposure to dinocap, a fungicide, induces deficits in swimming ability that are associated with malformation of the inner ears (124). Mice may exhibit behavioral strategies that are not conducive to locating the platform, such as wall-hugging or floating during a trial (10,77).

Spatial learning in the Morris water maze is dependent upon intact hippocampal function (37,38). TMT, which has neurotoxic effects on hippocampal granule cells, completely disrupts hidden platform learning in treated mice, without altering performance on the cued visible platform task (125). The impaired learning observed following neonatal exposure to MeHg, alone or in combination with PBDE 99, may be due to neurotoxic effects on nicotinic receptors in the hippocampus and cerebral cortex (39). Although the water maze test is most often used to assess reference memory (memory for information that does not change over time), the procedure can be modified to test other types of learning. For example, following acquisition, the platform can be moved to a different location and mice tested for perseveration of the learned pattern versus the ability to learn the new placement, termed reversal learning. The location of the platform can also be altered on a daily basis in order to assess working memory, or memory for information necessary for short-term choices (126).

The Barnes maze provides a way to assess spatial learning without the stressful physical demands of swimming and exposure to water (127). The maze consists of a large, brightly-lit circular table, with twenty to forty holes around the rim. One of the holes lead to a hidden escape box. Mice learn to escape from the surface of the table by entering the correct hole. For each trial, measures are taken of latency to find the correct hole and the number of errors (visits to incorrect holes). To encourage mice to enter the escape box, additional motivation may be needed (such as lining the box with paper nesting material). Interestingly, exposure to an environmental estrogen early in development leads to *improved* performance on the Barnes maze in female mice (128). The altered spatial learning may be due to the masculinizing effects of the estrogenic compound on brain development.

Y-maze and T-maze

Spatial alternation on a Y-maze or T-maze can be used to assess working memory. Spontaneous alternation procedures rely on intrinsic motivation to explore novel locations and do not require food or water deprivation. Exposure to CO or TMT (117) or inoculation with LP-BM5 [a murine leukemia retrovirus mixture (129)], leads to decreases in spontaneous alternation in the Y-maze. In appetitively motivated alternation tasks, mice are typically food-deprived to 85–90% of their free-feeding weight. Some mouse strains, such as C57BL/6J, may need to be deprived to 80% of free-feeding weight for adequate motivation in the maze task. Mice are rewarded with a small reinforcer, such as a sucrose pellet, for entering alternate arms of a T-maze. The alternation test may be made more difficult by increasing the time between arm presentations. This type of procedure, known as a delayed non-matching-to-sample test, has been used to demonstrate impaired function of the frontal cortex in genetic mouse models of schizophrenia (130). One study found that treatment with delta$_9$-tetrahydrocannabinol (delta$_9$-THC) led to deficits in the T-maze working memory task (126). However, this drug also had effects on activity in the test, shown by longer latencies to arm choice. The researchers noted that the water maze procedure was more resistant to confounding changes in activity, as shown by a lack of delta$_9$-THC effects on swim speed or escape latencies in the first trials.

Radial Arm Maze

The radial arm maze can be used to evaluate both reference and working memory. The maze consists of a central hub with eight arms radiating outward. Food rewards are placed at the end of selected arms. For reference memory tasks, the same subset of reinforced arms is presented each day, and the mice learn which arms to enter for a reward. Entries into incorrect (non-reinforced) arms are considered reference memory errors, while re-entries into arms that have already been visited and therefore no longer contain the rewards are errors of working memory. A more simple version of the task presents the mice with all eight arms baited and measures how many arms are visited until the first error (returning to a previously explored arm), and the total number of arm entries until all of the rewards have been found. Both MeHg and PBDE 99 can increase the number of errors in the radial arm maze, with greater effects on memory from co-exposure to the two neurotoxins (39).

MULTI-DOMAIN BEHAVIORAL PHENOTYPING IN NEONATAL MICE

Neonatal Behavioral Phenotyping

The neonatal period in the mouse (birth to one month) is the least investigated period for neurotoxicity, in part because of the pup's extremely small size and physiologic fragility, yet it is a critical period to examine for alterations in normal ontogeny, emergence of aberrant behaviors, and developmental origins of later deficits. The current importance of the mouse as a platform for the study of genetic function now underscores the need to reliably evaluate neurobehavioral ontogeny across postnatal development. Whether a gene is completely absent from conception or conditionally absent or over-expressed at a particular time in perinatal life, its perturbation may affect early functions and maturation of the behavioral phenotype. Subclinical central nervous system (CNS) teratogenic effects may manifest as subtle alterations in early reflex, sensory, or motor capacities. Such early effects can alter the induction of the normal behavioral repertoire as the animal matures, potentially confounding the interpretation of tests conducted only in adulthood. Developmental deficits can be categorized as retardation (delayed maturation), regression (loss of previously established function), and intrusions (presence of stereotypic or self-injurious behaviors) (44).

An observational screen for physical, reflex, and sensorimotor development in mice was published by William Fox in 1965 (131). Investigators have worked more recently to clarify these scales and expand test batteries to assess multiple domains for neonatal mice (132–136). However, measurement scales and analysis approaches for neonatal mouse behavior are not well standardized in comparison to those for adult screening batteries. Procedures to examine developmental milestones and trajectories and adaptations to measures described above for adults to the neonatal period are reviewed in the sections that follow. Specific tests for behavioral ontogeny in juveniles are a research priority.

Physical and Reflex Screening Profiles for Developing Mice

In his original study, Fox (131) tested over 45 litters from six different strains of mice to establish a composite profile of normal neonatal mouse development. Fox describes five developmental periods based on the dominant reflex response sets present: perinatal (PND 0–3), neonatal (PND 3–9), postnatal transitional (PND 9–15), postnatal infantile (PND 15–26), and juvenile (PND 26 to sexual maturity). Detailed procedures for administration of neonatal screening measures were given most recently by Heyser (132). Examiners should have established interrater reliability prior to beginning experimental data collection. An experienced examiner should be able to test a pup in ~7 min. In general, maneuvers are ordered from least to most stressful, although the order can vary as what is most stressful changes across the stages of neonatal development.

A brief description of the procedures and variables used in our current work to further develop methods for testing neonatal mice follows. First, observations of litter location and mothering behaviors in the home cage are recorded before disturbing the nest. Three to four pups are taken from the litter at a time to a small warmed and bedding-lined holding box, to reduce disturbances of the nest and yet maintain pup warmth and needed physical contact. An individual pup is placed in a testing arena with a padded surface and observed freely moving for kinesthetic response (generalized coarse and fine tremors). Scores are assigned for development of posture, quadruped locomotion and gait (pivot/crawl versus linear walk). Emergence of other motor behaviors (sit, groom, rear, run, jump, leap) and intrusive behaviors are noted throughout the test session. Reflex and sensory responses are tested as appropriate for postnatal age (crossed extensor, rooting, surface righting, cliff aversion, placing responses, fore- and hind-limb grasp, aerial righting, auditory startle, nociception). Negative geotropism (time to orient head-up on a 30° tilted wire mesh screen), and strength (bar holding or hanging on an inverted 4mm mesh wire screen) are tested to a maximum of 15 and 30 seconds, respectively.

Physical developmental characteristics are recorded (e.g., presence of vibrissa, eyelid opening, pinna separation, external ear canal opening, incisor eruption, hair type and distribution, presence of milk band, weight and tail length). Finally, a score for over-generalized response to handling is given. The animal is returned to a warmed holding cage while notes are recorded. After all pups are tested, the group is returned to the home cage.

Measurement Issues

Neonatal observational screening data present several challenges for measurement, analysis, and interpretation, including the level of measurement, inter-correlation of variables, and the changing capacities of the mice. Many different variables comprise the modified Fox battery (10 physical development, 12 reflex, 13 sensorimotor, and 2 generalized variables). Most of these variables are measured on ordinal or categorical scales; few generate interval/ratio data (e.g., weight, length, time to head-up orientation, and time holding onto bar). Fox's original ordinal scores ranged from 0 to 9, but data were reported only at anchors of 0 (unelicitable), 1 (barely present), 5 (moderate response), and 9 (adult-like response) (131). Others have modified the profile content slightly and altered the scoring of the ordinal scale to 0–3 (137,138) or varied the levels of descriptive specificity enough to utilize interval level analysis techniques (0.0–1.0), (133,136). Others have dichotomized the scales by scoring absence/presence of adult-like behaviors, thus reporting percent of animals at criterion by a specific postnatal age (134,135). Current work is being undertaken to extend and more precisely anchor the measurement of ordinal variables (Brunssen, unpublished observations) to achieve greater precision, facilitate inter-rater reliability, and to quantitate responses that are abnormally heightened, modeled after ordinal scoring of observational batteries in adults (49).

Another issue that needs to be considered during analysis is that individual items are highly correlated with each other, as several measures tap functional maturation of the same underlying neurological systems. Appropriately, scores for reflexes and behaviors that disappear over postnatal development, such as rooting and crossed extensor reflexes, correlate negatively with those that emerge over time.

As when testing human infants, the mouse pup must be in an aroused, awake state for reliable and valid test results. Young mice (<one week old) may shut down their responses when stressed, as do human premature infants. Thus, examiners may need to test a reflex several times, or interrupt a test session until the pup recovers in the warmed holding box. Similarly, changes in the animals' developing capacities, particularly onset of vision (~PND 13–15), further complicates scoring of the timed variables. Comparing the number of seconds the animal can support its weight in bar-holding is meaningless when sighted animals chose to let go and drop to the surface, or when they are able to climb over the edge of the wire screen in less than 30 sec. Similarly, older pups may abandon the protective value of the reflexive head-up orientation on a slope when testing negative geotropism in favor of slope exploration once sight matures. Separate descriptive variables scored on ordinal scales can be added to better describe these behaviors when testing late neonatal and juvenile pups. While there is general agreement as to the reflexes, motor skills, and milestones to assess in the neonate, differences in scoring metrics and study designs require different approaches to analysis of the data (see analysis section below).

Interpretation Issues

Interpretation of neonatal screening data is facilitated by examining variables within major domains of function: physical growth and maturation; retardation or regression of reflex ontogeny, sensory responses, and motor capacities; level of response to testing; and presence of intrusive behaviors. Ontogenic profiles of individual variables comprising a domain should be examined for group variation prior to analyzing combined domain scores. For instance, an exposure may delay onset of function in one sensory system with associated early maturation of another; resulting in a combined sensory score that looks identical and leads to the wrong interpretation (139). As interpretation of behavioral results relies heavily on motor responses, one question that must be considered is whether motivational or sensory systems are differentially affected (140). In other words, does the motor response reflect heightened sensation (or motivation), or accelerated or delayed maturation of these systems? Is the difference in sensation (or motivation) a persistent feature of the phenotype or does transient alteration in early life program later motor responses? Finally, are motor capacities inherently altered by the exposure of interest? More specific tests of motor, sensory, affective and cognitive functions are generally added to neonatal observational test batteries to aid in characterizing the development of the phenotype.

Modifications for Assessment of Motor Development

A comprehensive discussion of motor developmental assessment and the timing of selected milestones is provided by Kallman (141).

Rotarod and Other Measures of Coordination

Neonatal mice show progression of grip strength and coordination on the rotarod over the third week of life (142). The ability to remain on the rod at both 10 and 20 rpm increases significantly at PND 20–22 and is not changed by training at earlier ages. Thus, the emergence of coordinated postural control appears to reflect the maturation of underlying neural substrates. No alteration to the rotarod paradigm is needed other than proportionate reductions in the diameter of the rod. Similarly, the balance beam, inverted screen, and other coordination tests described for the adult mouse can be used in neonates, with significant skill acquisition over the third week of life.

Locomotion and Activity

As in adult mice, spontaneous exploration of an open field is a common test for motor activity and development of locomotor skills. Beyond tracking the ontogeny of gait, pivoting to linear movement, and crawling to walking locomotion that is a component of all neonatal observational screening batteries, complementary assessments can be helpful to interpret behavioral teratogenesis. The figure-eight maze (attached to the litter box) can be used to assess ontogeny of spontaneous exploration after about PND 12, for the entire litter. For the individual pup, the radial maze can be used to assess exploratory activity by simply measuring the number of arms entered in a time block (141).

Automated activity testing can be of utility about PND 17, depending on the size of pups in the strain. Very young pups are too small to reliably trigger photocells on most automated equipment, especially in the vertical plane, nor can they tolerate the length of testing (40–60 min) used for adults. Simple manual observations scaled to the size of young pups can suffice for short periods using real-time or videotaped recording. For example, motor activity, speed, and patterns of locomotion can be assessed for 10 min in an 18 × 18 cm clear plastic box overlying a 1.5 × 1.5 cm grid by simply counting the number of squares the pup crosses in 2-min bins, or by using video tracking programs. Other motor behaviors such as grooming episodes and sniffing behaviors can be quantified. Wall climbing can be an indication of catecholaminergic (and later, cholinergic) activity between PND 7 and 17 in rats (143,144). The behavior has also been identified in mice. Failure to habituate, escalating motor activity, or unusual motor behaviors seen at an early age may indicate perturbed maturation of specific neurotransmitter systems. The estrogenic endocrine disruptors bisphenol A (BPA; a component in plastics) and methoxychlor (an insecticide) in low-dose exposure eliminated normal male-female differences in exploratory and emotional behaviors both in juvenile and adult offspring (145). BPA effects relate to reduction in the normally higher number of tyrosine hydroxylase neurons in the rostral periventricular preoptic area in female mice (146).

Modifications for Assessment of Sensory Ontogeny

Olfaction

Olfaction is the first of the senses to develop and perhaps the most important ethologically for the survival of the pup in the first two weeks of postnatal life. The effect of the olfactory toxin 3-methylindole was successfully modeled in neonatal mice by measuring suppression of ultrasound vocalizations (USV) of pups in response to the odors of male adults compared with clean bedding (147). By adapting specialized equipment (olfactometer) and co-opting a developmentally specific behavior (USV) that can be reliably and precisely measured, the restricted motor response capacities of the neonatal mouse were no longer an obstacle to investigating early olfactory function. Paradigms using the positive olfactory value of nesting materials, or unusual odors such as lemon oil as cues to the home nest, have formed the basis of social and cognitive paradigms described below.

Vision

Eyelid opening occurs about PND 13, varying somewhat by strain, followed by maturation of visual placing responses about PND 15 (132). In the intervening 24 to 48 hours, pups may exhibit greater freezing or hesitancy on such tests as cliff aversion, negative geotaxis, or inverted screen climbing, as they integrate visual stimuli. The Morris water maze visible platform test (described for adults above) may be used to evaluate visual function in pups as soon as swimming ability is established.

Audition

The auditory startle response is apparent about PND15 (±3 days by strain), following opening of the external ear canal. The cochlea and hair fibers normally mature under the influence of Insulin-like Growth Factor (IGF)-1 by PND 14 (148). Delayed inner ear maturation and neuronal loss has been detected in IGF-1 knockout mice by auditory evoked brainstem response (ABR) at PND 28 (149). ABR is a sensitive but technically difficult technique that measures hearing threshold by the primary reflex response at the level of the brainstem. The characteristic adult organized sensorimotor "startle reflex" behavior (described in the section on adult tests above) matures over the third and fourth postnatal weeks, dependent on myelination of brain stem reflex motor pathways for speed (latency in msec), and modulation by developing forebrain circuits (inhibition by prepulse stimulus and habituation). The importance of myelination was demonstrated by measuring latency to compound action potential in thigh muscle following a 120 dB stimulus in PND 15 mice treated with docosohexanoic acid (150).

Less sophisticated automated systems described for the adult above can be adapted to test mice as early as PND17 by increasing the sensitivity of the piezoelectric force transducer for the lower weight of the pup and/or decreasing the height and length of the animal chamber to keep the pup centered over the transducer on all four extremities. In recent experiments, pups over 9 gm total body weight could be reliably tested using the SR-Lab ABS system (San Diego Instruments) with attention to the following considerations given that the instrument is not designed for such small animals. (*i*) Pups should be tested for threshold of hearing first to remove animals with poor acuity. (*ii*) Set the equipment to record only the first 25 msec following initiation of the startle impulse, as later recovery movement may be of higher force in young pups, thus falsely elevating the latency of response. (*iii*) Consider dropping pups generating less than 100 relative force units on startle trials (115–120 dBs). Such results may reflect a disorganized motoric response, and likely will confound interpretation of inhibition trials. (*iv*) Establish each animal's characteristic latency to respond to startle trials in msec. Identify outlier trials, examine the full tracing of that trial, and correct or eliminate those trials with errors introduced by the scoring algorithm. Cleaning the data thus reduces variance and enhances conclusion validity.

Modifications to Tests for Anxiety-Like Behavior

Light/dark place preference and elevated plus maze paradigms can be useful measures of anxiety without significant modification to the test paradigm in neonatal mice after vision is established about PND 15. Minor modifications may be necessary to the scale of the equipment. For instance, a smaller dark box can be installed inside the dark side of a light/dark place preference chamber to enhance the attraction for a young pup that is perhaps one-third the size of a mature adult.

Early Social Behavior

One paradigm for the early development of emotion and communication in mice is the isolation calling response which measures USV with frequencies between 30 and 90 kHz (151), believed to be a transient protective response in the neonate (152). USV is detectable in pups isolated from the nest as early as PND 1, peaks in frequency at PND 9, and is nearly extinguished by PND 14 in C57BL/6 mice (147). Neonatal mice may be assessed for the effect of isolation, for quieting to the presence of the dam or littermates, for potentiation following a prior maternal contact, and for suppression of vocalizations in response to the scent of an unfamiliar adult male. Measurement of USV may be particularly relevant in studies of anxiogenic and anxiolytic compounds. Ricceri and colleagues (153) have suggested that measures of USVs may provide an early index of social behavior deficits in mouse models for autistic-like behaviors.

Tests of Early Cognitive and Spatial Learning

Most tests of cognitive function and spatial learning in adults require extensive shaping or appetitive restriction over time and therefore are of limited use in assessing the maturation of memory and learning capacities in neonatal or juvenile mice. Simple choice paradigms that rely on olfactory or visual cues, habituation to novel environments or acoustic startle, or tests relying on avoidance and fear conditioning have been adapted to young mice.

Avoidance and Fear Conditioning

Passive avoidance can be reliably elicited by the end of the third postnatal week (PND 21) 24 hours after exposure to a single trial involving a mild scrambled foot shock as described for adult animals above. The response appears during the second postnatal week and is dependent on maturation of central cholinergic pathways (40–43). For studies of maturation of the response, the acquisition phase of the paradigm is usually modified by giving subjects up to 10 trials to reach the behavioral criterion of withholding entry into the dark chamber for 120 sec for two consecutive trials. To control for differences in habituation to the apparatus that might accrue from varying numbers of trials to reach criterion in test mice, an additional group of mice are exposed to the chamber without reinforcement of foot shock. Learning and memory are assessed 24 hours later, as usual, by latency to step into the dark chamber during a single trial without shock. Calamandrei, et al. (154) provide details of the procedure and statistical methods.

Simplified Spatial Learning Tasks

Swimming behavior develops in mice from essentially floating with the nose submerged to gradual progression of coordinated movement from PND 6 to 15 (141). Once it is established that the test pups can swim and eyes are open, the Morris water maze may be used to test spatial learning. A variant on adult testing procedures is to use the cued visible platform trial first to establish that the pups have adequate visual capacity, followed by several days of hidden platform trials to test spatial learning.

Cognitive functions can be assessed in neonatal mice by constructing tasks that make use of appropriate sensory and motor capacities and meaningful rewards for the pups, such as milk, odors associated with the litter, or contact with the dam or littermates. In rat pups, variants on the T maze that rely on olfactory function have been administered successfully as early as the beginning of the second week of postnatal life for simple position discrimination; while the ability to perform tests relying on recent spatial memory (delayed alternation paradigms) does not emerge until the third postnatal week (140).

DESIGN CONSIDERATIONS FOR ASSESSING NEONATAL MICE

Litter Effects

In neonatal testing, the litter, not the individual pup, is the unit of analysis. Physical growth and behavior of birth littermates are correlated based on their shared heredity, and intrauterine and rearing environments. Hereditary and intrauterine factors can be randomly redistributed in experimental litters by constructing rearing litters on PND 1–2 from timed pregnancies. This strategy is useful for designs involving postnatal exposure of pups to the compound of interest, especially where the goal is to represent all experimental conditions and postnatal endpoints within a rearing litter.

Factors in the rearing litter (birth or reconstructed) are also a source of variation in postnatal development. Maternal behavior and litter size are two such factors. When birth litters are too small, hormonal stimulation of maternal behavior and milk production may be insufficient. Conversely, very large litters affect pup access to nursing sites, and this in turn affects rate of growth and can affect the achievement of developmental milestones. Wainwright, et al. (155) demonstrated slower growth and behavioral retardation as a function of litter size. Thus, it may be prudent to restrict behavioral testing to birth litters within a specific size range (e.g., 6–12 pups) and/or from previously successful mothers.

In studies in which pups must be maintained in birth litters throughout the testing (such as fetal ± lactation exposures via the dam), litters can be culled to an equal number of pups (usually 8–10). This number should remain constant until weaning and across all litters in a given study protocol. Variations in maternal behaviors and nutrition contribute significantly to litter effects, even when strategies such as random reconstruction of rearing litters and culling have been employed. For example, in a series of CD-1 litters randomly reconstructed on PND1 to rearing litters of 10 male pups, by PND4 the weight of pups varied within any one litter by only 1 gm, but varied across litters from 2 to 4.5 gm (137). Maternal nesting and nursing behaviors should be recorded daily throughout testing. At a minimum, these observations should include presence of milk bands in young pups (<PND7) and the rate of pup growth as internal experimental controls.

Effects of Litter Manipulation on Maternal Behavior

Manipulating litters for testing inherently disrupts the normal rearing environment, which also affects behavior. For example, maternal nesting behaviors alter later responses of offspring to stress via epigenetic changes in histone acetylation, gene transcription, and DNA methylation of glucocorticoid receptor gene promoters in the hippocampus that emerge during the first week of life (156). Thus, maternal behaviors should be considered in study design, in protocols for handling litters, and in the interpretation of results.

Maternal tolerance of repeated disturbance of the nest for early developmental testing should be considered in strain selection, the value of cross-fostering, and creation of testing schedules. Daily manipulations of the litter may be necessary when the study purpose is to characterize differences in developmental trajectories. Various strategies to minimize stress to the dam include maintaining rearing litters in a quiet vivarium, delaying cage changes until after test sessions, removing test litters to a holding room at least an hour prior to testing, testing at nearly the same time every day, and testing half of litters on odd days while half are tested on even days. The latter strategy permits assessment of trajectories while disturbing each litter only once every 48 hours, but it is not as useful if the investigator wishes to determine the time to onset/loss of specific behaviors when that developmental window is typically only 12 to 24 hours. The dams of pre-mobile pups continue nesting behaviors when only half the pups are removed from the litter at a time. The first half of the pups are returned as a group outside the nest and, while the dam retrieves them, the second half of the pups can be removed for testing. The cage is disturbed three times, instead of twice, but the dam never stops nesting behaviors.

Repeated testing of pups from the same litter raises the issue of techniques to identify the pups. Short of neonatal tail or paw tattoos, which some view as too disturbing for early behavioral testing, pups can be identified by bands made with permanent colored markers around the tail. More distal, circumferential marks are more resistant to maternal grooming. Marks need to be replaced at least every 48 hours. At about PND 5, the pinna of normally developing mice is sufficiently detached from the scalp to make small clips to enhance identification. By PND 8, permanent identification is possible by clipping the final digit of one toe (allowing identification of up to 8 pups per rearing litter). Clipping a single toe does not interfere with testing neonatal motor skills.

Correlation of pup behavior with birth or rearing littermates is a powerful confound that is known to persist well into adulthood. Therefore, studies that test juvenile animals (beyond weaning) should also consider the birth or rearing litter as a potential variable confounding the assumption of independence as described in the section on analysis considerations below.

Study Designs for Neonatal Assessment

Application of basic principles of good research design is essential in developmental studies to support conclusion validity. Neonatal studies may use prospective repeated measures, cross-sectional, or hybrid designs. In all neonatal studies, random prospective assignment of the animal to the experimental conditions and to time of testing is an essential strategy to control bias and one that is infrequently addressed in published papers.

Repeated measures designs test animals in a litter more than once to determine the timing and pattern of acquisition of behaviors within and across litters. This design accounts for variance both within the litter and the animal, thus reducing the number of litters needed, especially if all experimental conditions can be represented within the litter. Potential confounds of repeated testing include learned responses, handling stress, and negative effects on growth rate. Growth restriction is associated with delayed neurological development. If, however, the purpose of the study is to examine effects of a toxin at a particular time or over a short set of timepoints, a cross-sectional design may be more appropriate. In this case, a single animal may be chosen to represent the litter, tested once, identified, returned to the litter, and never tested again during the neonatal period. A disadvantage of this design is that a randomly selected pup may not be representative of the litter as a whole; therefore, more litters are needed within each factor to achieve a representative average response. Hybrid designs use features of both repeated measures and cross-sectional designs; individual pups may be first tested at different pre-assigned postnatal ages, but then continue to be tested until weaning. This strategy attempts to control for effects of repeated testing (handling stress, learned responses) while ensuring that all the individual variance in a litter is represented in the data.

Issues in Testing Genetically Altered Neonates

Studies to characterize gene effects present particular challenges. Genetically altered strains may reproduce more sporadically with smaller litters, and maternal behaviors may not be robust. Fostering the pups to a wild-type dam or one from a different strain may be necessary to distinguish the potential confound of an altered rearing environment from the effects of the genetic manipulation itself. Since the genotypes of individual pups are unknown at birth, litter reconstruction to represent all "experimental conditions" within the litter is usually not feasible when testing will begin within a few days of birth. Culling birth litters to a standard size can help reduce variance, but effects on development from variable representation of genotypes and gender within the rearing environment may need to be addressed in the analysis. Additional concerns for lack of genetic independence may exist when litters share the same sires, as might occur in a small colony of genetically altered mice. Detailed records of mating and ancestry are essential to avoid this potential confound. Maternal behavior may also drift over several generations if "wildtype" females selected for breeding were reared in genetically altered litters, were themselves tested as neonates, or carry necessary latent constructs.

PRINCIPLES OF STATISTICAL ANALYSIS OF BEHAVIORAL DATA

Choice of the proper statistical analysis method depends on the research hypothesis, the study design, and the type of data collected. Therefore, it is necessary to consider each of these features before choosing an appropriate statistical test. The initial step in designing a behavioral study is to formulate a testable hypothesis. The experiment should then be designed to adequately test the hypothesis. As discussed previously, considerations governing the design include strain, stock, sex, age, genetic modifications, exposures, timing of interventions, and timing of behavioral assessments. Developmental changes, for example, are more reliably detected by making repeated observations on the same pups than by observing pups over the same time span, but observing each pup only once.

Additional considerations include desired significance level, expected direction of the effect, statistical power and the efficient use of animals. While the significance level is often arbitrarily set at 0.05, some thought should be given to whether the statistical test should be one-sided or two-sided. If an increase (or decrease) in response is hypothesized, a one-sided test is preferable because it has more statistical power, thus requiring fewer animals than a two-sided test having the same power. A drawback to performing one-sided tests, however, is that occasionally the results are in the opposite direction from that expected, and the hypothesis is not supported even if the effect is large, but in the opposite direction. If the direction of the effect can not be confidently predicted, a two-sided test should be used.

Power, Effect Size, and Sample Size

Statistical power is the probability of detecting, at a given level of significance, an effect of a pre-specified size when it is present. Power, sample size, significance level, effect size, and error variance are interrelated, and their relationships depend on the experimental design. The researcher exercises the most control over power through selection of the sample size. In determining the required sample size for an experiment, it is important to use formulas that are appropriate for the design selected (157,158). Sample size formulas for a repeated measures design with a control and two dosed groups, for example, are quite different from formulas for a cross-sectional design with two groups. If an experiment lacks adequate power, it is unlikely that effects will be observed (false negatives), and furthermore, if they are not observed, it will not be clear whether the effects are not present or the sample was simply too small to detect the effects. On the other hand, if an experiment involves too many animals, the extra animals are wasted along with time and resources. This can also yield statistically significant effects that are not biologically significant.

Experimental Design

Experimental designs can be broadly categorized as cross-sectional or prospective (i.e., repeated measures) designs. Cross-sectional studies provide a snap-shot of a phenomenon at a particular time; this is useful for assessing the prevalence of a characteristic but does not provide information about how it changes over time. Prospective studies follow animals over time, making assessments on each animal two or more times during an experiment. Hybrid studies contain elements of cross-sectional and prospective designs. For example, a battery of tests may be administered for which several tests are performed only once, while other tests are repeated on the same animals.

Confounding occurs when the effects of one treatment or factor can not be separated from the effects of a second treatment or factor. This is a problem when the treatment or factor is of central interest. Thoughtful design of the experiment may avoid, or at least ameliorate, problematic confounding. In the example given in the section on passive avoidance learning in neonatal mice above, differences in habituation to the environment would be expected to accrue based on the different numbers of trials needed to reach the behavioral criterion in young pups. An extra group of mice was added to the design that was exposed only to the environment, to control for this confound. On the other hand, confounding is sometimes built into the experimental design, as in a Latin Square design (159). In this variant of a randomized block design, each animal (one random source of variance) is exposed to each treatment condition (the second random source of variance) in a unique random order. The number of treatment conditions equals the number of animals (e.g., 5×5 square) and it may be replicated to enhance power (160). Two-way interactions of treatment and order, treatment and animal, and animal and order, as well as the three way interaction of treatment, order, and animal are confounded and can not be tested or estimated in this design.

Clustering and the Unit of Analysis

Data often have some amount of clustering which may induce correlations among members of the cluster. For example, as discussed above, littermates are more similar to each other than to other pups as a result of genetic similarities, intrauterine environment and position, and rearing behaviors of the dam, in addition to exposure of the mother to test chemicals prior to or during pregnancy and lactation. When such clustering may be present, its magnitude should be assessed and, if it is nontrivial, accommodations for it should be made in the data analysis, either by treating the litter as the unit of analysis or by building in a correlation structure among littermates that accounts for the clustering. Failure to accommodate these positive correlations will increase the false positive rate (1).

A major goal of statistical analysis is to identify and assess sources of variability. Generally, researchers are interested in quantifying sources of variability attributable to treatment and/or genetic differences. Such differences are read over *intra*-animal variation, as well as *inter*-animal variation in behavioral measurement. Behavioral measurements on a group of animals include variation among different animals within the same group, or *inter*-animal variation, and are reported as the standard deviation, standard error of the mean, or inter-quartile range for each group. However, an animal may be highly variable in its behavioral responses upon repeated testing (*intra*-animal variation) even during the same test session. One strategy to enhance the ability to detect group differences given high intra-animal variance is to average the results of multiple trials as the best estimate of a given animal's response on that day. For example, in analysis of the amplitude of acoustic startle response, say that each animal is tested at given sound levels (e.g., 75,85,95,105 dBs) for six repeated trials within a single session. After reviewing the data and discarding failed trials, the mean response is calculated for each mouse at each sound level and entered into the analysis of group values. Thus the number of degrees of freedom reflects the N of mice, not the N of trials. If the data are not interval/ratio level, other strategies can be applied such as taking the median intra-animal response, after eliminating trials that fail.

Measurement Issues

Reliability and Validity

Reliability is an indicator of how repeatable a measurement is, or how precise. Validity indicates how accurate a measurement is. A good measure is both valid and reliable, i.e., it has both accuracy and precision. In non-automated systems, it is typical to have two or more raters evaluating behavior. In this case, inter-rater reliability should be established before conducting the actual study. If the same rater evaluates the same behavior over an extended data collection period, intra-rater reliability should also be measured regularly to establish lack of drift in scoring. These reliability measures should always be reported in descriptions of the experimental methods. Creating a set of scored videotaped sessions as a standard for training and monitoring intra- and inter-rater reliability is one useful technique.

Bias

Measurement bias reduces validity. There are many forms of bias, or a tendency to over-estimate or under-estimate the true response. Some bias may arise from poor study designs, measurement error, researcher subjectivity, or statistical estimation techniques. Whatever the source, attempts should be made to minimize the effects of bias, by correcting design flaws, calibrating equipment properly, masking observers, using unbiased statistical techniques, or estimating the magnitude of bias.

Missing Data

While data missing completely at random does not bias results, systematically missing data may induce bias. Most statistical software discards data from units of analysis (e.g., animals, litters) having missing data, though some software allows the imputation of missing values. Using the mean to impute missing values artificially reduces variation. More complex methods, such as regression imputation, are better as they preserve more of the variation (161). Data sets that have values missing by design (e.g., repeated measures of the same variable at different ages for different animals) can be analyzed with statistical methods such as general linear mixed modeling (162) or generalized estimating equations (163,164).

Random Effects Versus Fixed Effects

In analysis of variance, factors may be assigned to be fixed or random. The choice depends on what generalizations the researcher wants to make. For example, "animal" is typically a random effect to enable generalization to all animals, not just the ones in the study. Sex is typically a fixed effect because there are only two sexes and the desire is to generalize only about males and females. Treatment and genotype are usually fixed effects as conclusions can be drawn only on the particular treatment and genotype involved in the experiment.

Analysis Strategies

As described earlier in this chapter, behavior may be measured on many different scales. It is important to choose a statistical analysis that tests the research hypothesis, is appropriate for the experimental design, and fits the nature of the variables. Variables may be classified as nominal, ordinal or interval/ratio level. Some of the more common tests are indicated in Table 1. The analysis of developmental patterns from the neonatal behavioral screening scales discussed earlier provides an interesting illustration. Scores on these scales for any developmental variable at a given postnatal age are highly skewed and not normally distributed. Each item has a sudden maturational off-set or on-set, usually over a period of 24 hours, which graphs as a quadratic or ogive curve with a steep linear segment. The typical developmental study hypothesizes a shift in the timing and/or slope of this segment. Thus, the researcher is looking for a method to estimate the average curve for each

Table 1 Selected Statistical Analysis Approaches by Level of Data and Hypothesis

	Level of variables			
Hypothesis	Nominal	Ordinal	Interval/ratio (not normally distributed)	Interval/ratio (normally distributed)
Difference between groups	Chi-square test	Chi-square test Mann-Whitney U Kruskal-Wallis H	Mann-Whitney U Kruskal-Wallis H	Student's *t*-test ANOVA
Paired/repeated measures	McNemar's chi-square test	Nonparametric correlation (e.g., Spearman's rho) Wilcoxon's signed-rank test	Nonlinear modeling (e.g., cyclical patterns) Wilcoxon's signed-ranks test	Paired *t*-test, Repeated measures, ANOVA Pearson correlation
Developmental trend	Categorical data modeling	Generalized estimating equations	Generalized estimating equations Nonlinear modeling	General linear mixed models
Time to event			Cox regression Log-rank test	

treatment group in order to predict group differences in development attributable to a treatment, exposure, or genetic modification and its interaction with age. The assumptions of independence, normal distribution, and homogeneity of variance required for a standard repeated measures ANOVA approach do not hold in this case. Furthermore, the data are clustered within litter and correlated over time. Mixed effects modeling approaches address these issues (165).

The interval/ratio level variables of weight and length, which are typically normally distributed, can be analyzed using general linear mixed models (162) while ordinal and categorical variables may require categorical data modeling (166) or generalized estimating equations (163,164). Predicted average developmental curves with 95% confidence intervals are generated for each treatment group. For hierarchical data clustered within litter, analyses should first explore litter and litter by age effects. If litter effects are not significant, the models can be reduced to treatment and age factors and interaction effects. Linear and quadratic terms for age can be included to model the shape of the response over postnatal maturation (e.g., length may be best fit with a straight line, while rate of weight gain may be best described by a curvilinear function over the first three postnatal weeks).

Time to event data may require special statistical methods, particularly if the event is not observed for one or more animals. For example, if an animal in the Morris water maze fails to locate the escape platform within the allotted time, the exact time to locate the platform is not known, but it is known to be at least as long as the allotted time. Measurements such as these are considered "censored" in the sense that only partial information is available for the statistical analysis. A number of survival analysis methods have been developed for dealing with time to event data with censored measurements (167); the log-rank test (168) and Cox regression modeling (169) are the most common in practice.

SUMMARY AND CONCLUSIONS

Neurobehavioral assessments shed light on the functional consequences of exposure to neurotoxic substances, to nutrients, and to the effects of altered gene expression across the age spectrum. Screening test batteries for adult and neonatal mice (Tier 1 tests) should be supplemented further with tests for specific sensorimotor, affective, and cognitive capacities, some of which were briefly summarized here. Given the importance of the mouse as a model for human health conditions, considerable work is still needed to develop test paradigms for cognition and learning that are most appropriate to the mouse behavioral repertoire and to document differences across mouse strains. Particular attention is needed to devise standard longitudinal test batteries appropriate to the neonatal and juvenile periods in the mouse as models to investigate the susceptibilities of the human nervous system from mid-gestation through adolescence, as we strive to understand the developmental basis of health and disease.

Behavioral outcomes increasingly are a necessary complement to biomolecular and structural studies in neurotoxicity and neurogenetics. However, behavioral testing requires a major commitment of space, time, and well-trained personnel from multiple disciplines to yield quality data that is not confounded or biased. The approach to testing across multiple domains of behavior has been stressed here as a necessary strategy to inform interpretation of the specific findings in any one domain within the integrated neurological function of the whole animal. Such an approach should better inform hypotheses for studies in neurotoxicity or of the mechanisms of gene functions in the nervous system.

REFERENCES

1. Crawley JN. What's Wrong with My Mouse? New York: Wiley-Liss, John Wiley & Sons, Inc., 2000.
2. Crawley JN, Belknap JK, Collins A, et al. Behavioral phenotypes of inbred mouse strains: Implications and recommendations for molecular studies. Psychopharm 1997; 132: 107–24.
3. Crawley JN, Paylor R. A proposed test battery and constellations of specific behavioral paradigms to investigate the behavioral phenotypes of transgenic and knockout mice. Horm Behav 1997; 31: 197–211.
4. Rogers DC, Jones DNC, Nelson PR, et al. Use of SHIRPA and discriminant analysis to characterise marked differences in the behavioural phenotype of six inbred mouse strains. Behav Brain Res 1999; 105: 207–17.
5. Tarantino LM, Bucan M. Dissection of behavior and psychiatric disorders using the mouse as a model. Human Mol Genetics 2000; 9: 953–65.
6. Ho M, Post CM, Donahue LR, et al. Disruption of muscle membrane and phenotype divergence in two novel mouse models of dysferlin deficiency. Hum Mol Genetics 2004; 13: 1999–2010.
7. McFadyen MP, Kusek G, Bolivar VJ, et al. Differences among eight inbred strains of mice in motor ability and motor learning on a rotarod. Genes Brain Behav 2004; 2: 214–19.
8. Moy SS, Nadler JJ, Young NB, et al. Mouse behavioral tasks relevant to autism: phenotypes of 10 inbred strains. Behav Brain Res 2007; 176: 4–20.
9. Owen EH, Logue SF, Rasmussen DL, et al. Assessment of learning by the Morris water task and fear conditioning in inbred mouse strains and F1 hybrids: implications of genetic background for single gene mutations and quantitative trait loci analyses. Neurosci 1997: 80: 1087–99.
10. Wahlsten D, Cooper SF, Crabbe JC. Different rankings of inbred mouse strains on the Morris maze and a refined 4-arm water escape task. Behav Brain Res 2005; 165: 36–51.
11. Logue SF, Owen EH, Rasmussen DL, et al. Assessment of locomotor activity, acoustic and tactile startle, and prepulse inhibition of startle in inbred mouse strains and F1 hybrids: implications of genetic background for single gene and quantitative trait loci analyses. Neurosci 1997; 80, 1075–86.
12. Johnson KR, Zheng QY, Erway LC. A major gene affecting age-related hearing loss is common to at least ten inbred strains of mice. Genomics 2000; 70: 171–80.
13. Zheng QY, Johnson KR, Erway LC. Assessment of hearing in 80 inbred strains of mice by ABR threshold analyses. Hearing Res 1999; 130: 94–107.
14. The Jackson Laboratory. Genetic background effects: can your mice see? JAX/Notes 2002; 485: 2.
15. Bolivar VJ, Walters SR, Phoenix JL. Assessing autism-like behavior in mice: variations in social interactions among inbred strains. Behav Brain Res 2007; 176: 21–6.
16. Brodkin ES, Hagemann A, Nemetski SM, et al. Social approach-avoidance behavior of inbred mouse strains towards DBA/2 mice. Brain Res 2004; 1002: 151–7.
17. Guillot P-V, Chapouthier G. Intermale aggression and dark/light preference in ten inbred mouse strains. Behav Brain Res 1996; 77: 211–13.
18. Crabbe JC, Cotnam CJ, Cameron AJ, et al. Strain differences in three measures of ethanol intoxication in mice: the screen, dowel and grip strength tests. Genes Brain Behav 2003; 2: 201–13.
19. McLin JP, Thompson LM, Steward O. Differential susceptibility to striatal neurodegeneration induced by quinolinic acid and kainate in inbred, outbred and hybrid mouse strains. Eur J Neurosci 2006; 24: 3134–40.
20. Padilla A, Govezensky T, Possani LD, et al. Experimental envenoming of mice with venom from the scorpion Centruroides limpidus limpidus: differences in mortality and symptoms with and without antibody therapy relating to differences in age, sex, and strain of mouse. Toxicon 2003; 41: 959–65.
21. Sedelis M, Hofele K, Auburger GW, et al. MPTP susceptibility in the mouse: behavioral, neurochemical, and histological analysis of gender and strain differences. Behav Genet 2000; 30: 171–82.
22. Boyd JD, Jang H, Shepherd KR, et al. Response to 1-methyl-4-phenyl-1,2,3,6-tetrahydropyridine (MPTP) differs in mouse strains and reveals a divergence in JNK signaling and COX-2 induction prior to loss of neurons in the substantia nigra pars compacta. Brain Res 2007; 1175: 107–16.

23. McLaughlin P, Zhou Y, Ma T, et al. Proteomic analysis of microglial contribution to mouse strain-dependent dopaminergic neurotoxicity. Glia 2006; 53: 567–82.
24. Bayer SA, Altman J, Russo RJ, et al. Timetables of neurogenesis in the human brain based on experimentally determined patterns in the rat. Neurotoxicology 1993; 14: 83–144.
25. Back SA, Luo NL, Borenstein NS, et al. Late oligodendrocyte progenitors coincide with the developmental window of vulnerability for human perinatal white matter injury. J Neurosci 2001; 21: 1302–12.
26. Craig A, Luo NL, Beardsley DJ, et al. Quantitative analysis of perinatal rodent oligodendrocyte lineage progression and its correlation with human. Exp Neurology 2003; 181: 231–40.
27. Bachevalier J, Beauregard M. Maturation of medial temporal lobe memory functions in rodents, monkeys and humans. Hippocampus, 1993; 3: 191–202.
28. Rice D, Barone S. Critical periods of vulnerability for the developing nervous system: evidence form human and animal models. Envir Health Persp 2000; 108: 511–33.
29. Voorhees CV. Handbook of Behavioral Teratology. New York: Plenum Press, 1986.
30. Ben-Ari Y. Excitatory actions of GABA during development: the nature of the nurture. Nat Rev Neurosci 2002; 3: 728–39.
31. Tokumo K, Hirai T, Nishino H. Effects of perinatal stress on the emotion-related behavior of the adolescent mouse. Biol Pharm Bull 2006; 29: 1036–8.
32. Bouret SG, Simerly RB. Developmental programming of hypothalamic feeding circuits. Clin Genet 2006; 70: 295–301.
33. Eriksson P, Ankarberg E., Viberg H, et al. The developing cholinergic system as a target for environmental toxicants nicotine and polychlorinated biphenyls (PCBs): implications for neurotoxicological processes in mice. Neurotox Res 2001; 3: 37–51.
34. Ammassari-Teule M, Passino E, Restivo L, et al. Fear conditioning in C57BL/6 and DBA/2 mice: variability in nucleus accumbens function according to the strain predisposition to show contextual- or cue-based responding. Euro J Neurosci 2000; 12: 4467–74.
35. Calandreau L, Trifilieff P, Mons N, et al. Extracellular hippocampal acetylcholine level controls amygdala function and promotes adaptive conditioned emotional response. J Neurosci 2006; 26: 13556–66.
36. Gerlai R. Contextual learning and cue association in fear conditioning in mice: a strain comparison and a lesion study. Behav Brain Res, 1998; 95: 191–203.
37. Schimanski LA, Nguyen PV. Mouse models of impaired fear memory exhibit deficits in amygdalar LTP. Hippocampus 2005; 15: 502–17.
38. Morris RGM, Garrud P, Rawlins JNP, et al. Place navigation impaired in rats with hippocampal-lesions. Nature 1982; 297: 681–3.
39. Fischer C, Fredriksson A, Eriksson P. Co-exposure of neonatal mice to a flame retardant PBDE 99 (2,2′,4,4′, 5-pentabromodiphenyl ether) and methylmercury enhances developmental neurotoxic defects. Toxicol Sci 2008; 244: 157–65.
40. Alleva E, D'Udine B. Early learning capability in rodents: a review (Rattus norvegicus and Mus musculus). Int J Comp Psychol 1988; 1: 107–25.
41. Calamandrei G, Ricceri L, Valanzano A. Systemic administration of anti-NGF antibodies to neonatal mice impairs 24-h retention of an inhibitory avoidance task while increasing ChAT immunoreactivity in the medial septum. Behav Brain Res 1996; 78: 81–91.
42. Ray D, Nagy MZ. Emerging cholinergic mechanisms and ontogeny of response inhibition in the mouse. J Comp Physio Psychol 1978; 92: 335–49.
43. Stehouwer DJ, Campbell BA. Ontogeny of passive avoidance: role of task demands and development of species-typical behaviors. Dev Psychobiol 1980; 13: 385–9.
44. Wagner, GC, Reuhl KR, Cheh M, et al. A new neurobehavioral model of autism in mice: pre- and postnatal exposure to sodium valproate. J Autsim Dev Disord 2006; 36: 779–93.
45. National Research Council. Environmental Neurotoxicology. Washington, DC: National Academy Press, 1992.
46. Bushnell PJ, Moser VC, Samsam TE. Comparing cognitive and screening tests for neurotoxicity: effects of acute chlorpyrifos on visual signal detection and a neurobehavioral test battery in rats. Neurotox Teratol 2001; 23: 33–44.
47. Ehman KD, Moser VC. Evaluation of cognitive function in weanling rats: a review of methods suitable for chemical screening. Neurotox Teratol 2006; 28: 144–61.
48. Moser VC. Approaches for assessing the validity of a functional observational battery. Neurotox Teratol 1990; 12: 483–8.
49. Moser VC. The functional observational battery in adult and developing rats. Neurotoxicol 2000; 21: 989–96.
50. Ohl F, Keck ME. Behavioural screening in mutagenised mice—in search for novel animal models of psychiatric disorders. Eur J Pharmacology 2003; 480: 219–28.

51. Takao K, Miyakawa T. Investigating gene-to-behavior pathways in psychiatric disorders. The use of a comprehensive behavioral test battery on genetically engineered mice. Ann N Y Acad Sci 2006; 1086: 144–59.
52. Onishchenko N, Tamm C, Vahter M, et al. Developmental exposure to methylmercury alters learning and induces depression-like behavior in male mice: effects of developmental MEHG exposure. Toxicol Sci 2007; 97: 428–37.
53. McIlwain KL, Merriweather MY, Yuva-Paylor LA, et al. The use of behavioral test batteries: effects of training history. Physiology Behav 2001; 73: 705–17.
54. Paylor R, Spencer CM, Yuva-Paylor LA, et al. The use of behavioral test batteries. II: effect of test interval. Physiology Behav 2006; 87: 95–102.
55. Rogers DC, Fisher EMC, Brown SDM, et al. Behavioral and functional analysis of mouse phenotype: SHIRPA, a proposed protocol for comprehensive phenotype assessment. Mamm Genome 1997; 8: 711–13.
56. Irwin S. Comprehensive observational assessment: Ia. A systematic, quantitative procedure for assessing the behavioral and physiologic state of the mouse. Psychopharmacologia (Berl) 1968; 13: 222–57.
57. Luvisetto S, Rossetto O, Montecucco C, et al. Toxicity of botulinum neurotoxins in central nervous system of mice. Toxicon 2003; 41: 475–81.
58. Sills RC, Valentine WM, Moser V, et al. Characterization of carbon disulfide neurotoxicity in C57BL6 mice: behavioral, morphologic, and molecular effects. Toxicol Pathol 2000; 28: 142–8.
59. Kozak W, Conn CA, Kluger MJ. Lipopolysaccharide induces fever and depresses locomotor activity in unrestrained mice. Am J Physiol Regulatory Integrative Comp Physiol 1994; 266: 125–35.
60. Lacosta S, Merali Z, Anisman H. Behavioral and neurochemical consequences of lipopolysaccharide in mice: anxiogenic-like effects. Brain Res 1999; 818: 291–303.
61. Kim SH, Johnson VJ, Sharma RP. Oral exposure to inorganic mercury alters T lymphocyte phenotypes and cytokine expression in BALB/c mice. Arch Toxicol 2003; 77: 613–20.
62. Santarelli L, Bracci M, Mocchegiani E. In vitro and in vivo effects of mercuric chloride on thymic endocrine activity, NK and NKT cell cytotoxicity, cytokine profiles (IL-2, IFN-gamma, IL-6): role of the nitric oxide-L-arginine pathway. Int Immunopharmacol 2006; 6: 376–89.
63. Herbert G, Arsaut J, Dantzer R, et al. Time-course of the expression of inflammatory cytokines and matrix metalloproteinases in the striatum and mesencephalon of mice injected with 1-methyl-4-phenyl-1,2,3,6-tetrahydropyridine, a dopaminergic neurotoxin. Neurosci Lett 2003; 349: 191–5.
64. Harry JG, Bruccoleri A, d'Hellencourt LC. Differential modulation of hippocampal chemical-induced injury response by ebselen, pentoxifylline, and TNFalpha-, IL-1alpha-, and IL-6-neutralizing antibodies. J Neurosci Res 2003; 73: 526–36.
65. Dantzer R. Cytokine-induced sickness behavior: mechanisms and implications. Ann N Y Acad Sci 2001; 933: 222–34.
66. Dantzer R, Bluthe R-M, Gheusi G, et al. Molecular basis of sickness behavior. Ann N Y Acad Sci, 1998; 840: 586–90.
67. Hart BL. Biological basis of the behavior of sick animals. Neurosci Biobehav Rev 1988; 12: 123–37.
68. Ross JF. Tier 1 neurological assessment in regulated animal safety studies. In: Massaro EJ, ed. Handbook of Neurotoxicology, Volume II. Totowa, NJ: Humana Press, 2002; 461–505.
69. Masuda Y, Suzuki M, Akagawa Y, et al. Development and pharmacological features of mouse emotional piloerection. Exp Anim 1999; 48: 209–11.
70. McFadyen MP, Kusek G, Bolivar VJ, et al. Differences among eight inbred strains of mice in motor ability and motor learning on a rotarod. Genes Brain Behav 2003; 2: 214–19.
71. Miana-Mena FJ, Munoz MJ, Yague G, et al. Optimal methods to characterize the G93A mouse model of ALS. Amyotroph Lateral Scler 2005; 6: 55–62.
72. Shin JH, Cho SI, Lim HR, et al. Concurrent administration of Neu2000 and lithium produces marked improvement of motor neuron survival, motor function, and mortality in a mouse model of amyotrophic lateral sclerosis. Mol Pharmacol 2007; 71: 965–75.
73. Wedyt P, Hong SY, Kliot M, et al. Assessing disease onset and progression in the SOD1 mouse model of ALS. NeuroReport 2003; 14: 1051–4.
74. Carter RJ, Lione LA, Humby T, et al. Characterization of progressive motor deficits in mice transgenic for the human Huntington's disease mutation. J Neurosci 1999; 19: 3248–57.
75. Richter F, Hamann M, Richter A. Chronic rotenone treatment induces behavioral effects but no pathological signs of Parkinsonism in mice. J Neurosci Res 2007; 85: 681–91.
76. Paylor R, Zhao Y, Libbey M, et al. Learning impairments and motor dysfunctions in adult Lhx5-deficient mice displaying hippocampal disorganization. Physiol Behav 2001; 73: 781–92.
77. Wahlsten D, Metten P, Phillips TJ, et al. Different data from different labs: lessons from studies of gene-environment interaction. J Neurobiol 2003; 54: 283–311.

78. Fernagut PO, Hutson CB, Fleming SM, et al. Behavioral and histopathological consequences of paraquat intoxication in mice: effects of -synuclein over-expression. Synapse 2007; 61: 991–1001.
79. Fleming SM, Salcedo J, Fernagut P-O, et al. Early and progressive sensorimotor anomalies in mice overexpressing wild-type human -synuclein. J Neurosci 2004; 24: 9434–40.
80. Bespalov A, Jongen-Relo AL, van Gaalen M, et al. Habituation deficits induced by metabotropic glutamate receptors 2/3 receptor blockade in mice: reversal by antipsychotic drugs. J Pharmacol Exp Ther 2007; 320: 944–50.
81. Mohn AR, Gainetdinov RR, Caron MG, et al. Mice with reduced NMDA receptor expression display behaviors related to schizophrenia. Cell, 1999; 98: 427–36.
82. Nabeshima T, Ishikawa K, Yamaguchi K, et al. Phencyclidine-induced head-weaving observed in mice after ritanserin treatment. Eur J Pharmacol 1987; 139: 171–8.
83. Ralph RJ, Paulus MP, Geyer MA. Strain-specific effects of amphetamine on prepulse inhibition and patterns of locomotor behavior in mice. J Pharmacol Experimental Therapeutics 2001; 298: 148–55.
84. Fox MW. The visual cliff test for the study of visual depth perception in the mouse. Animal Behav 1965; 13: 232–3.
85. Dulawa SC, Geyer MA. Psychopharmacology of prepulse inhibition in mice. Chinese J Physiology 1996; 39: 139–46.
86. Paylor R, Crawley JN. Inbred strain differences in prepulse inhibition of the mouse startle response. Psychopharm 1997; 132: 169–80.
87. Carlson S, Willott JF. The behavioral salience of tones as indicated by prepulse inhibition of the startle response: relationship to hearing loss and central neural plasticity in C57BL/6J mice. Hearing Res 1996; 99: 168–75.
88. Bast T, Feldon J. Hippocampal modulation of sensorimotor processes. Progress Neurobio 2003; 70: 319–45.
89. Larrauri J, Schmajuk N. Prepulse inhibition mechanisms and cognitive processes: a review and model. EXS 2006; 98: 245–78.
90. Braff DL, Geyer MA. Sensorimotor gating and schizophrenia. Arch. Gen. Psychiatry, 1990; 47: 181–8.
91. Geyer MA, Swerdlow NR, Mansbach RS, Braff DL. Startle response models of sensorimotor gating and habituation deficits in schizophrenia. Brain Res Bull 1990; 25: 485–98.
92. Perry W, Minassian A, Lopez B, et al. Sensorimotor gating deficits in adults with autism. Biol Psychiatry 2006; 61: 482–6.
93. Varty GB, Walters N, Cohen-Williams M, et al. Comparison of apomorphine, amphetamine and dizocilpine disruptions of prepulse inhibition in inbred and outbred mice strains. Eur J Pharm 2001; 424: 27–36.
94. Duncan GE, Moy SS, Perez A, et al. Deficits in sensorimotor gating and tests of social behavior in a genetic model of reduced NMDA receptor function. Behav Brain Res 2004; 153: 507–18.
95. Del Punta K, Leinders-Zufall T, Rodriguez I, et al. Deficient pheromone responses in mice lacking a cluster of vomeronasal receptor genes. Nature 2002; 419: 70–4.
96. Leypold BG, Yu CR, Leinders-Zufall T, et al. Altered sexual and social behaviors in trp2 mutant mice. Proc Natl Acad Sci U S A 2002; 99: 6376–81.
97. Liebenauer LL, Slotnick BM. Social organization and aggression in a group of olfactory bulbectomized male mice. Physiol Behav 1996; 60: 403–9.
98. Starkey ML, Barritt AW, Yip PK, et al. Assessing behavioural function following a pyramidotomy lesion of the corticospinal tract in adult mice. Exp Neurology 2005; 195: 524–39.
99. Christianson JA, Ryals JM, McCarson KE, et al. Beneficial actions of neurotrophin treatment on diabetes-induced hypoalgesia in mice. J Pain 2003; 4: 493–504.
100. Hargreaves K, Dubner R, Brown F, et al. A new and sensitive method for measuring thermal nociception in cutaneous hyperalgesia. Pain 1988; 32: 77–88.
101. Kayser V, Elfassi IE, Aubel B, et al. Mechanical, thermal and formalin-induced nociception is differentially altered in 5-HT1A-/-, 5-HT1B-/-, 5-HT2A-/-, 5-HT3A-/- and 5-HTT-/- knock-out male mice. Pain 2007; 130: 235–48.
102. Lister RG. The use of a plus-maze to measure anxiety in the mouse. Psychopharm 1987; 92: 180–5.
103. Nieto-Fernandez FE, Ruiz A, Ntukogu N, et al. Short term lead exposure induces a stress-like response in adult mice. Med Sci Monit 2006; 12: BR325–9.
104. Crawley JN. Neuropharmacological specificity of a simple animal model for the behavioral actions of benzodiazepines. Pharmacol Biochem Behav 1981; 15: 695–9.
105. Engeland CG, Neilsen DV, Kavaliers M, et al. Locomotor activity changes following lipopolysaccharide treatment in mice: a multivariate assessment of behavioral tolerance. Physiol Behav 2001; 72: 481–91.
106. Swiergiel AH, Dunn AJ. Effects on interleukin-1 and lipopolysaccharide on behavior of mice in the elevated plus-maze and open field tests. Pharmacology Biochem Behav 2007; 86: 651–9.

107. Frenois F, Moreau M, O'Conner J, et al. Lipopolysaccharide induces delayed FosB/DeltaFosB immunostaining within the mouse extended amygdala, hippocampus and hypothalamus, that parallel the expression of depressive-like behavior. Psychoneuroendocrinology 2007; 32: 516–31.
108. Mori A, Ohashi S, Nakai M, et al. Neural mechanisms underlying motor dysfunction as detected by the tail suspension test in MPTP-treated C57BL/6 mice. Neurosci Res 2005; 51: 265–74.
109. Ricceri L, Markina N, Valanzano A, et al. Developmental exposure to chlorpyrifos alters reactivity to environmental and social cues in adolescent mice. Toxicol Applied Pharm 2003; 191: 189–201.
110. Ricceri L, Venerosi A, Capone F, et al. Developmental neurotoxicity of organophosphorous pesticides: fetal and neonatal exposure to chlorpyrifos alters sex-specific behaviors at adulthood in mice. Toxicol Sci 2006; 93: 105–13.
111. Venerosi A, Calamandrei G, Ricceri L. A social recognition test for female mice reveals behavioral effects of developmental chlorpyrifos exposure. Neurotoxicol Teratology 2006; 28: 466–71.
112. Lin KI, Johnson DR, Freund GG. LPS-dependent suppression of social exploration is augmented in type 1 diabetic mice. Brain Behav Immun 2007; 21: 775–82.
113. Dluzen DE, Kreutzberg JD. 1-Methyl-4-phenyl-1,2,3,6-tetrahydropyridine (MPTP) disrupts social memory/recognition processes in the male mouse. Brain Res 1993; 609: 98–102.
114. Ferguson JN, Young LJ, Hearn EF, et al. Social amnesia in mice lacking the oxytocin gene. Nat Genet 2000; 25: 284–8.
115. Robledo P, Trigo JM, Panayi F, et al. Behavioural and neurochemical effects of combined MDMA and THC administration in mice. Psychopharmacology (Berl) 2007; 195: 255–64.
116. Weiss B, Stern S, Cox C, Balys M. Perinatal and lifetime exposure to methylmercury in the mouse: behavioral effects. Neurotoxicology 2005; 26: 675–90.
117. Maurice T, Phan V-L, Noda Y, et al. The attenuation of learning impairments induced after exposure to CO or trimethyltin in mice by sigma ( ) receptor ligands involves both 1 and 2 sites. Brit J Pharmacol 1999; 127: 335–42.
118. Kassed CA, Willing AE, Garbuzova-Davis S, et al. Lack of NF- B p50 exacerbates degeneration of hippocampal neurons after chemical exposure and impairs learning. Exp Neurol 2002; 176: 277–88.
119. Valentinuzzi VS, Kolker DE, Vitaterna MH, et al. Automated measurement of mouse freezing behavior and its use for quantitative trait locus analysis of contextual fear conditioning in (BALB/cJ × C57BL/6J)F2 mice. Learn Mem 2006; 5: 391–403.
120. Davis JA, Kenney JW, Gould TJ. Hippocampal 4 2 nicotinic acetylcholine receptor involvement in the enhancing effect of acute nicotine on contextual fear conditioning. J Neurosci 2007; 27: 10870–7.
121. Gould TJ. Ethanol disrupts fear conditioning in C57BL/6 mice. J Psychopharmacol 2003; 17: 77–81.
122. Weitemier AZ, Ryabinin AE. Alcohol-induced memory impairment in trace fear conditioning: a hippocampus-specific effect. Hippocampus 2003; 13: 305–15.
123. Wood SC, Fay J, Sage JR, Anagnostaras SG. Cocaine and Pavlovian fear conditioning: dose-effect analysis. Behav Brain Res 2007; 176: 244–50.
124. Gray LE, Jr, Rogers JM, Ostby JS, et al. Prenatal dinocap exposure alters swimming behavior in mice due to complete otolith agenesis in the inner ear. Toxicol Appl Pharmacol 1988; 92: 266–73.
125. Halladay AK, Wilson DT, Wagner GC, Reuhl KR. Trimethyltin-induced alterations in behavior are linked to changes in PSA-NCAM expression. Neurotoxicology 2006; 27: 137–46.
126. Varvel SA, Anum E, Niyuhire F, et al. Delta9-THC-induced cognitive deficits in mice are reversed by the GABAA antagonist bicuculline. Psychopharm 2005; 178: 317–27.
127. Barnes CA. Memory deficits associated with senescence: a neurophysiological and behavioral study in the rat. J Comp Physiol Psychol 1979; 93: 74–104.
128. Ryan BC, Vandenbergh JG. Developmental exposure to environmental estrogens alters anxiety and spatial memory in female mice. Horm Behav 2006; 50: 85–93.
129. Iida R, Yamada K, Mamiya T, et al. Characterization of learning and memory deficits in C57BL/6 mice infected with LP-BM5, a murine model of AIDS. J Neuroimmunol 1999; 95: 65–72.
130. Kellendonk C, Simpson EH, Polan HJ, et al. Prenatal dinocap exposure alters swimming behavior in mice due to complete otolith agenesis in the inner ear. Toxicol Appl Pharmacol 1988; 92: 266–73.
131. Fox WM. Reflex-ontogeny and behavioural development of the mouse. Anim Behav 1965; 13: 234–41.
132. Heyser CJ. Assessment of developmental milestones in rodents. Curr Protoc Neurosci 2003; 8.18.1–15.
133. Wainwright PE. Methodological issues in the assessment of behavioral development in laboratory mice. In: Crusio WE, Gerlai RT, eds. Handbook of Molecular-Genetic Techniques for Brain and Behavior Research. New York: Elsevier Science, 1999: 710–21.
134. Calamandrei G, Venerosi A, Branchi I, et al. Effects of prenatal AZT on mouse neurobehavioral development and passive avoidance learning. Neurotoxicol Teratol 1999; 21: 29–40.
135. Santucci D, Cagiano R, Calamandrei G. IGF-1 and IGF-1 (24–41) but not IGF-1 (57–70) affect somatic and neurobehavioral development of newborn male mice. Brain Res Bull 1994; 35: 376–81.

136. Wahlsten DA. Developmental time scale for postnatal changes in brain and behavior in B6D2F2 mice. Brain Res 1974; 72: 251–64.
137. Brunssen SH. Perinatal exposure to hyper-interleukin-6 affects neurodevelopment in the mouse Abstract . University of North Carolina at Chapel Hill, Dissertation Abstracts International, 2002.
138. Tremml P, Lipp HP, Müller U, et al. Neurobehavioral development, adult openfield exploration and swimming navigation learning in mice with a modified -amyloid precursor protein gene. Behav Brain Res 1998; 95: 65–76.
139. Wainwright PE. The role of nutritional factors in behavioural development in laboratory mice. Behav Brain Res 2001; 125: 75–80.
140. Stanton M. Assessment of learning and memory in developmental neurotoxicology. In: Harry GJ, ed. Developmental Neurotoxicology. Boca Raton: CRC Press, 1994: 123–55.
141. Kallman MJ. Assessment of motoric effects. In: Harry GJ, ed. Developmental Neurotoxicology. Boca Raton: CRC Press, 1994: 103–22.
142. Chapillon P, Lalonde R, Jones N, et al. Early development of synchronized walking on the rotarod in rats: effects of training and handling. Behav Brain Res 1998; 93: 77–81.
143. Scalzo FM, Burge LJ. The ontogeny of phencyclidine-induced wall climbing and locomotor activity. Dev Psychobiol 1992; 25: 597.
144. Linville DG, Spear LP. Cholinergic influences on wall climbing and its ontogenetic decline. Psychopharmacol 1988; 95: 200.
145. Gioiosa L, Fissore E, Ghirardelli G, et al. Developmental exposure to low-dose estrogenic endocrine disruptors alters sex differences in exploration and emotional responses in mice. Horm Behav 2007; 52: 307–16.
146. Rubin BS, Lenkowski JR, Shaeberle CM, et al. Evidence of altered brain sexual differentiation in mice exposed perinatally to low, environmentally relevant levels of bisphenol A. Endocrinology 2006; 147: 3681–91.
147. Lemasson M, Delbe C, Gheusi G, et al. Use of ultrasonic vocalizations to assess olfactory detection in mouse pups treated with 3-methylindole. Behav Processes 2005; 68: 13–23.
148. Camarero G, Avendano C, Fernandez-Moreno C, et al. Delayed inner ear maturation and neuronal loss in postnatal Igf-1-deficient mice. J Neurosci 2001; 21: 7630–41.
149. Cediel R, Riquelme R, Contreras J, et al. Sensorineural hearing loss in insulin-like growth factor I-null mice: a new model of human deafness. Eur J Neurosci 2006; 23: 587–90.
150. Haubner L, Sullvan J, Ashmeade T, et al. The effects of maternal docosahexanenoic acid intake on rat pup myelin and the auditory startle response. Dev Neurosci 2007; 29: 460–7.
151. Branchi I, Santucci D, Alleva E. Ultrasonic vocalization emitted by infant rodents: a tool for assessment of neurobehavioural development. Behav Brain Res 2001; 125: 49–56.
152. Hofer MS, Shair HN, Brunelli SA. Ultrasonic vocalizations in rat and mouse pups. Curr Protoc Neurosci 2002; Ch8: Unit 8.14.
153. Ricceri L, Moels A, Crawley J. Behavioral phenotyping of mouse models of neurodevelopmental disorders: relevant social behavior patterns across the life span. Behav Brain Res 2007; 176: 40–52.
154. Calamandrei G, Venerosi A, Branchi I, et al. Effects of prenatal zidovudine treatment on learning and memory capacities of preweanling and young adult mice. Neurotoxicology 1999; 20: 17–26.
155. Wainwright PE, Pelkman C, Wahlsten D. The quantitative relationship between nutritional effects on preweaning growth and behavioral development in mice. Dev Psychobiol 1989; 22: 183–95.
156. Weaver ICG, Cerbvoni N, Champagne FA, et al. Epigenetic programming by maternal behavior. Nat Neurosci 2004; 7: 1–8.
157. Dell RB, Holleran S, Ramakrishnan R. Sample size determination. ILAR J 2002; 43: 207–13.
158. Winer BJ. Statistical Principles in Experimental Design, 2nd edn. New York: McGraw-Hill, 1971.
159. Box GEP, Hunter WG, Hunter JS. Statistics for Experimenters. New York: John Wiley & Sons, 1978.
160. Armitage P, Berry G, Matthews JNS. Statistical Methods in Medical Research, 4th edn. Oxford: Blackwell, 2001.
161. Rubin DB. Multiple imputation after 18+ years. JASA 1996; 91: 473–89.
162. Neter J, Kutner MH, Nachtsheim CJ, et al. Applied Linear Statistical Models. Boston: WCB McGraw-Hill, 1996.
163. McCullagh P, Nelder JA. Generalized Linear Models, 2nd edn. London: Chapman and Hall, 1989.
164. Zeger SL, Liang K-Y. Longitudinal data analysis for discrete and continuous outcomes. Biometrics 1986; 42: 121–30.

165. Wainwright PE, Leatherdale ST, Dubin JA. Advantages of mixed effects models over traditional ANOVA models in developmental studies: a worked example in a mouse model of fetal alcohol syndrome. Dev Psychobiol 2007; 49: 664–74.
166. Agresti A. Categorical Data Analysis, 2nd edn. New York: John Wiley & Sons, 2002.
167. Hosmer DW, Lemeshow S. Applied Survival Analysis: Regression Modeling of Time to Event Data. New York: John Wiley & Sons, 1999.
168. Peto R, Peto J. Asymptotically efficient rank invariance test procedures (with discussion). J R Stat Soc Ser A 1972; 135: 185–206.
169. Cox DR. Regression models and life tables (with discussion). J R Stat Soc Ser B, 1972; 34: 187–220.

Index